MOLECULAR BIOLOGY OF LIFE

ENCYCLOPAEDIA OF MOLECULAR BIOLOGY-V

MOLECULAR BIOLOGY OF LIFE

By

Dr. M.Prakash

Dept. of Zoology

M.M.H. Post Graduate College

Ghaziabad

(U.P.)

DISCOVERY PUBLISHING HOUSE PVT. LTD.

NEW DELHI-110 002

First Published-2008

ISBN 978-81-8356-270-6

Published by

DISCOVERY PUBLISHING HOUSE PVT. LTD.
4831/24, Ansari Road, Prahlad Street,
Darya Ganj, New Delhi-110002 (India)
Phone: 23279245 • Fax: 91-11-23253475
E-mail: dphbooks@rediffmail.com
dphtemp@indiatimes.com

Printed at:
Sachin Printers, Delhi

Preface

The Present title **Molecular Biology of Life** is a fast growing area of research from majors or careers in physics, chemistry, mathematics and engineering as well as animal, plant, cell biology and medicine. The overall objective of this publication is to provide a professional level reference work with comprehensive coverage of the molecular basis of life and the application of that knowledge in genetics, evolution, medicine, and agriculture. It deals with the life processes at a molecular level genetic disease diagnosis and genetic therapy; the theory and techniques for understanding manipulating, and synthesizing biological molecules and their aggregates; and the application of biological process to make or modify products to improve plants or animals or to develop microorganisms for specific uses.

Teachers and professors in schools and universities will use this publication for course preparation, and members of the press will find useful background information on new development in biotechnology and genetic medicine. Efforts have been made to prevent a concise treatment of their field of expertise at a level useful to both colleagues and researchers who are experts in related fields, as well as to university students requiring an introduction to a specific molecular biology discipline.

There can be no claim to originality except in the manner of treatment and much of the information has been obtained from the books and scientific journals available in the different libraries.

The author expresses his thanks to his friends and colleagues whose continue inspirations have initiated him to bring out this book.

The author expresses his gratitude to Mr. Wasan and staff of M/s Discovery Publishing House for their whole hearted co-operation in the publication of this book.

Author

CONTENTS

1

INTRODUCTION

In 1985, scientists discovered that the high-altitude ozone layer, which shields organisms from harmful ultraviolet radiation, had thinned greatly over Antarctica. Ozone (O_3) is produced by the action of sunlight on atmospheric oxygen (O_2) in tropical regions and is transported to high latitudes, where it is destroyed. Chlorine compounds, produced mainly by humans, appear to be the main cause of the unusually high rates of ozone destruction. Ozone is now seriously depleted at very high latitudes, where conditions favour its destruction.

Because ultraviolet radiation damages DNA, it increases the incidence of mutation and cancer. A 1 percent decrease in atmospheric ozone is estimated to result in a 6 percent increase in the incidence of skin cancer. Average ozone concentrations have decreased about 3 percent, enough to have caused a 20 percent increase in skin cancers at midlatitudes. Skin cancer, which is sometimes lethal, is a serious human health problem.

The modification of Earth's atmosphere by human activity is just the latest of a series of atmospheric changes caused by organisms. The most dramatic of these changes was the generation of O_2 by photosynthetic prokaryotes 2.5 billion years ago. Organisms have also changed Earth's waters and soils. Conditions on Earth, in turn, have influenced the evolution of life. In this chapter we examine how conditions on Earth 3.8 billion years ago influenced where and how life evolved and how life in turn, during the early phases of its evolution, changed Earth.

The phenomenon of reproduction distinguishes living organisms from nonliving and was central to discussions at many points in Parts One and Two of this book. How could something that is self-replicating have existed before there were cells? What was the earliest self-replicating unit, and how did it reproduce? Remarkably, scientific evidence is good enough that we can suggest answers to these questions.

Earth's oldest rocks contain no fossils, showing that for at least several hundreds of thousands years after its formation, Earth had no life. Today we are confident that all organisms come from other organisms. The first life, however, must have come from nonliving matter. How did this happen? Under what conditions did life originate on Earth?

Is Life Forming from Nonlife Today?

Until the nineteenth century, people believed in *spontaneous generation*—the regular formation of living organisms from nonliving matter. Flies and maggots were believed to arise from rotting meat and barnyard manure, eels and fish from sea mud, and frogs and mice from moist soil. For more than 2,000 years, belief in spontaneous generation was accepted by most scholars, clergy, and plain folk, although attempts to disprove it had been made.

In 1862, the great French scientist Louis Pasteur performed a series of meticulous experiments showing that microorganisms come only from other microorganisms and that a genuinely sterile solution remains lifeless indefinitely unless contaminated by living creatures. His most elegant experiment relied on swan-necked flasks that were open to the air. Pasteur filled the flasks with nutrient medium, heated them to kill any microorganisms present, then cooled them slowly. No new growth appeared in these flasks. The shape of the necks kept any new organisms or minute dust particles from falling into the medium. However, in flasks without the narrow swan necks, microorganisms entered and grew rapidly.

As a result of Pasteur's experiments, similar experiments by other scientists, and other discoveries of nineteenth-century science, most people accepted that all life comes from existing life. These experiments suggested that life was not forming from nonliving matter under the current conditions on Earth, but they shed no light on when or how life did form on Earth 3.8 billion years ago.

How Did Life Evolve from Nonlife?

To understand how life arose, we need to know the physical conditions that prevailed on Earth before there was life. And we need to know when life arose. The early conditions on Earth determined which types of chemical reactions could take place. If we know what types of reactions took place, we can suggest how those reactions could have led to the appearance of life.

For its First Billion Years, no Life Existed on Earth

How did the universe form? Scientists now believe that between 10 billion and 20 billion years ago there was a mighty explosion. The matter of the universe, which had been highly concentrated, began to spread apart rapidly. Eventually clouds of gases collapsed on themselves through gravitational attraction, forming the galaxies-great clusters of hundreds of billions of stars.

Somewhat less than 5 billion years ago, toward the outer edge of our galaxy (the Milky Way), our solar system (the sun, Earth, and our

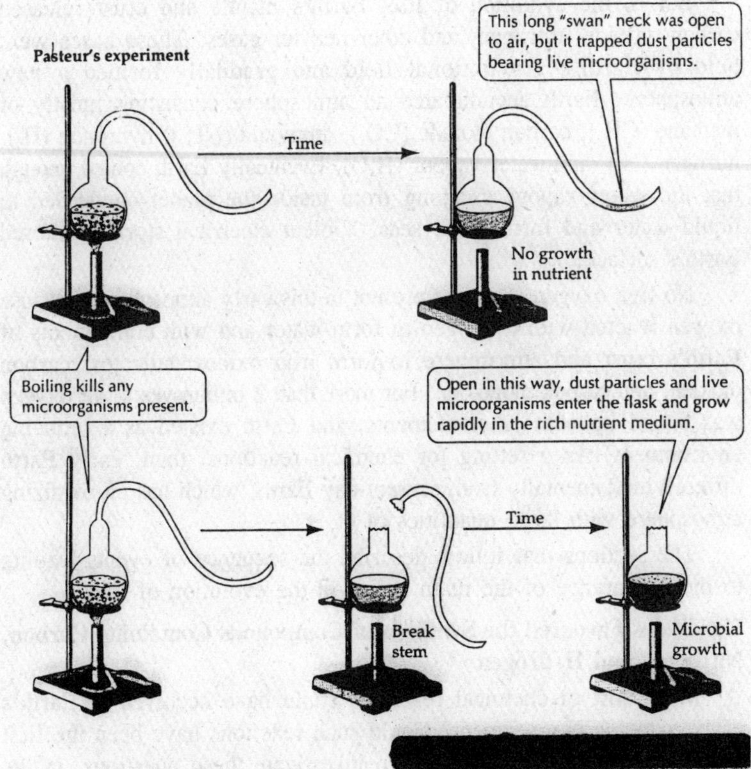

Fig. 1.1. Experiments disproved the spontaneous generation of life. Louis Pasteur's classic experiments showed that a genuinely sterile solution remains lifeless indefinitely; only "contamination" by living organisms could cause life to appear in the flasks.

sister planets) took form. Earth probably formed by gravitational attraction of rocks of various sizes. As Earth grew by this process over millions of years, the weight of the outer layers compressed the interior of the planet. The resulting pressures, combined with the energy from radioactive decay, heated the interior until it melted. Within this viscous liquid, the heavier elements settled to produce a fluid iron and nickel core with a radius of approximately 3,700 km that persists to this day. Around the core lies a mantle of dense silicate materials that is 3,000 km thick. Over the mantle is a lighter crust, more than 40 km thick under the continents but as little as 5 km thick in some places under the oceans.

Conditions on Earth 3.8 Billion Years Ago Differed from Those of Today

Before the evolution of life, Earth's mantle and crust released carbon dioxide, nitrogen, and other heavier gases. These gases were held by Earth's gravitational field and gradually formed a new atmosphere. Earth accumulated an atmosphere consisting mostly of methane (CH_4), carbon dioxide (CO_2), ammonia (NH_3), hydrogen (H_2), nitrogen (N_2), and water vapour (H_2O). Eventually Earth cooled enough that the water vapour escaping from inside the planet condensed to liquid water and formed the seas. Violent electrical storms battered Earth's surface.

No free oxygen (O_2) was present in this early atmosphere, because oxygen reacted with hydrogen to form water and with components of Earth's crust and atmosphere to form iron oxides, silicates, carbon dioxide, and carbon monoxide. For more than 2 billion years, all oxygen was bound up with other elements, and Earth existed as a reducing environment. As a setting for chemical reactions, then, early Earth differed fundamentally from present-day Earth, which has an oxidizing atmosphere with large quantities of O_2.

The sections that follow describe the sequence of events leading to the appearance of the main stages in the evolution of life.

Conditions Favoured the Synthesis of Compounds Containing Carbon, Nitrogen, and Hydrogen

What sort of chemical reactions could have occurred in Earth's early reducing environment? Could such reactions have been the first step toward the origin of life? To investigate these questions, in the 1950s Stanley Miller studied chemical reactions proceeding under conditions that resembled those believed to exist on Earth 4 billion years ago.

Within a closed system of glass tubes, he established a reducing atmosphere of hydrogen, ammonia, methane gas, and water vapour. Through these gases, he passed a spark to simulate lightning. Within a few hours the system was found to contain numerous simple organic compounds, including hydrogen cyanide, formaldehyde, cyanogen, acetaldehyde, cyanoacetylene, and propionaldehyde. In water, these compounds dissolved and were rapidly converted into amino acids, simple acids, purines, and pyrimidines—the building blocks of life.

The same or similar compounds are produced under a variety of conditions, provided that free oxygen is absent—that is, if the environment is a reducing one. Thus, once Earth cooled enough for water to condense and form oceans, molecules of many kinds formed, and they probably accumulated until they reached relatively high concentrations. From such a prebiotic ("before life") molecular soup, life emerged from nonliving matter.

Polymerization Provided Macromolecules with Diverse Properties

The next stage in the sequence leading to life was the generation of large molecules by *polymerization* of small molecules. The important large molecules of which organisms are composed (polysaccharides, proteins, and nucleic acids) are polymers formed by the combination of subunits called monomers. The polymerization reactions that generate these large molecules belong to a class of reactions called *condensations*, or *dehydrations*. During a condensation reaction, water forms. Large molecules are assembled through repeated condensations of monomers, and each condensation reaction requires energy. In the prebiotic soup, polymers that formed faster or were more stable would have come to predominate.

Some polymers can direct the synthesis of molecules identical to themselves. Which of the molecules on prebiotic Earth were most likely to reproduce themselves? Nucleic acids, the basis of today's genetic code, are clearly capable of self-copying, and the purine and pyrimidine constituents of nucleotides are formed under conditions similar to those believed to have prevailed on early Earth.

Sugars, another basic component of living cells, are generated by the polymerization of formaldehyde, another molecule formed in Miller's experiments. Solutions of formaldehyde spontaneously polymerize to form several different five- and six-carbon sugars, but these sugars are unstable in aqueous solutions and break down to alcohols and carboxylic acids. However, three and four-carbon sugars are very stable and can accumulate for hundreds of years in aqueous solutions.

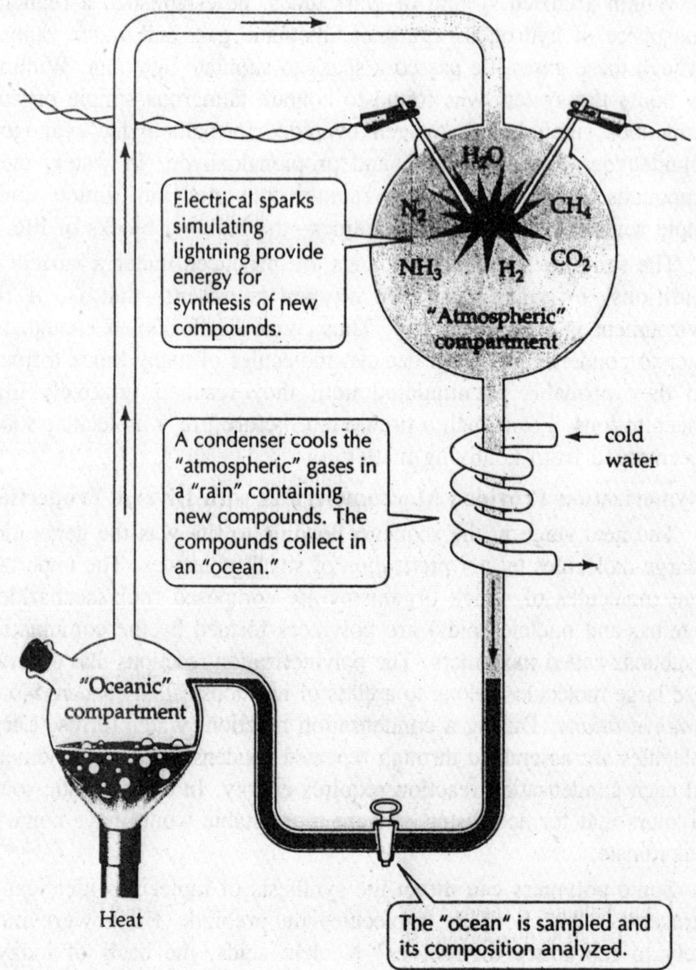

Fig. 1.2. Synthesis of molecules in an experimental atmosphere.

Over millions of years, these organic molecules would have accumulated in the oceans. They would have reached even higher concentrations in drying ponds. High concentrations of polymers, in turn, would have stimulated further polymerization and chemical reactions leading to the synthesis of sugars, which were important as components of nucleotides and for energy storage in early cells.

Phosphate-based Polymerizations Occurred in the Prebiotic Soup

Monomers formed under prebiotic conditions can polymerize by phosphorylation (the addition of a phosphate group). Phosphorylated monomers are stable enough to accumulate in solution but reactive enough to polymerize further. The first biologically active polymers of nucleic acid bases may have been compounds formed from three- and four-carbon sugars. They could have self-replicated by forming complementary double-stranded molecules, just as pentose-based nucleic acid polymers do today. At some point, by as yet unidentified processes, polymers based on small sugars were replaced by those based on larger sugars, such as ribose, the sugar in the nucleotides of RNA.

RNA was Probably the First Biological Catalyst

The enzymes that control the types and rates of reactions within organisms are proteins. Proteins are synthesized by a process that begins with transcription of information from DNA to an RNA molecule that has a base sequence complementary to that of one strand of the DNA. This information is then translated into mRNA and is eventually used to synthesize a specific polypeptide from amino acids. Amino acids are brought to a ribosome by specific tRNA molecules and are attached sequentially to the growing polypeptide. But how could such a system have evolved if protein catalysts needed nucleic acids for their formation but nucleic acids needed proteins to catalyze their own replication? Which came first, if both are necessary?

The inability to solve this dilemma held up research on the origin of life for several decades. The discovery that provided a solution came in 1981 from researchers working with the unicellular protist *Tetrahymena thermophila*. These scientists were studying the excision of introns and the splicing together of exons. They found entirely contrary to expectation—that excision is catalyzed in the absence of enzymes. The intron itself a 400-nucleotide sequence of RNA-carries out the excision and splicing. In addition, *Tetrahymena* ribosomes, which contain several molecules of RNA and a variety of proteins, have a catalytic RNA that operates in protein synthesis. RNAs that catalyze chemical reactions, called *ribozymes*, have now been found in many organisms.

The current system of macromolecular synthesis (DNA → RNA → protein) probably evolved gradually from much simpler processes. Biochemists believe that the first information-carrying molecules were short strands of RNA that replicated themselves without the help of enzymes. Evidence that RNAs can replicate themselves came first from experiments conducted by Manfred Eigen in the late 1970s. Eigen added RNA molecules to solutions containing monomers for making more RNA and found that sequences of five to ten nucleotides formed.

If he added a simple inorganic molecule such as zinc, much longer sequences were copied.

Although these experiments showed that RNA could replicate itself, they did not demonstrate that RNA could catalyze the synthesis of other molecules as true enzymes do. Within two years, however, studies of a tRNA-processing enzyme that contains RNA and a protein (ribonuclease P) showed that the RNA alone can cut the pre-tRNA molecule at the correct spot, whereas the protein cannot.

Many scientists believe that the first genetic code was based on RNA that catalyzed both its own replication and other chemical reactions. In such conditions a high concentration of RNA would have been needed, so that it could participate in many different chemical reactions. The accumulated products of RNA-catalyzed reactions could then participate in other reactions to form structures—for example, the production and accumulation of lipidlike molecules to form cell membranes, and the synthesis of proteins. However, after proteins evolved, they eventually took over most enzymatic functions because they are better catalysts than RNA and are capable of more diverse specificities.

In order to replicate, different RNAs would have competed with one another for monomers. Some RNA molecules would have been better at replicating in certain environments because their base sequences produced the most stable configurations (folded structures) under the particular conditions of temperature and salinity. With their higher rates of replication and greater stability, these molecules would have come to dominate the populations, of RNA in the corresponding environments.

In the precellular world, RNA that contained both genetic information and catalytic capacity evolved. Molecules with these capacities have been produced in the laboratory by investigators who started with completely random-sequence RNA. These ribozymes ligate (bind together) two RNA molecules in a reaction similar to that employed by the enzymes that synthesize RNA.

The investigators simulated the "evolution" of RNA molecules by selecting in test tubes for ribozymes with high ligation ability. By this method, they produced ribozymes with reaction rates 7 million times faster than the uncatalyzed reaction rate. These experiments suggest that ribozymes could have evolved rapidly when conditions on Earth became suitable for the formation of nucleic acids. But ribozymes in solution, even active ones, do not constitute life. Life required a barrier

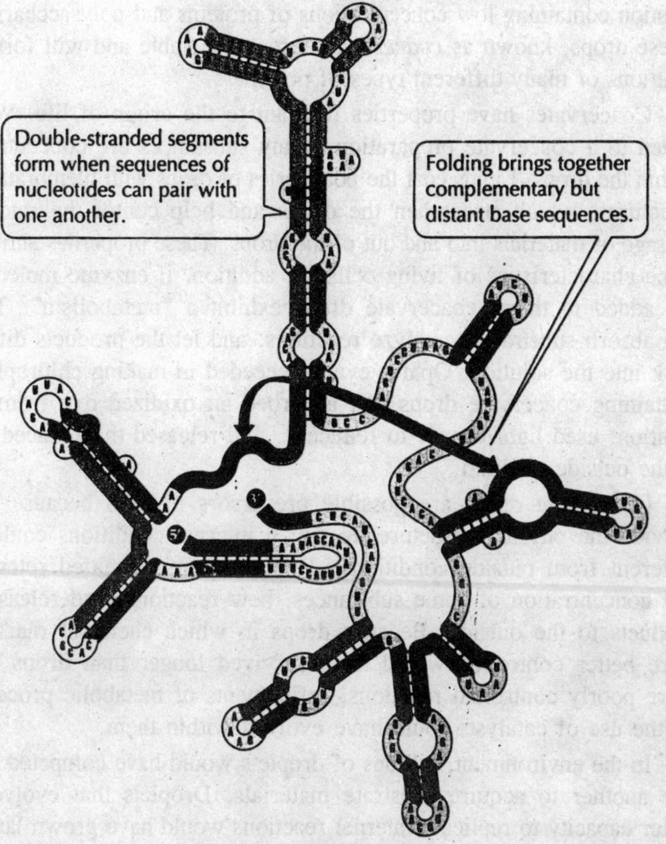

Fig. 1.3. A ribozyme from a protist.

that permitted the homeostatic control of internal conditions; life could not appear until cells enclosed in membranes evolved. The prebiotic RNA world still lacked cells.

Membranes Permitted a Stable Internal Environment to Form

The way in which ribozyme-based systems in solution might have evolved into cells was first suggested in the 1920s by the experiments of the Russian scientist Alexander Oparin, who spent much of his career studying complex solutions. Oparin observed that if he shook a mixture of a large protein and a polysaccharide, the drops that formed were divided into two "phases": an interior separated from but in contact with an exterior. These interiors, which were primarily protein and polysaccharide, with some water, were surrounded by an aqueous solution containing low concentrations of proteins and polysaccharides. These drops, known as *coacervates*, are quite stable and will form in solutions of many different types of polymers.

Coacervates have properties relevant to the origin of life. When added to a coacervate preparation, many substances are concentrated within the drops. Lipids coat the boundaries of drops with membranelike structures, which strengthen the drops and help contol the rates of passage of materials into and out of the drops. These properties simulate some characteristics of living cells. In addition, if enzyme molecules are added to them, coacervate drops exhibit a "metabolism": They can absorb substrates, catalyze reactions, and let the products diffuse back into the solution. Oparin even succeeded in making chlorophyll-containing coacervate drops that absorbed an oxidized dye from the solution; used light energy to reduce it, and released the reduced dye to the outside medium.

Coacervate drops are possible precursors to cells because they provide the physical structure by which internal conditions could be different from outside conditions. This structure permitted retention and concentration of some substances, new reactions, and release of products to the outside. Because drops in which chemical reactions were better controlled would have survived longer than drops with more poorly controlled reactions, refinements of metabolic processes by the use of catalysts could have evolved within them.

In the environment, billions of droplets would have competed with one another to acquire substrate materials. Droplets that evolved a better capacity to replicafe internal reactions would have grown larger. But when coacervates become large, they are less stable, and physical agitation breaks them apart into smaller droplets. During this early form of reproduction, thèy would have passed on at least rough copies of their molecules to daughter droplets. At some point, such droplets may have accumulated enough RNA to display the key property we associate with living cells—the use of energy to maintain homeostasis.

DNA Evolved from An RNA Template

If the first cells used RNA as their hereditary molecule, then RNA must have provided the template for the synthesis of DNA. In solution, DNA is less stable than RNA. Therefore, DNA probably did not evolve until RNA-based life became contained in membrane-enclosed cells where water concentrations were lower than in the surrounding environment. Because DNA is a more stable storage location for genetic information than is RNA, once the appropriate environments were available, DNA probably evolved rapidly, replacing RNA as the genetic code for most organisms. But by then, RNAs had assumed their current roles as intermediaries in the translation of genetic information into proteins.

Early Anaerobic Cells

Earth's atmosphere lacked the gas O_2 for about a billion years after life evolved. Therefore, the earliest organisms must have processed energy chemically without using O_2 as an electron acceptor. No traces of the metabolic pathways of those early organisms have been preserved, but some living anaerobic prokaryotes still employ similar pathways. By studying them, we can make reasoned guesses about the metabolism of ancient prokaryotes.

Three major types of anaerobic photosynthetic prokaryotes live today in sediments that lack oxygen: green sulfur bacteria, purple sulfur bacteria, and purple nonsulfur bacteria. In fact, oxygen is poisonous to them, as it must have been to organisms that evolved in a reducing atmosphere. These bacteria contain types of chlorophyll, called bacteriochlorophyll *a, c,* and *d*. Anaerobic photosynthetic bacteria also contain red and yellow carotenoids. These pigments absorb wavelengths of light that are not absorbed by chlorophyll and pass the absorbed energy along to chlorophyll for conversion into chemical energy.

The photosynthetic system of these bacteria is embedded in membrane complexes that also possess the electron transport chains and enzymes by which captured solar energy is converted to chemical energy and used to generate ATP. Photosynthetic prokaryotes similar in their metabolism to today's forms were so abundant about 3.4 billion years ago that their partly decomposed fossil remains formed extensive deposits of carbon, resembling the coal produced by vascular plant fossils 3 billion years later.

To reduce carbon dioxide (CO_2), a photosynthetic cell needs a source of electrons (hydrogen atoms). Many photosynthetic prokaryotes use light energy to generate ATP and NADPH + H^+. The particular waste product that they liberate depends on the source of hydrogen atoms they use. The green and purple sulfur photosynthetic bacteria obtain their hydrogen atoms from hydrogen sulfide (H_2S) and generate sulfur as a waste product. The purple nonsulfur bacteria typically obtain hydrogen atoms from organic compounds such as ethanol, lactic acid, or pyruvic acid, or directly from hydrogen gas (H_2).

In some environments today, the H_2 is produced by other prokaryotes as the end product of their fermentations. Under the anaerobic conditions of early Earth, hydrogen sulfide and other compounds containing hydrogen were more abundant than they are now. Today O_2 directly oxidizes H_2S, H_2, pyruvic acid, and similar compounds into water, carbon dioxide, and sulfur oxides.

Aerobic Photosynthesis is the Source of Atmospheric O_2

The evolution of aerobic photosynthesis, slightly more than 3 billion years ago, changed the course of evolution and changed Earth. The key change was the acquisition of the ability to use water as the source of hydrogen: $2\ H_2O - 4\ H + O_2$. The chemical splitting of H_2O produced O_2 as a waste product and made available hydrogen atoms for reducing CO_2.

This ability appeared first in certain sulfur bacteria that evolved into cyanobacteria. Remains of these bacteria are abundantly preserved in fossil *stromatolites*. Cyanobacteria are still forming stromatolites in a few very salty places on Earth.

The ability to split water was doubtless the cause of the extraordinary success of the cyanobacteria. The O_2 they liberated opened the way for the evolution of oxidation reactions as the energy source for the synthesis of ATP. Their success made possible the evolution of the full respiratory chain of reactions now carried out by all aerobic cells. The evolution of life irrevocably changed the nature of our planet. Life created the O_2 of our atmosphere, and it removed most of the carbon dioxide from the atmosphere by transferring it to ocean sediments.

Oxygen was poisonous to the anaerobic organisms that lived on Earth, but the prokaryotes that evolved a tolerance to it were able to live successfully in environments empty of other, organisms. Also, aerobic (oxygenated) metabolism could proceed much more rapidly and efficiently than the anaerobic metabolism that had dominated life until then.

Eukaryotic Cells: Increasing Complexity

The generation of atmospheric O_2 by cyanobacteria probably set the stage for the evolution of eukaryotes. Small, prokaryotic cells can obtain enough oxygen by simple diffusion even when oxygen concentrations are very low. Larger cells, however, have a lower surface area-to-volume ratio and need higher environmental O_2 concentrations to meet their needs.

About 1.5 billion years ago oxygen concentrations became high enough for larger cells to flourish and diversify. Some of these cells grew big enough to consume smaller cells; they were the first predators. Most of the prey died after they were engulfed, but some of them survived and eventually evolved into the mitochondria and chloroplasts that now maintain mutually beneficial relationships with their host.

Chromosomes Organize Large Amounts of DNA

All early organisms were haploid. They reproduced by dividing, and each division was accompanied by a duplication of their circular chromosomes. The single circular DNA molecule of prokaryotes fits into a small cell, and it can be duplicated rapidly. Over time, the amount of DNA in the cells of early organisms increased, as it proved advantageous to encode information for the enzymatic catalysis of new reactions and to pass this information to progeny cells.

But replicating a long DNA molecule would take many hours if it could be started only at one spot. Replication can proceed much more rapidly if genes are separated into chromosomes, each of which has multiple initiation sites for duplication. Division of the genome into chromosomes also makes it less likely that the DNA molecules will become tangled when they divide. For these reasons, biologists believe that chromosomes evolved soon after the appearance of eukaryotes, about 2.5 billion years ago.

Variable Offspring are Advantageous in Variable Environments

All early organisms reproduced by simple fission, and each division was accompanied by a duplication of their chromosomes. Occasionally and accidentally, the duplication of DNA was not accompanied by a cell division, and the result was a diploid cell. A diploid cell has certain advantages. It can repair more kinds of chromosome damage than a haploid cell because the undamaged duplicate copy can guide the repair. But even if repair was not possible, a duplicate copy of the genome offered assurance that the life of the cell could continue.

Today, chromosome breakage can be repaired only by diploid cells. Because chromosome damage is especially likely in stressful environments, diploidy is advantageous in those conditions. Many present-day unicellular organisms that live in a haploid state most of the time produce diploid cells when they are environmentally stressed. Diploid cells also have protection from point mutations, most of which alter only one copy of a gene. The unaltered copy on the homologous chromosome continues to function normally.

A diploid cell can also result if two haploid cells fuse. The chromosomes of two haploid cells are unlikely to have the same

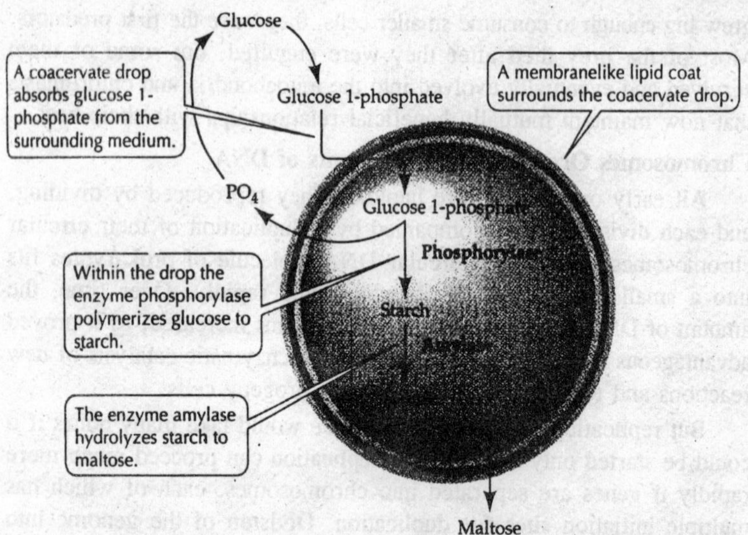

Fig. 1.4. "Metabolism" of a coacervate drop.

damages. Thus, a diploid cell formed by acquiring chromosomes from another cell would be able to repair all its damaged chromosomes. This ability may have provided the original advantage of sexual recombination, but we can only speculate. Whatever its original advantages, sex appeared early during the evolution of life, and nearly all living organisms engage in sexual recombination at least occasionally during their life cycles. Those that do not are believed to have evolved recently from sexual ancestors; they are destined to have short evolutionary lives.

Sex is maintained in multicellular organisms today by a benefit different from the ones that initially favoured it. Whereas the original benefits accrued directly to the individuals accepting genetic material, sex is maintained because of its effects on offspring. Whereas in unicellular organisms genes are not exchanged during reproduction (cell division), in multicellular organisms genes are nearly always exchanged at the time of reproduction, which is the only time during the life cycle when a single-cell stage exists.

The offspring of sexual multicellular organisms are genetically highly variable. In an environment that varies in space and time, the genetically uniform offspring of an asexual organism are less likely to find suitable conditions than are the variable offspring of a sexual organism.

Matter, Time, and Inevitable Life

Scientists have been able to gather information that provides many insights into the origin of life on Earth. In laboratory experiments they have studied chemical reactions under conditions similar to those believed to have prevailed on early Earth. Under laboratory conditions, they have witnessed the "evolution" of ribozymes with catalytic activity from random-sequence RNA.

Taken together, this information suggests that the evolution of life as we know it was highly probable under the conditions that prevailed on Earth 3.8 billion years ago. The molecules on which life is based form readily under such conditions, as does the organization of those molecules into larger units. Much remains to be learned about the early evolution of the complex metabolism of living organisms, but techniques are available to help us.

If the origin of life was almost inevitable, why is new life not still being assembled from nonliving matter on today's Earth? The reason is that simple biological molecules released into today's environment are quickly consumed by existing life. They cannot accumulate to the densities that characterized the prebiotic "soup," even in anaerobic environments. In aerobic environments these molecules are quickly oxidized to other forms. They would not accumulate even if they were not consumed. Life was generated from nonlife on Earth, but it was an event of the remote past. Once life had evolved, it prevented the formation of other life from nonlife.

2

History of Life

In 1967, the U.S. Surgeon General, William H. Stewart, announced that the time had come "to close the book on infectious diseases." Stewart and other world health officials were so confident of the power of their medical arsenal that they believed they were on the threshold of totally eradicating infectious diseases. However, the future did not conform to their prediction. In 1975, an era of new bacterial diseases began with legionnaires' disease, followed by Lyme disease, Brazilian purpuric fever, epithelioid angiomatosis, toxic shock syndrome, and others.

Since 1967, several hundred newly described bacteria have been associated with human disease, and dozens of other bacteria species that were thought to be harmless have been found to cause human disease. In addition, some historical diseases that had been nearly wiped out have resurged—some to epidemic proportions. Infectious diseases killed more than 16.5 million people in 1993, more than were killed by noninfectious diseases such as cancer and heart disease.

Health officials erred in their predictions because they did not understand the evolutionary capabilities of disease-causing organisms. Hippocrates described the symptoms of malaria, mumps, diphtheria, tuberculosis, and influenza 2,300 years ago, and human fossils show that our species has been plagued by infectious diseases throughout prerecorded history. Science, by making possible the rapid increase of Earth's human population, has helped create ideal conditions for the evolution of new pathogens.

Evolutionary changes are taking place all around us, and they have powerful implications for human welfare. Our attempts to control populations of undesirable species and increase populations of desirable species make us powerful agents of evolutionary change. In addition to producing the results we desire, we often cause undesirable outcomes, such as the evolution of resistance to medicines and pesticides and the evolution of greater virulence on the part of human pathogens. Medicine and agriculture can respond creatively to the evolutionary changes they are causing only if we understand why those changes happen.

What is biological evolution? *Biological evolution* is a change over time in the genetic composition of members of a population. Changes that take effect over a small number of generations constitute *microevolution*. Changes that take centuries, millennia, or longer to be completed are called *macroevolution*. The fossil record documents macroevolutionary changes among organisms. Many of these changes are dramatic. The goals of this part of the book are to document the history of life on Earth and to describe the processes of evolutionary change and the agents that cause them.

We begin with an overview of the history of life on Earth. Because the genetic material and basic cellular metabolic pathways are very similar or identical in all organisms, biologists believe that all living organisms have descended from a single ancestral lineage. How biologists determine the evolutionary histories of organisms, how they study the mechanisms of evolutionary change, and how the millions of species that live today (as well as those that became extinct) formed from a single common ancestor are addressed in subsequent chapters. We examine how life probably arose from nonliving matter several billion years ago.

To understand the long-term patterns of evolutionary change that we will document in this chapter, we must think in time frames spanning many millions of years and imagine events and conditions very different from those we now observe. The Earth of the distant past is, to us, a foreign planet inhabited by strange organisms. The continents were not where they are today, and climates were different. One of the remarkable achievements of modern science has been the development of techniques that enable us to infer past conditions and to date them with some precision.

In this chapter we first examine how events in the distant past can be dated. Then we review the major changes in physical conditions on Earth during the past 4 billion years, look at how those changes affected life, and discuss the major patterns in the evolution of life.

Determining How Earth Has Changed

Geologists divide Earth's history into four *eons*: the Hadean eon, the Archean eon, the Proterozoic eon, and the Phanerozoic eon. The Phanerozoic eon is subdivided into *eras*, which are further subdivided into *periods*. The boundaries between these divisions are based on major differences in the fossils contained in successive layers of rocks. The fossils of organisms, not the rocks themselves, provided the information geologists first used to order the sequence of events, because evolving life provided a record that can be ordered through time. There is no directional evolution of rocks; a rock of a particular type can be formed at any time. However, scientists do have ways of determining the date a rock was formed.

Radioactivity Provides a Way to Date Rocks

Radioactivity can be used to date rocks precisely because in successive, equal periods of time, an equal fraction of the remaining radioactive material of any radioisotope decays, becoming the corresponding stable isotope. For example, in 14.3 days, one-half of any sample of phosphorus-32 (^{32}P), a radioactive isotope of phosphorus, decays. During the next 14.3 days, onehalf of the remaining half decays, leaving one-fourth of the original sample of ^{32}P. After 42.9 days, three halflives have passed, so one-eighth (that is, $1/2 \times 1/2 \times 1/2$) of the original radioactive material remains, and so forth.

Each radioisotope has a characteristic half-life. Tritium (^{3}H) has a half-life of 12.3 years, and carbon-14 (^{14}C) has a half-life of about

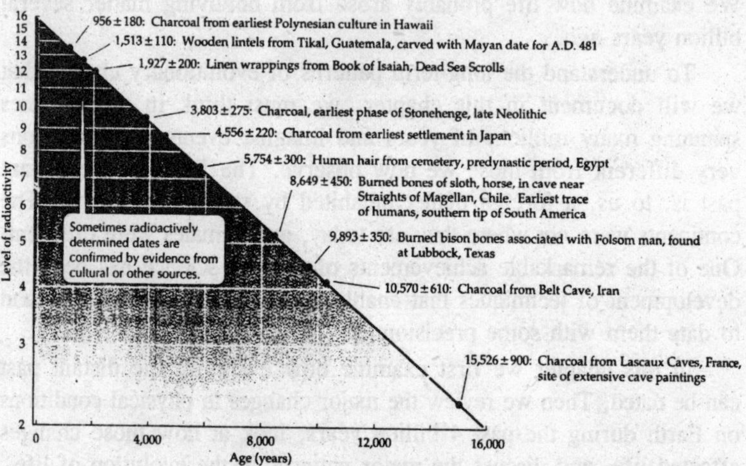

Fig. 2.1. Radioactive clocks tell time.

5,700 years. Some radioisotopes have much longer half-lives: The half-life of potassium-40 (^{40}K) is 1.3 billion years; that of uranium-238 (^{23}SU) is about 10 billion years. Which isotope is used to estimate the ages of some ancient material depends on how old the material is thought to be. The decay of potassium-40 to argon-40 has been used to date most of the ancient events in the evolution of life that we will describe in this chapter.

To use radioisotopes to date past events, we must know the concentrations of the isotopes at the time of those events. In the case of carbon, we know the initial amounts of ^{14}C because the production of new ^{14}C in the upper atmosphere (by the reaction of neutrons with ^{14}N) just balances the natural radioactive decay of ^{14}C. Therefore a steady state exists.

The ratio of radioactive ^{14}C to nonradioactive ^{12}C in a living creature is always the same as that in the environment because carbon is constantly being exchanged between the environment and organisms. However, as soon as a tree or any other living thing dies, it ceases to equilibrate its carbon compounds with the rest of the world. Its decaying ^{14}C is not replenished from outside, and the ratio of ^{14}C to ^{12}C decreases. By measuring the fraction of ^{14}C in a carbon specimen, we can easily calculate how much time has elapsed since it died, but ^{14}C can be used to date events only within the last 30,000 years. Some radiocarbon dates of archaeological objects.

Armed with information on the ages of mileposts in Earth's history, we can analyze what caused the changes in types of organisms preserved in the rocks that led geologists to identify boundaries between geological periods.

Unidirectional Changes in Earth's Atmosphere

The atmosphere of early Earth was a reducing one; that is, it lacked free oxygen. Perhaps the most important environmental change since Earth cooled enough for water to condense on its surface is the largely unidirectional increase in atmospheric O_2 concentrations that began in the Phanerozoic eon. The oxygen concentration increased because certain sulfur bacteria evolved the ability to use water as the source of hydrogen during photosynthesis. The cyanobacteria that evolved from these sulfur bacteria became very abundant. They liberated enough O_2 to open the way for the evolution of oxidation reactions as the energy source for the synthesis of ATP.

An oxygenated atmosphere also made possible larger cells and more complicated organisms. Small, unicellular aquatic organisms can obtain enough O_2 by simple diffusion even when O_2 concentrations are very low. Larger unicellular organisms have lower surface area-to-volume ratios. In order to obtain enough O_2 by simple diffusion, they must live in an environment where concentrations of O_2 are higher than those that can support small prokaryotic cells. Bacteria can thrive on 1 percent of current atmospheric O_2 levels, but eukaryotic cells require oxygen levels that are at least 2 to 3 percent of current atmospheric concentrations.

About 1,500 million years ago (mya), O_2 concentrations became high enough for large eukaryotic cells to flourish and diversify. Further increases in atmospheric O_2 levels some 700 to 570 mya enabled multicellular organisms to evolve. The fact that it took millions of years for Earth to develop an oxygenated atmosphere probably explains why only unicellular prokaryotes lived on Earth for more than a billion years.

Processes of Major Change on Earth

Unlike the largely unidirectional change in O_2 concentrations in Earth's atmosphere, most major changes on Earth have been characterized by irregular oscillations in the planet's internal processes, such as the activity of volcanoes and the shifting and colliding of the continents. External events, such as collision with meteorites, have also left their mark, sometimes causing major disruptions in the history of life.

Continents have Changed Position

The maps and globes that adorn our walls, shelves, and books give an impression of a static Earth. It is easy for us to believe that the continents have always been where they are, but this conclusion would be quite incorrect. Earth's crust consists of solid plates approximately 40 km thick that float on a fluid mantle. The mantle fluid circulates because heat produced by radioactive decay sets up convection cells. The plates move because the seafloor spreads along ocean ridges where material from the mantle rises and pushes the plates aside. Where plates come together, either they move sideways or one plate moves under the other, creating mountain ranges. The movement of the plates and the continents they contain—a process known as *continental drift*—has had enormous effects on climate, sea levels, and the distribution of organisms.

At times, the drifting of the plates has brought the continents together; at other times the continents have drifted apart. The positions and sizes of the continents influence ocean circulation patterns and sea levels. Mass extinctions of species, particularly marine organisms, have usually accompanied major drops in sea level. Later in this chapter we will discuss how the positions and movements of the continents, vulcanism, and large meteorites influenced the major events in the evolution of life.

Earth's Climate Shifts between Hothouse and Icehouse Conditions

Through much of its history, Earth's climate was considerably warmer than it is today, and temperatures decreased more slowly toward the poles. At other times, however, Earth was colder than it is today. Large areas were covered with glaciers during the late Proterozoic, the Carboniferous, the Permian, and the Quaternary, but these cold periods were separated by long periods of milder climates. Because we live in one of the colder periods in the history of Earth, it is difficult for us to imagine the mild climates that were found at high latitudes during much of the history of life.

Usually climates change slowly, but major climatic shifts have taken place over periods as short as 5,000 to 10,000 years, primarily as a result of changes in Earth's orbit around the sun. A few shifts appear to have been even more rapid. For example, during one interglacial period, the Antarctic Ocean changed from being ice-covered to being nearly ice-free in less than 100 years. Such rapid changes are usually caused by sudden shifts in ocean currents. Climates have some times changed rapidly enough that extinctions caused by them appear "instantaneous" in the fossil record.

Volcanoes Disrupt the Evolution of Life

On the morning of August 27, 1883, Krakatau, an island the size of Manhattan located in the Sunda Strait between Sumatra and Java, was devastated by a series of volcanic eruptions. Tidal waves caused by the eruption hit the shores of Java and Sumatra, demolishing towns and villages and killing 40,000 people. The 1883 explosion at Krakatau was not the largest in recorded history. The 1815 eruption of Tambora on the Indonesian island of Sumbawa was five times greater. An even larger eruption on Sumatra 75,000 years ago created the 65-km-long Lake Toba. Impressive though these eruptions were, their effects were either local or short-lived. They did not cause major changes in patterns of the evolution of life. But the even larger volcanic eruptions that occurred several times during the history of Earth, did have major consequences for life on Earth.

The collision of continents during the Permian period (260 to 250 mya) to form a single, gigantic land mass called Pangaea, caused massive volcanic eruptions. The ash the volcanoes ejected into Earth's atmosphere reduced the penetration of sunlight to Earth's surface, lowering temperatures and triggering the massive glaciation of that time. Massive volcanic eruptions also occurred as the continents drifted apart during the late Triassic period and at the end of the Cretaceous.

External Events have Triggered other Changes on Earth

At least 30 meteorites between the sizes of baseballs and soccer balls hit Earth each year, but collisions with large meteorites are rare. A large meteorite, weighing 5,000 kg, fell in Norton County, Kansas, on February 18, 1948. A prehistoric gigantic meteorite formed Canyon Diablo in Arizona.

The hypothesis that the mass extinction of life at the end of the Cretaceous period, about 65 mya, might have been caused by the collision of Earth with a large meteorite was proposed in 1980 by Luis Alvarez and several of his colleages at the University of California, Berkeley. These scientists based their hypothesis on the finding of abnormally high concentrations of the element iridium in a thin layer separating rocks deposited during the Cretaceous from those of the Tertiary. Iridium is abundant in some meterorites but is exceedingly rare on Earth's surface.

To account for the estimated amount of iridium in this layer, Alvarez postulated that a meteorite 10 km in diameter collided with Earth at a speed of 72,000 km per hour. The force of such an impact would have ignited massive fires, created great tidal waves, and sent up an immense dust cloud that encircled Earth, blocking the sun and cooling the planet. The settling dust would have formed the iridium-rich layer.

This hypothesis generated a great deal of controversy, which continues today. Many paleontologists (scientists who study fossils) believe that the fossil record shows that the mass extinction took place over a period of millions of years, not instantaneously as the meteorite theory demands. Also, because some volcanoes emit substantial quantities of iridium, the massive vulcanism of the time could have generated the iridium layer.

This controversy has stimulated much activity. Some scientists have tried to locate the site of impact of the supposed meteorite. Others have worked to improve the precision with which events of that age could be dated. Still others have tried to determine more exactly the speed with which extinctions occurred at the Cretaceous-Tertiary boundary. Progress on all three fronts has tended to support the meteorite theory.

The theory was supported by the discovery of a circular crater 180 km in diameter buried beneath the north coast of the Yucatan Peninsula, Mexico, thought to have been formed by an impact 65 mya. Some new fossil evidence also suggests that there may have been a sudden extinction of organisms 65 mya, as required by the meteorite theory. Therefore, many scientists accept that the collision of Earth with a large meteorite contributed importantly to the mass extinctions at the boundary between the Cretaceous and Tertiary periods, but concensus has not yet been reached.

Fossil Record of Life

Geological evidence is a major source of information about changes on Earth during the remote past. But the preserved remains of organisms that lived in the past, not the rocks themselves, are what have enabled geologists to order those events in time. What are these remains and what do they tell us about the influence of physical events on the evolution of life on Earth? After examining the conditions that preserve remains, we will consider the completeness of the fossil record, and how that record reveals patterns in life's history.

Much of what we know about the history of life is derived from *fossils*—the preserved remains of organisms or impressions of organisms in materials that formed rocks. An organism is most likely to be preserved if it dies or is deposited in an environment that lacks O_2. However, most organisms live in oxygenated environments and therefore decompose when they die. Thus many fossil assemblages are collections of organisms that were transported by wind or water to sites that lacked O_2. Occasionally, however, organisms are preserved where they lived. In such cases-especially if the environment in question was a cool, anaerobic swamp, where conditions for preservation were excellent—we obtain a picture of communities of organisms that lived together.

How Complete is the Fossil Record?

About 300,000 species of fossil organisms have been described, and the number is growing steadily. However, this number is only a tiny fraction of the species that have ever lived. We do not know how many species lived in the past, but we have ways of making reasonable estimates. Of the present-day biota-that is, the species in all kingdoms (bacteria, archaea, protists, plants, fungi, and animals)—approximately 1.5 million species have already been named. The actual number of living species is probably at least 10 million (and possibly as high as 50 million) because most species of insects, the richest animal group, have not yet been described. Thus the number of known fossil species is less than 2 percent of the probable total of living species.

Because life has existed on Earth for at least 3.5 billion years, and because species last, on average, less than 10 million years, Earth's biota must have turned over many times during geological history. The total number of species over evolutionary time greatly exceeds the number living today.

The sample of fossils, although small in relation to the total number of extinct species, is better for some groups than for others. The record is especially good for marine animals that have hard skeletons. Among the nine major animal groups that have hard-shelled members, approximately 200,000 species have been described from fossils, roughly twice the number of living marine species in these same groups. Paleontologists lean heavily on these groups in their interpretations of the evolution of life in the past. Insects, although much rarer as fossils, are also relatively well represented in the fossil record.

Fossil Record Demonstrates Several Patterns

Despite its incompleteness, the fossil record reveals several patterns that are unlikely to be altered by future discoveries. First, great regularity exists: Organisms are not mixed together at random, but rather appear sequentially. Second, as we pass from ancient periods of geological time toward the present, fossils increasingly resemble species living today. The fossil record also tells us that extinction is the eventual fate of all species.

The fossil record contains many good series that demonstrate gradual change in lineages of organisms over time. A good example is the series of fossils showing the pathway by which whales evolved from hoofed terrestrial mammals, beginning about 50 mya. The intermediate fossils illustrate the major changes by which ancestors of whales became adapted for aquatic existence and lost their hind limbs. Interestingly, whales retain the genetic potential for developing legs; occasionally living whales have been found with small hind legs that extend outside their bodies. The claim made repeatedly by creationists that the fossil record does not contain examples of intermediates is false. Intermediates abound, and more and more of them are being discovered.

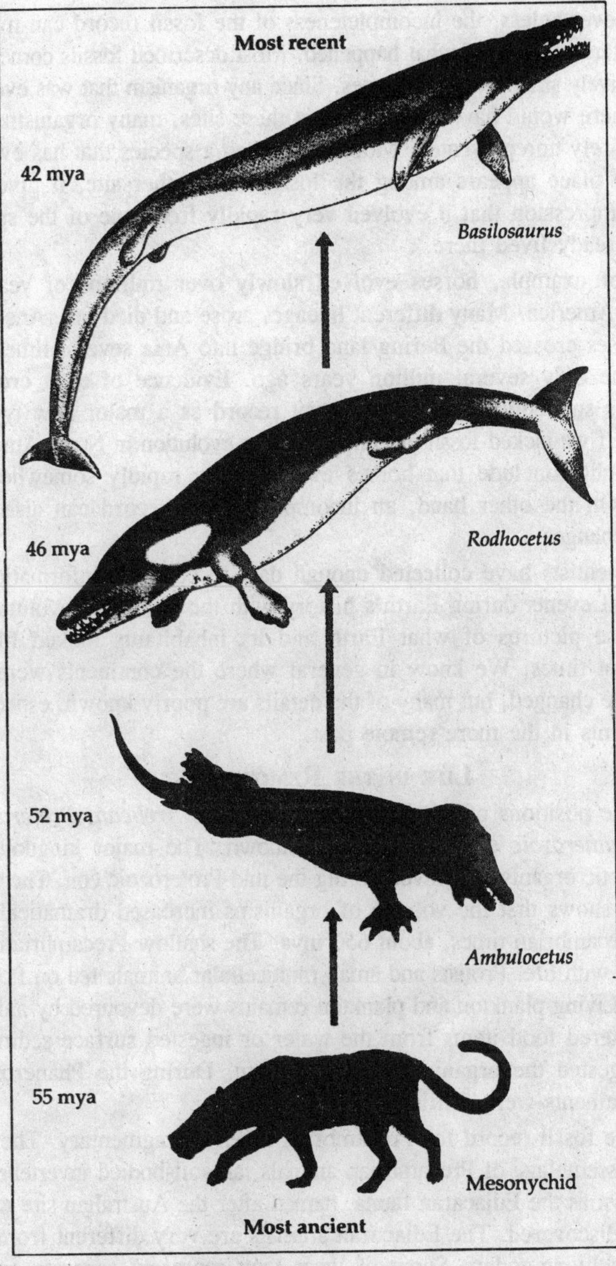

Fig. 2.2. From terrestrial to aquatic life.

Nevertheless, the incompleteness of the fossil record can mislead our interpretations of what happened. Most described fossils come from a relatively small number of sites. Since any organism that was evolving elsewhere would have been absent at these sites, many organisms may be entirely unrepresented. Moreover, when a species that has evolved in one place appears among the fossils of another site, it gives the false impression that it evolved very rapidly from one of the species that already lived there.

For example, horses evolved slowly over millions of years in North America. Many different lineages arose and died out. Ancestors of horses crossed the Bering land bridge into Asia several times, the last one only several million years ago. Evidence of each crossing appears suddenly in the Asian fossil record as a major new type of horse. If we lacked fossil evidence of horse evolution in North America, we might conclude that horses evolved very rapidly somewhere in Asia. On the other hand, an incomplete fossil record can also hide rapid changes.

Scientists have collected enough data to combine information of physical events during Earth's history with the course of evolution to compose pictures of what Earth and its inhabitants looked like at different times. We know in general where the continents were and how life changed, but many of the details are poorly known, especially for events in the more remote past.

Life in the Remote Past

The positions of the continents during the *Archean*, *Proterozoic*, and *Phanerozoic* eons are not well known. The major kingdoms of eukaryotic organisms evolved during the mid-Proterozoic eon. The fossil record shows that the volume of organisms increased dramatically in late Precambrian times, about 650 mya. The shallow Precambrian seas teemed with life. Protists and small multicellular animals fed on floating algae. Living plankton and plankton remains were devoured by animals that filtered food items from the water or ingested surface sediments and digested the organic remains in them. During the Phanerozoic, the continents were drifting apart.

The fossil record for Precambrian times is fragmentary. The best fossil assemblage of Precambrian animals, all soft-bodied invertebrates, is known as the Ediacaran fauna, named after the Australian site where it was discovered. The Ediacaran animals are very different from any animals living today. Some of them may represent separate animal lineages that have no living descendants. In the discussions that follow

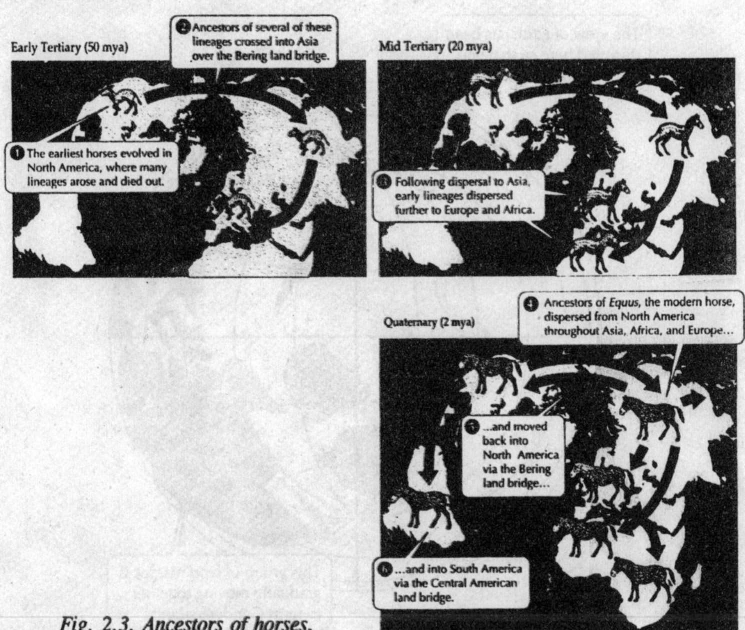

Fig. 2.3. Ancestors of horses.

in this section of the chapter, we will examine the changes that occurred in the Paleozoic, Mesozoic, and Cenozoic eras following the enormous proliferation of life in the Cambrian period.

Life Exploded during the Cambrian Period

During the *Cambrian period* (540 to 500 mya), O_2 levels in Earth's atmosphere approached their current concentration, and drifting of the plates resulted in several continental masses, the largest of which was Gondwana. Thus, O_2 concentration was no longer *a* constraint on the evolution of large, multicellular organisms. All animal phyla that have species living today had already evolved, as revealed by the exceptionally well preserved fossils in the Burgess Shale in British Columbia, which were deposited in an equatorial sea. The evolution of hard skeletons in representatives of so many phyla about 540 mya suggests that predation became very intense during this period when life "exploded."

Paleozoic Era was a Time of Major Changes

Ordovician

During the *Ordovician period* (500 to 440 mya) the continents were located primarily in the Southern Hemisphere. No evidence exists of land or shallow marine environments in the Northern Hemisphere

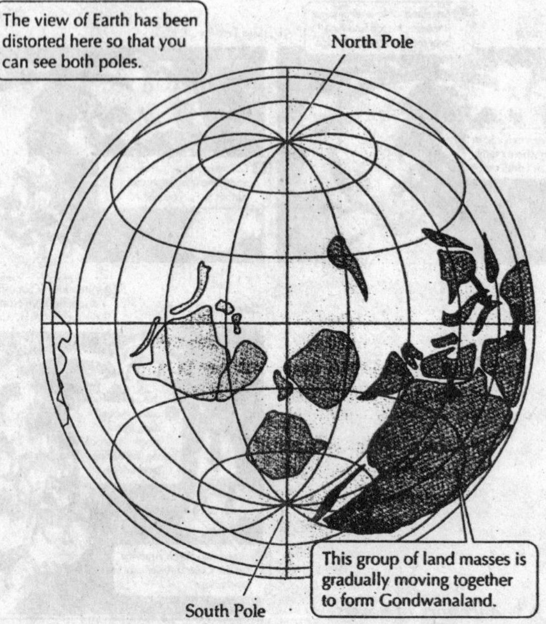

Fig. 2.4. Positions of the continents during mid-Cambrian times.

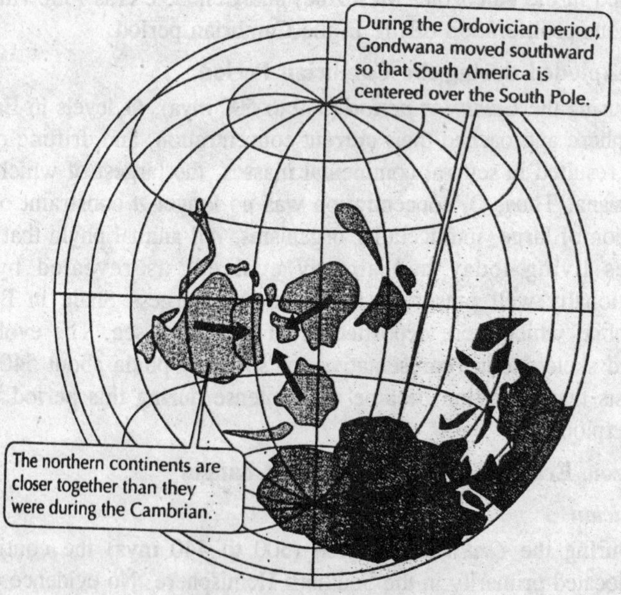

Fig. 2.5. Ordovician continents.

north of the Tropics during the Ordovician. Early during the period, the number of kinds of animals that filter small prey from the water, including brachiopods and mollusks, increased greatly. Floating graptolites, members of a now extinct phylum, were abundant. Ancestors of club mosses and horsetails colonized wet terrestrial environments, but they were still relatively small. At the end of the Ordovician, sea levels dropped about 50 m as massive glaciers formed over Gondwana. Much of the continental shelf was exposed, and ocean temperatures dropped. About 85 percent of the species of marine animals became extinct, probably because of these major environmental changes.

Silurian

During the *Silurian period* (440 to 400 mya), the northern continents, which had been relatively close to one another during the Ordovician, coalesced, but the general positions of the continents did not change much. Marine life rebounded from the massive extinction at the end of the Ordovician, but no major new groups of marine organisms evolved. The tropical sea was uninterrupted by land barriers, and most marine genera were widely distributed. More significant changes happened on land where the first terrestrial arthropods—scorpions and millipedes—appeared.

Devonian

Rates of evolutionary change accelerated in many groups of organisms during the *Devonian period* (400 to 345 mya). Earth continued to be divided into northern and southern land masses, both of which were slowly moving northward.

There was a great evolutionary radiation of corals and shelled squidlike cephalopods. Fishes diversified as jawed forms replaced jawless ones, and the heavy armor that had characterized most earlier fishes gave way to the less rigid outer coverings of modern fishes.

Terrestrial communities also changed markedly during the Devonian period. Land plants became common, and some reached the size of trees. Most of them were club mosses and horsetails, along with some tree ferns. Toward the end of the period the first gymnosperms appeared. Distinct floras evolved on the two land masses. The first known fossils of centipedes, spiders, pseudoscorpions, mites, and insects date to this period, and the first fishlike amphibians began to occupy the land.

The end of the Devonian was marked by the extinction of about 80 percent of all marine species. Paleontologists disagree on the cause of this major extinction. Some believe that it was triggered by the

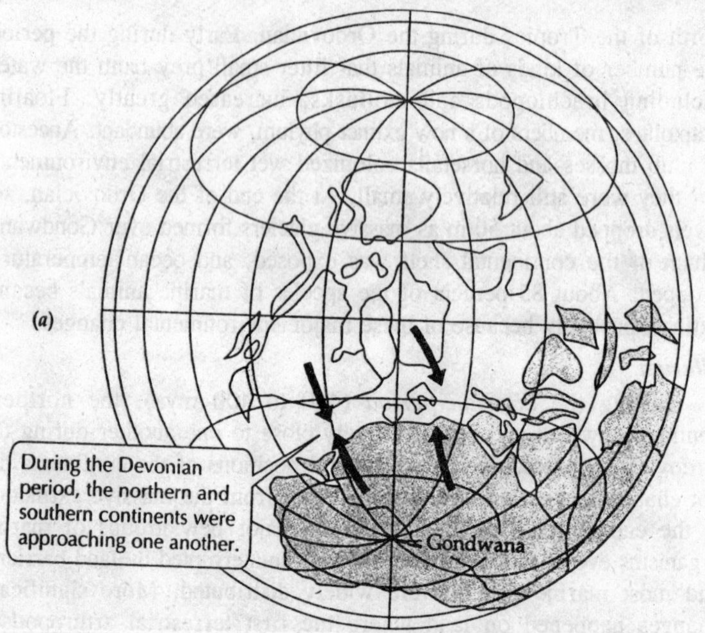

Fig. 2.6. Devonian continents.

collision of the two continents, which destroyed much of the existing shallow, warm-water marine environment. This hypothesis is supported by the fact that extinction rates were much higher among tropical than among cold-water species. Other paleontologists believe that the extinction was caused by collision of a large asteroid with Earth.

Carboniferous

Extensive forests grew on the tropical continents during the *Carboniferous period* (345 to 290 mya). The compressed remains of trees that grew in swampy forests where they fell into deep, anaerobic mud that preserved them from biological degradation are the coal we now mine for energy. Carboniferous beds are rich in fossils, many of which retain traces of the fine details of their structure. The diversity of terrestrial animals increased greatly in the Carboniferous period. Snails, scorpions, centipedes, and insects were present in great abundance and variety. Insects evolved wings, which enabled them to move readily among tall and structurally complex plants. Many Carboniferous plant fossils show evidence of damage by feeding insects. Amphibians became better adapted to terrestrial existence. Some of them were large animals more than 5 m long, quite unlike any surviving today. From one amphibian stock, the first reptiles evolved late in the period.

Permian

Deposits from the *Permian period* (290 to 245 mya) contain representatives of most modern groups of insects, including dragonflies with wingspreads that measured 2 feet, the largest insects that ever lived. By the end of the period, reptiles greatly outnumbered amphibians. These reptiles included a variety of terrestrial forms, as well as species that were major predators in marine and freshwater environments. The lineage leading to mammals diverged from one reptilian lineage late in the period. In fresh waters, the Permian period was a time of extensive radiation of bony fishes.

At the end of the Permian period, about 90 percent of all species, both terrestrial and marine, became extinct. What caused this mass extinction? There was massive volcanic activity in Siberia at that time, but there was no collision of a large meteorite with Earth. The

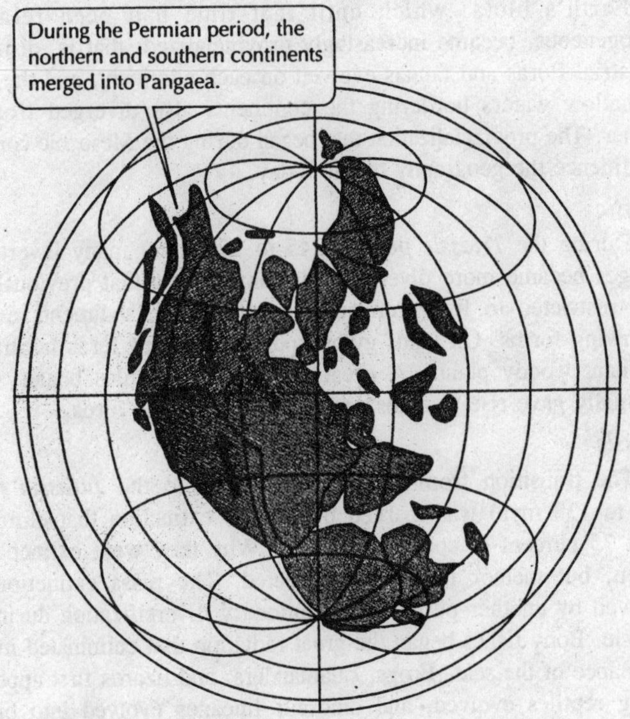

Fig. 2.7. Pangaea formed in the Permian period.

Permian extinctions may have happened slowly, perhaps over more than 10 million years, but the time frame is uncertain. The most probable cause was the coalescing of the continents into the supercontinent Pangaea. The interior of Pangaea, far removed from the oceans, experienced very harsh climates. The largest glaciers in Earth's history formed, causing sea levels to drop, drying out many of the shallow seas where most marine organisms lived.

Geographic Differentiation Increased in the Mesozoic Era

At the start of the **Mesozoic** era, the few surviving organisms found themselves in a relatively empty world. As Pangaea slowly separated into individual continents, the glaciers melted and the oceans rose and reflooded the continental shelves, forming huge, shallow inland seas. Life again diversified, but lineages different from those of the Permian period came to dominate Earth. The trees that dominated the great coal-forming forests were replaced by cycadeoids, cycads, ginkgos, and conifers.

Earth's biota, which until that time had been relatively homogeneous, became increasingly provincialized; that is, distinctive terrestrial floras and faunas evolved on each continent, and the biotas of shallow waters bordering the continents also diverged from one another. The provincialization that began during the Mesozoic continues to influence the geography of life today.

Triassic

During the *Triassic period* (245 to 195 mya) many invertebrate lineages became more diverse, and many groups that previously had been restricted to living on surfaces of bottom sediments evolved burrowing forms. On land, gymnosperms and seed ferns became the dominant woody plants. A great radiation of reptiles began, which eventually gave rise to dinosaurs, crocodilians, and birds.

Jurassic

The transition from the Triassic period to the *Jurassic period* (195 to 138 mya) was marked by a mass extinction that eliminated about 75 percent of species on Earth. Why they went extinct is not known, but meteor impact is suspected. The mass extinction was followed by another period of evolutionary diversification during the Jurassic. Bony fishes began the great radiation that culminated in their dominance of the seas. Frogs, salamanders, and lizards first appeared. Flying reptiles evolved, and dinosaur lineages evolved into bipedal predators and large quadrupedal herbivores. Several groups of mammals also evolved.

Cretaceous

By the early *Cretaceous period* (138 to 66 mya), the northern continents had completely separated from the southern ones and a continuous sea encircled the Tropics. Sea levels were high, and Earth existed in a hothouse state. Biological production was high both on land and in the oceans. Marine invertebrates increased in variety and number of species. On land, dinosaurs continued to diversify. The first snakes appeared during the Cretaceous, but their lineage did not radiate until much later.

By the end of the period, many groups of mammals had evolved, but these mammals were generally small. Early in the Cretaceous period, possibly somewhat earlier, flowering plants evolved from gymnosperm ancestors and began the radiation that led to their current dominance on land. And insects added angiosperms to the list of plants they ate. Another mass extinction took place at the end of the Cretaceous period. On land, all vertebrates larger than about 25 kg in body weight apparently became extinct. In the seas, many planktonic organisms and bottomdwelling invertebrates became extinct. As we

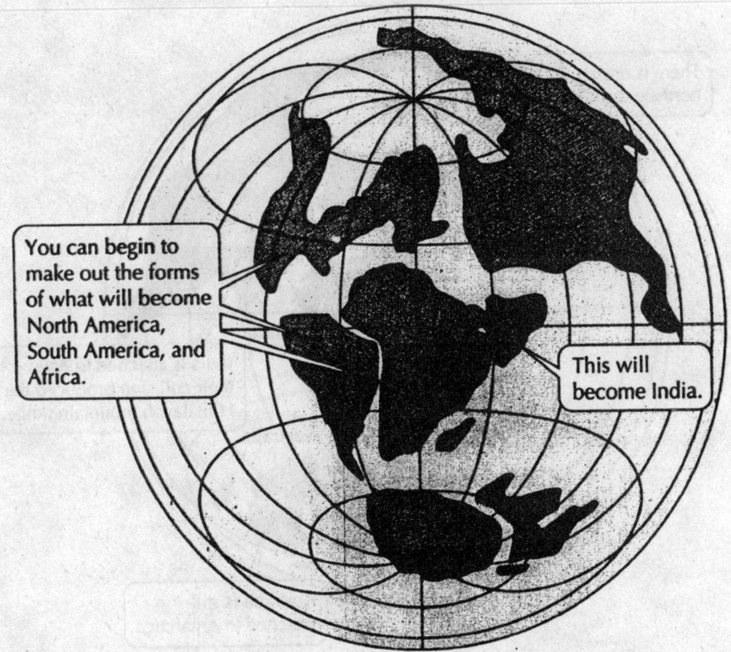

Fig. 2.8. Continents during the Cretaceous period.

discussed earlier, this extinction may have been caused by a large meteorite that collided with Earth off the Yucatan Peninsula.

Modern Biota Evolved during the Cenozoic Era

By the early *Cenozoic era* (66 mya), the positions of the continents had begun to resemble those of today, but Australia was still attached to Antarctica, the Atlantic Ocean was much narrower, and the northern continents were connected. The Cenozoic era was characterized by an extensive radiation of mammals, but other taxa were also undergoing important changes. Flowering plants diversified extensively and dominated world forests, except in cool regions.

Tertiary

During the *Tertiary period* (66 to 2 mya), Australia began its northward drift, and by 20 mya it had nearly reached its current position. The map of the world during the Tertiary for the first time looks familiar to us.

In the middle of the Tertiary, when the climate became considerably drier and cooler, many lineages of flowering plants evolved herbaceous (nonwoody) forms; grasslands, with and without scattered

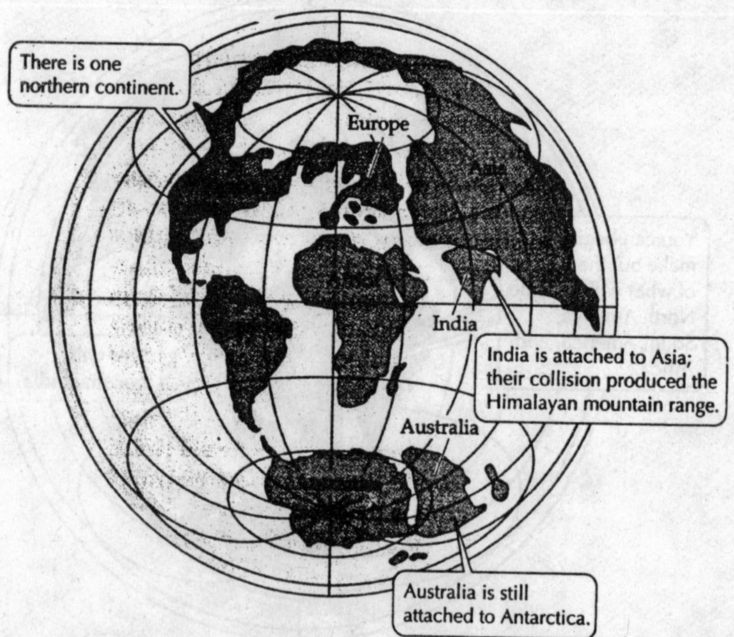

Fig. 2.9. Continents during early Tertiary period.

trees, spread over much of Earth. By the beginning of the Cenozoic era, invertebrate faunas were already modern in most respects. It is among the vertebrates that evolutionary change during the Tertiary period was most rapid. Living groups of reptiles, such as snakes and lizards, underwent extensive radiations during this period, as did birds and mammals.

Quaternary

The present geological period, the *Quaternary period*, began with the Pleistocene epoch about 2 mya. The Pleistocene was a time of drastic cooling and climatic fluctuations. During four major and about 20 minor glacial episodes, Earth became much cooler and distributions of animals and plants shifted toward the equator. The last glaciers retreated from temperate latitudes less than 10,000 years ago.

Organisms are still adjusting to these changes. Many high-latitude ecological communities have occupied their current locations no more than a few thousand years. Interestingly, most Pleistocene climatic fluctuations resulted in few extinctions. But many large birds and mammals became extinct in North and South America and in Australia when *Homo sapiens* arrived on those continents. Human hunting may have caused these extinctions, although existing evidence does not convince all paleontologists.

It took several million years for life to recover from mass extinctions

Fossil evidence shows that the number of species on Earth returned to the value it had before a mass extinction event within I to 7 million years. This recovery rate is rapid in geological time but slow relative to human life spans. Therefore, we cannot rely on evolution to replace the biological diversity we are currently losing as a result of human modification of Earth.

Although in the past the number of species has recovered within a few million years. after each mass extinction, the kinds of organisms that have diversified have differed from those that dominated life on Earth before the extinctions. How have these different kinds of organisms arisen and achieved dominance?

Patterns of Evolutionary Change

Major new features, such as the feathers of birds or the legs of terrestrial vertebrates, that adapt organisms to a special way of life are called *evolutionary innovations*. How such novelties arise has been the subject of much debate from Darwin's time to the present. The variety of sizes and shapes among living organisms seems almost limitless, but when we take a closer look, the number of truly novel structures is remarkably small.

As fiction writers often do, we can imagine unusual vertebrates with wings sprouting from their shoulders, but in reality the wings of vertebrates are always modified front legs. Modern mammals are highly varied in their shapes, but all of their structures are modifications of structures found in ancestral mammals. As we saw earlier, transforming a terrestrial mammal into a whale did not require a drastic reorganization of the mammalian body plan. Only a few evolutionary innovations, such as the notochord of chordates, do not appear to be modifications of a preexisting structure.

In the following discussions, we will see that minor developmental changes can provide major morphological changes. Then we will look at faunal patterns in time, in organism size, and in predator capacity.

Changes in Growth Rate Alter Shapes Dramatically

A major reason why evolution typically proceeds via modifications of existing plans is that major changes in shape can result from minor changes in developmental processes. As first demonstrated by D'Arcy Thompson in his 1917 book *On Growth and Form*, dramatically different

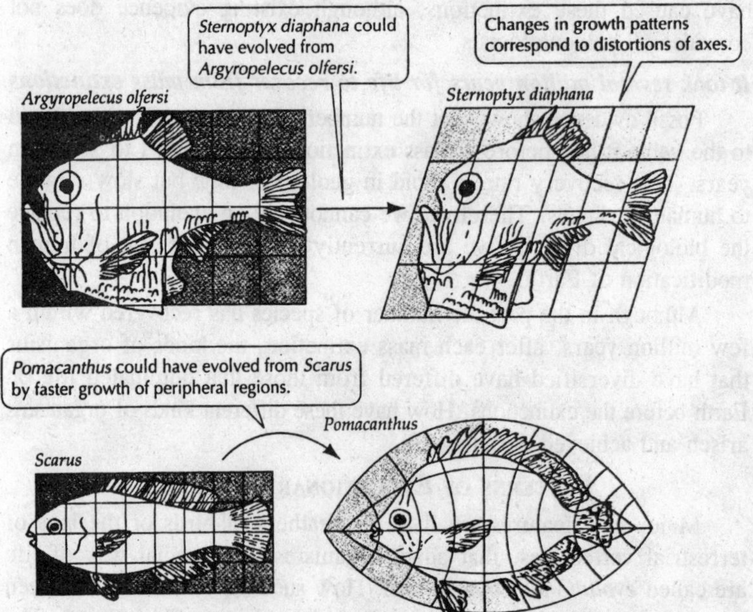

Fig. 2.10. Changes in growth rate alter shapes.

shapes of organisms can result simply from changes in rates of growth of different parts of the body. Changes in shape caused by differential growth rates are probably easy to evolve because they are often under very simple genetic control and because the sizes of all body parts are adjusted appropriately.

Three Major Faunas have Dominated Animal Life on Earth

Only three events during the evolution of life have resulted in the evolution of major new faunas. The first one, known as the Cambrian explosion, took place about 500 mya. The second, about 60 million years later, resulted in the Paleozoic fauna. The great Permian extinctions 300 million years later were followed by the third event, the Triassic explosion, which led to our modern fauna.

During the Cambrian explosion all the major groups of present-day organisms appeared, along with a number of phyla that subsequently became extinct. The Paleozoic and Triassic explosions greatly increased the number of families, genera, and species, but no new *phyla* of organisms evolved. These later explosions resulted in many new species of organisms, but all these species had modifications of the body plans already present when the great biological diversifications began.

Biologists have long puzzled over the striking differences between the Cambrian explosion and the two later explosions. The most commonly accepted theory is that because the Cambrian explosion took place in a world that contained few species of organisms, all of which were small, the ecological setting was favourable for the evolution of many new body plans and different ways of life. Many types of organisms were able to survive initially in this world, but as competition intensified and new types of predators evolved, many forms were unable to persist.

Although Earth was relatively poor in species at the time of the other two evolutionary explosions as well, the species that were already present included a wide array of body plans and ways of life. Therefore, new major innovations were less likely to evolve at these times than in the Cambrian period.

Size and Complexity of Organisms have Increased

The earliest organisms were small prokaryotes. A modest increase in size and structural complexity accompanied the evolution of eukaryotes 2.5 billion years ago. Since then, the maximum sizes of organisms in many lineages have increased, irregularly to be sure. E. D. Cope noted in 1885 that sizes of vertebrates often increase with time within lineages, a trend that has since been verified in many other groups. The most striking exception to this trend is insects, which have remained relatively small throughout their evolutionary history, although some were much larger during the Carboniferous.

The overall increase in body size is the result of two opposing forces. Within a species, selection often favours larger size because larger individuals can dominate smaller ones. But larger species on average survive for less time than small species do, which is one reason why Earth is not populated primarily by large organisms.

Predators have Become More Efficient

During the Cretaceous period (138 to 66 mya) many species of crabs with powerful claws evolved, and carnivorous marine snails able to drill holes in shells began to fill the seas. Skates, rays, and bony fishes with powerful teeth capable of crushing mollusk shells also evolved, and large, powerful marine reptiles-the placodonts—fed heavily on clams. The increasing thickness and narrowing openings of snail shells during the Cretaceous is evidence that predation rates intensified. Other evidence of heavy predation pressure is the increase in the percentage of fossil shells that show signs of having been repaired following an attack that did not kill its owner.

Although shell thickness provided some protection from predators, predators were so effective that clams disappeared from the surfaces of most marine sediments. The survivors in those environments were *species* that burrowed into the substrate, where they were more difficult to capture.

Speed of Evolutionary Change

Following each mass extinction, the diversity of life rebounded. How fast did evolution proceed during those times? Were rates too high to be accounted for by the speed at which current evolutionary changes are happening? Why did some lineages evolve rapidly while others did not? Scientists have made enough progress in studying evolution to give at least tentative answers to these questions.

Fossil evidence can be used to estimate rates of change of size and shape in many lineages of organisms. One of the fastest known rates of evolution has been in the skeletons of house sparrows that were introduced into the United States from Europe in the nineteenth century, but that rate is many times slower than rates achieved by scientists performing evolutionary experiments in the laboratory. Nonetheless, house sparrows diverged from their European ancestors at rates about 20 times faster than the highest rate of change that has been measured in characters among rapidly evolving fossil mammals. Rates of increase in sizes of dinosaurs were ten times slower than that. Thus, the most rapid rates of evolutionary change are compatible with the microevolutionary mechanisms. In considering the speed of evolution, we will look at variations in the rates of both evolution and extinctions.

Why are Evolutionary Rates Uneven?

The fossil record shows that many species have experienced times of *stasis*, during which they change very little over long periods. For example, many marine lineages evolved very slowly during the Silurian period. The horseshoe crabs that lived 300 mya are almost identical in appearance to those living today. The chambered nautiluses of the late Cretaceous are indistinguishable from living species.

"Living fossils" are found today in environments that have changed relatively little for millennia. These environments tend to be harsh in one way or another. The sandy coastlines where horseshoe crabs spawn have extremes in temperature and in salt concentration that are lethal to many organisms. Chambered nautiluses spend their days in deep, dark ocean waters, ascending to feed in food-rich surface waters only under the protective cover of darkness. Their intricate shells provide little protection against today's visually hunting fishes.

Periods of stasis may be broken by times during which changes, either in the physical or biological environment, create conditions that favour new traits. How new conditions favour rapid evolutionary changes is illustrated by the sizes of spines on the three-spined stickleback (*Gasterosteus aculeatus*). This widespread marine fish has repeatedly invaded fresh water throughout its long evolutionary history. Sticklebacks are quite tiny (usually less than 10 cm long); all marine and most freshwater populations have well-developed pelvic girdles with prominent spines that make it difficult for other fishes to swallow them. However, large predatory insects can readily grasp the stickleback's spines. These insects prey selectively on stickleback individuals with the largest spines. When sticklebacks invade freshwater habitats where predatory fish are absent but predatory insects are present, the lineages rapidly evolve smaller spines. Populations with reduced spines are found primarily in young lakes that were covered by ice during the last glaciation. These lakes do not have large predatory fishes.

The extensive fossil record of sticklebacks shows that spine reduction evolved many times in different populations that invaded fresh water. In addition, molecular data reveal that each freshwater population is most closely related to an adjacent marine population, not to other freshwater populations. Therefore, spine reduction has evolved rapidly many times in different places in response to the same ecological situation: the absence of predatory fish.

Extinction Rates Vary Over Time

More than 99 percent of the species that have ever lived are extinct. However, we know relatively little about causes of extinctions, except for species that have become extinct in historical times. Species have become extinct throughout the history of life, but extinction rates have fluctuated dramatically. In some periods, some groups have had high extinction rates while others were proliferating. Paleontologists distinguish between normal or background extinction rates and rates during mass extinctions.

Each mass extinction changed the flora and fauna of the next period by selectively eliminating some types of organisms, thereby increasing the relative abundance of others. For example, among planktonic foraminifera, important marine protists, the only survivors of mass extinction were relatively simple species with broad geographic ranges. Among the seashells of the Atlantic coastal plain of North America, species with a broad geographic range were less likely to become extinct during normal periods than were species with a small geographic range.

On the other hand, during the mass extinction of the late Cretaceous, *groups of closely related species* with large geographic ranges survived better than those with small ranges, even if the *individual species* had small ranges. Similar patterns are found in other molluscan groups elsewhere, suggesting that traits favouring long-term survival during normal times are often different from those that favour survival during times of mass extinctions.

At the end of the Cretaceous period, extinction rates on land were much higher among large vertebrates than among small ones. The same was true during the Pleistocene mass extinction, when extinction rates were high only among large mammals and large birds. In addition, as we have seen, during some mass extinctions marine organisms were heavily hit while terrestrial organisms survived well. Other extinctions affected organisms living in both environments. These differences are not surprising, given that major changes on land and in the oceans did not always coincide.

Future of Evolution

The agents of evolution are operating today as they have been since life first appeared on Earth, but major changes are under way as a result of the dramatic increase of Earth's human population. The loss of large vertebrates that dominated human-caused extinctions until recently is being supplemented by increasing extinctions of small species, driven primarily by changes in Earth's vegetation. Deliberately or inadvertently, people are moving thousands of species around the globe, reversing the provincialization of Earth's biota that evolved during the Cenozoic era.

Humans have taken charge of the evolution of certain valuable species by means of artificial selection. Our ability to modify species has been enhanced by modern molecular methods that enable us to move genes at will among species, even among distantly related ones. In short, humans have become the dominant evolutionary agent on Earth today. How we handle our massive influence will powerfully affect the future of life on Earth.

3

Chemistry of Life

One of the oldest arguments in the history of biology concerned the basic nature of living things. Many biologists in the past thought that the substances composing living things are fundamentally different from those in nonliving things. They thought there is a special life chemistry, governed by its own set of laws different from those governing nonliving matter. They also proposed that there is a special force or energy, "the vital force," that is unique to life. Despite all efforts to detect it, however, no such vital force has ever been found, and chemical laws have been found to operate the same whether the chemical processes occur in living or nonliving things.

Today, biologists recognize that while all matter—living and nonliving—consists of atoms and molecules, many of the molecules in living organisms are larger and more complex than those commonly found in nonliving systems. As we will see, these relatively large and complex molecules are arranged into highly organized systems such as membranes and cells. Because of the high degree of organization among their molecules, living things have properties not found in nonliving systems.

In order to be able to explore the properties of living things, we need to understand some characteristics of the substances that constitute them.

Elements and Compounds

All matter is composed of basic substances called *elements*. An element cannot be broken down into simpler units by chemical reactions; it contains only one kind of atom. An *atom* is the smallest characteristic unit of an element.

Atoms of most elements do not occur in isolation. Usually they are combined with other atoms to form *molecules*. A molecule is a combination of like or different atoms. For example, we do not find individual oxygen atoms in the atmosphere. Rather, oxygen in air is in the form of oxygen molecules, each of which contains two oxygen atoms.

There are ninety-two naturally occurring elements and a number of artificial elements that can be made in laboratories. Each element has a symbol consisting of a one- or two-letter abbreviation of its English or Latin name. For example, the symbols for hydrogen (H) and oxygen (O) are derived from their English names, while the symbols for sodium (Na) and potassium (K) are derived from their Latin names, *natrium* and *kalium*.

A *compound* is a substance that can be split into two or more elements. Water is a compound because it can be split into its components, hydrogen and oxygen. The *formula* of a compound gives information about the kinds and numbers of atoms that make up each molecule of that compound. A formula contains the symbols (abbreviations) of the kinds of atoms in each molecule and subscripts that indicate the number of each kind of atom in the molecule. For example, the formula for water, H_2O, indicates that a water molecule contains two hydrogen atoms and one oxygen atom; and a molecule of the sugar glucose, $C_6H_{12}O_6$, contains six carbon atoms, twelve hydrogen atoms, and six oxygen atoms.

ATOMIC STRUCTURE

While atoms are the smallest units that possess all the properties of the particular elements, each atom is composed of still smaller particles.

Protons, Neutrons, and Electrons

There are several kinds of subatomic particles, but only three are crucial to the properties of elements we will consider here: protons, neutrons, and electrons. *Protons* and *neutrons* are located in a central area of the atom called the *nucleus*. *Electrons* are located at various distances from the nucleus. Protons have a positive electric charge, and electrons have a negative charge; thus, protons attract electrons. Neutrons have no electric charge.

While protons and neutrons have nearly the same mass, their mass is much greater than the mass of electrons (more than 1,800 times as great). Chemists customarily use the *Atomic Mass Unit* (AMU), or *dalton*, to describe the mass of subatomic particles. Both the proton and the neutron have masses of approximately one dalton.

Each atom of an element has a characteristic number of protons in its nucleus, and this number is known as the element's *atomic number*. For example, because a carbon atom has six protons in its nucleus, its atomic number is six. You can see the atomic numbers of some biologically important elements in table 3.1.

Table 3.1. Some elements important in biology.

Element	*Symbol*	*Atomic number*	*Atomic weight*
Hydrogen	H	1	1.01
Carbon	C	6	12.01
Nitrogen	N	7	14.01
Oxygen	O	8	16.00
Phosphorus	P	15	30.97
Sulfur	S	16	32.06
Sodium	Na	11	23.00
Magnesium	Mg	12	24.31
Chlorine	Cl	17	35.45
Potassium	K	19	39.10
Calcium	Ca	20	40.08
Iron	Fe	26	55.85
Copper	Cu	29	63.54
Zinc	Zn	30	65.37

Normally, the number of electrons outside the nucleus of the atom is equal to the number of protons in the nucleus. Because the positive charges of the protons equal the negative charges of the electrons, the charges cancel one another and the atom is neutral.

The *mass number* of an element indicates how many protons and neutrons are in the nucleus of one of its atoms. For example, most carbon atoms contain six protons and six neutrons. Carbon's mass number, 12, is symbolized with a superscript numeral preceding the element's symbol: ^{12}C (read "carbon-12"). The *atomic weight* is the *actual measured weight* of an element and is usually very close to, but not exactly the same as, the mass number.

Isotopes

Use of the word isotope has become common in our society. We hear almost daily about the use of isotopes, particularly radioactive isotopes (or simply "radioisotopes") in medical diagnosis and cancer treatment. We also hear these terms in discussions of nuclear power plants and nuclear waste disposal. What is meant by the term "isotope," and what are radioactive isotopes?

All atoms of a given element have the same number of protons, but different atoms of that element can have different numbers of neutrons. Recall that the mass number of an element is equal to the number of protons plus the number of neutrons in each of its atomic nuclei. The atoms of different forms, or *isotopes*, of a given element have different numbers of neutrons and different mass numbers.

As we saw above, the most common form of carbon atom, with six protons and six neutrons in its nucleus, has a mass number of 12 (^{12}C). A carbon isotope with six protons and seven neutrons is ^{13}C, while a carbon isotope with six protons and eight neutrons is ^{14}C.

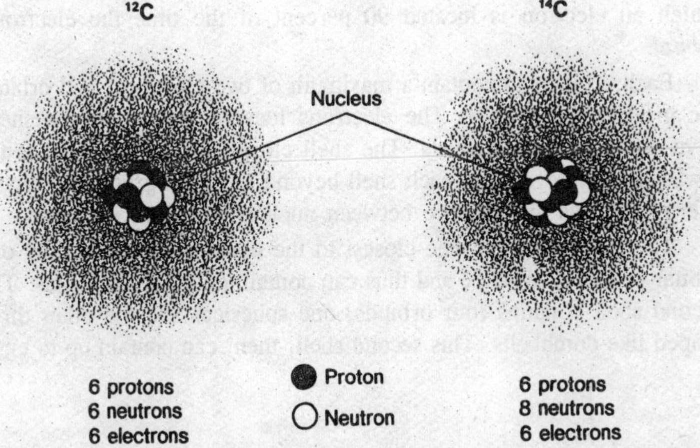

Fig. 3.1. Nuclei of carbon isotopes differ only in the number of neutrons present.

Some isotopes are stable, while others are unstable and tend to decay (break down) by spontaneously emitting small particles. These unstable isotopes are *radioactive*. ^{12}C and ^{13}C are stable isotopes, while ^{14}C is radioactive.

Radioactive isotopes of some elements emit particles with a great deal of energy. Some of the radioisotopes produced in nuclear reactors emit very high energy radiation and are extremely dangerous to handle and difficult to store safely.

Certain isotopes that emit lower-energy radiation are very useful in biological research and medical diagnosis. Radioactive isotopes can be used as chemical "labels" to mark certain compounds in cells or tissues. Because the different isotopes of an element react chemically in the same way, radioactive isotopes may be substituted for stable isotopes to make chemical substances under observation easier to detect.

Electrons and Orbitals

Although electrons have a much smaller mass than protons and neutrons, they are by no means less important in determining the chemical properties of atoms. The location of electrons in relation to the nucleus has a great influence on the way atoms and molecules react with each other.

Because electrons are continually in motion around the nucleus in every atom, we cannot say exactly where a particular electron is located at any one moment. We can say, however, where electrons are most likely to be located most of the time. We call the space in which an electron is located 90 percent of the time the electron's *orbital*.

Each orbital can contain a maximum of two electrons, and orbitals are grouped into *shells*. The electrons located in the various shells have different energy levels. The shell closest to the nucleus has the lowest energy level, and each shell beyond has a progressively higher energy level as the distance between nucleus and shell increases.

The first shell, the one closest to the nucleus, contains only one orbital (a spherical one), and thus can contain only two electrons. The second shell contains four orbitals, one spherical and the other three shaped like dumbbells. This second shell, then, can contain up to eight

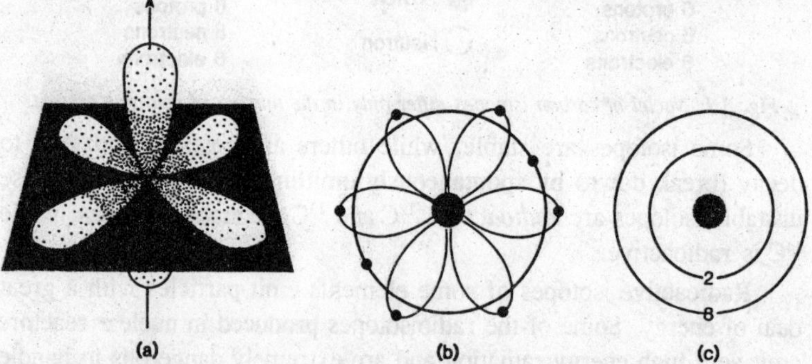

Fig. 3.2. Electron orbitals. (a) The three dumbbell-shaped orbitals of the second shell are at right angles to each other. (b) All four orbitals of the second shell on a flat plane. (c) A simpler way of representing the orbital shells with the number of electrons contained in each

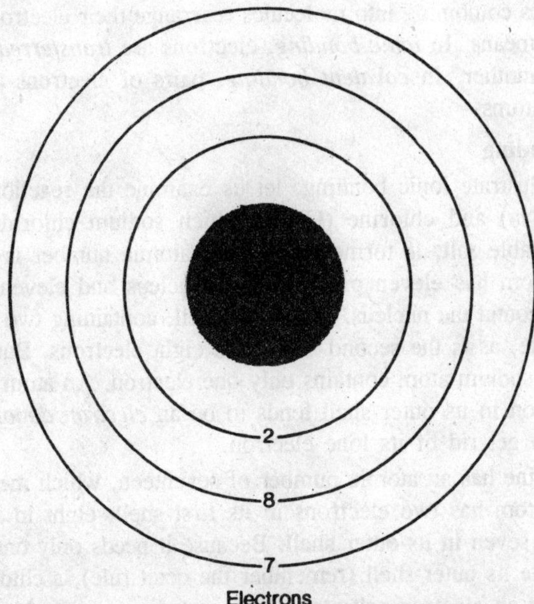

Fig. 3.3. Electron orbital shells of a chlorine atom.

electrons. Although the third shell and those beyond can hold more than eight electrons, these shells are in a particularly stable condition when they do hold exactly eight electrons.

Under normal circumstances, electrons will be in the innermost shells in which orbital space is available. In this way, each lower-energy shell is filled before the next shell is. When an atom's outermost shell contains eight electrons, the atom is very stable and does not react readily with other atoms. This *octet* (eight) *rule*, as it is called, is very important in the associations of atoms to form molecules.

Chemical Bonding

Free atoms are quite rare in nature—at least in the earth's crust and in living things. Instead, atoms usually bind together in molecules such as O_2 or H_2O, or in larger aggregates, such as crystals which contain enormous numbers of atoms and molecules.

How do atoms bind together to become molecules? Atoms are held together in molecules by *chemical bonds*. By forming chemical bonds, atoms tend to achieve a more stable electron configuration than they have by themselves. A stable molecule is formed when the uniting atoms can reshuffle their electrons so that each atom has a complete, or at least more stable, outer shell.

Atoms combining into molecules rearrange their electrons through different means. In *ionic bonding*, electrons are *transferred* from one atom to another. In *covalent bonding*, pairs of electrons are *shared* between atoms.

Ionic Bonding

To illustrate ionic bonding, let us examine the reaction between sodium (Na) and chlorine (Cl) in which sodium chloride (NaCl), common table salt, is formed. Sodium's atomic number is eleven—a sodium atom has eleven protons in its nucleus and eleven electrons orbiting around the nucleus. The inner shell, containing two electrons, is complete, as is the second shell with eight electrons. But the third shell of a sodium atom contains only one electron. An atom with only one electron in its outer shell tends to be an *electron donor*—that is, it tends to get rid of its lone electron.

Chlorine has an atomic number of seventeen, which means that a chlorine atom has two electrons in its first shell, eight in its second shell, and seven in its outer shell. Because it needs only one electron to complete its outer shell (remember the octet rule), a chlorine atom tends to be an *electron acceptor*. When a sodium atom and a chlorine atom come together, an electron is transferred from the sodium atom

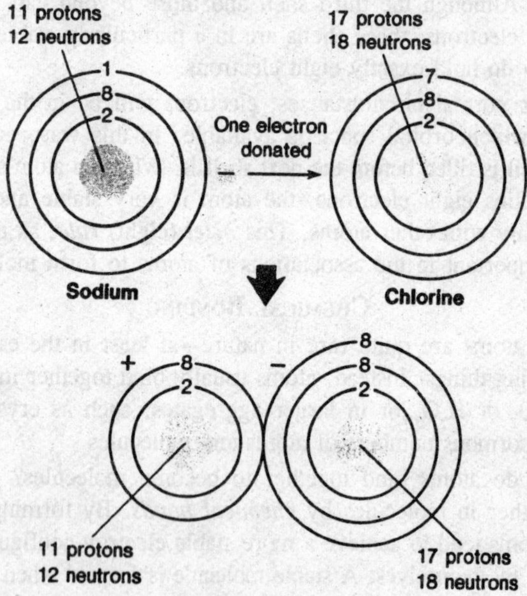

Fig. 3.4. Ionic bonding

to the chlorine atom. But this electron transfer changes the balance between the protons and electrons within each of the two atoms. The sodium atom ends up with one more proton than it has electrons, and the chlorine atom with one more electron than it has protons. The sodium atom is left with a net charge of +1 (symbolized Na^+), while the chlorine atom's net charge is –1 (symbolized Cl^-). Charged atoms are called *ions* and are attracted to one another by their opposite charges. Thus, an ionic compound such as sodium chloride (NaCl) is held together by the attraction between its positive and negative ions; that is, by an *ionic bond.*

The ionic bonds of sodium chloride are quite strong when the salt exists as a dry solid, but when NaCl is dissolved in water, it separates into Na^+ and Cl^- ions. Ionic compounds are most commonly found in this *dissociated* (*ionized*) form in the watery environment inside living things.

Covalent Bonding

In *covalent bonds* electrons are not donated from one atom to another, but are shared between atoms. Thus, a covalent bond consists of a pair of electrons that are shared between two atoms and that occupy two stable orbitals, one of each atom. If a single pair of electrons is shared between a pair of atoms, it is called a *single bond.* Sometimes two atoms share two or three pairs of electrons to form *double* or *triple bonds.*

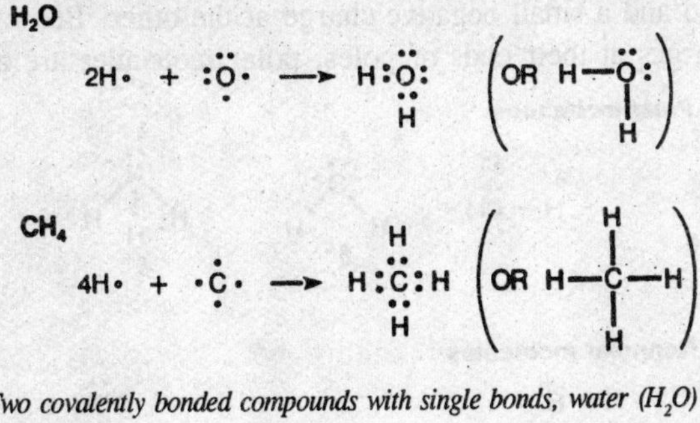

H_2O

2H• + :Ö• —► H:Ö: (OR H—Ö: with H below O)

CH_4

4H• + •Ċ• —► H:C:H with H above and below (OR H—C—H with H above and below C)

Fig. 3.5. Two covalently bonded compounds with single bonds, water (H_2O) and methane (CH_4).

Weaker Interactions

In addition to ionic and covalent bonds, there are also several types of weaker interactions among atoms and molecules. These weaker interactions are largely responsible for the three-dimensional shapes of many large biological molecules, such as proteins, and for some of

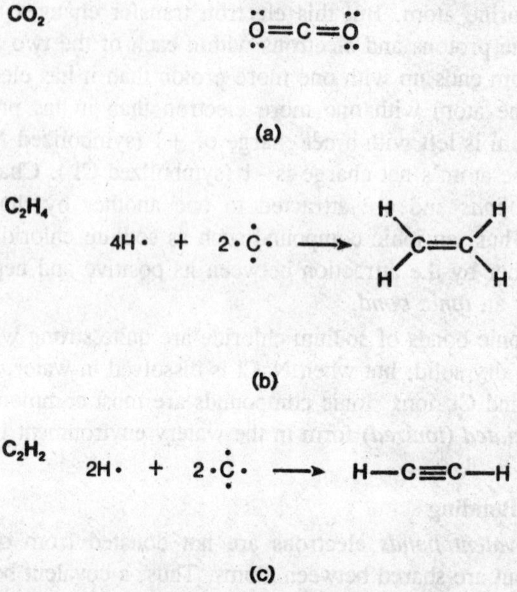

Fig. 3.6. Some compounds with multiple covalent bonds.

their functional properties. In some molecules formed by covalent bonds, the electrons tend to be attracted more toward one atom than the other. This unequal amount of attraction among the electrons is called *polarity*. Polarity results in molecules that have a small positive charge at one end and a small negative charge at the other. Because of these slight charges at their ends or poles, polar molecules are attracted to

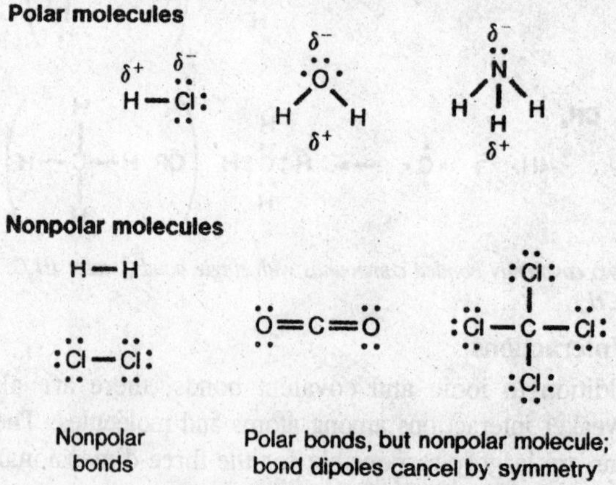

Fig. 3.7. Examples of polar and nonpolar molecules.

each other. The weak electrical attractions between opposite charges on polar molecules are known as *dipole forces*.

Hydrogen bonding

An example of this attraction between polar molecules is *hydrogen bonding*. Hydrogen bonding occurs between a hydrogen atom that is already covalently bonded to nitrogen or oxygen and a *different* nitrogen or oxygen atom. This type of bond can be formed between different molecules or between different parts of the same molecule.

Both types of hydrogen bonding are important in biological processes. For instance, hydrogen bonding between different molecules is crucial to the role of water in living systems. Hydrogen bonding within molecules is especially important in the threedimensional organization of protein and nucleic acid molecules.

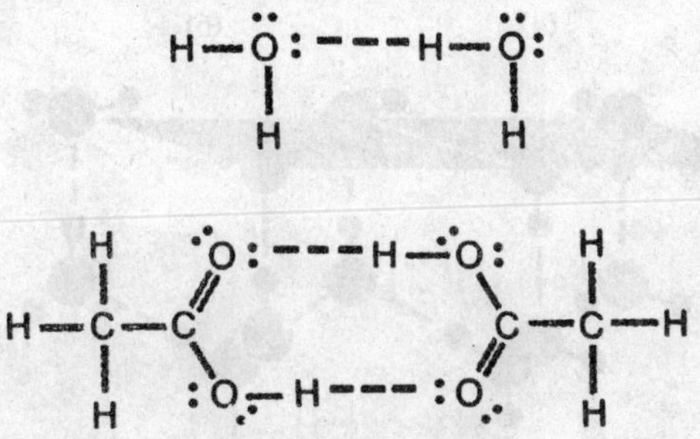

Fig. 3.8. Hydrogen bonding.

London dispersion forces

London dispersion forces (also called *van der Waals forces*) are weak forces resulting from the interaction of temporary dipoles. They are only about one-fifth as strong as hydrogen bonds but, weak as they are, they assume great importance in large molecules because there are so many of them per molecule. London dispersion forces also help hold aggregates of smaller molecules together. For instance, the phospholipids that are major components of cell membranes are held together mainly by these forces.

Water in Living Systems

Water is the most abundant material in living things, which are 60 to 95 percent water by weight. Understanding the properties of water is essential to understanding virtually all processes that occur in living systems.

Some Properties of Water

The structure of the water molecule is nonlinear (bent) with a bond angle of 104.5° between the HOH (hydrogen-oxygen hydrogen) atoms. Water is a polar molecule, with the oxygen atom forming the

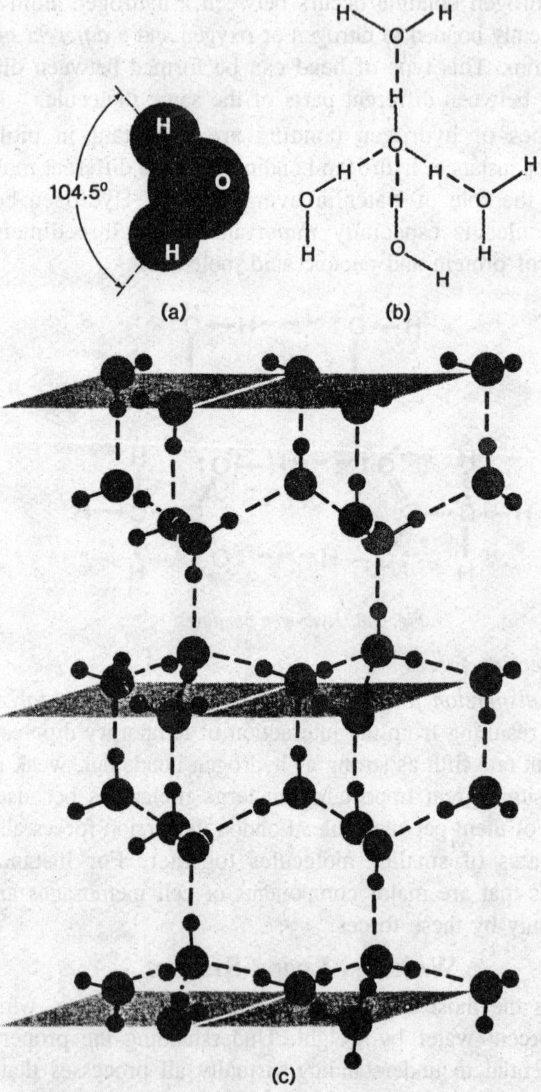

Fig. 3.9. (a) Structure of a water molecule. (b) Tetrahedral hydrogen bonding of water molecules in ice. (c) The open structure of ice.

negative end and the hydrogen atoms the positive end. Therefore, hydrogen bonding occurs between water molecules. We can see the importance of hydrogen/ bonding in water most strikingly by examining the structure of ice. In ice, each oxygen atom is surrounded tetrahedrally by four hydrogen atoms, two close (covalent bonds) and two distant (hydrogen bonds). Hydrogen bonding in ice produces a very open structure, resulting in a low density (or a large volume per molecule). The density of ice is, in fact, less than the density of water in its liquid state. Water reaches its maximum density at 4°C. This explains why ice floats on cold water and why lakes freeze at the top first, leaving water below.

When a body of water freezes at the surface, the ice acts as an insulator, preventing much of the water below it from freezing. If ice were denser than water, lakes and ponds would freeze solid, from the bottom up, and most aquatic organisms would die during the winter.

Liquid water has several other physical properties that are important to its role in living things. Because of the strong hydrogen bonds holding its molecules together, water has much higher melting and boiling points than would otherwise be expected for a molecule of its size. It also has a distinctively high *heat capacity* (the amount of heat energy that must be added or subtracted to raise or lower its temperature). The high heat capacity of water helps to stabilize and protect organisms from rapid fluctuations in temperature. On a larger scale, it stabilizes the temperature of aquatic environments such as lakes and oceans and moderates the climates of land areas near large bodies of water.

Water also has a high *heat of vaporization*. Heat of vaporization is the amount of heat energy required to convert water from a liquid to a gaseous state. Water's high heat of vaporization makes cooling by evaporation an efficient process; for example, the evaporation of a single gram of water cools 540 grams of water by 1°C. This property gives organisms in hot environments an efficient cooling method, as processes such as perspiration can help to dissipate excess body heat.

The hydrogen bonding that takes place among water molecules also contributes to the large amount of *surface tension* in water. Surface tension is the *force per unit length* (or the amount of work per unit area) needed to expand a surface. Put simply, the molecules of water making up the surface film of water resist being separated. Water's surface tension is directly exploited by insects such as the water strider, which can "skate" on a water surface, its small weight supported by surface tension. The mutual attraction of water molecules is also important in many other biological processes, including the movement of fluids in plants.

Water as a Solvent

Major parts of all living things consist of watery solutions. In a *solution*, there is a uniform mixture of the molecules of two or more substances. In living things, water is by far the most important *solvent*, the dissolving substance that is present in the greatest amount. Thus, molecules of various *solutes* (dissolved substances) are dispersed among molecules of water in solutions in living things.

How does water function as a solvent, and what determines which substances will or will not dissolve in water? Water's properties as a solvent relate to the polarity of water molecules. Recall that one end (pole) of each water molecule has a slight positive charge, and the other a slight negative charge. Many molecules that are important in living things are polar or have polar regions. The positive regions of such molecules are attracted to the negative ends of water molecules and vice versa, and water molecules are similarly attracted to them. The attractive forces involved are called *hydrophilic* ("water-loving") interactions. Because of these hydrophilic interactions with water molecules, such polar molecules become interspersed among water molecules. Sugars, salts, some alcohols, and ammonia are examples of water-soluble substances.

Other molecules, however, are nonpolar or have large nonpolar parts, and those molecules do not dissolve in water. This is because the forces attracting water molecules to one another are stronger than those acting between water and nonpolar molecules. Thus, the nonpolar molecules are "squeezed out" from between water molecules. Because nonpolar molecules do not interact with water molecules, they cannot become dispersed among water molecules to form a solution. Fats and oils are good examples of such substances. No matter how vigorously you stir oil in water, the oil remains in droplets and does not go into solution. The interactions between water molecules and nonpolar molecules or nonpolar parts of molecules are referrerd to as *hydrophobic* ("water-hating") interactions.

When an ionic compound, such as sodium chloride (NaCl), is dissolved in water, its ions separate because the charged poles of the water molecules attract the ions more than the ions attract each other. Not only are the ions pulled apart in the process, but they become surrounded by layers of water molecules called *hydration spheres*. The

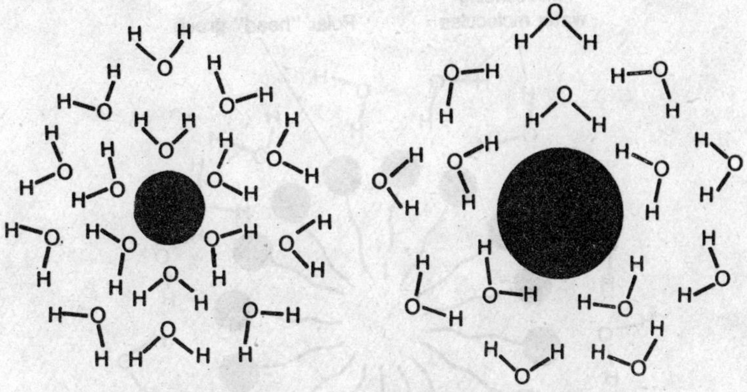

Fig. 3.10. Hydration of ions, using Na^+ and Cl^- ions as examples.

water molecules around the ions are oriented so that the negative (O) end of each water molecule is adjacent to the positive ion (Na^+). Similarly, the positive (H) end of each water molecule is attracted to the negative ion (Cl^-). Hydration spheres effectively increase the size of ions and affect the movement of ions in biological systems, including the passage of ions through cell membranes. The hydration process also affects the structure of water, because the water molecules in each sphere are oriented in one direction—with either their positive or negative ends attracted toward the ion. Because of their fixed orientation, the water molecules cannot, for example, take part so readily in ice formation. This explains why the higher the salt content of water, the more resistant it is to freezing—solutes depress the freezing point of water.

What happens when a molecule with both a small, polar (hydrophilic) portion and a large, nonpolar (hydrophobic) portion is added to water? When molecules with polar "heads" and long, nonpolar "tails" are mixed with water, they form structures known as *micelles*. In a micelle the nonpolar tails turn toward each other and away from the H_2O molecules. However, hydrophilic attraction does occur between the polar heads and adjacent H_2O molecules. Detergent action is a familiar, everyday example of micelle formation. A detergent disperses nonpolar grease by incorporating the grease molecules into the interior of micelles.

The phenomenon of micelle formation reveals much about the structure of biological membranes because biological membranes are largely made up of molecules that have both hydrophilic and hydrophobic ends.

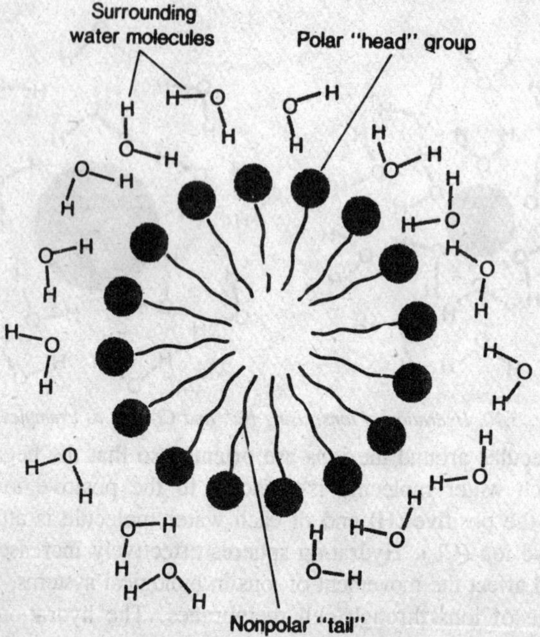

Fig. 3.11. The formation of a micelle by molecules that have a polar "head" and a nonpolar "tail."

Concentrations of Solutions

The properties of a solution depend not only on the natures of the solvent and solute, but also on their concentrations. We can express the concentrations of solutions in several ways. Stating the concentration as a *percentage* (normally, weight of solute to volume of solvent) is one simple approach. Physiological saline solution, for example, is 0.9 percent (weight-to-volume) NaCl in water.

In describing chemical reactions, however, the *mole* is a more useful unit of measurement for the amounts of different substances. A mole is the mass of substance, expressed in grams, that is numerically equal to the *molecular* (or *formula*) *weight*. The formula weight of any molecule is equal to the sum of the atomic weights of the atoms in the molecule. Thus, the formula weight of NaCl is 58.5, and one mole of NaCl is 58.5 grams:

23.0 (atomic weight of Na) + 35.5 (atomic weight of Cl) = 58.5

The mole is useful for expressing quantities of substances. One mole of any substance contains Avogadro's number (6.02×10^{23}) of molecules of the substance.

Molarity (*M*) is the number of moles of solute per liter of solution. Physiological saline solution contains 9 grams of NaCl per liter of solution, and thus, it is a 9/58.5 or 0.154 M NaCl solution. Moles and molarity can be used to describe accurately the small concentrations of many materials in living systems. For example, in table 2.2, the concentrations of ions in seawater and in the body fluids of several organisms are compared in terms of millimoles (1 millimole = 1/1,000 mole) per liter.

Acid-Base Phenomena

In biological systems, one of the most important ions is also the simplest: H^+, the hydrogen ion. The concentration of hydrogen ions strongly affects the rates of many reactions, the shape and function of enzymes and other protein molecules, and even the integrity of the cell itself. The powerful effect of the H^+ ion on other molecules is due to its positive charge and its extremely small size.

A hydrogen atom is the simplest of all atoms. It contains one proton in the nucleus and one electron in an orbital outside the nucleus. When the electron is lost (forming the H^+ ion), all that remains is the proton in the nucleus. In fact, the H^+ ion is often simply called a proton.

In the internal environment of organisms, the most abundant molecule is water, which carries two pairs of electrons that are not involved in bonding:

```
  :Ö:
  / \
 H   H
```

A proton (H^+) is strongly attracted to one of these pairs of electrons in a water molecule, thus forming a *hydronium ion* (OH_3^+).

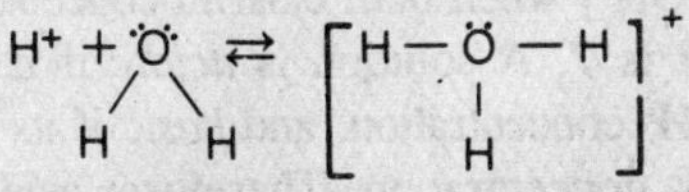

```
                       +
H+ + :Ö:  ⇌  [ H—Ö—H ]
     / \        |
    H   H       H
```

In fact, this proton attraction to H_2O is so strong that *no free* H^+ ions exist in aqueous solution: the H^+ ion is always attached to a water molecule. For convenience, however, we will continue to represent the hydrogen ion as H^+.

H^+ concentration and pH

At any given time, a small proportion of the molecules in pure water are ionized to produce hydrogen ions (H+) and hydroxide (OH^-) ions:

$$H_2O \rightleftharpoons H^+ + OH^-$$

This reaction is reversible, as indicated by the arrows pointing in both directions. We say that this reaction is at *equilibrium* because in pure water, the rate of ionization of water molecules

$$H_2O \rightarrow H^+ + OH^-$$

is equal to the rate of reaction among the ions.

$$H_2O \leftarrow H^+ + OH^-$$

Thus, at equilibrium there is no further *net* change in concentration, even though both reactions occur continually.

In pure water, this ionization occurs to such an extent that

$$[H^+] = 10^{-7} \text{ M and } [OH^-] = 10^{-7} \text{ M}$$

(Square brackets are conventionally used to denote concentrations.) Of course, the concentration of H^+ ions in pure water must equal the concentration of OH^- ions because neither can be formed without the other.

A convenient way of designating the H^+ concentration in a solution is provided by the *pH scale*, which is expressed as the negative logarithms of hydrogen ion concentrations. A *logarithm* is the exponent that indicates the power to which one number must be raised to obtain another number. (For example, the logarithm of 100 to the base 10 is 2.) Thus,

$$pH = \log_{10} \frac{1}{[H^+]}$$

$$(\text{or, } pH = -\log_{10} [H^+])$$

As an example, if $[H^+] = 10^{-3}$ M (.001 M) then pH = 3.

A solution is *neutral* if its H^+ and OH^- concentrations are equal. For water, $[H^+] = [OH^-]$ when both exist in concentrations of 10^{-7} M; that is, when the pH is 7. A solution is *acidic* if it has a higher H^+ concentration than OH^- concentration, and *basic* if its H^+ concentration is lower than its OH^- concentration. Therefore, acidic solutions have pH values below 7, and basic solutions have pH values above 7.

Acids and bases

Substances that change the concentrations of H^+ or OH^- ions when they dissolve in water are known as *acids* and *bases*. An *acid* is a *proton* (*hydrogen ion*) *donor*; it causes an increase in H^+ ion concentration when it ionizes in solution. For example, hydrochloric acid (HCl) is ionized quite completely by the reaction:

$$HCl \rightarrow H^+ + Cl^-$$

Adding hydrochloric acid to a solution increases the relative number of H^+ ions and, therefore, lowers the pH of the solution. The solution of gastric juices in the human stomach, for example, contains a great deal of hydrochloric acid and sometimes may have a pH as low as 1.0. This is such an acidic solution that stomach contents coming in contact with skin could cause a serious acid burn.

A *base* is a *proton acceptor* and has the opposite effect on a solution. For example, sodium hydroxide (NaOH) ionizes almost completely in solution and strongly increases the relative concentration of OH^- ions. Hydroxide ions (OH^-) combine with H+ ions to form water, and thereby decrease the H^+ ion concentration (thus raising the pH of the solution).

Buffers

Even small pH changes may have great effects on biological processes. Because many processes are sensitive to pH changes, it is important that the pH of body fluids remains stable. Cell interiors have an average pH of 7.0 to 7.3; and in humans, for example, blood and tissue fluids have a pH of 7.4 to 7.5. These values remain very stable despite the fact that many biochemical reactions either release or incorporate H^+ ions.

How does pH remain so constant in living things? Such pH stability is possible because organisms have built-in mechanisms to prevent pH change. The most important of these mechanisms are *buffers*. A buffer resists pH change by removing H^+ ions when the H^+ ion concentration rises, and releasing H^+ ions when the H^+ ion concentration falls.

An example of a buffer is the carbonic acid-bicarbonate ion buffer system, which is involved in the buffering of human blood:

$$\underset{\text{Carbonic acid}}{H_2CO_3} \rightleftarrows H^+ + \underset{\text{Bicarbonate ion}}{HCO_3^-}$$

In this example, if H^+ ions are added to the system, they combine with HCO_3^- to form H_2CO_3. This reaction removes extra H^+ ions and keeps the pH from changing. If H^+ ions are removed from the system (for example, if OH^- ions are added and H_2O is formed), more H_2CO_3 will ionize and replace the H^+ ions that were used. Again, pH stability is maintained. This and other buffer systems prevent potentially harmful pH changes in living things.

Organic Molecules

Most of the thousands of kinds of chemical compounds in living organisms are *organic compounds*, compounds that contain carbon. The

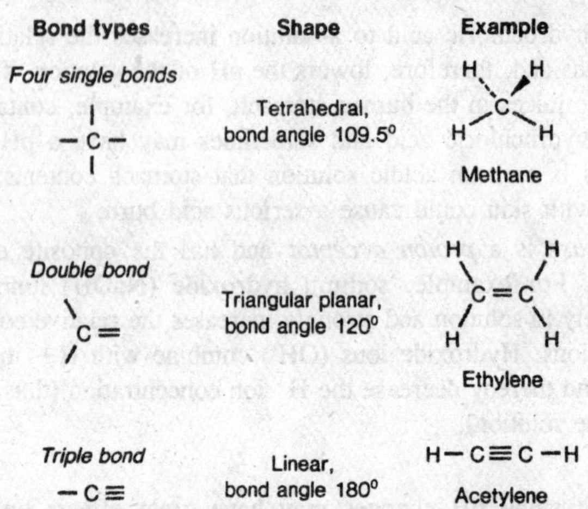

Fig. 3.12. Possible arrangements of bonds around the carbon atom.

most important characteristics of organic compounds depend on properties of this key element, carbon.

The carbon atom has an atomic number of six, with six protons and six electrons. Since such an atom has two electrons in the first shell and four electrons in the second shell, it can form covalent bonds with as many as four other atoms. A carbon atom can bond to a variety of elements, but it most commonly bonds to hydrogen, oxygen, nitrogen, and carbon.

The carbon-to-carbon bonding ability of carbon atoms makes possible carbon chains (sometimes called carbon skeletons) of various lengths and shapes. The carbon skeleton establishes the overall framework of an organic molecule. Carbon atoms can form single, double, or triple covalent bonds with one another, and such bonding patterns determine the characteristics of organic compounds.

Many organic compounds consist of just carbon and hydrogen, but many others contain *functional groups* as well. Functional groups are characteristic patterns of atoms other than those involved in carbon-carbon and carbon-hydrogen bonds. These functional groups are the parts of organic molecules that are most often involved in chemical reactions.

One reason there are so many organic compounds is that carbon forms chains of atoms better than any other element. An unusually long noncarbon chain might contain ten atoms. On the other hand, organic compounds with chains of more than fifty carbon-carbon bonded atoms can be found in living systems, and chains containing carbon

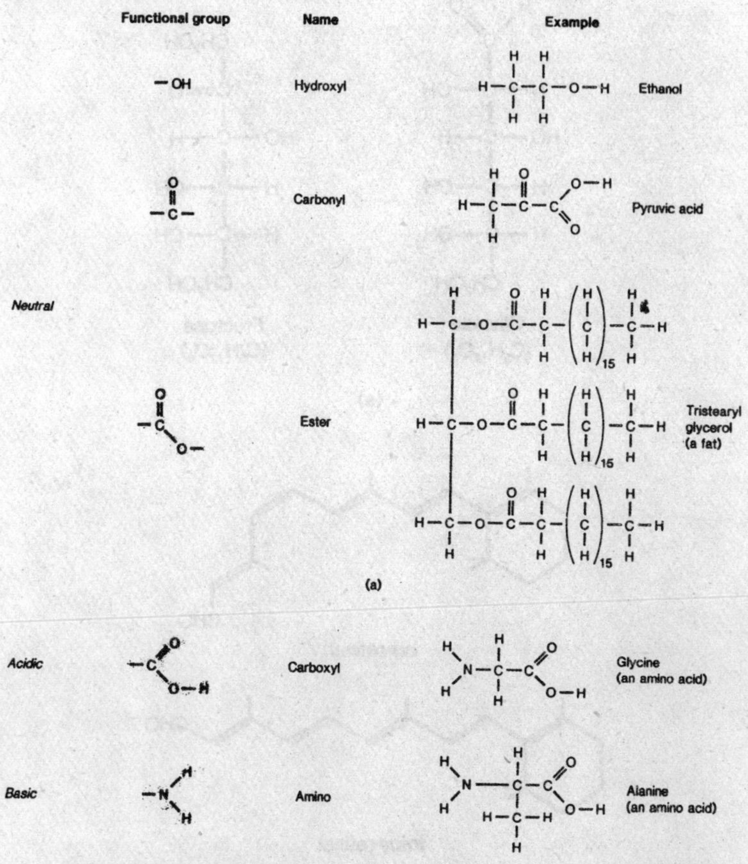

Fig. 3.13. Some examples of functional groups found in organic molecules. (a) Neutral functional groups. (b) Acidic and basic functional groups.

along with other elements such as nitrogen and oxygen are sometimes thousands of atoms long.

Another factor contributing to the diversity of organic compounds is the presence of numerous *isomers*. Isomers are compounds having the same molecular formulas but different atom arrangements. *Structural isomers* are completely different chemical compounds; they share the same molecular formula but have different pairs of atoms connected by bonds. *Geometric isomers* differ from each other only in the orientations of bonds between particular pairs of atoms. These differences are possible in the presence of certain rings or C=C double bonds. Although rotation around C—C single bonds occurs readily,

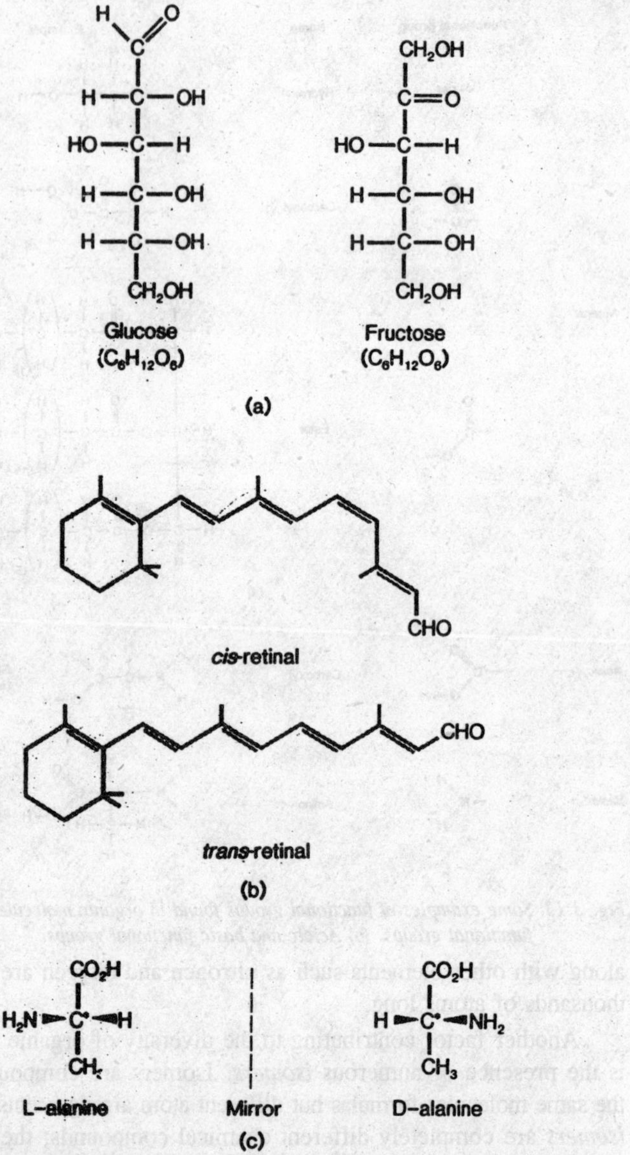

Fig. 3.14. Some examples of isomers. (a) Structural isomers are molecules with identical molecular formula, but different pairs of atoms connected by bonds. (b) Geometric isomers are molecules that have different orientations of bonds in space, but that are not minor images of each other. (c) Optical isomers are pairs of molecules that are nonsuperimposable mirror images of each other.

rotation around C=C double bonds does not. *Optical isomers* occur as pairs of nonsuperimposible mirror images, similar to the relationship between your right and left hands. Optical isomers have at least one asymmetric carbon atom, a carbon atom attached to four different groups. This makes it possible for the four different groups to occupy two different arrangements in space. These mirror-image forms are designated L- and D-. Such differences can be biologically important. For example, it is interesting and a little curious that all of the amino acids that occur in proteins in living things are of the L- form.

4

LIFE ON OTHER PLANET

The evolution of stars and the evolution of living organisms appear to be completely dissimilar processes. The differences between the two can be explained in terms of how the particles of matter are bound together and how energy is exchanged among them. Indeed, the evolution of living organisms represents one outcome of stellar evolution. It was the steady flow of energy from the sun for four or five billion years that brought about the biological developments on earth, culminating in the emergence of intelligent organisms able to contemplate the whole remarkable story. If the sun had a different history, life would not have appeared in its immediate vicinity.

Astronomical evidence acquired in recent years indicates that what has happened here is probably not unique. Two decades ago it was thought that the solar system might have originated in a near-collision of the sun and another star, which event supposedly pulled away enough matter from the sun to from the planets. Because such encounters must be rare events, they would give rise to few stars with a company of planets. Today most astronomers believe that a star is formed by the condensation of a cloud of dust and gas. This hypothesis much more readily explains the origin of the solar system and is supported by the observation that more than half of the stars in our galaxy are double or multiple systems. In fact, it now appears that most stars are accompanied by other stars or by planets, though the latter must be so small as to escape sure detection by the present instruments of astronomy. Thus the appearance of life—even the appearance of mind—may be far from unusual events in the universe.

On the other hand, certain critical conditions must be satisfied if life processes are to be initiated and maintained. In the first place the star must shine long enough and steadily enough to permit life to evolve. The star must also be hot enough to warm up a habitable zone deep enough to offer a reasonable chance that a planetary orbit will fall within it. And the planet must ply a stable orbit within this zone. The number of stars that have given rise to life must therefore be considerably smaller than the immediately suggested by the dust-cloud hypothesis.

Taking account of all these factors, what is the probability that man will ever be able to visit the life-bearing planet of another star, or that the earth will receive a visitor from such a planet? Does the possibility justify taking measures now-in advance of a visit—to put existing technology to the task of scanning the sky for signals from intelligent organisms outside the solar system, or for transmitting signals in the hope they may be heard? If so, what sort of signals should be listened for or sent?

A reliable answer to these questions calls for a somewhat closer consideration of the conditions critical for life and an estimate of how often they are likely to be satisfied in the evolution of stars. The first condition is a steady and prolonged flow of energy. How long it takes life to evolve may be judged from the single instance available: Here on earth rational animals evolved from inanimate matter in about three billion years. Biological evolution proceeds by the purely random process of mutation, that is, the unpredictable occurrence of novel chemical processes among the fantastically numerous reactions that constitute life. Since the process is a random one, the laws of probability suggest that the time-scale of evolution on earth should resemble the average time-scale for the development of higher forms of life anywhere. The rate at which mutations occur is of course a variable in the calculation. It is affected by the electromagnetic and corpuscular radiation from the parent star, and the amount of such radiation that reaches the surface of any planet is governed in turn by the magnetic field of the planet, the depth and composition of its atmosphere and other factors. But since mutation itself is a random process, the introduction of a few more variables does not affect the calculation greatly. Furthermore, a higher mutation rate does not necessarily accelerate evolution, because most mutations are harmful. The more frequently they occur, the greater the chance that an individual will suffer an injurious mutation. In order to favour natural selection, mutations should be rare, perhaps as rare as in the evolution of life on earth.

It is somewhat easier to estimate the time-scale of stellar evolution. In contrast to the random nature of biological evolution, stellar evolution is governed by the universal law of gravitation and by a relatively small number of thermonuclear reactions. When a star begins to form in a cloud of dust and gas, gravitational attraction among the gas and dust particles causes the cloud to condense until the pressure raises the temperature within it to the point at which the thermonuclear reactions that convert hydrogen to helium begin. The tremendous quantities of energy liberated by these reactions now set up a counterpressure from the center of the star that exactly balances the force of gravitational contraction. In this state of equilibrium the star shines for a much longer time than that required for its condensation out of dust and gas.

The vast majority of the visible stars are in this phase of their evolution. They are called "main-sequence" stars because their luminosity (energy output per unit of time) plotted on graph against their surface temperature places them in sequence in a narrow band. What the chart shows is that the hottest stars are also the most luminous. Luminosity depends upon mass, and so the point at which a star appears on the main sequence depends primarily upon the mass of material incorporated in it during condensation. The hottest stars are designated by the letter O, followed in descending order by stars classified B, A, F, G, K and M; these classifications are usually called spectral type. The adjectives "early" and "late," which have nothing to do with the age of the star, are often used before the spectral-type designation to denote further relative temperature differences. An early F-type star is hotter than a late one, which in turn has a higher temperature than an early G-type star such as our sun. The classification is further refined by a number from 0 to 9 written after the letter. Thus the class of early B-type stars includes those from B0 to B4; and the late B-type stars, those from B5 to B9.

The length of time a star remains in the equilibrium state on the main sequence can be calculated from the total mass of hydrogen in its core and the rate at which the hydrogen is consumed. Both factors in the calculation are determined by the position of the star on the main sequence. Luminosity, which is an index of the rate of fuel consumption, increases as the fourth power of the star's mass. The most massive stars thus use up their substance most rapidly and so have the shortest lifetimes in the equilibrium state. In general a star will evolve away from the main sequence when the core in which the hydrogen has been consumed has a mass of about 12 per cent of that of the entire star. With the exhaustion of fuel in the central furnace, gravitational contraction takes over again, heating up the interior until the thermonuclear reaction spreads to outer layers. The star now leaves the main sequence and in a comparatively brief period of time evolves into a red giant or supergiant. Its evolution thereafter cannot presently be predicted in detail. But somehow, perhaps through the rapid loss of mass by ejection, it ends up as a hot, faint, dense object known as a white dwarf. A majority of the intrinsic variable stars, of novae and of nova-like objects are apparently in the stage between red giants and white dwarfs.

In leaving the main sequence a star releases so much energy that it would destroy life on any of its planets. Thus the main-sequence stage of the star's evolution is the only important one so far as life is concerned. The O and early B stars are the most massive, but since they burn much more rapidly than the smaller, less luminous stars, their life on the main sequence lasts only a million to 10 million years. The small M stars, in contrast, remain on the main sequence for more than 100 billion years. None of the early stars in the O, B and A groups has a stable lifetime longer than three billion years; they cannot therefore sustain biological evolution long enough for intelligent organisms to appear on any of their planets. Closer calculation shows that only the stars farther down the sequence than F4 maintain their equilibrium for a sufficient length of time to bring biological evolution to its culmination.

If time were the only critical condition, then the late-K and M stars would stand the greatest chance of having life-bearing planets. But these stars have low luminosity, and the habitable zones in which planets might travel about them must be quite narrow. A more luminous star can obviously warm up a larger space than a less luminous one. It is like a fire in a field on a cold night: the bigger the fire, the wider the zone around it in which the temperature will be neither too hot nor too cold. Thus the chance of finding a planet and intelligent life within the habitable zone of any particular M or even late-K star is quite small. However, since there are 10 times as many M stars as G stars, the total number of life-bearing planets traveling around the M stars may not be negligibly small.

On the basis of their lifetimes and the depth of their habitable zones, stars of the late-F, G and early-K types seem to offer the most favourable environments for life. Approximately 10 per cent of stars fall in these types. Out of the 200 billion stars in the Milky Way, therefore, some 20 billion might foster intelligent life.

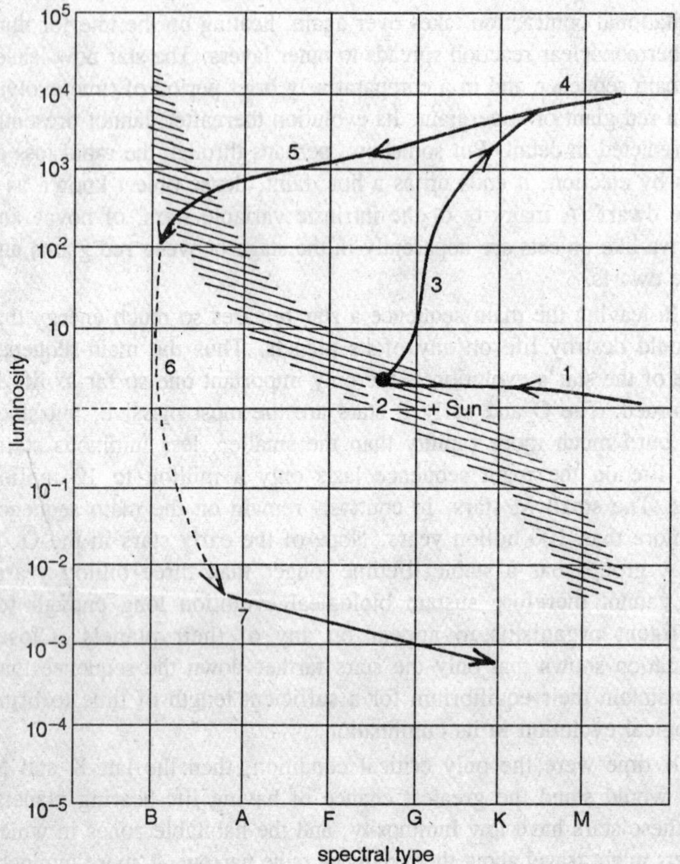

Fig. 4.1. Evolutionary path of a star with a mass 1.2 times that of the sun is traced.

This number is reduced considerably upon consideration of the third critical condition; the maintenance of stable planetary orbits. The dust-cloud mechanism that increases the likelihood of planets has also brought most stars into existence as double or multiple systems. It is obvious that the presence of two or more stars in a system will profoundly perturb the orbits of planets in the system. A few double-star systems have members that are so far apart that if they are sufficiently near to us, the stars can be distinguished by a telescope or even by the naked eye. Such systems are called visual binaries. Much more common are the "spectroscopic binaries." In these systems the stars are so close together that even if they are relatively near to us, they cannot be separated by the largest telescope. We can tell that

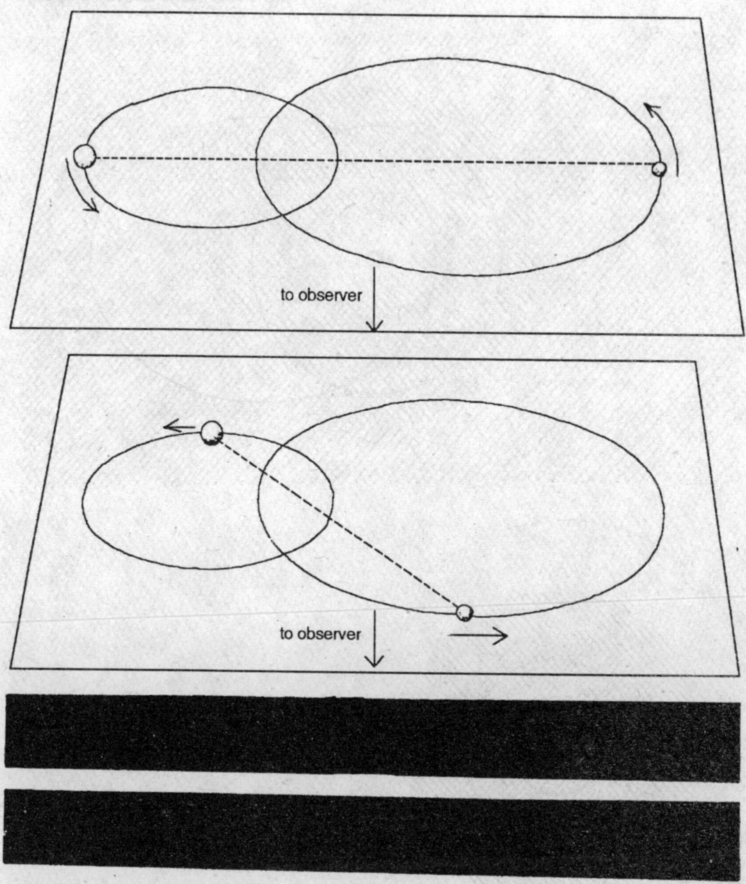

Fig. 4.2. Spectroscopic detection of close binaries is depicted in these diagram and spectra.

they are double stars only by regular changes in their spectra. These changes also enable us to calculate their orbits.

The orbits of planets in such systems are so complicated that astronomers have been able to work them out for only a few idealized cases. A life-bearing planet would have to travel on an obrbit close to one of the stars in a widely separated system, or on an orbit at a large distance from the stars in a close system. In the case of a hypothetical binary composed of two stars with the same luminosity as our sun and revolving around a common center in a nearly circular orbit, a planet would find the thermally habitable zone dynamically stable only if the stars were more than 10 astronomical units or less than .05 astronomical unit apart (an astronomical unit is the distance

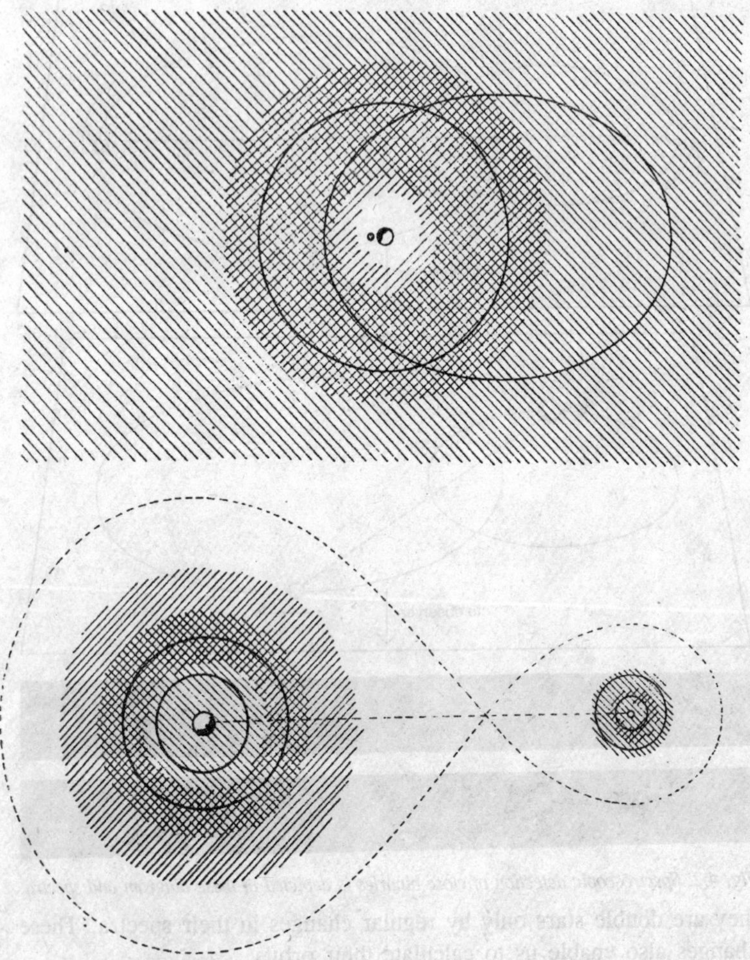

Fig. 4.3. Planetary orbits around binary stars are depicted here in an idealized manner.

between the earth and the sun). With a separation between the stars of .5 to 2 astronomical units there is no overlapping of the habitable zone and the dynamically stable zone. Therefore no inhabitable planets can exist in such systems.. Taking everything into consideration, only 1 to 2 per cent of all double and multiple stars may possess inhabitable planets, and perhaps 3 to 5 per cent of all the stars in our galaxy have such planets.

For the present there is no hope of detecting on a photographic plate the existence of a planet of another star. Such planets as may attend even the nearest star are completely lost to view in the brilliance of the star's light. Some-day there may be a telescope on a platform in space, free of the interfering effects of the earth's atmosphere. As has been suggested by Nancy G. Roman of the National Aeronautics and Space Administration, the instrument would produce a sharp star image that could be blocked out so that a planet near the star could be detected by means of a long photographic exposure. Even the subtle methods used to detect spectroscopic binaries are not refined enough to find a planet, because the effect of a planet upon the motion of its parent star is so slight.

There is nonetheless other well-established evidence to support the contention that planets are common. In the first place, no sharp distinction can be drawn between binary or multiple stars and stars with planetary systems. According to Gerard P. Kuiper of the Yerkes Observatory, the mean distance of separation between the components of all binaries so far investigated is about 20 astronomical units. This is of the same order of magnitude as the distance between the sun and its major planets (Jupiter, Saturn, Uranus and Neptune). Kaj Aa. Strand at the U.S. Naval Observatory in Washington has studied small perturbations in the orbital motion of a star called 61 Cygni, which

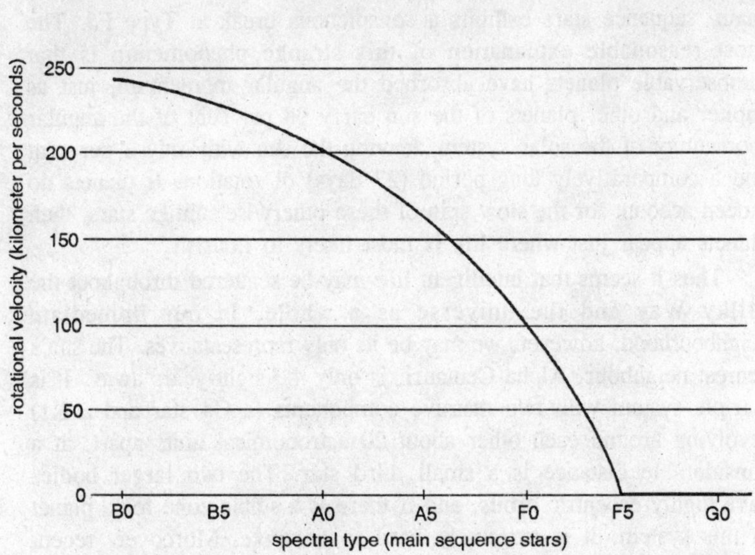

Fig. 4.4. Rotational velocity of stars on the main sequence is diagrammed here according to spectral type.

has a companion that is too faint to be directly observed. He has found that the mass of this unseen object is about a hundredth of that of the sun. This mass lies between that of stars and of Jupiter. It is therefore reasonable to believe that the masses of small stars in binary systems grade continuously down to the masses of planets. Since binaries are so common in our galaxy, it would seem that many stars that now appear to be alone actually possess planets.

Harold C. Urey of the Scripps Institution of Oceanography has found additional evidence for planets in a certain kind of meteorite. According to Urey, diamonds embedded in these objects show that they must at one time have been under high pressure in a body the size of the moon. Such moonlike objects, known as "pestellar nuclei," would enhance the formation of both stars and planets from a dust cloud. Any irregularity in the motions of a dust cloud should be expected to produce more than one such nucleus, and the formation of planets or multiple stars would follow as a normal consequence.

Finally, measurements of the angular momenta of many stars give every indication that planets exist outside the solar system. Otto Struve of the National Radio Astronomy Observatory has pointed out that main-sequence stars more massive than Type F5 usually rotate rapidly, but starting with this type the rotation of stars slows down abruptly. In other words, the average angular momentum per units mass of the main- sequence stars exhibits a conspicuous break at Type F5. The most reasonable explanation of this strange phenomenon is that unobservable planets have absorbed the angular momentum, just as Jupiter and other planets of the sun carry 98 per cent of the angular momentum of the solar system, leaving the sun with only 2 per cent and a comparatively long period (27 days) of rotation. If planets do indeed account for the slow spin of these otherwise sunlike stars, then planets appear just where life is most likely to flourish.

Thus it seems that intelligent life may be scattered throughout the Milky Way and the universe as a whole. In our immediate neighbourhood, however, we may be its only representatives. The sun's nearest neighbour, Alpha Centauri, is only 4.3 light-years away. It is a triple system with two massive components (a G4 star and a K1) revolving around each other about 20 astronomical units apart; at a considerable distance is a small third star. The two larger bodies have highly eccentric orbits, and if there is a stable zone for a planet in this system, it is extremely hard to compute. Moreover, recent investigations indicate that the system may be much younger than the sun, so that higher forms of life might not have had time to evolve even if a habitable planet does exist in it.

Forty other stars are located within five parsecs (16.7 light-years) of the sun. Only two-Epsilon Eridani (a K2 type) and Tau Ceti (G4) —seem to fulfil the conditions for the existence of advanced forms of life, and Epsilon Eridani may not be exactly on the main sequence. Tau Ceti is 10.8 light-years distant, has an apparent visual magnitude of 3.6, is located on the celestial sphere about 16 degrees south of its equator and appears above the horizon in the northern sky only in the winter.

Since intelligent life is probably not a rare phenomenon, and since at least one star in our vicinity meets the specifications of a life-fostering star, it may seem odd that we have had no visitors from other worlds. The idea would have drawn ridicule 20 years ago, but today it deserves consideration. There are, however, several reasons for believing that we have had no visitors from outer space. For one thing, the 10.8 light-years that separate us from Tau Ceti—astronomically a short distance—is an extremely long distance in terms of human experience. Even traveling at the speed of the artificial satellites that man has launched, space voyagers would need hundreds of thousands of years to traverse it. It is possible that organisms from Tau Ceti might have a far longer life-span than man, but this supposition invokes radical assumptions that cannot be supported by present knowledge. Furthermore, if a meeting is to occur, our high technological civilization would have to be contemporary with that created by intelligent organisms on other planets. The cultural evolution that brought mankind to its present technical competence began only a few centuries ago. And even if human civilization endures for hundreds of thousands of years, it would be a brief episode in the time-scale of biological evolution. Therefore the chance that advanced civilizations might be flourishing at the present time on the planets of one or two nearby stars is excessively small.

Some workers have concluded, however, that the chance is good enough and certainly intriguing enough—to institute radio surveillance of signals originating outside the solar system. Guiseppe Cocconi and Philip Morrison of Cornell University have pointed out that the most favourable wavelength would be one close to but outside of the 21-centimeter line emitted by hydrogen in space, because in this region of the spectrum the galactic noise and the noise produced in the earth's atmosphere are at a minimum. Moreover, this important wavelength would be of as much interest to astronomers on another planet as to those on earth. Cocconi and Morrison have urged that radiation picked up by the 600-foot radio telescope, now being built by the Navy in West Virginia, be analyzed for the presence of signals. According to them, that telescope will be capable of detecting signals generated 10 light-years away by a technology no more advanced than our own. Meanwhile Frank D. Drake of the National Radio Astronomy Observatory, who makes a more generous estimate of the distance at which signals sent by intelligent organisms could be detected, is in charge of an actual project that is employing a smaller telescope to detect any radio signals transmitted by living beings in other "solar systems."

What kind of signals may we expect to receive or should we send out? Probably the most abstract and the most universal conception that any intelligent organisms anywhere would have devised is the sequence of cardinal numbers: 1, 2, 3, 4 and so on. The most likely signal would be a series of pulses indicating this sequence repeated at regular intervals. Such a signal may upon first consideration appear to be too simple for the sophisticated task of communicating with other beings far away among the stars. It would sound like baby talk. But after-all interstellar communication is surely still in the baby-talk stage.

Search for Life on Mars

Is there life on Mars? The question is an interesting and legitimate scientific one, quite unrelated to the fact that generations of science-fiction writers have populated Mars with creatures of their imagination. Of all the extraterrestrial bodies in the solar system Mars is the one most like the earth, and it is by far the most plausible habitat for extraterrestrial life in the solar system. For that reason major objective of the Viking mission to Mars was to search for evidences of life.

The two Viking spacecraft were launched from Cape Canaveral in the summer of 1975. Each spacecraft consisted of an orbiter and an attached lander. When the spacecraft arrived at Mars in July and August of 1976, each was put in a predetermined orbit around the planet, and the search for a landing place began. Cameras aboard the orbiters were the principal source of information on which the choice of the landing sites was based; important data also came from infrared sensors on the orbiters and from radar observatories on the earth. The sole consideration in the final selection of the sites was the safety of the spacecraft. It would be a mistake to suppose, however, that the sites were therefore without biological interest. Biological criteria dominated the initial decisions as to the latitude at which each spacecraft would land. Once the latitudes had been chosen there was relatively little difference between sites at different longitudes.

On command from the earth each lander separated from its orbiter. With the help of its retroengines and parachute it dropped to the surface of Mars. Both orbiters continued to circle the planet, operating their own scientific instruments and relaying to the earth data transmitted from the landers. Both landings were in the northern hemisphere of Mars, and the Martian season was summer. (Mars has seasons like those on the earth, but each season lasts approximately twice as long. The Martian year is 687 Martian days; each Martian day, named a sol by the Viking team to distinguish it from a terrestrial day, is 24 hours 39 minutes long). On July 20, 1976, the *Viking I* lander came to rest in the Chryse Planitia region of Mars, some 23 degrees north of the equator, Six weeks later the *Viking* 2 lander settled down in the Utopia Planitia region, some 48 degrees north of the equator. In longitude the two landers are separated by almost exactly 180 degrees, thus placing them on opposite sides of the planet. Since the instrumentation of the two landers is identical, the difference in their landing sites is the only distinction between them.

The first biologically significant task carried out by each lander was the analysis of the Martian atmosphere. Life is based on the chemistry of light elements, notably carbon, hydrogen, oxygen and nitrogen. To be suitable as an abode of life a planet must have those elements in its atmosphere. Spectroscopic observations from the earth and from spacecraft that had flown past Mars in previous years had already shown that carbon dioxide was the principal component of the Martian atmosphere. Small quantities of carbon monoxide, oxygen and water vapour had also been detected. Nitrogen had not been detected in any form, however, and atmospheric theory suggested that Mars had lost most of its nitrogen in the past.

Each Viking lander analyzed the atmosphere by means of two mass spectrometers. One spectrometer, operating during the descent to the surface, sampled and analyzed the atmospheric gases every five seconds. The second spectrometer operated on the ground. The results showed that the atmosphere near the ground was approximately 95 per cent carbon dioxide. 2.5 per cent nitrogen and 1.5 per cent argon, and that it also held traces of oxygen, carbon monoxide, neon, krypton and xenon. At both landing sites the atmospheric pressure was 7.5 millibars. (The atmospheric pressure at sea level on the earth is 1.013 millibars).

Since the Viking spacecraft revealed that nitrogen is indeed present in the Martian atmosphere, we can say that the elements necessary for life are available on Mars. Missing from the list of gases, however is one critically important compound: water vapour. Although earlier measurements had shown that traces of water vapour are present in the Martian atmosphere, the quantity varies with season and place. The Viking orbiters carried out a survey of water vapour over the entire planet with infrared spectrometers. The results showed that the highest concentration of atmospheric water vapour was at the edge of the north polar cap (the summertime hemisphere), and that the concentration fell off toward the south (the opposite of what is found on the earth). In the polar region the amount of water vapour in the atmosphere would form a film only a tenth of a millimeter thick if all of it were to be condensed on the planet's surface. At the landing sites the concentration of water vapour ranged between 10 and 30 per cent of the concentration at the pole.

These numbers put into quantitative terms a long-known fact about Mars: It is a very dry place. Mars has ice at its poles, but nowhere on its surface are there oceans or lakes or any other bodies of liquid water. The absence of liquid water is related to the dryness of the atmosphere through a fundamental law of physical chemistry: the phase rule. The phase rule states that for liquid water to exist on the surface of a planet the pressure of the water vapour in the atmosphere must at some times and in some places be at least 6.1 millibars. The Viking measurements imply that the vapour pressure of water at the surface of Mars in the northern hemisphere is at most .05 millibar, even if all the water vapour is concentrated in the lower atmosphere. At that low pressure liquid water cannot remain in the liquid phase: depending on the temperature, it must either freeze or evaporate. By the same token raindrops cannot form in the Martian atmosphere and ice cannot melt on the Martian surface.

The extreme dryness presents a difficult problem for any Martian biology. Liquid water is essential for life on the earth. All terrestrial species have high and apparently irreducible requirements for water; none could live on Mars. If there is life on Mars, it must operate on a different principle as far as water is concerned. If Mars had a more favourable environment in the past, however, and if the planet did not dry up too fast, species may have had time to evolve and adapt to present conditions. Pictures made by the *Mariner 9* spacecraft, which went into orbit around Mars in 1971, suggested that Mars may indeed have had running water on its surface in the past. The pictures from the Viking orbiters have confirmed that impression. The evidence consists of channels in the Martian desert that resemble dry riverbeds. There seems to be little doubt that the channels were carved by rapidly flowing liquid, and there is widespread agreement that the most probable liquid is water.

If liquid water once existed on Mars, could life have arisen on the planet? If the life evolved to meet changing conditions, could it exist there still? There is no way to settle these questions by deductive reasoning or even by experimentation in laboratories on the earth. They can be answered only by the direct exploration of Mars, and that is what the Viking spacecraft did.

Five different types of instrument on each Viking lander were involved in the search for evidences of life: two cameras for photographing the landscape, a combined gas chromatograph and mass spectrometer for analyzing the surface for organic material and three instruments designed to detect the metabolic activities of any microorganisms that might be present in the soil engineers and managers whose joint efforts made all the Viking projects possible. They work in universities, industrial laboratories and the National Aeronautics and Space Administration and its field centers. Their names are recorded in the growing technical literature dealing with this historic mission.

Each of the Viking landers carried two cameras of the facsimile type, which built up a picture of the scene by scanning it in a series of narrow strips. Such cameras make pictures slowly, but they are rugged and versatile. Their resolution was moderately high: a few millimeters at a distance of 1.5 meters. They produced pictures in black and white, in colour and in stereo. The two cameras on each lander could between them survey the entire horizon around the spacecraft.

As life-seeking tools cameras have inherent advantages and disadvantages. Their chief advantage lies in the fact that a picture contains a large amount of information. In principle it would be possible to prove unequivocally the existence of life on Mars with a single photograph. For example, if a line of trees were visible on the horizon or if foot-prints appeared on the ground in front of the spacecraft one morning, there would be no room for doubt that there is life on Mars. Another advantage lies in the fact that pictorial evidence is independent of all assumptions about the chemistry and physiology of Martian organisms. The organisms need not respond in certain ways to certain substances or treatments in order to be recognized. The cameras could identify, say, a mushroom made of titanium as a form of life if one were to sprout up from under a rock in the course of the mission. Of course, reliance on pictorial evidence rests on its own set of assumptions about the morphology of living things. The most obvious disadvantage of the camera as a life-seeking instrument is the fact that an entire world of life can exist below the camera's limit of resolution.

Of all the results of the Viking mission the wonderful photographs of the Martian desert at the two landing sites are the most impressive. The photographs have been eagerly scanned by alert and hopeful eyes, but no investigator has yet seen anything suggesting a living form.

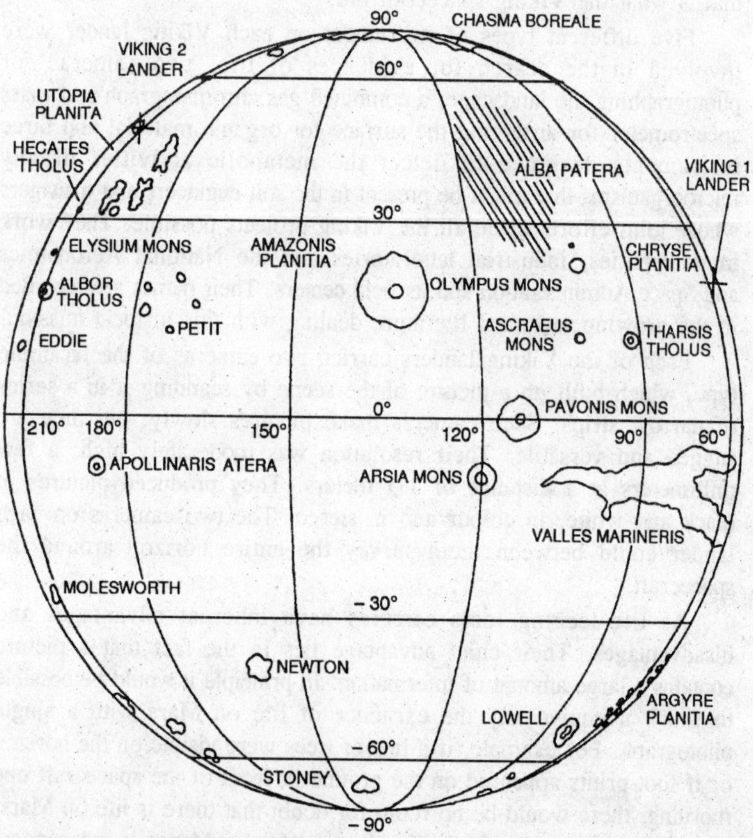

Fig. 4.5. Locations of the two Viking Landers are indicated on this map of Mars, which shows some of the major geological features of the planet.

The next step was to analyze the soil for any organic constituents. Among the elements carbon is unique in the number, variety and complexity of the compounds it can form. The special properties of carbon that enable it to form large and complex molecules arise from the basic structure of the carbon atom. The structure enables the carbon atom to form four strong bonds with other atoms, including other carbon atoms. The molecules thus formed are very stable at ordinary temperatures, so stable, in fact that there seems to be no limit to the size they can attain. The connection between life and organic chemistry (that is, the chemistry of carbon) rests on the fact that the attributes by which we identify living things—their capacity to replicate themselves, to repair themselves, to evolve and to adapt—originate in properties that are unique to large organic molecules. It is the highly complex information-rich proteins and nucleic acids that endow all the living things we know, even "simple" ones such as bacteria and viruses, with their essential nature. No other element, including that favourite of science-fiction writers silicon has the capacity carbon has to form large and complex structures that are so stable. It is no accident that even though silicon is far more abundant than carbon on the earth, it has only minor and nonessential roles in biochemistry. Biochemistry is largely a chemistry of carbon.

Such fundamental facts lead to the conclusion that wherever life arises in the universe it will most likely be based on carbon chemistry. That view has been strengthened by the discovery of organic compounds of biological interest in meteorites and in clouds of dust in interstellar space. Although these compounds are nonbiological in origin, they are closely related to the amino acids and the nucleotides that are the respective building blocks of proteins and of nucleic acids. The fact that they are formed in settings remote from the earth implies that carbon chemistry gives rise to familiar organic compounds throughout the universe. This fact in turn suggests that life elsewhere in the universe will be based on an organic chemistry similar to our own, although not necessarily identical with it.

Such considerations led to the decision to include an organic-analysis experiment *aboard* the Viking landers. The instrument used in the experiment was the mass spectrometer that had analyzed the atmosphere combined with a gas chromatograph and a pyrolysis furnace. A sample of the Martian soil was first heated in the furnace through series of steps up to a temperature of 500 degrees Celsius. Any volatile materials released were passed through the gas chromatograph. Since each of the different compounds has a different molecular weight, composition and polarity, among other properties, it passed through the columns of the gas chromatograph at a unique rate, and so the compounds were separated from one another. As each compound emerged from the chromatographic column it was directed into the mass spectrometer for identification. Since essentially all organic matter is cracked, or decomposed, into smaller fragments at 500 degrees C., the method is capable of detecting organic compounds that have a wide range of molecular weights.

Two soil samples were analyzed at each landing site. The only organic compounds detected were traces of cleaning solvents known to have been present in the apparatus. The fact that the solvents were detected shows the instruments were functioning properly. The heated samples gave off carbon dioxide and a small amount of water vapour; nothing else was found.

This result is surprising and weighs heavily against the existence of biological processes on Mars. The combined gas chromatograph and mass spectrometer aboard each Viking lander is a sensitive instrument, capable of detecting organic compounds at a concentration of a few parts per billion, a level that is between 100 and 1,000 times below their concentration in desert soils on the earth. Even if there is no life on Mars, it has been supposed the fall of meteorites onto the Martian surface would have brought enough organic matter to the planet to have been detected. Because Mars is near the asteroid belt, from which meteorites originate, it is believed to receive a much larger number of meteorite impacts than either the earth or the moon. Indeed, a question that was frequently discussed before the Viking spacecraft were launched was whether or not it would be possible to distinguish biological organic matter on Mars from the meteoritic organic matter that was expected to be present. The absence of organic matter at the parts-per-billion level, however, suggests that on Mars organic compounds are actively destroyed, probably by the strong ultraviolet radiation from the sun.

The other experiments aboard the Viking landers searched not just for organic matter in the soil but for living organisms. On the earth microorganisms such as bacteria, yeasts and molds are the hardiest of all species. There are few places on the earth where microbial forms do not live; they are the last survivors in environments of extreme temperature and aridity. The reasons for their hardiness are interesting but need not detain us here. Suffice it to say that if there is life on

Mars, the chance of detecting it would be maximized by searching for microorganisms in the Martian soil.

Each Viking lander carried three instruments designed to detect the metabolic activities of soil microorganism. First, the gas-exchange experiment was designed to detect changes in the composition of the atmosphere caused by microbial metabolism. Second, the labeled-release experiment was designed to detect decomposition of organic compounds

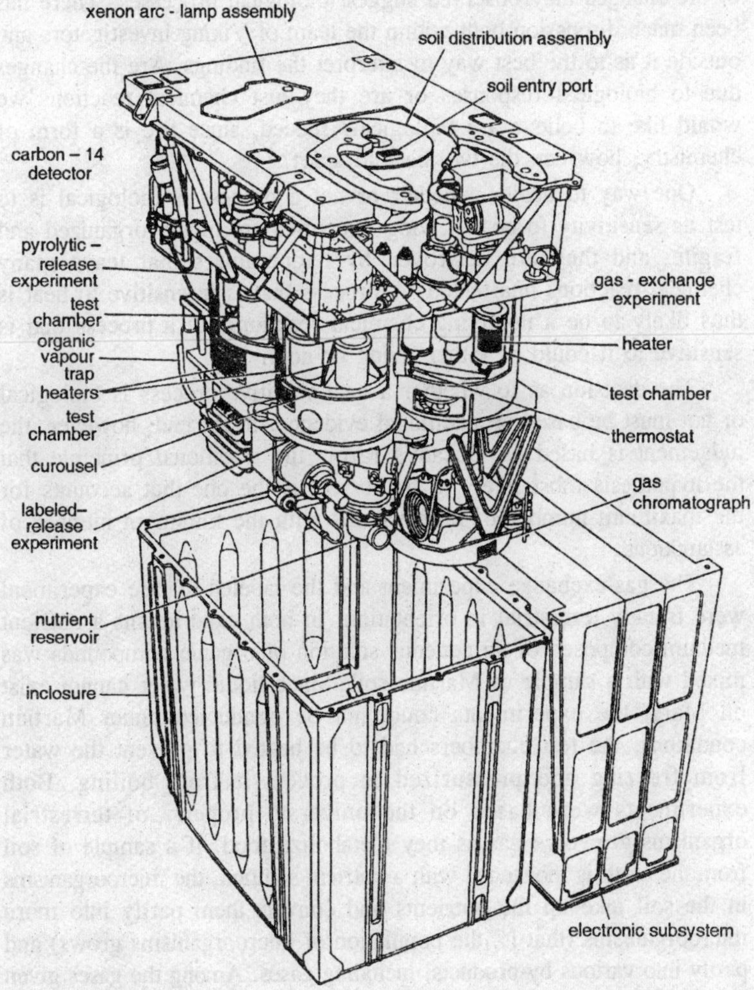

Fig. 4.6. Biological laboratory aboard both Viking spacecraft occupies a volume of only one cubic foot.

by soil microbes when they were fed with a nutrient. Third, the pyrolytic-release experiment was designed to detect the synthesis of organic matter in Martian soil from gases in the atmosphere by either photosynthetic or non-photosynthetic processes. All three experiments analyzed portions of each sample of Martian soil.

All the experiments detected chemical changes of one kind or another in the soil. All the experiments are now completed, and some of the changes they observed suggest biological processes. There has been much discussion both within the team of Viking investigators and outside it as to the best way to interpret the findings. Are the changes due to biological responses or are they just chemical reactions we would like to believe are biological? Indeed, since life is a form of chemistry, how can the two be told apart?

One way to decide whether or not a process is biological is to test its sensitivity to heat. Living structures are highly organized and fragile, and they are destroyed by temperatures that leave many chemical reactions unaffected. A process that is insensitive to heat is thus likely to be a nonliving chemical reaction, but a process that is sensitive to it could be either living or nonliving.

The decision as to whether a heat-sensitive process is biological or not must be based on additional evidence. In the end, however, the judgement is based on Occam's razor: the traditional principle that the hypothesis most likely to be correct is the one that accounts for the maximum number of observations with the minimum number of assumptions.

The gas-exchange experiment and the labeled-release experiment were frankly terrestrial in orientation. In both experiments a nutrient medium composed of an aqueous solution of organic compounds was mixed with a sample of Martian soil. Since liquid water cannot exist on Mars, the experiments could not be conducted under Martian conditions; the test chambers had to be heated to prevent the water from freezing and pressurized to prevent it from boiling. Both experiments were based on the universal property of terrestrial organisms to evolve gas as they metabolize food. If a sample of soil from the earth is moistened with a nutrient solution, the microorganisms in the soil take up the nutrients and convert them partly into more microorganisms (that is, the population of microorganisms grows) and partly into various by-products, including gases. Among the gases given off in microbial metabolism are carbon dioxide, methane, nitrogen, hydrogen and hydrogen sulfide. On the earth gases evolved by one species of organisms are eventually consumed by other species of organisms. In that way the light elements at the earth's surface are continually cycled through the biosphere and the atmosphere.

In the gas-exchange experiment a complex nutrient solution was added to a sample of Martian soil in a closed chamber, and the gases were analyzed periodically by means of a gas chromatograph. The experiment proceeded in two stages. In the first stage a small volume of the nutrient solution was introduced into the soil chamber in such a way that it humidified the chamber without actually wetting the soil, and the resulting gases were analyzed several times. In the second stage a large volume of the nutrient was poured into the chamber, saturating the soil. With the soil now in direct contact with the medium the main part of the experiment began. The soil was incubated for nearly seven months, so that whatever microorganisms might be in the sample had enough time to signal their presence by producing or consuming gases. During the period of incubation the atmosphere in the chamber was periodically analyzed.

The findings of the first stage of the experiment were both surprising and simple. Immediately after the soil sample was humidified

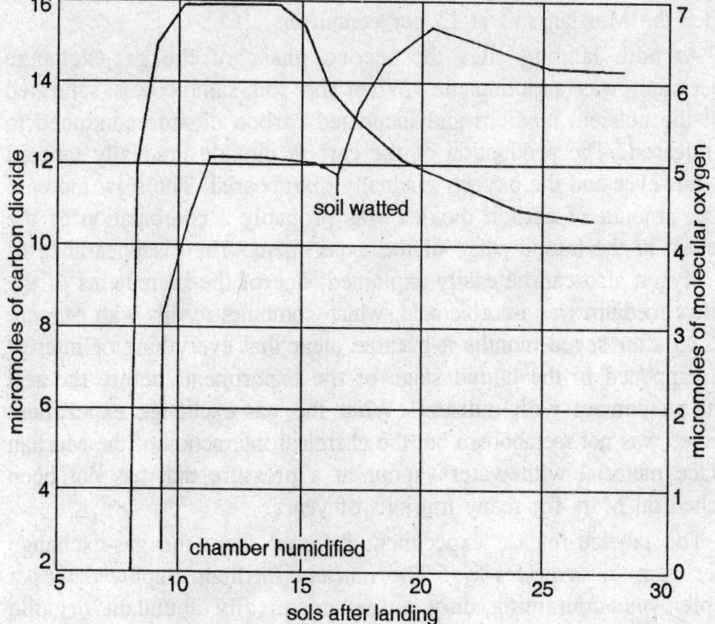

Fig. 4.7. Results of the gas-exchange experiment, according to data of Vance I.

carbon dioxide and oxygen were rapidly released. The release of the gases ceased soon after it had begun but not before the pressure in the chamber had risen measurably. At the Chryse site in a period of little more than one sol the quantity of carbon dioxide in the incubation chamber of the *Viking I* lander increased by a factor of five and the quantity of oxygen increased by a factor of 200. At the Utopia site the increases were less, but they were still considerable.

The rapidity and the brevity of the response recorded by both landers clearly suggested that the process observed was a chemical reaction not a biological one. The appearance of the carbon dioxide is readily explained. Carbon dioxide gas would be expected to be adsorbed on the surface of the dry Martian soil; if the soil was exposed to a very humid atmosphere, the gas would be displaced by water vapour. The appearance of the oxygen is more complex. The production of so much oxygen seems to require an oxygen-generating chemical reaction, not just a physical liberation of pre-existing gas. It is likely that the oxygen was released when the water vapour decomposed an oxygen-rich compound such as a peroxide. Peroxides are known to decompose if they are exposed to water in the presence of iron compounds, and according to the X-ray fluorescence spectrometer aboard each Viking lander the Martian soil is 13 per cent iron.

At both landing sites the second phase of the gas-exchange experiment was anticlimactic. When the soil sample was saturated with the nutrient medium and incubated carbon dioxide continued to be released. The production of the carbon dioxide gradually tapered off, however and the oxygen gradually disappeared. The slow increase in the amount of carbon dioxide was probably a continuation of the reaction in the humid stage of the experiment. The disappearance of the oxygen also can be easily explained: one of the ingredients of the nutrient medium was ascorbic acid, which combines readily with oxygen. And so after seven months it became clear that everything of interest had happened in the humid stage of the experiment, before the soil came in contact with nutrient! What the gas-exchange experiment detected was not metabolism but the chemical interaction of the Martian surface material with water vapour at a pressure that has not been reached on Mars for many millions of years.

The labeled-release experiment differed from the gas-exchange experiment in several ways. The nutrient medium employed was a simpler one containing only a few cosmically abundant organic compounds such as formic acid (HCOOH) and the amino acid glycine

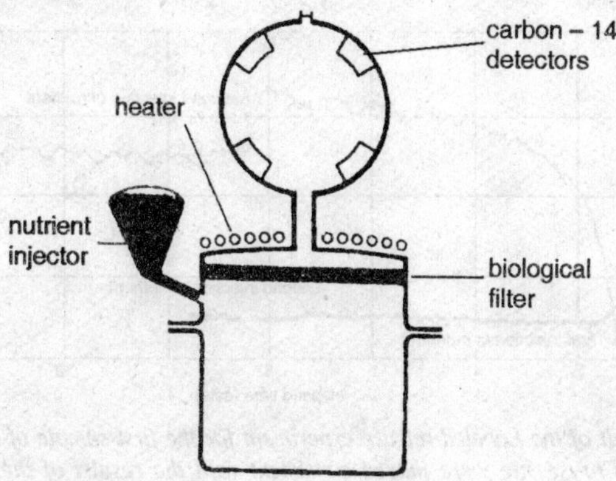

Fig. 4.8. Labeled-release experiment on the Viking landers tested the Martian soil for microorganisms that could metabolize simple organic, or carbon, compounds.

(NH_2CH_2COOH). All the compounds were labeled with atoms of the radioactive isotope carbon·14. The labeled-release instrument was designed to detect radioactive gases, principally carbon dioxide, released when the nutrient medium was added to a sample of soil. The number of radioactive disintegrations in gases can be counted quite efficiently, so that the labeled-release experiment is faster and more sensitive than the gas-exchange experiment in detecting microbial activity in terrestrial soil. The labeled-release experiment's sequence of operations did not include a humid stage as such, but it attempted to accomplish the same end by injecting a volume of nutrient medium that was insufficient to wet the entire soil sample but sufficient to humidify the chamber. If the experiment worked on Mars as planned, subsequent injections of the medium, which were controlled by commands sent from the earth, brought the medium into contact with some soil that had previously been wetted and with other soil that had been humidified but not wetted.

As in the gas-exchange experiment, immediately after the nutrient medium was added to the soil in the labeled-release experiment, gas surged into the chamber. The release of gas tapered off soon after the first sol. The gas, undoubtedly carbon dioxide, was radioactive, showing that it had been formed from the radioactive compounds of the medium and not from compounds in the Martian soil. Non-radioactive gases, which also must have formed when the aqueous medium came in contact with the soil were not detectable in the experiment.

The production of radioactive carbon dioxide in the labeled-release experiment is understandable in the light of the evidence from the

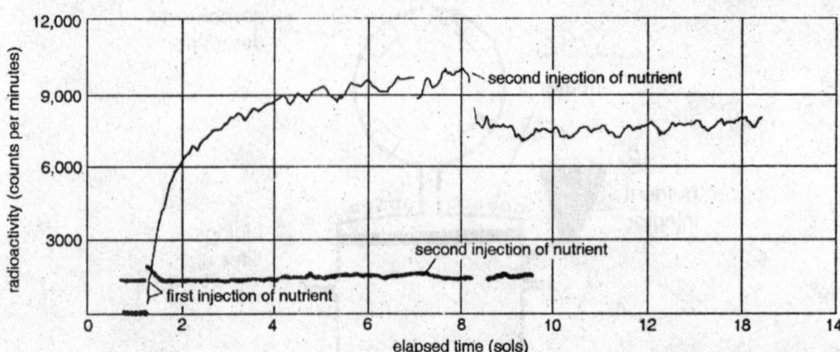

Fig. 4.9. Result of the Labeled-release experiment for the first sample of soil analyze at the Chryse site were indeed consistent with the results of the gas-exchange experiment.

gas-exchange experiment suggesting that the surface material of Mars contains peroxides. Formic acid, one of the compounds of the labeled-release nutrient medium, is oxidized with particular ease: if a molecule of formic acid (HCOOH) reacts with one of hydrogen peroxide (H_2O_2), it will form a molecule of carbon dioxide (CO_2) and two mɔlecules of water ($2H_2O$). The amount of radioactive carbon dioxide given off in the labeled release experiment was only slightly less than what would have been expected if all the formic acid in the medium had been oxidized in this way.

If the source of the oxygen released in the humid stage of the gas-exchange experiment was indeed peroxides in the soil decomposed by water vapour, then in the labeled-release experiment all the peroxides should also have been decomposed by the first injection of nutrient. Thus the next injection should have evolved no additional radioactive gas inspite of the fact that part of the sample presumably had not yet been wetted by the medium. That proved to be the case. When a second volume of medium was injected into the chamber, the amount of the gas in the chamber was not increased; indeed, it decreased. The decrease is explained by the fact that carbon dioxide is quite soluble in water; when fresh nutrient medium was added to the chamber, it absorbed some of the carbon dioxide in the head space above the sample.

This result was obtained with all the samples tested by the labeled-release experiment at both Viking sites. In that respect the results of the labeled-release experiment did not parallel the results of the gas-exchange experiment. At both sites both experiments tested soil gathered from the ground's exposed surface; at the Utopia site the experiments also tested soil gathered from under a rock. Although the labeled-release experiment found essentially no difference in the amount of gas released by any of the samples, the gas-exchange experiment recorded about three-fourths as much carbon dioxide from the surface samples at the Utopia site as it had from the surface samples at the Chryse site, and it recorded even less carbon dioxide from the sample from under the rock.

The gas-exchange experiment also recorded less oxygen from the samples from the Utopia region, but the interference of the ascorbic acid in the complex nutrient medium of that experiment makes it difficult to quantify the difference. In every case, however the gas-exchange experiment detected considerably more gas than the labeled-release experiment did with portions of the same sample. Those results, however, do not contradict the thesis that the production of oxygen detected by the gas-exchange experiment and the production of radioactive carbon dioxide detected by the labeled-release experiment are simply different measurements of the same surface chemistry. The gas-exchange experiment measures the total amount of oxidant in the surface; the labeled-release experiment measures only a fraction of it.

The labeled-release experiment also tested the stability of the reaction to heat. When the soil was reheated to 160 degrees C. for three hours before incubation, the reaction was abolished. When it was heated to 46 degrees for the same length of time, the magnitude of the reaction was reduced by about half. These results have been regarded by some as evidence in favour of the hypothesis that the reaction is biological. The results are of course consistent with such a hypothesis, but they are also consistent with a chemical oxidation in which the oxidizing agent is destroyed or evaporated at relatively low temperatures. A variety of both inorganic peroxides and organic peroxides could probably have produced the same results.

The third microbiological experiment, the pyrolytic-release experiment, differed from the gas-exchange and labeled-release experiments in two respects. First, it attempted to measure the synthesis of organic matter from atmospheric gases rather than its decomposition. Second, it was designed to operate under the conditions of pressure, temperature and atmospheric composition that actually obtain on Mars, since those are the conditions under which any form of Martian life must exist. In practice the conditions in the chamber were a reasonably good approximation of Martian conditions except for the temperature, which stayed warmer than the outside temperature because of heat sources within the spacecraft.

A sample of Martian soil was sealed in a chamber along with some Martian atmosphere. A quartz window in the chamber admitted simulated Martian sunlight from a xenon arc lamp. Into this Martian microcosm small amounts of radioactive carbon dioxide and radioactive carbon monoxide were introduced. Both gases are present in the Martian atmosphere but not in radioactive form. After five days the lamp was turned off, the atmosphere was removed from the chamber and the soil was analyzed for the presence of radioactive organic matter.

First the soil was heated in the pyrolysis furnace to a temperature high enough to crack any organic compounds into small volatile fragments. The fragments were swept out of the chamber by a stream of helium and passed through column that was designed to trap organic molecules but allow carbon dioxide and carbon monoxide to pass through. The radioactive organic molecules were thus transferred from the soil to the column and at the same time were separated from any remaining gases of the incubation atmosphere. The organic molecules were released from the column by raising the column's temperature. Simultaneously the radioactive organic molecules were decomposed into radioactive carbon dioxide by copper oxide in the column. The carbon dioxide was then carried by the stream of helium into a radiation counter. If organic compounds had been synthesized in the soil, they would be detected as radioactive carbon dioxide; if no organic compounds had been synthesized, no radioactive carbon dioxide would have been formed.

Surprisingly, seven of the nine pyrolytic-release tests executed on Mars gave positive results. The two negative results were obtained at the Utopia site, but a third sample tested at Utopia was positive. This third sample was actually incubated in the dark, implying that light may not be required for the reaction. The amount of carbon fixed in the soil by the experiment was small: enough to furnish organic matter for between 100 and 1,000 bacterial cells. The quantity is so small, in fact, that it could not have been detected by the organic analysis experiment. The quantity is nonetheless significant; it was surprising that in such a strongly oxidizing environment even a small amount of organic material could be fixed in the soil.

Even more significant, the pyrolytic-release instrument had been rigorously designed to eliminate nonbiological sources of organic compounds. During the development of the experiment it had been found that in the presence of short-wavelength ultraviolet radiation, carbon monoxide spontaneously combined with water vapour to form

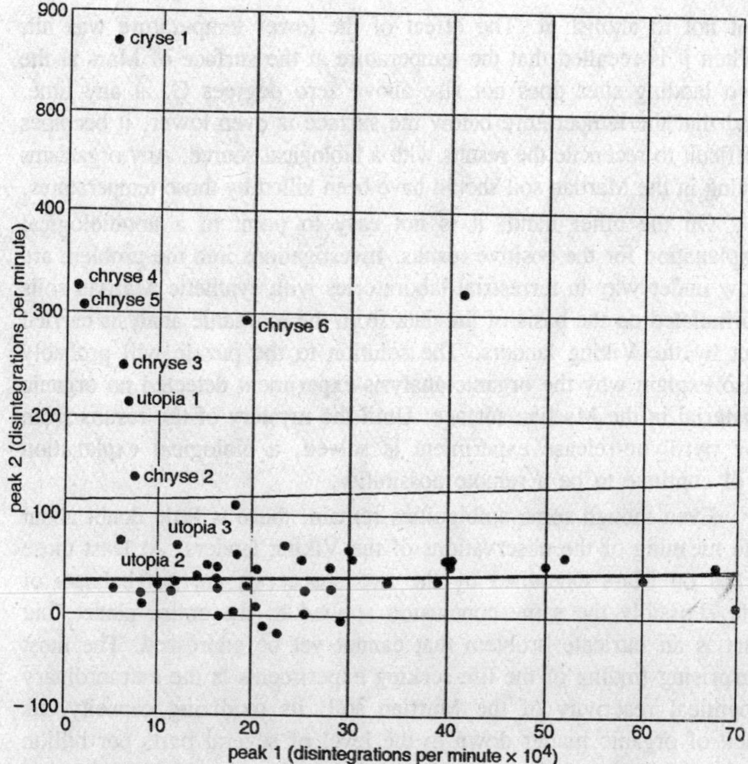

Fig. 4.10. Results of Pyrolytic-release experiment are shown for all the samples tested on Mars.

organic molecules on glass, quartz and soil surfaces in the experimental chamber. In order to avoid those reactions and the confusion they would have caused, the short-wavelength ultraviolet was filtered out of the radiation allowed to enter the incubation chamber. To receive positive results from the soil on Mars in spite of that precaution was startling.

Nevertheless, it appears that the findings of the pyrolytic-release experiment must also be interpreted nonbiologically. The reason is that the reaction detected was less sensitive to heat than one would expect of a biological process. In two of the nine pyrolytic-release experiments performed on Mars the soil sample was heated before the radioactive gases were injected and the incubation was begun. In one case the sample was held at 175 degrees C. for three hours and in the other it was held at 90 degrees for nearly two hours. The effect of the higher temperature was to reduce the reaction by almost 90 per cent but not to abolish it. The effect of the lower temperature was nil. When it is recalled that the temperature at the surface of Mars at the two landing sites does not rise above zero degrees C. at any time, and that the temperature below the surface is even lower, it becomes difficult to reconcile the results with a biological source. Any organisms living in the Martian soil should have been killed by those temperatures.

On the other hand, it is not easy to point to a nonbiological explanation for the positive results. Investigations into the problem are now under way in terrestrial laboratories with synthetic Martian soils formulated on the basis of the data from the inorganic analysis carried out by the Viking landers. The solution to the puzzle will probably also explain why the organic-analysis experiment detected no organic material in the Martian surface. Until the mystery of the results from the pyrolytic-release experiment is solved, a biological explanation will continue to be a remote possibility.

Even though some ambiguities remain, there is little doubt about the meaning of the observations of the Viking landers: At least those areas on Mars examined by the two spacecraft are not habitats of life. Possibly the same conclusion applied to the entire planet, but that is an intricate problem that cannot yet be addressed. The most surprising finding of the life-seeking experiments is the extraordinary chemical reactivity of the Martian soil: its oxidizing capacity, its lack of organic matter down to the level of several parts per billion and its capacity to fix atmospheric carbon (presumably into organic molecules) at a still lower level. It seems Mars has a photochemically activated surface that, due to the low temperature and the absence of water, is maintained in a state far from chemical equilibrium.

These conclusions drawn from the results of the life-seeking experiments on the Viking landers are undeniably disappointing. The discovery of life would have been much more interesting, to say the least. There are doubtless some who, unwilling to accept the notion of a lifeless Mars, will maintain that the interpretation I have given is unproved. They are right. It is impossible to prove that any of the reactions detected by the Viking instruments were not biological in origin. It is equally impossible to prove from any result of the Viking experiments that the rocks seen at the landing sites are not living organisms that happen to look like rocks Once one abandons Occam's razor the field is open to every fantasy. Centuries of human experience warn us, however, that such an approach is not the way to discover the truth.

5

EARLIEST CELLS ON EARTH

When *On the Origin of Species* appeared in 1859, the history of life could be traced back to the beginning of the Cambrian period of geologic time, to the earliest recognized fossils, forms that are now known to have lived more than 500 million years ago. A far longer prehistory of life has since been discovered: it extends back through geologic time almost three billion years more. During most of that long Precambrian interval the only inhabitants of the earth were simple microscopic organisms, many of them comparable in size and complexity to modern bacteria. The conditions under which these organisms lived differed greatly from those prevailing today, but the mechanisms of evolution were the same. Genetic variations made some individuals better fitted than others to survive and to reproduce in a given environment, and so the heritable traits of the better-adapted organisms were more often represented in succeeding generations. The emergence of new forms of life through this principle of natural selection worked great changes in turn on the physical environment, thereby altering the conditions of evolution.

One momentous event in *Precambrian* evolution was the development of the biochemical apparatus of oxygen-generating photosynthesis. Oxygen released as a by-product of photosynthesis accumulate in the atmosphere and effected a new cycle of biological adaptation. The first organisms to evolve in response to this environmental change could merely tolerate oxygen; later cells could actively employ oxygen in metabolism and were thereby enabled to extract more energy from foodstuff.

A second important episode in Precambrian history led to the emergence of a new kind of cell, in which the genetic material is aggregated in a distinct nucleus and is bounded by a membrane. Such nucleated cells are more highly organized than those without nuclei. What is most important, only nucleated cells are capable of advanced sexual reproduction, the process whereby the genetic variations of the parents can be passed on to the off spring in new combinations. Because sexual reproduction allows novel adaptations to spread quickly through a populationists development accelerated the pace of evolutionary change. The large, complex, multicellular forms of life that have appeared and quickly diversified since the beginning of the *Cambrian period* are without exception made up of nucleated cells.

The history of life in its later phases, since the start of the *Cambrian period*, has been reconstructed mainly from the study of fossils preserved in sedimentary rocks. In the 18th and 19th centuries it gradually became apparent that the fossil record has appreciable chronological and geographical continuity. The fossil deposits form recognizable layers which can be identified in widely separated geological formations. Boundaries between such layers, where one characteristic suite of fossils gives way to another, provide the basis for dividing geologic time into eras, periods and epochs.

One of the most dramatic boundaries in the rock record is the one that separates the Cambrian period from all that came before. The 11 periods of geologic time since the start of the Cambrian are referred to collectively as the ***Phanerozoic era***, which might be translated from the Greek as the era of manifest life. The preceding era is called simply the Precambrian.

By itself the geologic time scale cannot provide dates for fossil deposits; it only lists their sequence. Ages can be calculated, however, from the constant rate of decay of radioactive isotopes in the earth's crust. By determining how much of an isotope has decayed since the minerals in a rock unit crystallized, a date can be assigned to that unit and to nearby strata containing fossils. Radioactive-isotope studies of this kind, carried out on rocks from many parts of the world, have established a rather well-defined date for the start of the Phanerozoic era: it began about 570 million years ago. The same method indicates that the earth itself and the rest of the solar system are 4.6 billion years old. Thus the Precambrian era encompasses some seven-eighths of the earth's entire history.

The boundary between the Precambrian era and the Cambrian period has traditionally been viewed as a sharp discontinuity. In Cambrian strata there are abundant fossils of marine plants and animals:

seaweeds, worms, sponges, mollusks, lampshells and, what are perhaps most characteristic of the period, the early arthropods called trilobites. It was thought from many years that fossils were entirely absent in the underlying Precambrian strata. The Cambrian fauna seemed to come into existence abruptly and without known predecessors.

Life could not have begun with organisms as complex as *trilobites*. In *On the Origin of Species* Darwin wrote: "To the question why we do not find rich fossiliferous deposits belonging to....periods prior to the Cambrian system. I can give no satisfactory answer... The case at present must remain inexplicable; and may be truly urged as a valid argument against the views here entertained." The argument is no longer valid, but it is only in the past 20 years or so that a definitive answer to it has been found.

One part of the answer lies in the discovery of primitive fossil animals in rocks below the earliest Cambrian strata. The fossils include the remains of jellyfishes, various kinds of worms and possibly sponges, and they make up a fauna quite distinct from that of the predominantly shelled animals of the Cambrian period. These discoveries, however, extend the fossil record by only about 100 million years, less than four per cent of the Precambrian era. It can still be asked: What came before?

Since the 1950's a far-reaching explanation has emerged. It has come to be recognized that not only are many Precambrian rocks fossil-bearing but also Precambrian fossils can be found even in some of the most ancient sedimentary deposits known. These fossils had escaped notice earlier largely because they are the remains only of microscopic forms of life.

An important clue in the search for Precambrian life was discovered in the early years of the 20th century, but its significance was not fully appreciated until much later. The clue came in the form of masses of thinly layered limestone rock discovered by *Charles Doolittle Walcott* in Precambrian strata from western North America. Walcott found numerous moundlike or pillar like structures made up of many draped horizontal layers, like, tall stacks of pancakes. These structures are now called stromatolites, from the Greek *stroma* meaning bed or coverlet, and *lithos*, meaning stone.

Walcott interpreted the stromatolites as being fossilized reefs that had probably been formed by various types of algae. Other workers were skeptical, and for many years the stromatolites were widely attributed to some non biological origin. The first convincing evidence substantiating Walcott's hypothesis came in 1954, when *Stanley A. Tyler* of the University of Wisconsin and *Elso S. Barghoorn* of Harvard University reported the discovery of fossil microscopic plants in an outcropping of Precambrian rocks called the Gunflint Iron formation near Lake Superior in Ontario. Most of the Gunflint fossils, which form the layers of dome-shaped and pillar-like stromatolites, resemble modern blue-green algae and bacteria. More recently, living stromatolites have been identified in several coastal habitats, most notably in a lagoon at Shark Bay on the western coast of Australia. They are indeed built up by communities of blue-green algae and bacteria, and they are strikingly similar in form to the fossilized Precambrian structures.

Today microfossils have been identified in some 45 stromatolitic deposits (All but three of these fossilized communities have been found in the past 10 years). The fossils are often well preserved, the cell walls being petrified in three-dimensional form, and they have become a prime source of documentation for the early history of life. In recent years the search for Precambrian microfossils in other kinds of sediments, such as shales deposited in offshore environments, has also been rewarded. These fossils are generally not as well preserved as the ones in stromatolites, most of them having been flattened by pressure; on the other hand, they supply information about Precambrian life in a habitat quite different from that of the shallow-water stromatolites.

A surprising amount of information can be derived from the fossil remains of a microorganism. Size, shape and degree of morphological complexity are among the most easily recognized features, but under favourable circumstances even details of the internal structure of cells can be discerned. In retracing the course of Precambrian evolution, however, there is no need to rely exclusively on the fossil record. An entirely independent archive has been preserved in the metabolism and the biochemical pathways of modern, living cells. No living organism is biochemically identical with its Precambrian antecedents, but vestiges of earlier biochemistries have been retained. By studying their distribution in modern forms of life it is sometimes possible to deduce when certain biochemical capabilities first appeared in the evolutionary sequence.

Still another independent source of information about the early evolutionary progression is based neither on living nor on fossil organisms but on the inorganic geological record. The nature of the minerals found there reflects physical conditions at the time the minerals were deposited, conditions that may have been influenced by biological innovations. In order to understand the introduction of oxygen into the early atmosphere, for example, all three fields of study must be called on to testify, the mineral record tells when the change took place the fossil record reveals the organisms responsible and the distribution of biochemical capabilities among modern organisms puts the development in its proper evolutionary context.

Since the 1960's it has become apparent that the greatest division among living organisms is not between plants and animals but between organisms whose cells have nuclei and those that lack nuclei. In terms of biochemistry, metabolism, genetics and intracellular organization, plants and animals are very similar; all such higher organisms, however, are quite different in these features from bacteria and blue-green algae, the principal types of non-nucleated life. Recognition of this discontinuity has been important for understanding the early stages of biological history.

Organisms whose cells have nuclei are called eukaryotes, from the Greek roots *eu-,* meaning well or true, and *karyon,* meaning kernel or nut. Cells without nuclei are prokaryotes, the prefix promeaning before. All green plants and all animals are eukaryotes. So are the fungi, including the molds and the yeasts, and protists such as *Paramecium* and *Euglena*. The prokaryotes include only two groups of organisms, the bacteria and the blue-green algae. The latter produce oxygen through photosynthesis like other algae and higher plants, but they have much stronger affinities with the bacteria than they do with eukaryotic forms of life. Later on several workers refer to blue-green algae by an alternative and more descriptive name, the *cyanobacteria.*

Several important traits distinguish eukaryotes from prokaryotes. In the nucleus of a eukaryotic cell the DNA is organized in chromosomes and is enclosed by an intracellular membrane, many prokaryotes have only a single loop of DNA which is loose in the cytoplasm of the cell. Prokaryotes reproduce asexually by the comparatively simple process of binary fission. In contrast, asexual reproduction in eukaryotic cells takes place through the complicated process of mitosis, and most eukaryotes can also reproduce sexually through meiosis and the subsequent fusion of sex cells. (The "parasexual" reproduction of some prokaryotes differs markedly from advanced eukaryotic sexuality). Eukaryotic cells are generally larger than prokaryotic ones, although the range of sizes overlaps, and almost all

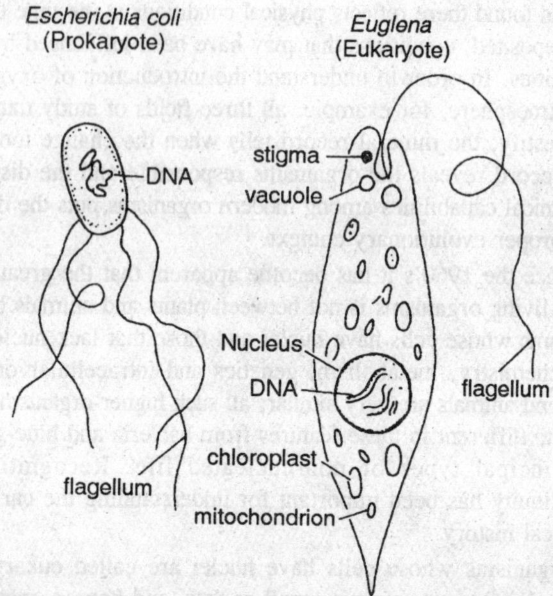

Fig. 5.1. Greatest division among organisms in the one separating cells with nuclei (eukaryotes) from those without nuclei (prokaryotes).

prokaryotes are unicellular organisms whereas the majority of eukaryotes are large, complex and many-celled. A mammalian animal, for example, can be made up of billions of cells, which are highly differentiated in both structure and function.

An intriguing feature of eukaryotic cells is that they have within them smaller membrane-bounded subunits, or organelles, the most notable being mitochondria and chloroplasts. Mitochondra are present in all eukaryotes, where they play a central role in the energy economy of the cell. Chloroplasts are present in some protists and in all green plants and are responsible for the photosynthetic activities of those organisms. It has been suggested that both mitochondria and chloroplasts may be evolutionary derivatives of what were once free-living microorganisms, an idea discussed in particular by Lynn Margulis of Boston University. The modern chloroplast, for example, may be derived from a cyanobacterium that was engulfed by another cell and that later established a symbiotic relationship with it. In support of this hypothesis it has been noted that both mitochondria and chloroplasts contain a small fragment of DNA whose organization is somewhat like that of prokaryotic DNA. In the past several years the testing of this hypothesis has generated a large body of data on the comparative biochemistry of modern microorganism, data that also provide clues to the evolution of life in the Precambrian.

One further difference between prokaryotes and eukaryotes is of particular importance in the study of their evolution: the extent to which the two types of organisms tolerate oxygen. Among the prokaryotes oxygen requirements are quite variable. Some bacteria cannot grow or reproduce in the presence of oxygen: they are classified as obligate anaerobes. Others can tolerate oxygen but can also survive in its absence; they are facultative anaerobes. There are also prokaryotes that grow best in the presence of oxygen but only at low concentrations, far below that of the present atmosphere. Finally, there are fully aerobic prokaryotes forms that cannot survive without oxygen.

In contrast to this variety of adaptations the eukaryotes present a pattern of great consistency: with very few exceptions they have an absolute requirement for oxygen, and even the exceptions seem to be evolutionary derivatives of oxygen-dependent organisms. This observation leads to a simple hypothesis: the prokaryotes evolved during a period when environmental oxygen concentrations were changing, but by the time the eukaryotes arose the oxygen content was stable and relatively high.

One indication that eukaryotic cells have always been aerobic is provided by mitotic cell division, a process that can be considered a definitive characteristic of the group. Many eukaryotic cells can survive temporary deprivation of oxygen and can even carry on some metabolic functions; it appears that no cell, however, can undergo mitosis unless oxygen is available at least in low concentration .

The pathways of metabolism itself—the biochemical mechanisms by which an organism extracts energy from foodstuff—provide more detailed evidence. In eukaryotes the central metabolic process is respiration, which in overall terms can be described as the burning of the sugar glucose with oxygen to yield carbon dioxide, water and energy. Some prokaryotes (the aerobic or facultative ones) are also capable of respiration, but many derive their energy solely from the simpler process of fermentation. In bacterial fermentation glucose is not combined with oxygen (or with any othersubstance from outside the cell) but is simply broken down into smaller molecules. In both respiration and fermentation part of the energy released through the decomposition of glucose is captured in the form of high-energy phosphate bonds, usually in molecules of adenosine triphosphate (ATP). The rest of the energy is lost from the cell as heat.

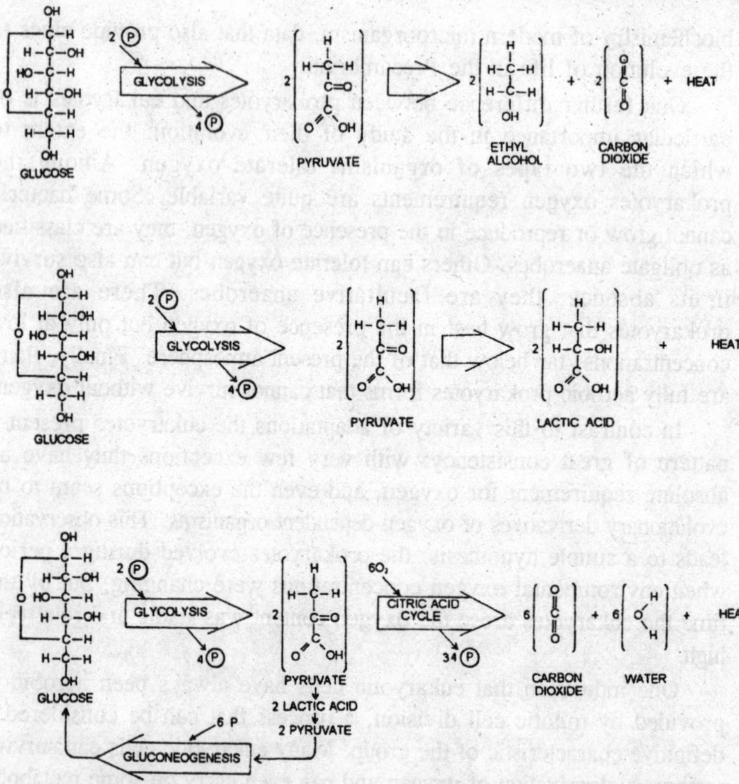

Fig. 5.2. Metabolic pathways by which cells extract energy from foodstuff apparently evolved in response to an increase in free oxygen.

Respiratory metabolism has two main components: a short series of chemical reactions, collectively called glycolysis, and a longer series called the citric acid cycle. In glycolysis a glucose molecule, with six carbon atoms, is broken down into two molecules of pyruvate, each having three carbon atoms. No oxygen is required for glycolysis, but on the other hand it releases only a little energy with a net gain of only two molecules of ATP.

The fuel for the citric acid cycle is the pyruvate formed by glycolysis. Through a series of enzyme-controlled reactions the carbon atoms of the pyruvate are oxidized and the oxidations are coupled to other reactions that result in the synthesis of ATP. For each two molecules of pyruvate (and hence for each molecule of glucose entering the sequence) 34 additional molecules of ATP are formed. The complete respiratory pathway is thus far more effective than glycolysis alone.

In respiration the proportion of energy released that can be recovered in useful form (as ATP) is higher than it is in fermentation, about 38 per cent instead of only some 30 per cent, and in respiration the net energy yield to the cell is some 18 times greater. By breaking down the glucose to simple inorganic molecules (carbon dioxide and water) respiration liberates virtually all the biologically usable energy stored in the chemical bonds of the sugar.

The metabolism of the prokaryotes immediately suggests an evolutionary relationship between them and the eukaryotes; up to a point fermentation is indistinguishable from glycolysis. In bacterial fermentation a molecule of glucose is split into two molecules of pyruvate, with a net yield of two molecules of ATP. As in glycolysis, no oxygen is required for the process. In anaerobic prokaryotes, however, the metabolic pathway essentially ends at pyruvate. The only further reactions transform the pyruvate into such compounds as lactic acid, ethyl alcohol or carbon dioxide, which are excreted by the cell as wastes.

The similarity of fermentation in prokaryotes to glycolysis in eukaryotes seems too close to be a coincidence, and the assumption of an evolutionary relationship between the two groups provides a ready explanation. It seems likely that anaerobic fermentation became established as an energy-yielding process early in the history of life. When atmospheric oxygen became available for metabolism, it offered the potential for extracting 18 times as much useful energy from carbohydrate: a net yield of 36 molecules of ATP instead of only two molecules. The oxygen-dependent reactions did not however, simply replace the anaerobic ones: they were appended to the existing anaerobic pathway.

Further evidence for this proposed evolutionary sequence can be found in the behaviour of some eukaryotic cells under conditions of oxygen deprivation. In mammalian muscle cells, for example, prolonged exertion can demand more oxygen than the lungs and the blood can supply. The citric acid cycle is then disabled, but the cells continue to function, albeit at reduced efficiency, through glycolysis alone. Under such conditions of oxygen debt pyruvate is not consumed in the cell, but in the liver it can be converted back into glucose (at a cost in energy of six ATP molecules). Significantly the pyruvate itself is not transported to the liver but instead is converted into lactic acid, which in the liver must then be returned to the form of pyruvate. This use of lactic acid may represent a vestige of an earlier, bacterial pathway that under aerobic conditions has been suppressed. Indeed, the oxygen-starved muscle cell seems to revert to a more primitive, entirely anaerobic form of metabolism.

The development of an oxygen-dependent biochemistry can also be traced through a consideration of reaction sequences in the synthesis of various biological molecules. Once again stages in the synthetic pathway that emerged early in the Precambrian can be expected to proceed in the absence of oxygen. Reaction steps nearer the end product of the pathway, which were presumably added at a later age, might with increasing frequency require oxygen. The distribution of the oxygen-demanding steps among various kinds of organisms could also have evolutionary significance. If only one pathway has evolved for the synthesis of a class of biochemical substances, then primitive forms of life might be expected to exhibit only the initial, anaerobic steps. Organisms that arose later might exhibit progressively longer. oxygen-dependent synthetic sequences.

In aerobic organisms it might seem at first that virtually all biochemical synthesis require oxygen; eukaryotic cells exhibit relatively little synthetic activity under anoxic conditions. For the most part, however, the oxygen requirement of such synthesis is simply for metabolism: the construction of biological molecules demands energy in the form of ATP, and most of the ATP is supplied through the oxygen-dependent citric acid cycle. If ATP is made available from some other source many synthetic pathways can proceed unimpaired.

Some synthesis, however have an intrinsic requirement for oxygen, quite apart from metabolic demands. Molecular oxygen is needed, for example, in the synthesis of bile pigments in vertebrates, of chlorophyll *a* in higher plants and of the amino acids hydroxyproline and, in animals tyrosine. The oxygen dependence of two synthetic pathways in particular has been determined in detail. One of these pathways controls the manufacture of a class of compounds that includes the sterols and the carotenoids and the other is concerned with the synthesis of fatty acids.

Sterols, such as cholesterol and the steroid hormones, are flat, platelike molecules derived from the compound squalene, which has 30 carbon atoms. Carotenoids are derived from the 40-carbon compound phytoene: they are pigments, such as carotene, the orange-yellow compound in carrots, and they are found in virtually all photosynthetic organisms. A common starting point for the synthesis of both groups of compounds is isoprene, a five-carbon molecule that is also the repeating unit in synthetic rubber. In the biological synthesis two isoprene subunits are joined head to tail then a third isoprene is added to form a 15-carbon polymer, farnesyl pyrophosphate. At this point there is a fork in the pathway. In one continuaion of the synthesis two farnesyl chains are joined to form squalene, the 30-carbon precursor of the sterols. In the other continuation a fourth isoprene subunit is added, and only then are two of the chains joined. The product in this case is phytoene, the 40-carbon precursor of the carotenoids and of other pigments derived from them, such as the xanthophylls.

Up to this step in the synthetic pathway none of the reactions requires the participation of molecular oxygen. The next step in the synthesis of sterols, however, is the conversion of the linear squalene molecule to a 30-carbon ring, and this transformation does require oxygen; so do most of the subsequent steps in sterol synthesis. On the other branch of the pathway there are a few more anaerobic reactions, and indeed carotenoids can be made from phytoene without oxygen. Several further modifications of the carotenoids, however, such as the production of the pigments called epoxy xanthophylls, are oxygen dependent.

Two observations about the evolution of these biosynthetic pathways are appropriate. Even in groups of organisms that have long been aerobic the first steps in the synthesis are independent of the oxygen supply; molecular oxygen enters the reaction sequence only at later stages. In a similar way the most primitive living organisms, the anaerobic bacteria, are capable only of the first segments of the pathway, the anaerobic segments. The more complex aerobic bacteria and the photosynthetic cyanobacteria have longer synthetic pathways, including some steps that require oxygen. Advanced eukaryotes, such as vertebrate animals and higher plants, have long, branched synthetic pathways, with many steps in which molecular oxygen is required.

A similar pattern can be discerned in the synthesis of fatty acids and their derivatives. The fatty acids are straight carbon-chain compounds that have a carboxyl group (COOH) at one end. A fatty acid is said to be saturated if there are no double bonds between carbon atoms in the chain: it is saturated with hydrogen, which fills all the available bonding positions. An unsaturated fatty acid has a double bond between two carbon atoms or it may have several such double bonds: for each double bond two hydrogen atoms must be removed from the molecule.

In the synthesis of fatty acids the molecule grows by the repeated addition of units two carbon atoms long. The first few steps in the synthesis are identical in all organisms, and they yield fully saturated fatty acids. The first branch in the pathway comes when the developing chain is eight carbons long. At that point many prokaryotes can introduce a double bond, which eukaryotes cannot. There is a second branch at the next step when the saturated chain is 10 carbons long: a double bond can similarly be introduced at that point by many prokaryotes but not by eukaryotes. No matter which branch is followed, elongation of the chain ends at 18 carbons. At that point the fatty acids produced by many prokaryotes contain a double bond, but in eukaryotes the product is always the fully saturated molecule, stearic acid. None of the steps in this sequence, whether in prokaryotes or eukaryotes, requires molecular oxygen.

If no subsequent transformations of fatty acids were possible, eukaryotic cells would be incapable of synthesizing any but the fully saturated forms. Actually extensive modifications can be accomplished through the process of oxidative desaturation, in which double bonds are formed by removing two hydrogen atoms and combining them with oxygen to form water, Oxidative desaturation can take place only in the presence of molecular oxygen (O_2). Through this mechanism cyanobacteria make unsaturated fatty acids with two, three and four double bonds, and eukaryotes form polyunsaturated fatty acids (with multiple double bonds).

As in the sterol-carotenoid synthesis, an analysis of the fatty-acid pathway argues for a pattern of biochemical evolution in which the increasing availability of atmospheric oxygen played a central role. The first steps in the synthetic sequence are common to all organisms capable of making fatty acids, and in the most primitive organisms those are the only steps. Hence the reactions that come first in the biochemical sequence apparently also developed early in the history of life; these first steps are all anaerobic. Organisms that presumably emerged somewhat later (such as aerobic bacteria and cyanobacteria) have longer pathways, including a few steps of oxidative desaturation. In advanced eukaryotes a substantial proportion of the steps are oxygen dependent.

Comparisons of the metabolism and biochemistry of prokaryotes and eukaryotes thus provide strong evidence that the latter group arose only after a substantial quantity of oxygen had accumulated in the atmosphere. Hence it is of interest to ask when eukaryotic cells first appeared. It seems apparent that an oxygen-rich atmosphere cannot have developed later than this signal evolutionary event.

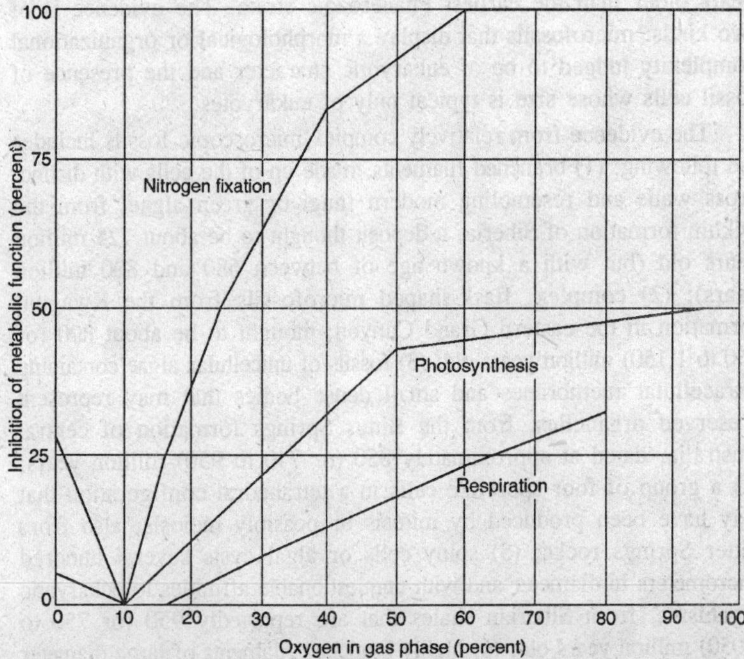

Fig. 5.3. Oxygen inhibition of metabolic functions in cyanobacteria suggests that these aerobic prokaryotes are adapted to an optimum oxygen concentration of about 10 percent, or roughly half the oxygen concentration of the earth's present atmosphere.

The primary means of assigning a date to the origin of the eukaryotes is through the fossil record. Because this field of study is so new, however, the available information is scanty and often difficult to interpret. It is rarely a straightforward task to identify a microscopic, single-cell organism as being eukaryotic merely from an examination of its fossilized remains. And even when a fossil has been identified as unequivocally eukaryotic the available radioactive-isotope methods of dating can rarely assign it a precise age. At best such methods have an accuracy of only about *plus* or minus 5 per cent. What is more, the age determinations are generally carried out on rocks that were once molten, such as volcanic lavas, whereas the fossils are found in sedimentary deposits. Consequently the stratum of the fossil itself usually cannot be dated: it is merely assigned an age somewhere between the ages of the nearest underlying and overlying datable rock units.

In spite of these difficulties there is now substantial evidence for the existence of eukaryotic fossils in rocks hundreds of millions of years older than the earliest Phanerozoic strata. The evidence is of two kinds: microfossils that display a morphological or organizational complexity judged to be of eukaryotic character and the presence of fossil cells whose size is typical only of eukaryotes.

The evidence from relatively complex microscopic fossils includes the following: (1) branched filaments, made up of the cells with distinct cross walls and resembling modern fungi or green algae, from the Olkhin formation of Siberia, a deposit thought to be about 725 million years old (but with a known age of between 680 and 800 million years); (2) complex, flask-shaped microfossils from the Kwagunt formation in the eastern Grand Canyon, thought to be about 800 (or 650 to 1,150) million years old; (3) fossils of unicellular algae containing intracellular membranes and small dense bodies that may represent preserved organelles, from the Bitter Springs formation of central Australia, dated at approximately 850 (or 740 to 950) million years; (4) a group of four sporelike cells in a tetrahedral configuration that may have been produced by mitosis or possibly meiosis, also from Bitter Springs rocks; (5) spiny cells or algal cysts several hundred micrometers in diameter and with unquestionable affinities to eukaryotic organisms, from Siberian shales that are reportedly 950 (or 750 to 1,050) million years old; (6) highly branched filaments of large diameter and with rare cross walls, similar in some respects to certain green or golden-green eukaryotic algae, from the Beck Spring dolomite of southeastern California (1,300, or 1,200 to 1,400 million years old) and from the Skillogalee dolomite of South Australia (850, or 740 to 867 million years old); (7) spheroidal microfossils described as exhibiting two-layered walls and having "medial splits" on their surface and which may represent an encystment stage of a eukaryotic alga from shales 1,400 (or 1,280 to 1,450) million years old in the McMinn formation of northern Australia; (8) a tetrachedral group of four small cells, resembling spores produced by mitotic cell division of some green algae, from the Amelia dolomite of northern Australia, approaching 1,500 (or 1,390 to 1,575) million years in age; (9) unicellular fossils that appear to be exceptionally well preserved and that are reported to contain small membrane-bounded structures that could be remnants of organelles, from the Bungle-Bungle dolomite in the same region as the Amelia dolomite and of approximately the same age.

Thus the earliest of these eukaryote-like fossils are probably somewhat less than 1,500 million years old. Numerous types of

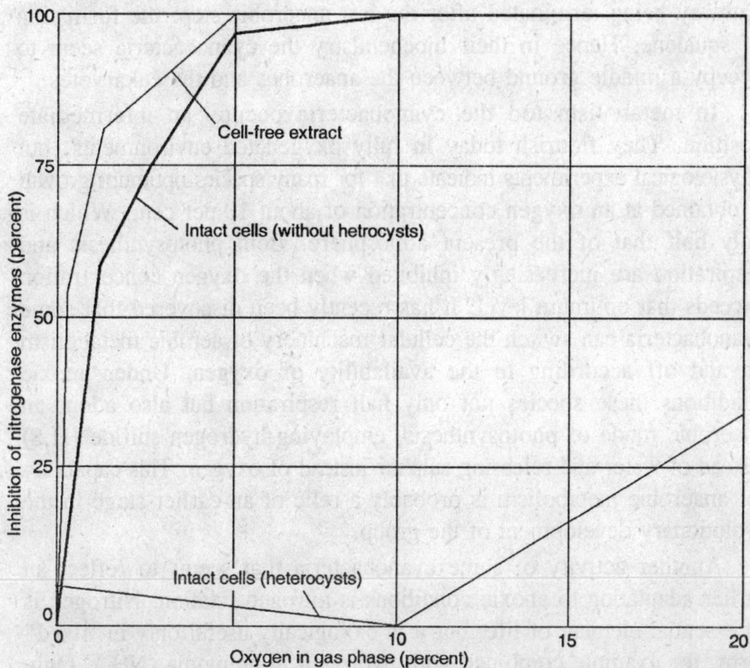

Fig. 5.4. Inhibition of nitrogen fixation in the presence of oxygen is caused by the deactivation of the nitrogenase enzymes.

microfossils have been discovered in older sediments, but none of them seems to be a strong candidate for that in overall effect (although not in mechanism) is the reverse of respiration. The energy of sunlight is employed to make carbohydrates from water and carbon dioxide, and molecular oxygen is released as a by-product. The cyanobacteria can tolerate the oxygen they produce and can make use of it both metabolically (in aerobic respiration) and in synthetic pathways that seem to be oxygen dependent (as in the synthesis of chlorophyll *a*). Nevertheless, the biochemistry of the cyanobacteria differs from that of green eukaryotic plants and suggests that the group originated during a time of fluctuating oxygen concentration. For example, although many cyanobacteria can make unsaturated fatty acids by oxidative desaturation some of them can also employ the anaerobic mechanism of adding a double bond during the elongation of the chain. In a similar manner oxygen-dependent synthesis of certain sterols can be carried out be some cyanobacteria, but the amounts of the sterols made in this way are minuscule compared with the amounts typical of eukaryotes. In other cyanobacteria those sterols are not found at all, the biosynthetic pathway being terminated after the last anaerobic step: the formation of squalene. Hence in their biochemistry the cyanobacteria seem to occupy a middle ground between the anaerobes and the eukaryotes.

In metabolism too the cyanobacteria occupy an intermediate position. They flourish today in fully oxygenated environments, but physiological experiments indicate that for many species optimum growth is obtained at an oxygen concentration of about 10 per cent. Which is only half that of the present atmosphere. Both photosynthesis and respiration are increasingly inhibited when the oxygen concentration exceeds that optimum level. It has recently been discovered that some cyanobacteria can switch the cellular machinery of aerobic metabolism on and off according to the availability of oxygen. Under anoxic conditions these species not only halt respiration but also adopt an anaerobic mode of photosynthesis, employing hydrogen sulfide (H_2S) instead of water and releasing sulphur instead of oxygen. This capability for anaerobic metabolism is probably a relic of an earlier stage in the evolutionary development of the group.

Another activity of some cyanobacteria that seems to reflect an earlier adaptation to anoxic conditions is nitrogen fixation. Nitrogen is an essential element of life, but it is biologically useful only in 'fixed" form, for example combined with hydrogen in ammonia (NH_3). Only prokaryotes are capable of fixing nitrogen (although they often do so in symbiotic relationships with higher plants). The crucial complex of enzymes for fixation, the nitrogenases, is highly sensitive to oxygen. In cell-free extracts nitrogenases are partially inhibited by as little as $\times 1$ per cent of free oxygen, and they are irreversibly inactivated in minutes by exposure to oxygen concentrations of only about 5 per cent.

Such a complex of enzymes could have originated only under anoxic conditions, and it can operate today only if it is protected from exposure to the atmosphere. Many nitrogen-fixing bacteria provide at protection simply by adopting an anaerobic habitat, but among the cyanobacteria a different strategy has developed; the nitrogenase enzymes are protected in specialized cells, called heterocysts, whose internal milieu is anoxic. The heterocysts lack certain pigments essential for photosynthesis, and so they generate no oxygen of their own. They have thick cell walls and are surrounded by a mucilaginous envelope that retards the diffusion of oxygen into the cell. Finally, they are equipped with respiratory enzymes that quickly consume any uncombined oxygen that may leak in.

Because of the thick cell walls heterocysts should be comparatively easy to recognize in fossil material. Indeed, possible heterocysts have been reported from several Precambrian rock units, the oldest being about 2.2 billion years in age. If these cells are indeed heterocysts, they may be taken as a sign that free oxygen was present by then, at least in small concentrations.

Nitrogen fixation has a high cost in energy, and the capability for it would therefore seem to confer a selective advantage only when fixed nitrogen is a scarce resource. Today the main sources of fixed nitrogen are biological and industrial, but biologically usable nitrate (NO_3-) is formed by the reaction of atmospheric nitrogen and oxygen. In the anoxic atmosphere of the early Pre-identification as eukaryotic for example, the well-studied Canadian fossils of the Gunflint and Belcher Island iron formations, which are about two billion years old, have been interpreted as exclusively prokaryotic.

The testimony of these as yet rare and unusual specimens can be checked through statistical studies of the sized of known Precambrian microfossils. The size ranges of prokaryotes and eukaryotes overlap, so that a particular fossil cannot always be classified unambiguously on the basis of size alone; by cataloguing the measured sized in a large sample of fossils, however, it may be possible to determine whether or not eukaryotic cells are present. Among modern species of spheroidal cyanobacteria about 60 per cent are very small, less than five micrometers in diameter: of the remaining species only a few are larger than 20 micrometers and none is larger than 60 micrometers. Unicellular eukaryotes, such as green or red algae, can be much larger. Typically they fall in the range between 5 and 60 micrometers, but several per cent of living species are larger than 60 micrometers and a few are larger than 1,000 micrometers (one millimeter).

Systematic size measurements have been made on some 8,000 fossil cells from 18 widely dispersed Precambrian deposits. On the basis of those data certain tentative conclusions can be drawn. Cells larger than 100 micrometers, and hence of distinctly eukaryotic dimensions, are unknown in rocks older than about 1,450 million years. Virtually all the unicellular fossils from rocks of that age, whether they grew in shallow-water stromatolites or were deposited in offshore shales, are of prokaryotic size.

Cells larger than modern prokaryotes (greater than 60 micrometers in diameter) first become abundant in rocks about 1,400 million years old. Algae of this type were apparently free-floating rather than mat-forming species, and they are therefore particularly common in shales, sediments deposited in deeper water. Such eukaryote-size fossils have been known for several years from shales of this age in China and in the U.S.S.R. Recently cells more than 100 micrometers in diameter have also been discovered in the Newland limestone of Montana, and cells more than 600 micrometers in size (10 times the size of the largest spheroidal prokaryote) have been found in the McMinn formation of Australia; the age of both of these fossil-bearing deposits is about 1,400 million years.

In somewhat younger Precambrian sediments there are still larger cells, fossils greater than one millimeter in diameter (with some as large as eight millimetres). They were first described in 1899 by Walcott, who discovered them in rocks from the Grand Canyon. They have since been found in nearly a dozen other rock units throughout the world. The oldest seem to be those from Utah and from Siberia, each about 950 million years old, and those from northern India, which could be even older (from 910 to 1,150 million years old).

Studies of both the morphology and the size of unicellular fossils therefore suggest that there is a break in the fossil record between 1,400 and 1,500 million years ago. Below this horizon cells with eukaryotelike traits are rare or absent; above it they become increasingly common. Moreover, the data suggest that the diversification of the eukaryotes began shortly after the cell type first appeared, apparently within the next few hundred million years. By a billion years ago there had been substantial increases in cell size, in morphological complexity and in the diversity of species. All these indicators also suggest, of course, that oxygen-dependent metabolism, which is highly developed even in the most primitive eukaryotes, had already become established by about 1.5 billion years ago.

The prokaryotes that must have held exclusive sway over the earth before the development of eukaryotic cells were less diverse in form, but they were probably more varied in metabolism and biochemistry than their eukaryotic descendants. Like modern prokaryotes, the ancient species presumably varied over a broad range in their tolerance of oxygen, all the way from complete intolerance to absolute need. In this regard one group of prokaryotes, the cyanobacteria, are of particular interest in that they were largely responsible for the development of an oxygen-rich atmosphere.

Like higher plants cyanobacteria carry out aerobic photosynthesis, a process cambrian the latter mechanism would obviously have been impossible. The lack of atmospheric oxygen would also have indirectly reduced the concentration of ammonia to very low levels. Ammonia is dissociated into nitrogen and hydrogen by ultraviolet radiation, most of which is filtered out today by a layer of ozone (O_3) high in the atmosphere; without free oxygen there would have been little ozone, and without this protective shield atmospheric ammonia would have been quickly destroyed.

It is likely that the capability for nitrogen fixation developed early in the Precambrian among primitive prokaryotic organisms and in an environment where fixed nitrogen was in short supply. The vulnerability of the nitrogenase enzymes to oxidation was of no consequence then, since the atmosphere had little oxygen. Later, as the photosynthetic activities of the cyanobacteria led to an increase in atmospheric oxygen, some nitrogen fixers adopted an anaerobic habitat and others developed heterocysts. By the time eukaryotes appeared, apparently more than half a billion years later, oxygen was abundant and fixed nitrogen (both NH_3 and NO_3^-) was probably less scarce, and so the eukaryotes never developed the enzymes needed for nitrogen fixation.

At present oxygen-releasing photosynthesis by green plants, cyanobacteria and some protists is responsible for the synthesis of most of the world's organic matter. It is not, however, the only mechanism of photosynthesis. The alternative systems are confined to a few groups of bacteria that on a global scale seem to be of minor importance today but that may have been far more significant in the geological past.

The several groups of photosynthetic bacteria differ from one another in their pigmentation, but they are alike in one important respect: unlike the photosynthesis of cyanobacteria and eukaryotes, all bacterial photosynthesis is a totally anaerobic process. Oxygen is not given off as a by-product of the reaction, and the photosynthesis cannot proceed in the presence of oxygen. Whereas oxygen appears to be required in green plants for the synthesis of chlorophyll *a*, oxygen inhibits the synthesis of bacteriochlorophylls.

The anaerobic nature of bacterial photosynthesis seems to present a paradox: photosynthetic organisms thrive where light is abundant, but such environments are also generally ones having high concentrations of oxygen, which poisons bacterial photosynthesis. These contradictory needs can be explained if it is assumed that anaerobic photosynthesis evolved among primitive bacteria early in the Precambrian, when the atmosphere was essentially anoxic. The photosynthesizers could thus have lived in matlike communities in shallow water and in full sunlight. Somewhat later such bacteria gave rise to the first organisms capable of aerobic photosynthesis, the precursors of modern cyanobacteria. For the anaerobic photosynthetic bacteria the moleculary oxygen released by this mutant strain was a toxin, and as a result the aerobic photosynthesizers were able to supplant the anaerobic ones in the upper portions of the mat communities. The anaerobic species became adapted to the lower parts of the mat, where there is less light but also a lower concentration of oxygen. Many photosynthetic bacteria occupy such habitats today.

Photosynthetic bacteria were surely not the first living organisms, but the history of life in the period that preceded their appearance is still obscure. What little information can be inferred about that early period, however, is consistent with the idea that the environment was then largely anoxic. One tentative line of evidence rests on the assumption that among organisms living today those that are simplest in structure and in biochemistry are probably the most closely related to the earliest forms of life. Those simplest organisms are bacteria of the clostridial and methanogenic types, and they are all obligate anaerobes.

There is even a basis for arguing that anoxic conditions must have prevailed during the time when life first emerged on the earth. The argument is based on the many laboratory experiments that have demonstrated the synthesis of organic compounds under conditions simulating those of the primitive planet. These syntheses are inhibited by even small concentrations of molecular oxygen. Hence it appears that life probably would not have developed at all if the early atmosphere had been oxygen-rich. It is also significant that the starting materials for such experiments often include hydrogen sulfide and carbon monoxide (CO), and that an intermediate in many of the reactions is hydrogen cyanide (HCN). All three compounds are poisonous gases, and it seems paradoxical that they should be forerunners of the earliest biochemistry. They are poisonous, however, only for aerobic forms of life: indeed, for many anaerobes hydrogen sulfide not only is harmless but also is an important metabolite.

It was argued above that oxygen must have been freely available by the time the first eukaryotic cells appeared, probably 1,400 to 1,500 million years ago. Hence the proliferation of cyanobacteria that released the oxygen must have taken place earlier in the Precambrian. How much earlier remains in question. The best available evidence bearing on the issue comes from the study of sedimentary minerals, some of which may have been influenced by the concentration of free-oxygen at the time they were deposited. In recent years a number of workers have investigated this possibility, most notably Preston E. Cloud, Jr., of the University of California at Santa Barbara and the U.S. Geological Survey.

One mineral of significance in this argument is uraninite (UO_2), which is found in several deposits that were laid down in Precambrian streambeds. In the presence of oxygen, grains of uraninite are readily oxidized (to U_3O_8) and are thereby dissolved. David E. Grandstaff of Temple University has shown that streambed deposits of the mineral probably could not have accumulated if the concentration of atmospheric oxygen was greater than about 1 per cent. Uraninite-bearing deposits of this type are found in sediments older than about two billion years but not in younger strata, suggesting that the transition in oxygen concentration may have come at about that time.

Another kind of mineral deposit, the iron-rich formations called red beds, exhibits the opposite temporal pattern: red beds are known in sedimentary sequences younger than about two billion years but not in older ones. The red beds are composed of particles coated with iron oxides (mostly the mineral hematite, Fe_2O_3), and many are thought to have formed by exposure to oxygen in the atmosphere rather than under water. It has been proposed that the oxygen may have been biologically generated. This hypothesis is consistent with several lines of evidence, but objections to it have also been raised. For example, most red beds are continental deposits rather than marine ones and are therefore susceptible to erosion; it is thus conceivable that red beds were formed earlier than two billion years ago as well as later but that the earlier beds have been destroyed. It is also possible that the oxygen in the red beds had a nonbiological origin; it may have come from the splitting of water by ultraviolet radiation. This has apparently happened on Mars to create a vast red bed across the surface of that planet, where there are only traces of free oxygen and there is no evidence of life.

Perhaps the most intriguing mineral evidence for the date of the oxygen transition comes from another kind of iron-rich deposit: the banded iron formation. These deposits include some tens of billions of tons of iron in the form of oxides embedded in a silica-rich matrix; they are the world's chief economic reserves of iron. A major fraction of them was deposited within a comparatively brief period of a few

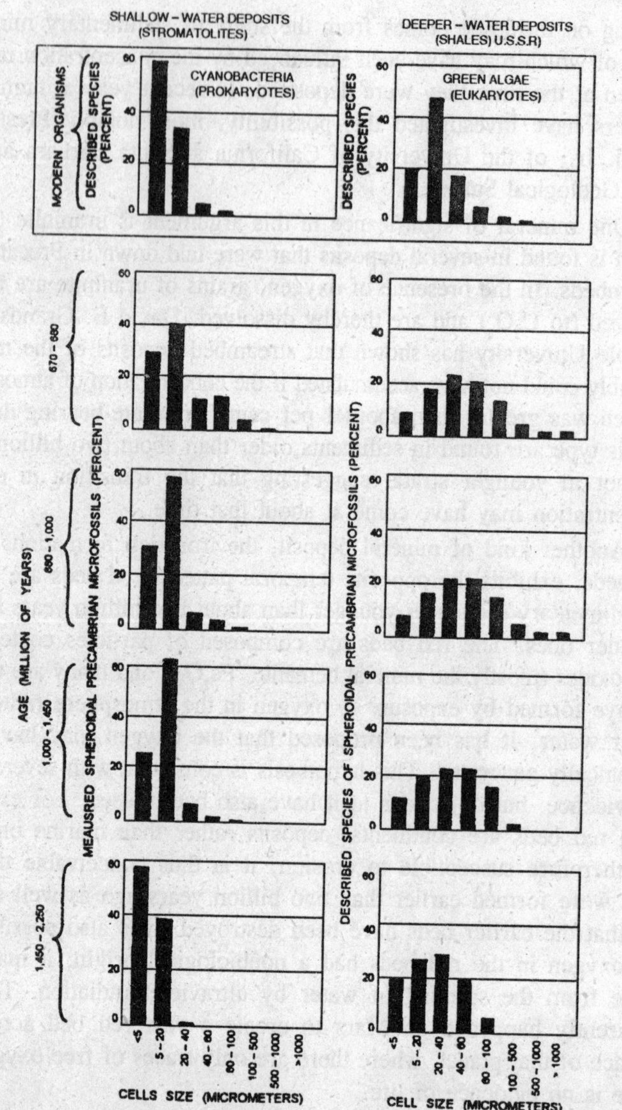

Fig. 5.5. Size of fossil cells provides evidence on the origin of the eukaryotes.

hundred million years beginning somewhat earlier than two billion years ago.

A transition in oxygen concentration could explain this major episode of iron sedimentation through the following hypothetical sequence of events. In a primitive, anoxic ocean, iron existed in the

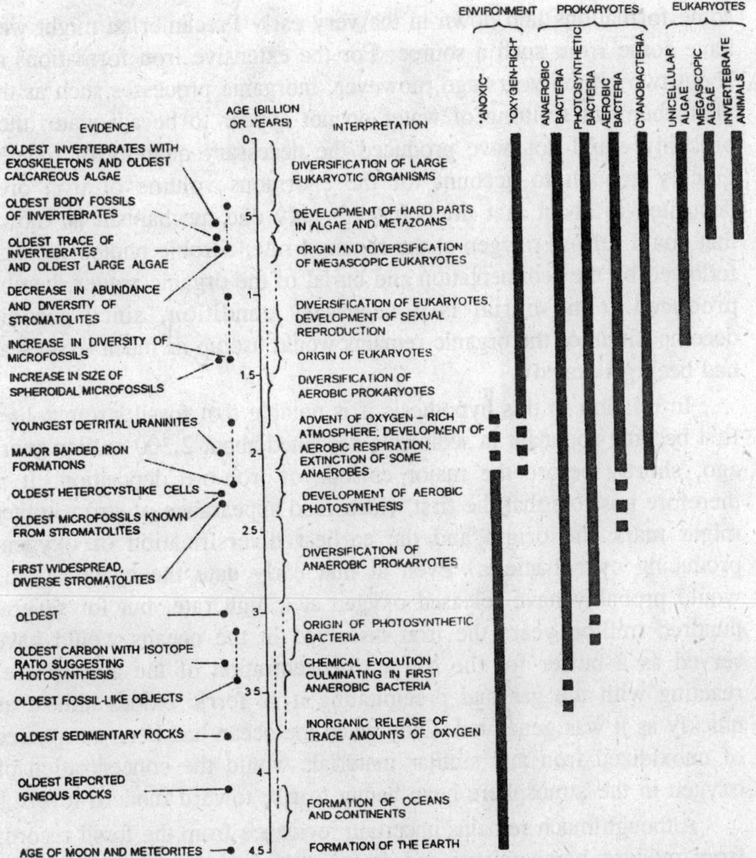

Fig. 5.6. Major events in Precambrian evolution are presented in chronological sequence based on evidence from the fossil record, from inorganic geology and from comparative studies of the metabolism and biochemistry of modern organisms.

ferrous state (that is, with a valence of +2) and in that form was soluble in seawater. With the development of aerobic photosynthesis small concentrations of oxygen began diffusing into the upper portions of the ocean, where it reacted with the dissolved iron. The iron was thereby converted to the ferric form (with a valence of +3), and as a result hydrous ferric oxides were precipitated and accumulated with silica to form rusty layers on the ocean floor. As the process continued virtually all the dissolved iron in the ocean basins was precipitated: in a matter of a few hundred million years the world's oceans rusted.

As in the deposition of red beds, an inorganic origin could also be proposed for the oxygen in the banded iron formations; the oxygen in some formations laid down in the very early Precambrian might well have come from such a source. For the extensive iron formations of about two billion years ago, however, inorganic processes such as the photochemical splitting of water do not appear to be adequate; they probably could not have produced the necessary quantity of oxygen quickly enough to account for the enormous volume of iron ores deposited at about that time. Indeed, only one mechanism is known that could release oxygen at the required rate: aerobic photosynthesis, followed by the sedimentation and burial of the organic matter thereby produced. (The burial is a necessary condition, since aerobic decomposition of the organic remains would use up as much oxygen as had been generated).

In relation to this hypothesis it is notable that fossil stromatolites first become abundant in sediments deposited about 2,300 million years ago, shortly before the major episode of iron-ore deposition. It is therefore possible that the first widespread appearance of stromatolites might mark the origin and the earliest diversification of oxygen-producing cyanobacteria. Even at that early date the cyanobacteria would probably have released oxygen at a high rate, but for several hundred million years the iron dissolved in the oceans would have served as a buffer for the oxygen concentration of the atmosphere, reacting with the gas and precipitating it as ferric oxides almost as quickly as it was generated. Only when the ocean had been swept free of unoxidized iron and similar materials would the concentration of oxygen in the atmosphere have begun to rise toward modern levels.

Although much remains uncertain, evidence from the fossil record, from modern biochemistry and from geology and mineralogy make possible a tentative outline for the history of precambrian life. The most primitive forms of life with recognizable affinities to modern organisms were presumably spheroidal prokaryotes, perhaps comparable to modern bacteria of the clostridial type. Initially at least they probably derived their energy from the fermentation of materials that were organic in nature but were of nonbiological origin. These materials were synthesized in the anoxic early atmosphere and were of the type that during the age of chemical evolution had led to the development of the first cells.

The first photosynthetic organisms apparently arose earlier than about three billion years ago. They were anaerobic prokaryotes, the precursors of modern photosynthetic bacteria. Most of them probably lived in matlike communities in shallow water, and they may have been responsible for building the earliest fossil stromatolites known, which are estimated to be about three billion years old.

The rise of aerobic photosynthesis in the mid-Precambrian introduced a change in the global environment that was to influence all subsequent evolution. The resulting increase in oxygen concentration probably led to the extinction of many anaerobic organisms, and others were forced to adopt marginal habitats, such as the lower reaches of bacterial mat communities. Nitrogen-fixing organisms also retreated to anaerobic habitats or developed heterocyst cells. With little competition for those regions having optimum light the cyanobacteria were able to spread rapidly and came to dominate virtually all accessible habitats. With the development of the citric acid cycle and its more efficient extraction of energy from foodstuff, the dominance of the biological community by aerobic organisms was confirmed. When the major episode of deposition of banded iron formations ended some 1,800 million years ago, the trend toward increasing oxygen concentration became irreversible.

By the time eukaryotic cells arose 1,500 to 1,400 million years ago a stable, oxygen-rich atmosphere had long prevailed. Adaptive strategies needed by earlier organisms to cope with fluctuations in the oxygen level were unnecessary for eukaryotes, which were from the start fully aerobic. The diversity of eukaryote cell types present by about a billion years ago suggests that some form of sexual reproduction may have evolved by then. Within the next 400 million years the rapid diversification of eukaryotic organisms had led to the emergence of multicellular forms of life, some of them recognizable antecedents of modern plants and animals.

In style and in tempo evolution in the Precambrian was distinctly different from that in the later, Phanerozoic era. The Precambrian was an age in which the dominant organisms were microscopic and prokaryotic, and until near the end of the era the rate of evolutionary change was limited by the absence of advanced sexual reproduction. It was an age in which the major benchmarks in the history of life were the result of biochemical and metabolic innovations rather than of morphological changes. Above all, in the Precambrian the influence of life on the environment was at least as important as the influence of the environment on life. Indeed, the metabolism of all the plants and animals that subsequently evolved was made possible by the photosynthetic activities of primitive cyanobacteria some two billion years ago.

6

LIFE OF UNICELLULAR ORGANISM

The classical division of the world of life into two kingdoms, animals and plants, has such a low informative and instructional value as to be almost useless for our purpose. At the level of unicellular organisms the distinction between animals and plants is based upon a single characteristic, the presence or absence of photosynthesis, and so is highly artificial. Much more instructive are the various modern schemes that recognize several kingdoms, and use several different characteristics for distinguishing them. That of *Whittaker* (1969), which recognizes five kingdoms, is most satisfactory. These kingdoms represent three grades of advancement; prokaryote (Monera), unicellular eukaryote (Protista), and multicellular or coenocytic eukaryote. The first two grades are each represented by a single kingdom, but at the multicellular level three kingdoms are recognized: plants, animals, and fungi. The reasons for doing this are discussed later in this chapter.

First, however, the historical reason for revising the biologist's conception of kingdoms must be reviewed. It results directly from the greatly increased knowledge of microorganisms acquired by biologists during the past century. From about 1870 to 1940 the relationships between different groups of unicellular eukaryotes or protista were intensively investigated and the modern discipline of protozoology was founded. These investigations confused rather than clarified the problem of the relationships between different kinds of microorganisms. In spite of the presence of cholorophyll and the ability to perform photosynthesis, many newly discovered unicellular organisms were recognized by zoologists as animals and classified as Protozoa because of their motility, their ability to ingest food particles, and their resemblance to other colourless Protozoa. These same organisms, of which the best known example is *Euglena*, were classified by botanists as plants, because of their photosynthetic ability. Each systematist could present logical arguments to support his claim. In this way a sort of "no man's land" of organisms arose, containing species and genera that were classified in one way, with one set of terms, by zoologists, and in an entirely different way by botanists.

During the same period the position and relationships of the bacteria became equally confused as a result of two conflicting tendencies. Since they do not ingest food and are very different from Protozoa, bacteria have never been recognized by zoologists as animals. They are, however, sufficiently plant-like to have attracted the attention of botanists, who have placed them in the plant kingdom as a separate division or phylum in most textbooks and general treatises. At the same time, a study of bacteria has become an essential part of medical science, and a large school of medical bacteriologists have devised their own methods of investigating bacterial relationships. In general, they have had little communication with botanists, even with mycologists, interested in bacteria. The resulting cross currents and lack of communication have led to a confusing situation.

The problem was clarified and progress was made toward its solution by two important developments in the middle of this century: the rise of molecular biology and the invention of the electron microscope. Modern research has revealed a previously unsuspected cleft between two major groups of organisms, one that cuts right across the preexisting classification of animal and plant kingdoms. With respect to their biochemistry, genetics, and cellular and nuclear fine structures, organisms can be divided into two sharply distinct groups, *eukaryotes* and *prokaryotes*. The prokaryotes include the bacteria and blue-green algae; all other cellulose organisms are eukaryotes. Table 6.1 summarizes the principal difference between these two groups. On the basis of biochemical differences, McLaughlin and Dayhoff (1970) have concluded that the evolutionary distance between prokaryotes and eukaryotes is twice that between animals, plants, and fungi.

The distinctive characteristics of prokaryotes are the absence of important structural features and reproductive cycles, such as mitosis, meiosis, and syngamy, which are present in most or all eukaryotes. That this absence in prokaryotes is a primary condition rather than a secondary loss is evident from the Precambrian fossil record, which shows that prokaryotic cells appeared long before eukaryotes. Consequently, evolutionists are now faced with a major problem: How did the first major ascent in evolutionary grade, the origin of eukaryotes from prokaryotes, take place? This problem is discussed later in this chapter.

Table 6.1. Principal differences between prokaryotes and eukaryotes

	Prokaryotes	*Eukaryotes*
Nucleus		
Nuclear membrane	No	Yes
Mitotic spindle	No	Yes (no)
Meiotic reduction of chromosome number	No	Yes (no)
Recombination by union of entire gametes	No	Yes (no)
Chromosomes		
Number of different nonhomologous chromosomes in basic genome	1	2-600+
Histones and acidic proteins complexed with DNA	No	Yes (no)
Regular condensation and relaxation of chromosomes during mitotic cycle	No	Yes (no)
Cytoplasm		
Mitochondria	No	Yes
Chloroplasts	No	Yes or no
Metabolic enzymes in cellular membrane	Yes	No
Cilia and flagella with 9-plus-2 fibre arrangement	No	Yes or no
Endoplasmic reticulum	No	Yes
Vacuoles	No	Yes (no)
Golgi apparatus	No	Yes or no
Phagocytosis, pinocytosis	No	Yes or no
Amoeboid movement, cytoplasmic streaming	No	Yes or no
Molecular synthesis		
Nature of cytoplasmic ribosomes	70S	80S
Nature of organellar ribosomes	No	70S

The separation of prokaryotes as a kingdom distinct from eukaryotes does not solve the problem of the distinction between plants and animals. *Whittaker's* solution is to recognize at the level of kingdoms a second advance in grade, that from unicellular to multicellular or coenocytic eukaryotes. This procedure localizes all the previous artificial separations between autotrophic and heterotrophic organisms either at the level of unicellular eukaryotes (Protista) or of prokaryotes (Monera), i.e., bacteria vs. blue-green algae. The difference at these levels between autotrophy and heterotrophy is no more significant than other cytological, physiological, and biochemical difference, and in most instances is useful chiefly for distinguishing between orders, families, or even genera within a family. The logic of recognizing Protista as a single and separate kingdom becomes clear when the distribution of autotrophy *vs*. heterotrophy among them is compared with that of other characteristics, such as the chemistry of the cell wall, the position and fine structure of flagellae, and the organization of the mitotic apparatus. For instance, with respect to these other characteristics the facultatively autotrophic *Euglena* resembles various heterotrophic flagellates and is completely different from the autotrophic flagellate *Chlamydomonas*. Calling *Euglena* a plant and placing it with *Chlamydomonas* rather than with the flagellates it resembles is a highly artificial procedure. Another course adopted by some botanists is to classify *Chlamydomonas* with plants because it resembles them in having a cellulose wall, and exclude *Euglena* because the chemical structure of its cell wall is completely different. This procedure creates other problems. Red and brown algae, which are plant-like in many respects, including photosynthetic ability, nevertheless have walls that do not contain cellulose or that differ considerably from those of typical green plants in other respects.

Unicellular eukaryotes, therefore, are here recognized as a single, probably monophyletic kingdom, the Protista, and kingdoms that are parallel branches or clades of major dimensions are recognized only at the grade of multicellular or coenocytic organisms. This procedure raises two additional problems: where to draw the boundaries between Protista and the kingdoms of multicellular organisms, and how many kingdoms to recognize at the multicellular or coenocytic grade. The first problem does not arise with respect to animals, since the gap between Protista and the animal kingdom, which according to the present treatment is synonymous with the Metazoa, is so great that the origin of Metazoa is still obscure. On the other hand, the different phyla of algae and fungi are connected to various groups of unicellular forms through a series of intermediates, which may be colonial cell aggregates, small coenocytes, or have low degrees of cellular differentiation. One might justly ask, therefore, why reject the artificial distinction between autotrophs and heterotrophs at the level of Protista, while erecting more or less artificial boundaries between unicellular and multicellular autotrophs?

Our first answer is that the world of life is not divided into a few sharply separated kingdoms, which can be easily delimited. If plants are to be separated from ciliate Protista, which are surely much more like animals than plants, artificial boundaries must be established somewhere. Second, the advantage of recognizing kingdoms at the somewhat indistinct boundary between the unicellular and the multicellular or coenocytic grade is admittedly heuristic. It calls attention to the great importance of this advance in grade for the evolution of life, and invites studies directed at learning how this advance came about.

Once the policy is adopted of recognizing several parallel kingdoms at the multicellular grade, the question arises: How many should be recognized? This question is not easy to answer. If one adopts the rule that each kingdom should be monophyletic, including only a single advance to the multicellular grade, two difficulties arise. First, the number of these advances is relatively large; an estimate of 16 is given later in this chapter. Second, their actual number is largely a matter of inference based upon comparisons between modern forms, since multicellular microorganisms lack a significant fossil record. Clearly the phylogenetic approach, because of its uncertainties, is a weak base upon which to found categories as broad as kingdoms.

The approach adopted here separates three kingdoms of multicellular or coenocytic organisms on the basis of fundamental differences in their mode of nutrition. Plants are autotrophic, except for a few saprophytes or parasites among seed plants, which are obviously of secondary origin. Fungi are heterotrophic and, like bacteria, absorb their food in a predigested form. Animals are heterotrophic ingesters. All three of these kingdoms are probably polyphyletic. Among animals, sponges have probably had a different origin from the remaining phyla. Among fungi, the water moulds, or oömycetes, probably arose from heterotrophic flagellates independently of terrestrial fungi. The plant kingdom is probably polyphyletic even if, following Margulis (1970), the brown and red algae are excluded from it and only green algae plus land plants are included. There are good reasons for believing that the coenocytic green algae were derived from Protista independently of the cellular forms. Even the filamentous green algae now appear to have had at least three separate origins from unicellular forms: the

Conjugales, Ulotrichales, and *Klebsormidium* group, which are the only green algae that form typical cross walls from a cell plate. Once polyphylesis is admitted, the addition of the brown and red algae to the plant kingdom as separate and independent phyla becomes reasonable because of the numerous parallelisms between their morphological and ecological evolution and that of green algae. In our view, kingdoms of multicellular organisms are aggregates of phyla that, considered together, have a heuristic value because they represent similar ways of achieving advances in grade.

The principal heuristic value inherent in this system is as follows. First, the origin of multicellular organisms having differentiation or division of labour among the parts of their bodies involved two major ascents in grade. The first was the evolution of the eukaryotic cell. Without the division of labour among the organelles and membranes evolved in this kind of cell, cellular differentiation such as exists in animals and plants could not evolve. Some of the essential characteristics of the eukaryotic cell evolved only once. Many writers have pointed out the similarity between all eukaryotic cells with respect to: cilia and flagella, when present; ribosomes and other protein-synthesizing machinery of nuclear origin; and with few exceptions, the mitotic apparatus and molecular organization of the chromatin of their chromosomes.

The second advance in grade, the origin of multicellular or multinucleate differentiated organizms, took place many times in various evolutionary lines, and followed the first advance only after tens or hundreds of millions of years had elapsed and extensive adaptive radiation had taken place at the unicellular level. The initial evolution of the eukaryotic type of cellular organization was a necessary, but not sufficient, condition for the origin of multicellularity and differentiation.

A second kind informational content inherent in this system is the recognition of the three principal types of nutrition: autotrophy, heterotrophy with absorption of soluble food, and heterotrophy with ingestion of solid particles and internal digestion.

The earliest true cells were probably heterotrophic, and surrounded by firm cell walls. They excreted enzymes into the surrounding medium and absorbed soluble food that had been externally digested. This form of nutrition is predominant in prokaryotes, since only the relatively advanced and, for prokaryotes, structurally complex blue-green algae plus a few groups of photosynthetic and sulfur bacteria have evolved autotrophy. On the other hand, all three forms of nutrition exist in unicellular eukaryotes or Protista. Not infrequently two different methods, such as autotrophy or ingestion, as well as ingestion or absorption, can be carried out by the same cell. At the unicellular level differentiation with respect to these three basic methods of nutrition does not require extensive differentiation with respect to structure, physiology, or gene content.

The situation is completely different in multicellular organisms. Both autotrophy and heterotrophy with absorption require extensive contact between cell surfaces and the external medium, whether it be water, air, or the cellular environment of a parasite's host. In a multicellular organism this can be achieved only by extending its external surface area more or less proportionately with its size. Large autotrophs and absorbers are, by their nature, sedentary, so that compactness of bodily form is not an advantage to them. Furthermore, because of their sedentary nature they can attain wide distributions and can colonize new habitats only with the aid of propagules: cells or multicellular structures especially adapted for dispersal. This explains the origin of spores, motile gametes, and, in the more advanced forms, seeds. Many sedentary ingesters have evolved similar propagules. The situation in heterotrophic ingesters is completely different. Since they find their food by active motion, they must remain relatively compact and symmetrical, no matter what their size. As size increases, each organ increases the surface area of its internal membranes or cellular layers relative to its size. If they are predators, efficiency of nutrition usually depends upon activity and keenness of the senses, hence the successive evolution of more efficient sense organs and locomotor apparatus. If they are small and nonpredatory, their compactness makes them the logical prey of predators. Increasing efficiency of sense organs and locomotor apparatus is adaptive for most multicellular ingesters, i.e., animals, the chief exceptions being those that acquire, secondarily, armoured coverings or noxious, unpalatable characteristics. At the same time, the sense organs and locomotor apparatus that are vital for nutrition and escape can be modified with relative ease to serve as aids in sexual union. Hence, the distinction between adaptations for survival and for reproduction is far less marked in animals than in plants.

At the multicellular level the distinction between plants and animals is sharp and clear. Moreover, fungi are almost as distinct from multicellular plants as are animals. They are, like bacteria, heterotrophic absorbers, so that with respect to their ecology and many of their biochemical properties the smaller fungi are more like bacteria than higher plants. In addition, the chitinous cell walls of most fungi are completely different from those of plants, and certain genetic characteristics, such as the heterokaryotic condition in Ascomycetes and the dikaryotic condition in Basidiomcetes, are also distinctive. The recognition of fungi as a kingdom separate from animals and plants, and of mycology as a discipline separate from botany, has both a phylogenetic basis and a justification in terms of its informational content.

Some groups of organisms do not fit into a system of five kingdoms based upon three grades. The best known of these are the slime molds, both cellular (Acrasiales) and plasmodial (Myxomycetes), and the Volvocales. The slime molds, during their stages of growth and metabolism, retain the properties of heterotrophic, ingesting unicellular Protista. Nevertheless, they have evolved the ability to form propagules, differentiated cells that enable them to reproduce in a manner essentially like that of fungi, by means of spores. The most highly evolved Volvocales, particularly the genus *Volvox*, have many properties of colonial autotrophic Protista, but their mature colonies have a polarized differentiation, and carry out a process of "gastrulation" not unlike that of animals.

Two ways of handling these anomalous groups seem reasonable. One way would be to stretch somewhat the definition of Protista, and treat the anomalous groups as unusual members of that kingdom. The other would be to erect little "principalities" for them, and thus to exclude them from all of the five kingdoms. The latter procedure has greater heuristic value, since it calls attention to the fact that multicellularity and differentiation have evolved many times, in response to immediate adaptive advantages in particular environments, and that only some of these multicellular organisms have undergone extensive further evolution.

The earliest known cellular organisms, bacteria of the African strata known as Fig Tree Chert, are about 3.3 billion years old. The first appearance of unicellular eukaryotes was "earlier than 1,300 million years ago and possibly earlier than 1,700 million years ago". Consequently, for more than 1,500 million years the only living organisms on the earth were prokaryotes. Since the oldest fossils that could be regarded as multicellular or multinucleate organisms are not more than 700 million years old, the period when evolution was exclusively at the prokaryote level was twice as long as that encompassing the entire evolution of multicellular eukaryotes. Since the evolution of prokaryotes is largely biochemical rather than structural, it remains, to a large extent, a closed book. Nevertheless, on the basis or reasonably reliable information about the nature of the earth during the era of prokaryote evolution, some inferences about the nature of prokaryote evolution can be made.

The first important point is that when life first arose the oxygen content of the atmosphere was much lower than in more recent times, so that aerobic metabolism, if it existed at all, had to be at a much lower rate than in modern aerobes. The greater part of the oxygen in the modern atmosphere was and is produced by the activity of photosensitizing autotrophs. The origin of a truly oxidizing atmosphere took place, probably, not much more than a billion years ago. The greater part of prokaryote evolution, therefore, was based upon anaerobic or weakly aerobic metabolism.

The most significant single event of this evolution was the origin of photosynthesis. It took place relatively soon after the origin of cellular life. The fossil record of blue-green algae is instructive in this connection. Their modern forms have molecules of chlorophyll that are identical with those of higher plants, and even the carotenoid pigment associated with their chlorophyll resembles that found in eukaryotes. Although some of them can carry out anaerobic metabolism, most of them are aerobic, and perform photosynthesis with H_2O as the hydrogen donor. Hence, by the time blue-green algae appeared, photosynthesis similar to that of higher plants either existed or was represented by a similar process. The oldest fossils of blue-green algae are calcareous deposits known as stromatolites, similar to the mineral layers that surround colonies of marine forms in some parts of the modern world, particularly Australia. The oldest stromatolites are in the Buluwaya formation of South Africa, 2.7 billion years old. They are 500 million years younger than the oldest known bacteria, but 1.4 billion years older than the earliest eukaryotes.

The complex, sophisticated kind of photosynthesis carried out by blue-green algae must have ben preceded by a whole series of more simple, primitive methods of autotrophy. The only methods known to us are those of three groups of photosynthetic bacteria: the green bacteria (*Cholorobium*, *Pelodictyon*) purple sulfur bacteria (*Thiospirillum*, *Lamprocystis*) and purple nonsulfur bacteria (*Rhodospirillum*, *Rhodopseudomonas*, *Rhodomicrobium*). All these forms have chlorophyll molecules that resemble those of other photosynthesizers with respect to the tetrapyrrole porphyrin ring surrounding a central atom of magnesium, but differ with respect to certain side chains. None of the forms has a strictly aerobic kind of photosynthesis characteristic of most blue-green algae and all higher plants. The green bacteria and purple sulfur bacteria are strictly anaerobic, and require sulphide, thiosulfate, or perhaps hydrogen as an electron donor. The purple nonsulfur bacteria carry out photosynthesis anaerobically, but use organic compounds as the principal electron donors, and can become aerobic and heterotrophic in the presence of oxygen and low illumination. Slow, inefficient methods of photosynthesis like that found in these bacteria must have preceded aerobic photosynthesis of the modern type. Although fossil evidence may never be adequate to solve the problem, the assumption that forms similar to photosynthetic bacteria evolved prior to blue-green algae is quite reasonable.

The total range of adaptive radiations carried out by bacteria before the evolution of eukaryotes may never be known. The ecology of the extinct species cannot be inferred, and one cannot expect to find among contemporary bacteria representatives of all groups that must have existed. If the relative diversity of modern forms is any kind of a guide, one must conclude that this radiation was much more restricted in its earlier stages, when oxygen was limited and only anaerobic forms could exist, than at later periods, when the principal radiants must have been aerobic. The contemporary anaerobic free-living bacteria are, besides the photosynthetic bacteria, principally of three groups. The methane bacteria inhabit the bottoms of ponds or quiet streams, where they obtain energy by the anaerobic oxidation of molecular hydrogen. The *Clostridium* group is characterized by the ability to form highly resistant spores, and live in moist soil. Finally, there is a large series of relatively little-known anaerobic bacteria living in the mud at the bottom of the sea, chiefly at depths of 1,000 to 2,000 metres. In addition, some pathogenic bacteria, including members of the *Closrtridium* group, as well as some of the bacteria that inhabit the mouth and digestive tract of animals, are obligate anaerobes. These latter groups, obviously, are relatively recent origin. In addition, many groups of bacteria, such as the lactic-acid bacteria, are facultative anaerobes, which are able to utilize oxygen for respiration in an aerobic atmosphere but which, in its absence, can derive energy from anaerobic metabolism.

The extent to which aerobic metabolism enabled bacteria to undergo additional adaptive radiation even before the evolution of eukaryotes may be estimated by the wide range of ecological niches in the contemporary world inhabited by aerobic prokaryotes. The blue-green algae have already been mentioned; their range of habitats extends from hot springs, at temperatures of 73° to 75°C, to the icy wastes of Antarctica. They are also much more tolerant of polluted water than are eukaryotic photosynthesizers. Derivatives of the blue-green algae that have reverted to heterotrophy are the filamentous gliders, such as *Beggiatoa* and *Thiothrix*, which live in sulfur springs. Other strictly aerobic free-living bacteria are the nitrogen-fixing soil bacteria (*Azotobacter*, *Beijerinckia*), the colonial actinomycetes, the highly salt-tolerant *Holobacterium*, and the free-living members of the group of spirochaetes. These organisms are remarkable for the long flagella-like structures contained in their cell wall, which enable them to carry out their characteristically rapid spiral motility. Judging from the diversity of these modern forms, the advent of oxygen in the earth's atmosphere greatly expanded the possibilities for adaptive radiation and evolution at the prokaryote level.

The advent of an oxygen-rich atmosphere required many readjustments of cellular metabolism. One of them must have consisted of ways by which cells could overcome the toxic effect of oxygen on obligate anaerobes. Of the many evolutionary strategies for acquiring this adaptation, one was particularly striking: the origin of bioluminescence in bacteria, and presumably in their protistan descendants. McElory and Seliger (1962) have produced a good argument to support their hypothesis that the biochemical reactions responsible for bioluminescence or fluorescence evolved from the vestiges of mechanisms that in the earliest aerobic organisms reduced molecular oxygen directly and quickly, thus preventing it from exerting toxic effects, which oxygen has when it accumulates in the cells that are normally anaerobic. The cells that produce luminescence produce a compound, luciferin aldehdye, which reacts with the potentially toxic peroxide produced by free oxygen to form the corresponding organic acid and water. The organisms responsible for the phosphorescent glow so often seen in seawater, the dinoflagellate *Noctiluca*, may have acquired its property via an evolutionary descent from a luminous prokaryote.

Evolution in Viruses

For many years viruses were not regarded as organisms at all, so that their evolution need not have been considered in connection with that of cellular organisms. Now, however, molecular biologists

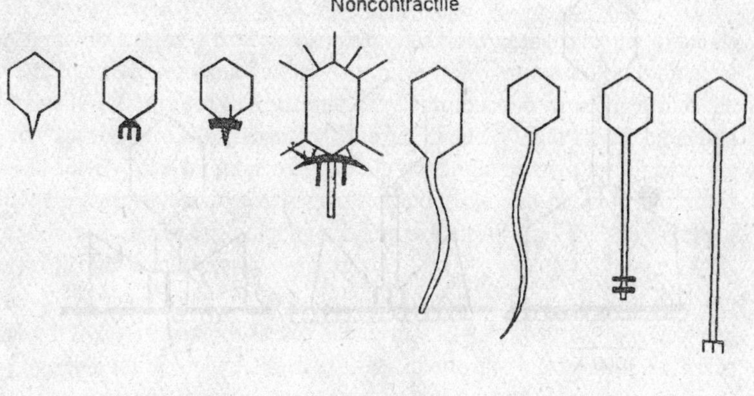

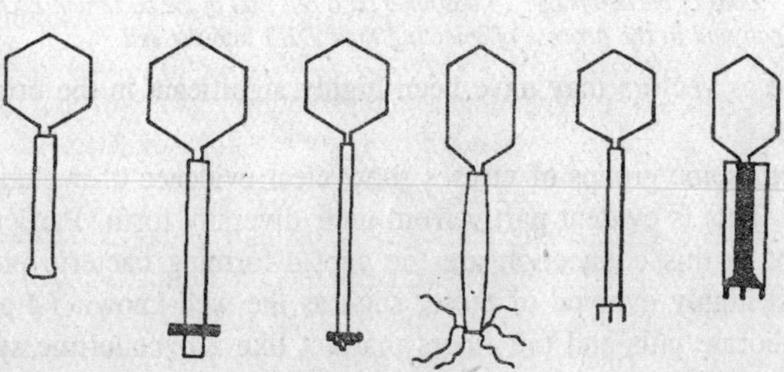

Fig. 6.1. Comparative morphology of bacteriophage viruses, showing various degrees of complexity.

generally accepted the hypothesis that viruses have been derived from detached portions of cellular organisms. They contain typical nucleic acids, either DNA or RNA or both, and with the aid of enzymes provided by the host synthesize proteins, just like cellular organisms.

Three further points established by Joklik (1974) in his review of virus evolution are significant to evolution in general. First, groups of viruses differ so radically from each other with respect to both the cross affinities of their nucleic acids and the immunological relationships of their proteins that they almost certainly have been derived separately from cellular organisms. Second, the members of the same viral group, with similar nucleic acids and proteins, include parasites of very different organisms. Among the closely knit group of Reoviridae, for instance, some attack mammals, other birds, while still others have plants and/or insects as hosts. In most of these groups with heterogeneous host affinities, insects appear to be the common denominators, so that

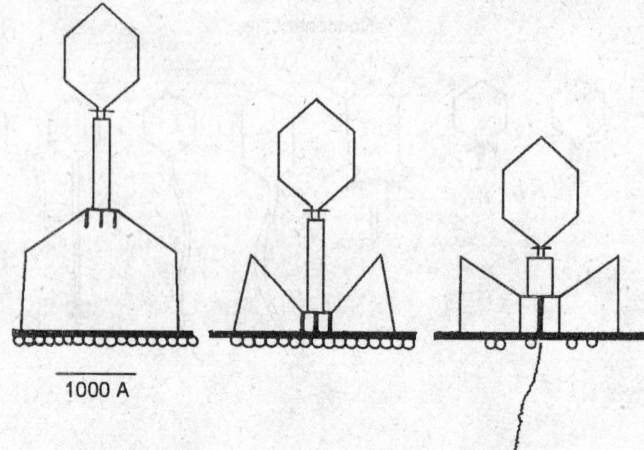

Fig. 6.2. Virions of bacteriophage T4 adsorbed on a cell wall of the bacterium Escherichia coli and in the process of injecting their DNA into the cell.

their role as vectors may have been highly significant in the origin of the group.

Third, some groups of viruses show clear evidence of evolutionary lineages. This is evident partly from their diversity form. Particularly important in this connection are the capsid-forming bacterio phages. The most highly evolved of these, such as the well-known *T4* phage, have elaborate tails and tail fibers that act like a hypodermic syringe in injecting DNA into the host cell. In different phages of this group a whole series of intermediate kinds of tail mechanisms have been recognized. These viruses have clearly undergone extensive evolution of the conventional sort after originating as detached portions of cells.

Origin of Eukaryotes

The first advance in grade after the origin of cellular life was that of the eukaryotic cell. Its origin can be regarded as the most significant single advance of organic evolution, since until cells had acquired the complexity and the degree of division of labour among membranes and organelles that are associated with the eukaryotic condition, multicellular organisms, having differentiated cells and tissues, could not evolve.

There is general agreement among biologists that the basic organization of eukaryotic cells evolved only once. This is because eukaryotic cells possess certain complicated organelles that are strikingly similar to each other in their supramolecular structure. The most conspicuous of these are the mitochondria, the plastids of autotrophic eukaryotes, and flagella or cilia having the 9+2 organization of fibrils. Among eukaryotes lacking these organelles, only the red algae can be postulated with reasonable probability as never having had such flagella. A second similarity is the mitotic apparatus in nearly all eukaryotes including red algae. Biochemical characteristics common to eukaryotes are the Embden-Meyerhof pathway of glucose metabolism, the universal presence of Krebs-cycle oxidations, and particularly the protein-synthesizing machinery. The ribosomes of eukaryotes are larger than those of prokaryotes, and their activity is insensitive to chloramphenicol, which inhibits protein synthesis in prokaryotes, while it is sensitive to cyclohexamide, to which prokaryotic cells are resistant.

Not unexpectedly, many biologists have speculated about the origin of the eukaryotic cell, a cardinal event of evolution. The most complete hypothesis is that of Margulis (1970), who has marshalled a great wealth of morphological, biochemical, and paleontological facts that, she believes, support the hypothesis that the cellular organelles of eukaryotes, even including flagella, centrioles, and spindle fibers, are derived from symbiotic prokaryotes that originally existed is a host cell and later became modified in various ways.

Raff and Mahler postulate that mitochondria originated from invaginations of the inner cellular membrane, which first formed respiratory vesicles bound by membranes. These vesicle became converted into mitochondria by the addition of a protein-synthesizing DNA plasmid, derived from another prokaryote. Raff and Malher say nothing about the origin of plastids, but reject outright the origin of flagella and the mitotic apparatus from symbiont microorganisms. They believe that the difference in modern eukaryotes between nuclear-directed and mitochondria-directed protein synthesis and other metabolic activities is due to divergent selective pressures that have acted since the acquisition of mitochondria.

Bogorad offers an alternative hypothesis for the origin of the eukaryote nucleus and the DNA contained in plastids and mitochondria. Emphasizing the well-known fact that some of the proteins of which these organelles consist are coded by nuclear genes, while others are coded by genes located in the organelles themselves, he has suggested the "cluster-clone" hypothesis. He believes that all of the DNA found in eukaryote cells originally existed in a nuclear area, perhaps surrounded by a common nuclear membrane. Later, clusters of genes derived from the proto-nucleus became separated and surrounded by their own membranes, thus forming the plastids and mitochondria.

At first sight, the hypothesis of symbiosis and the suggested alternatives appear to be diametrically opposed. Actually, they are not so contradictory when one examines the qualifications made in each theory to harmonize with the facts. An intermediate theory, containing

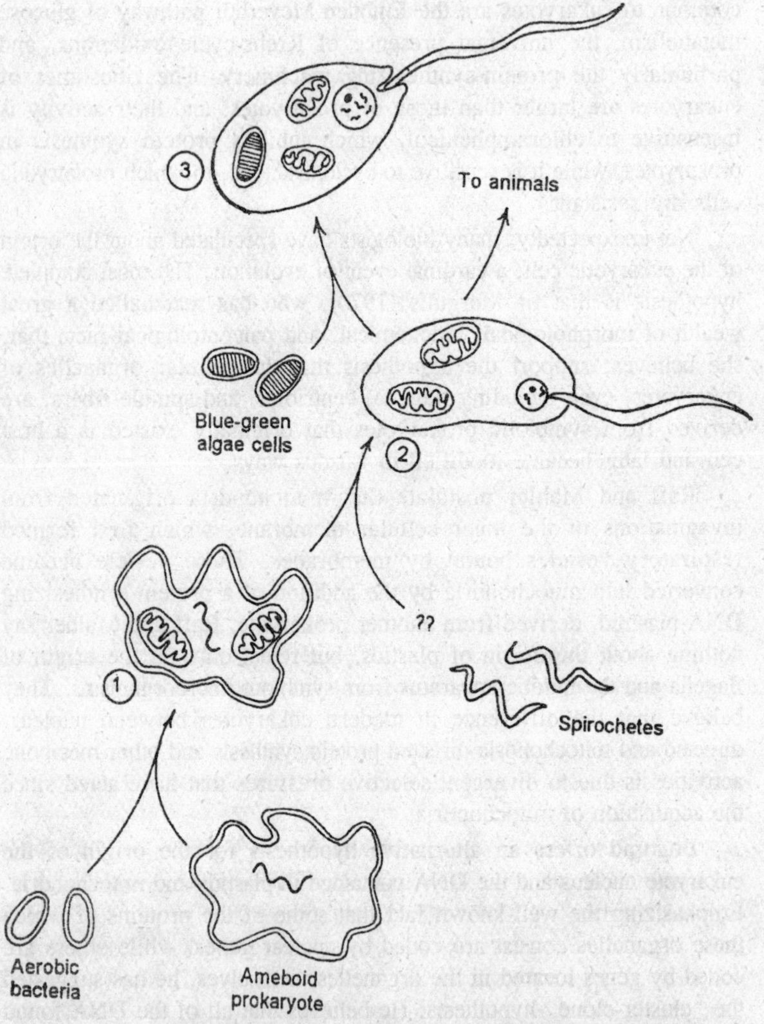

Fig. 6.3. The symbiosis theory of the origin of the eukaryotic cell. 1. A prokaryotic cell having a flexible cell wall, capable of ingesting bacterial cells. 2. An intermediate cell, in which ingested bacteria are evolving into mitochondria. 3. Blue green algal cells became incorporated into the colourless unicellular eukaryote to form a simple green flagellate.

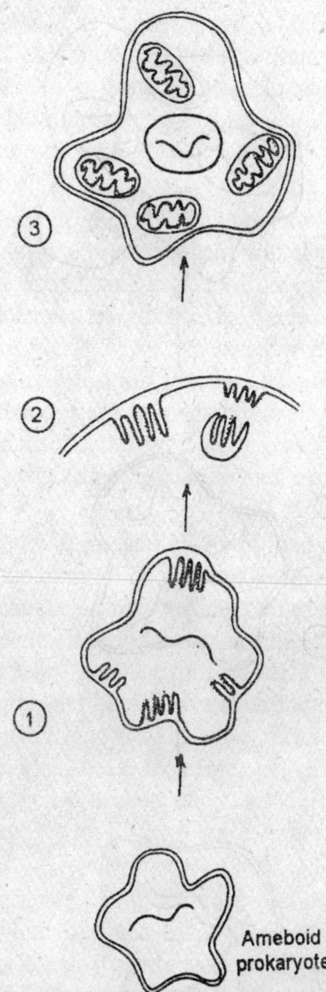

Fig. 6.4. The origin of mitochondria according to the invagination theory. 1. A prokaryotic cell. 2. Enzymes associated with glycolysis and respiration become associated with these invaginations. 3. The invaginated membranes lose contact with the cell membrane and become converted into mitochondria.

elements of both extreme hypotheses, may be acceptable. The biochemical similarities between plastids, mitochondria, and prokaryotic cells cannot be denied. One striking similarity is between the enzyme superoxide dismutase of chicken liver mitochondria and the corresponding enzymes in *Escherichia coli*: both are very different from the dismutases of nuclear origin found in bovine erythrocytes. Another similarity, which

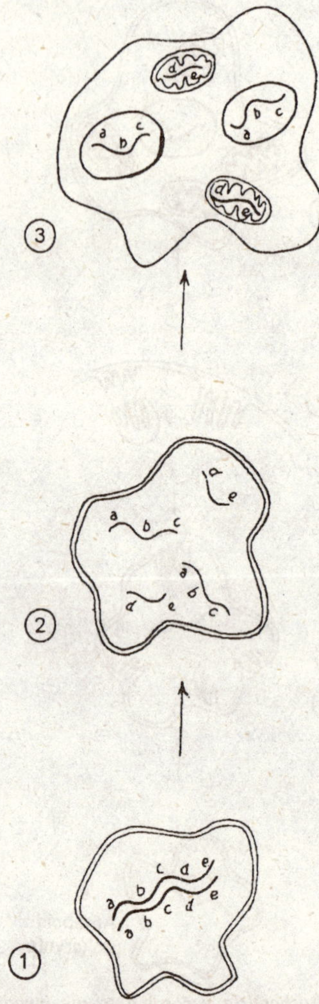

Fig. 6.5. Bogorad's cluster clone theory. 1. A prokaryote having a single duplicated strand of DNA on which are five genes. 2. A similar cell in which the DNA strand or chromosome has broken into two parts containing different genes. 3. Retention of some of these genes in a nucleus.

concerns the origin of chloroplasts, is the extensive homology between sequences of oligonucleotides found in the chloroplast DNA of the red alga *Porphyridium* and corresponding sequences recognized in prokaryotes, such as *E. coli*, *Bacillus subtilis*, and the blue-green alga *Anacystis nidulans*. Nevertheless, the similar sequences could have been introduced into the evolving prokaryotic cell in the form of plasmids, as Raff and Mahler have suggested. Whichever theory is accepted, evolutionary modification of both nuclear and organellular DNA during early eukaryote evolution must be postulated.

The strongest case for the symbiosis theory concerns the origin of plastids. The chlorophylls of eukaryotes are very similar to those of the prokaryotic blue-green algae; in addition, certain accessory pigments, particularly the phycobilins of red algae, are also similar. Furthermore, the cells of blue-green algae and the chloroplasts of red algae contain organelles known as phycobilisomes, which are so similar in their fine structure that they are almost certainly homologous. There are numerous other examples of unquestioned symbiosis between heterotrophic eukaryotes and blue-green algae. Two genera of unicellular flagellates, *Glaucocystis* and *Cyanophora*, which have typical eukaryotic nuclei, contain photosensitizing organelles (or perhaps symbionts) that in some respects are intermediate between blue-green algae cells and plastids. Furthermore, the chloroplast DNA of another autotrophic flagellate, *Euglena*, has been shown by hybridization techniques to be very similar to that of blue-green algae.

There is every reason to believe that some or all autotrophic eukaryotes have acquired their photosynthetic apparatus by evolution from blue-green algae. If one rejects the symbiosis hypothesis and assumes that plastids have originated by invagination, one must assume a direct connection via cell lineages from blue-green algae to autotrophic flagellates. This connection is very hard to accept, chiefly because of the great differences in the chemistry and the ultramicroscopic structure of cell walls between blue-green algae, the most primitive autotrophic eukaryotes, and typical green plants. The rigid walls of blue-green algae, like those of bacteria, have a thick network consisting of murein, which includes, covalently bonded together, N-acetylmuramic acid and a sugar, N-acetylglucoseamine. This complex is in no way related to the cellulose of higher plants, or to the mucilages based on galactose that form the cell walls of red algae. Furthermore, autotrophic eukaryotes described in the next section, such as dinoflagellates and *Euglena*, which have chromosomes and a mitotic apparatus intermediate between prokaryotes and eukaryotes, do not have rigid cell walls at all, but thin, somewhat gelatinous pellicles. One can logically postulate that organisms having thin flexible walls evolved from the lower grade of rigid-walled autotrophic prokaryotes and later gave rise to typical plants, algae as well as land plants.

Since in autotrophic organisms flexible cell walls are less adaptive than rigid ones, the loss of the rigid cell wall during the evolution of an autotrophic eukaryote would be very hard to explain. If, however, the first eukaryotic cell was heterotrophic and thin-walled, as postulated by the symbiosis hypothesis, the origin of the primitive autotrophic flagellates by ingestion of small blue-green algae and their conversion to plastids would be a plausible phenomenon, and one analogous to many well-documented evolutionary sequences. The acquisition of rigid cell walls having a different structure would certainly be adaptive in a number of ecological situations.

On the other hand, the internal membranes of eukaryotic cells, particularly the nuclear membrane and the endoplasmic reticulum, are best explained as invaginations. Since the prokaryote chromosomes are attached to the nuclear membrane, at least when they divide, an invagination of this part of the cellular membrane would automatically place the chromosomes inside the nucleus. The connections and molecular similarity between flagella and centrioles could be explained by assuming that in the earliest eukaryotes the nuclear membrane was still attached to the outer cell membranes at the time when both the mitotic apparatus and flagella were evolving simultaneously. The DNA of the basal body of the flagellum could then be explained by the addition of a plasmid, as suggested by Raff and Mahler.

The most plausible explanation of the complex series of changes required for the origin of the eukaryotic cell is therefore that both symbiosis and invagination were involved. The most disputed problem, that of the origin of mitochondria, may become settled when more biochemical data become available. As *Taylor* (1974) has maintained, the origin of flagella and the mitotic apparatus in symbiosis is highly improbable. Indeed, at the present state of knowledge no plausible hypothesis exists about the origin of flagella and the mitotic apparatus.

Both the invagination and symbiosis hypotheses can provide equally well a clue to the ecological conditions that prevailed during the prokaryote-eukaryote transition, and the type of organism that carried it out. Either of these methods would occur most easily in a relatively large cell having a thin wall and flexible cell membrane, i.e., an amoeboid prokaryote. The only organisms of this kind now alive are the mycoplasmas. It is highly unlikely these organisms are the ancestors of eukaryotes, since they inhabit chiefly the blood plasma of vertebrates, but wall-less forms of conventional bacteria, known as the L-phase, have been induced by a number of cultural techniques. Some of them breed true, apparently as a result of appropriate mutations. When derived from bacteria having the less specialized gram-positive type of cell wall, these forms are adapted to media having an osmotic strength of NaCl as high as 0.25 to 1.1 M. One might imagine that free-living gram-positive marine bacteria, living in the ooze at the bottom of seashore pools, would find a medium of the right osmostic strength, and a habit rich in available organic matter. If a prokaryote should begin occupation of this habitat, the ability to ingest the organic particles, derived from decaying marine organisms, would have a high adaptive value, so that the selective pressure for losing the murein mesh and acquiring the amoeboid form of nutrition would be strong. At the same time, a strong selective pressure would exist for transferring the enzymes responsible for glycolysis and respiration, which in bacteria are embedded in the external cellular membrane, to an internal position as associates of separate organells. This could be done efficiently either by invagination or by ingestion of smaller bacteria. Subsequently, selection would favour increased activity of enzymes located in the internal organelles and inactivation of those remaining on the external cellular membrane, so that the latter would become devoted solely to such functions as selective permeability and extensibility in connection with the formation of pseudopodia and the ingestion of food.

Since the marine environment postulated would have a low oxygen content, particularly if the atmosphere above it contained less oxygen than the contemporary atmosphere, these first amoeboid prokaryotes would be anaerobes. However, once they had acquired the division of labour between external and organelle membranes postulated in the last paragraph, they would be ready for further evolution toward more efficient metabolism by cellular activity. For this new phase of evolution, aerobic metabolism, motility by means of flagella, and autotrophy, either separately or combined with each other, would have a high adaptive value. If the ameboid prokaryotes already had mitochondria or proto-mitochondria, the number of these organelles per cell, and hence the rate of aerobic metabolism, could be increased by their differential reproduction relative to cell division. Flagella could be acquired by ingesting various other prokaryotes and by digesting all of their parts except the DNA of their chromosomes, which could be broken up into plasmids that could either be rejected or incorporated into the nucleus depending upon the adaptive value of the genes contained in them. As already mentioned, autotrophy would be acquired by ingestion of small unicellular blue-green algae and, their modification into plastid symbionts.

The suggestion of Sonea (1972), that bacterial plasmids derived from ingested cells contributed greatly to building up the chromosomal complements of eukaryotic cells, is one of the most plausible explanations yet offered for this fundamental part of the evolution of the eukaryotic nucleus. The transitional forms may well have evolved endonuclease enzymes specially adapted to the conversion of bacterial DNA's into such plasmids. The existence in viruses of endonucleases having this kind of specificity has recently been demonstrated by several workers. The partial endosymbiosis hypothesis, in the ecological version just proposed, is supported further by the reclassification of amoeboid Protozoa or Sarcodina proposed by Jahn, Bovee, and Griffith (1974). This greatly alters the phylogenetic position of the genus *Pelomyxa*, which Margulis singled out for particular attention because its cells contain large numbers of inclusions of various sorts. In earlier classifications of the Sarcodina this genus was regarded as relatively specialized, but according to Jahn, Bovee, and Griffith, it may be primitive.

The present version of the symbiosis hypothesis has a direct bearing upon transpacific evolution in general for the following reasons. First, it postulates that the event that triggered off this major advance in grade was one of many adaptive radiations carried out by anaerobic prokaryotes, and one that was distinctive because of the highly specialized and unusual environment the radiant entered. Second, it postulates that the one major way of food-getting that does not exist in free-living prokaryotes—ingestion of solid particles—was the primary method in eukaryotes. This is in accord with the fact that ingestion is more common among contemporary heterotrophic protists than is the bacterial method of external digestion and absorption. Finally, it postulates that the evolution of autotrophic organisms did not follow a direct line from blue-green algae to eukaryotic protists and algae, but involved the intermediate stage of heterotrophy followed by symbiosis.

These ideas are all completely compatible with the hypothesis of Margulis as well as with that of Raff and Mahler, at least concerning the mitochondria, the only organelles the latter authors considered. It is incompatible with the cluster-clone hypothesis of Bogorad, which postulates a direct continuity from blue-green algae to photosynthetic eukaryotes.

Origin of Mitosis and Sex

After the eukaryotes originated they evolved for at least 600 million years, and probably longer at the unicellular level. This evolution included the invasion of a large number of habitats, although present data do not permit definite conclusions as to what this adaptive radiation was like. Nevertheless, the perfection of the mitotic cycle and the origin of meiosis and the sexual cycle most certainly took place during this period.

Analyses of cell division in some group of flagellates give clues to the origin of the mitotic cycle, including chromosome splitting, condensation of chromosomes during prophase, relaxation at the end of telophase, and the formation of the mitotic spindle. Of greatest interest are the dinoflagellates, which have a kind of nucleus designated as mesokaryotic. Their nuclear membrane contains pores, and they have nucleoli like eukaryotes. The dinoflagellates contain several chromosomes rather than a single "genophore" as in typical prokaryotes. Their chromosomal fibres, however, are densely packed together in a very different way from those of eukaryote chromosomes. Furthermore, the metabolic nuclei, between mitosis, contain chromatin that is nearly or quite as condensed as that of metaphase and anaphase chromosomes; the cycle of condensation and elongation of characteristic of other eukaryote chromosomes is absent. They contain acidic proteins, but little or no basic protein or histone. During cell division, typical mitotic spindles are not formed, and the chromosomes remain within the nuclear membrane. The separation of the daughter chromosomes apparently results from elongation of this membrane, and so is somewhat similar to the separation of chromsomes or genophores in a dividing bacterial cell. The cytoplsmic organelles of dinoflagellates are typical of eukaryotes. The situation in *Euglena* is somewhat similar. On the basis of these observations, as well as a less anomalous but still aberrant situation in a parasitic protozoan, *Syndinium*, Ris and Kubi (1974) distinguish three successive stages in the origin of typical eukaryotic mitosis. Whether or not this scheme has general validity will not be clear until more transitional examples have been studied by means of modern electron-microscopic techniques.

With respect to the origin of sexual reproduction, two challenging questions present themselves. First, in what kinds of organisms did sex first arise? And second, what was the adaptive advantage that caused sexual reproduction to become predominant in higher organisms?

Until very recently no clear answer could be given to the first question. A survey of modern unicellular organisms conducted that those having the least-specialized cell structure—Chrysomonadina, Cryptomonadina, Euglenoidea, Dinoflagellata, and Amoebina—are not known to possess sex. Recently, however, earlier, doubtful records of a sexual cycle have been confirmed for the dinoflagellate genus *Crypthecodinium* (Gyrodinium) by clearcut morphological, cytological, and genetic evidence. Since this organism is mesokaryotic, its sexuality suggests strongly that sexual cycles evolved contemporaneously with the perfection of the mitotic cycle. Moreover, the sequence of meiotic divisions in this genus is not fully developed. Most probably, flagellate and ameboid protists with a fully evolved mitotic cycle but not known to possess sex either have an undetected sexual cycle or are asexual derivatives of sexual ancestors. The presence of fully developed sexual cycles in more specialized protists, such as *Chlamydomonas* and its relatives *Actinosphaerium* and *Actinophrys*, polymastigoid flagellates, Foraminifera, Sporozoa, and Ciliates, is in harmony with this conclusion. Paleontological data suggest that both mitosis and meiosis were present in microorganisms, probably algae, preserved in the Bitter Springs Formation of Australia, about 900 million years old.

The adaptive value of the sexual cycle has recently been debated, and the conventional statement, that it favours new adaptive combinations of genes, has been challenged. Williams points out that a well-adapted diploid genotype, in order to produce offspring that are equally well adapted to its particular environment, must pay a "50 per cent cost of meiosis." This is because each gamete produced by meiosis contains only 50 per cent of the parental alleles, including, presumably, those that compose the organism's adaptive gene combination. They are normally replaced in the zygote by genes derived from another individual, which may or may not be equally well adapted to the environment of the parent being considered. Obviously, if an organism occupies a constant, homogeneous environment, natural selection will favour those offspring that are genetically most like their parents. Since such offspring are always produced by asexual reproduction, the initial steps of evolution toward sexuality will lead to offspring with lowered reproductive fitness relative to the asexual offspring, and so will be eliminated by selection. The argument that sexuality can evolve because of its long-term advantages, particularly in a changing environment, is regarded as inadmissible by Williams, since natural selection is not predictive but operates from generation to generation.

According to Williams, the solution to this dilemma lies in the hypothesis that the first sexually reproducing organisms possessed an alternation between sexuality and asexuality, each process serving a somewhat different function. Asexuality served to maintain the population in a particular habitat, while sexuality made possible the exploration of new and slightly different habitats, thus increasing the ecological and geographic amplitude of the species. Examples of organisms having this duality are hydroid corals and molluscs among animals, and perennial plants, particularly those that, like the strawberry, reproduce efficiently by such vegetative structures as stolons, rhizomes, and bublets. Efficient exploration of the environment calls for cross-fertilization at random and high fecundity. This produces a great excess of gametes and zygotes with slightly different genotypes. If these can move about actively or passively, each habitat similar to the parental one can be colonized by a large number of genetically different zygotes or young offspring. Natural selection can then weed out those that are poorly adapted and favour those that are particularly well suited to the new environment. If the latter become capable of asexual reproduction, they can "capture" the new habitat and continue the cycle.

The early phylogenetic history of multicellular organisms support this hypothesis. Most of the phyla and classes of Metazoa, except for terrestrial vertebrates, arthropods, and molluscs, include some organisms that have efficient asexual reproduction, or at least the capacity to produce a great excess of gametes during their lifetime. Sponges (Porifera), colonial coelenterates, pelecypod molluscs, most algae, bryophytes, most spore-bearing vascular plants, and the great majority of seed plants are all good examples. Unfortunately, the exceptions—animals having either shorter lives or lower fecundity or both, as well as annual flowering pants—include nearly all of the multicellular organisms that have been studied intensively by geneticists. Hence, according to Williams, our knowledge of population genetics is biased toward those organisms with relatively low fecundity. Recent information about the gene pools of the more fecund organisms, based upon allozyme studies, should help to remove this bias.

Williams' hypothesis could explain why sexual cycles are rare or absent in free-living protists, such as *Euglena* and the majority of flagellates that make up the nannoplankton of the oceans. Once an organism has become adapted to the planktonic mode of life, exploration of new habitats is no longer necessary, and natural selection favours maximum production of genotypes similar to the successful parent and obtained by asexual reproduction.

Combining Williams' hypothesis and the one previously suggested for the origin of eukaryotes themselves, the following hypothesis may explain the very first origins of sexuality in primitive eukaryotes. These ameboid cells were asexually able to reproduce genetically similar ameboids, and thus could saturate the habitat of a particular tide pool. But without sex their future success depended either upon considerable phenotypic flexibility or the ability of passively dispersed propagules to find a new habitat sufficiently similar to the old one to make new colonization possible. Colonization might be aided by the evolution of active movements propelled by flagella. Actively motile propagules could also acquire the ability to attract each other by chemical means and so produce zygotes. If the latter then should settle down in a new pool and give rise by meiosis to four ameboid progeny, the chances that one of them could successfully colonize this new habitat by asexually produced progeny would be greatly increased. The hypothesis suggests that the first sexually reproducing eukaryote protists may have resembled the modern cellular slime molds.

It must be emphasized that the "cost of meiosis" argument applies only to diploid organisms. Protists, such as *Chlamydomonas*, as well as many green algae, lower fungi, and probably the earliest sexual organisms, such as *Crypthecodinium*, are strictly haploid or monoploid during their asexual or vegetative reproduction; the only diploid cell is the zygote. Such organisms can evolve mechanisms for genetic exploration of new habitats without paying any cost at all. Consequently, the really critical question to ask is: What was the initial advantage of diploidy, and why is it almost the only condition present among all phyla of Metazoa?

The most plausible answer to this question is that the first diploid organisms possessed marked heterosis or hybrid vigour. The advantage of delaying meiosis until many cell generations after fertilization (so that the somatic cells became diploid) was a particular rather than a general one. It existed only in the progeny of two parents having good combining ability. Diploidy helped to stabilize this highly adaptive heterotic condition. If diploidy and heterosis then became regularly associated in the population, reversions to haploidy would have been less fit and would have been eliminated. Later, the diploid condition made possible the enrichment of the gene pool through mutations, particularly those that gave rise to recessive alleles that were eliminated with difficulty or were retained indefinitely because of their contribution of superior heterozygotes. A high correlation exists between the diploid state and the existence of complex developmental patterns. This is most easily recognized by comparing parallel lines of evolution, one of them diploid (or predominantly so) and the other haploid. Examples are comparisons of the larger diploid brown algae with the haploid green algae; dikaryotic basidiomycete fungi with the monokaryotic ascomycetes; and the diploid vascular archegoniates ("Pteridophytes") with the haploid gametophytes of nonvascular ones ("Bryophytes"). These comparisons justify the conclusion of Raper and Flexer that the superiority of diploidy lies in the fact that two similar genomes, intimately associated, can exert subtle control upon minute steps of complex sequences of development and differentiation.

Another question raised by Williams (1975) is, "Why have most higher animals, which have acquired reduced fecundity and no longer perpetuate themselves by means of genetic exploration of the environment, nevertheless retained sexual reproduction, and are able to withstand the cost of meiosis?" He answers this question in part by noting that once sexuality has evolved into a highly complex and harmoniously integrated sequence of events, the pathway to successful asexuality passes through several intermediate stages that will usually be less adaptive. This adaptive "valley" is difficult to cross. A second answer for land vertebrates as well as many insects is that maternal care can provide a buffer that greatly reduces the significance of small genetic differences between sibling offspring. In birds and mammals the entire "cost of meiosis" is probably paid by parents in the form of feeding and otherwise taking care of their offspring.

Origins of Multicellular Organisms

Even a cursory review of the simpler multicellular organisms indicates that the grade of multicellularity was achieved many times independently, and probably for rather different reasons in different groups. Recent intensive studies—particularly of cell-wall structure and chemistry; cytoplasmic organelles, including cilia and flagella; the number of nuclei per cell or coenocyte; the chromosomal or sexual cycle; and the plastic pigments in autotrophic forms—have tended to increase our estimate of the number of separate origins. The following list is a minimal number (17) of simple multicellular kinds of organisms that we believe have arisen independently from unicellular Protista.

Autotrophic. Red algae; brown algae; filamentous golden algae; Vaucheriales; siphonalean green algae; *Acetabularia* and other Dasychadales; Volvocales; Conjugales; Ulotrichales and other uninucleate green algae; phragmoplast-forming green algae (*Klebsormidium* line).

Ideally, one would like to explain each of these separate origins as the outcome of a particular kind of adaptive radiation at the level of unicellular eukaryotes. Unfortunately, this is impossible, owing to the absence of fossils of the delicate organisms the earliest members of each line must have been, and the absence from the contemporary biota, probably through extinction, of most transitional forms. In the case of the earliest plant groups a reasonable hypothesis can be formed based upon two facts. First, the common denominator of differentiation with respect to the vegetative parts is polarity. All groups that evolved beyond diversification within a single order acquired filaments or other cellular aggregates in which one or more cells are adapted to anchoring the plant to a substrate and others are differentiated into the tip of a filament or the margin of a flat, sheet like structure. Second, the simpler aquatic algae having polarity reproduce asexually by means of motile cells that are themselves polarized and resemble unicellular protists in all cytological details. When these cells settle down to form a filament, their front end, bearing flagella, becomes attached to the substrate, and the first cross wall of the future filament is formed at right angles to the polarized axis of the motile cell. Hence, the polarization of the filament, which eventually in higher plants became transferred to the polarization root-shoot, did not arise *de novo*, but was derived directly from that of the motile unicellular ancestor.

On the basis of these facts a plausible hypothesis for the origin of the multicellular filament has been suggested. Unicellular autotrophs could colonize turbulent of flowing water only if they became attached to a firm substrate, to avoid being swept away. The flaggella of many unicelluair forms are viscid and adhesive, and so would be the natural organelles for attaching the cell to the substrate. Natural selection under these conditions would favour cells having the most viscid flagella, and mutations that replaced typical flagella with more elongate, branching structures similar to the rhizoids of algal cells would have a particularly high adaptive value. Once this happened, reproduction would depend upon the ability to produce, by mitosis, cells capable of differentiating into motile zoospores, which would have a very high selective value. On the other hand, the most successful attached filaments would be those in which a considerable amount of photosynthetic surface was produced before zoospores were initiated. This could be acquired by cellular elongation before the onset of the first cell division, or by mitotic divisions that produced cylindrical, photosynthesizing filament cells before producing cells capable of differentiating into zoospores. Ever since the nineteenth-century observations of Julius Sachs, botanists have known that when a plant cell becomes relatively long and narrow, the mitotic spindle becomes oriented parallel to its long axis and the first cross wall is formed at right angles to this axis. Hence, selection for increased photosynthetic surface would result directly in the evolution of a polarized filament. Further evolution forming branched filaments, flat plates, or three-dimensional structures would depend upon processes that as yet are not understood.

There is no reason to believe that the origin of multicellular organisms, any more than the origin of the eukaryotic cell, depended upon processes other than natural selection of genetic combinations fitting the earliest multicellular plants and animals to particular environmental niches. The most general cause favouring the later success of multicellular organisms was the adaptive value of increased size.

7

METAZOAN LIFE

Metazoa may be defined as animals that share a common multicellular ancestry with coelenterates and flatworms. Although there is a good fossil record of the major groups that have well-mineralized skeletons, the origins and earliest evolution of the metazoan phyla cannot be documented from fossil evidence. Nevertheless, from the fossils that we do have, from comparative studies among living organisms, and from interpretations of the functional significance of primitive characters, it is possible to produce models of metazoan origins and diversifications.

As noted already that there are several types of fossil remains, and each has a different probability of preservation. We shall place them in three classes here: trace fossils (trails, burrows, tracks), soft-bodied fossils (chiefly impressions or carbon films); and rigidly skeletonized fossils, which are usually impregnated with minerals (chiefly carbonates or phosphates). Although soft-bodied animals frequently posses hydrostatic skeletons, fossil skeletons are of mineralized or protein-tanned structures. Both soft-bodied and rigidly skeletonized fossil are called *body fossils*. So far as the record is now understood, trace fossils appear earliest, then soft-bodied forms, and finally mineralized skeletons.

Traces left by the activities of organisms are difficult to discriminate from sedimentary structures left by many inorganic processes. Although there are occasional claims of very old animal fossils, these can usually be clearly ascribed to inorganic causes; a few remain doubtful. The earaliest traces that appear likely to be valid organic traces were recently discovered in Zambia. They are probably near 1,000 million years old. They are horizontal burrows, up to about 10 mm in diameter, that have been back-filled with sediment, presumably by detritus-feeders that ingested the sediment itself. The next earliest known burrows may be less than 700 million years old. The long barren interval between these fossils is puzzling. Sediment deposits from this interval, confidently interpreted as shallow marine deposits, have been searched for fossils without success. The burrows that appear sometime after 7 million years ago (their ages are not closely determined as yet) include horizontal and vertical burrows and sea-bottom trails. The abundance and diversity of traces increase progressively in younger rocks until, during the Cambria Period (beginning 570 million years ago), they reach a richness in shallow water that they maintain for hundreds of millions of years.

Among the earliest authenticated records of soft-bodied invertebrate fossils are those of the Ediacaran fauna, named for an unusually large assemblage from the Ediacara formation, South Australia. This fauna seems to have lived between at least 680 and 580 million years ago and is found on several continents that are at present widely dispersed: Europe, Africa, North America, and Australia. The fauna is principally composed of medusae, pennatuloids, and other forms that appear also to have been coelenterates, and a small variety of worms of uncertain affinities, but perhaps including annelid-like forms.

The next younger assemblage of soft-bodied fossils of importance is from the Burgess Shale, British Columbia, Canada. It is of Middle Cambrian age and postdates the appearance of mineralized skeletons by a few tens of millions of years. For over 65 years this fauna has been known to exist and is now undergoing a much needed reappraisal. The new results are most impressive. About a quarter of the fauna represents groups that have already appeared by mid-Cambrian time, many with mineralized skeletons. These include coelenterates, molluscs, trilobite arthropods, brachiopods, echinoderms, and hemichordates. Another quarter is represented by non-trilo-bite arthropods, including sorts that are otherwise unknown; sponges constitute another quarter. Of the remainder, many specimens represent the earliest undisputed records of living phyla: Annelia (which may be represented in the Ediacara fauna), Priapuloida, Nemertea, and Chordata. Finally, a few specimens belong to groups that are distinct from living phyla; they represent extinct phyla that do not seem ancestal to any living animals. There are at leas three of these; only one has as yet been adequately described and is interpreted as a lophophorate coelomate.

The contrast between this soft-bodied assemblage, teeming with invertebrate phyla, and the depauperate late Precambrian fauna thus far known, is striking. There are only a few other occurrences that provide us with important glimpses of some of the soft-bodied associations of later times. An important one is in the lower carboniferous of Illinois, which has yielded a form that is not readily assignable to any living phylum.

The earliest mineralized skeletons appear slightly earlier than the generally accepted base of the Cambrian, below the Tommotian Stage of the Russian platform. The fossils are minute and of uncertain affinities. Additional skeletons appear in the Tommotian, including small coiled shells that are probably molluscs, and small conical elements that may be associated with tentacle of primitive suspension feeders. In the next stage many more phyla appear, so that within the first two Cambrian stages, seven skeletonized phyla are known. The last skeletonized phylum to appear, the Ectoporocta, does so in the earliest Ordovician.

During the Cambrian and Ordovician, radiations among the durably skeletonized phyla led to the establishment of numerous classes and orders, so that the number of taxa in those categories was considerably greater than at present for these same phyla. There is no reason to believe that the phyla for which records are poor—the soft-bodied groups—were not represented also by large numbers of classes and orders at that time.

To summarize, the earliest animal fossil (traces) may date from one billion years. Body fossils appear later and include coelenterates, possibly annelids, and a number of enigmatic forms. Then shortly after 570 million years ago, durably skeletonized forms that include several phyla of coelomate metazoans appeared in quantity. The next younger soft-bodied assemblage known is also rich in advanced metazoan types. By the early Ordovician, all living phyla that have durable skeletons are present; a large number of phyla that have no hard parts at all, including some extinct ones, also have been discovered. It appears that among marine invertebrates diversity of anatomical architecture as represented by phyla, and major modifications of these basic body plans as represented by classes and orders, reached higher levels in the early Paleozoic than ever again.

Evolution of the Tissue Grade

The earliest metazoans are unknown; the earliest yet described had already attained a tissue grade of organization. The adaptive and phylogenetic pathways that led to this grade have been inferred from functional and morphological studies of living organisms, which can be interpreted in a number of ways. As a result, there are a number of different historical models to explain both the evolution of metazoan grades and the radiations of body plans within grades.

It is widely accepted that metazoans evolved from protozoans, though there is a dispute as to the ancestral type. The first significant advance towards metazoans must have been the development of a multicellular organism. As we have seen, the step to multicellularity occurred several times. The ancestral type usually suggested for the metazoans is a flagellate protozoan, which developed a colonial habit not unlike living *Volvox*, leading eventually to obligate multicellularity. Hadzi (1963) suggests that the protozoan ancestors were ciliates and that multicellular metazoans arose from the division of a multinucleate syncytium. Among the difficulties faced by this hypothesis is the highly specialized nuclear behaviour of living ciliates, which is unlike that of metazoan cells.

The advantages of multicellularity per se flow chiefly from the sizes, shapes, and repetitions of cellular machinery that are achieved through a modular construction. Among them are increased homeostasis, longevity, and reproductive potential. Like the primitive eukaryotes, the metazoans must have evolved in the sea, where there would be numerous ecological opportunities for these primitive animals. For example, important adaptive zones in the sea contain detritus- or suspension-feeders, which utilize particles in the water column by sweeping them from the water or by receiving a rain of descending particles on the sea floor. In shallow water particles are likely to be somewhat patchily distributed, so that a large surface area for particle capture is advantageous. One way the primitive organisms can increase their areas is for their replicating cells to remain in contact. Uneven distribution of food items can then be corrected by transferring ingested materials or their degraded products across cell walls; in this way each cell has access to a more stable food supply, enhancing homeostasis.

The adaptation of the colony could be enhanced if growth patterns and morphogenetic movements were specified within the genome of the colony to control its size and shape. To accomplish this, the regulatory repertory would have to enlarge from control of individual cell functions to the integration of many genetically identical cells. Evolution of multicellular control could proceed through the intercellular exchange of substances, such as hormones or larger molecules originally employed in internal cell regulation but which become inductors. Once some genetic multicellular integration was achieved, variations in the shapes of colonies could have arisen through recombination and mutation, and natural selection would establish an array of shapes, each adapted to a distinctive range of habitats. For example, a dome-like or hummocky shape can be advantageous in benthic colonies because it increases the surface area of the colony. Furthermore, water currents flowing over irregular surfaces become turbulent, so that a hummocky colony living in fairly quiet water is washed by a disproportionately larger volume of water than a flat sheet of cells; of course, increased current flow increases the probability of food. Cylindrical or cup shapes also permit colonies to obtain food particles from a larger segment of the water column. Strands of cells extending above the colony proper serve a similar purpose.

Establishment of even simple shapes, while making for more efficient exploitation of resources by a colony, often results in an imbalance in food-uptake at different locations within a colony. In domical shapes, for example, the cells towards the center tend to receive more food than peripheral ones. Natural selection would promote differentiation to increase the feeding efficiency of the central cells, those chiefly responsible for nourishing the colony. Reproduction might well be suppressed in such cells but selected in well-nourished cells at neighboring localities, while peripheral cells would be differentiated for strength in support of the colony. Patterns of differential selective pressures on cells would vary among colonies of different shapes. Suspension-feeding colonies would be modified in shape to promote efficient current patterns for feeding. Other forms, especially those feeding on benthic detritus, would be required to creep along in order to exploit new areas of the substrate. A rim or some pattern of ventral cells could be specialized for locomotion, while other cells specialized in ingestion of food. In some lineages feeding cells came to be located in partial invaginations in order to increase cell number and localize digestive capability. Floating or swimming modes of life are often postulated for primitive multicellular colonies, with food items trapped by secretions or swept into pockets or locations specialized for accepting them.

During the evolution of intercellular communication, regulation, and differentiation, a point is reached where the cells no longer form a colony, but an integrated individual organism. We can imagine an array of simple organisms. The early forms at thus grade are all extinct. Although reconstructions are speculative, they indicate the selective advantages of small steps along an adaptive pathway from unicellular to truly multicellular organisms. Perhaps the most interesting hypotheses concerning the earliest metazoans are those by Haeckel (1874) and by Metschnikoff (1833). Haeckel postulated that metazoans arose through a hollow spherical layer of cells, the *blastaea*. The posterior cells of this sphere are inferred to have become specialized for feeding and to have invaginated, thus producing a two-layered *gastraea*. The interior cells became endoderm and the hollow of the invagination became a primitive gut. The evolution of a jellyfish-like coelenterate from such a primitive form can easily be imagined. Metschnikoff, however, believed that the primitive metazoan was solid, and that digesting (which is intracellular in lower metazoans) occurred in cells that filled the interior by ingression from the external cell layer. The planula larvae of most coelenterates is in fact a living analog of such a form. Invaginating to produce an endoderm and a primitive gut space is generally rare in metazoan ontogeny. To the extent that such ontogenies indicate phylogenetic pathways, the evidence favour Metschnikoff. However, if the very early ontogenetic stages evolved simply as effective developmental steps rather than reflecting phylogenetic development, then we are left without direct evidence of the precise phylogenetic and adaptive pathways during earliest metazoan evolution.

There are three living multicellular animal groups with especially primitive features; the Porifera (sponges), the Coelenterata (jellyfish, corals), and the Ctenophora (comb jellies). It is worthwhile to speculate briefly upon plausible pathways leading from simple multicellular forms to these groups.

The multicellular sponges display cellular differentiation, although they lack nervous systems or well-defined organs. Some are solitary, while others are considered "colonial" because they have numerous openings or osculae and can be considered as collections of individuals. However, there seems to be no differentiation of function whatsoever among individuals within a "colony." The body walls are composed of outer and inner cell layers separated by a gelatinous substance containing amoebocytes. The cell layers are not homologous with the ectoderm and endoderm of other animals; sponge development is quite unique. Most sponges secrete mineralized skeletal spicules, which are first found as fossils in the Lower Cambrian.

Since cells of the inner layer closely resemble certain flagellate protozoans, it is commonly suggested that sponges have evolved from such ancestors. Since their unique development suggests that they evolved independently of other animal-groups, they are often classified as Parazoa, separate from the Metazoa. It is quite possible that they evolved well after the Metazoa, which appear in the fossil record at least 100 million years earlier. Sponges illustrate that animals are perfectly viable at the simpler multicellular grades of organization.

Coelenterates possess two cell layers, ectoderm and endoderm—the *diploblastic* condition. Space between the layers is filled with a fibrous, gelatinous substance, the *mesoglea*, the jelly of jellyfish. Cell differentiation from the cell layers to form tissues is well advanced; gonads are present, and a relatively simple nerve net is developed. A number of evolutionary routes could lead to be coelenterate condition. A conical, detritus-feeding, multicellular animal with either internal or invaginated digestive tissues, which we have previously considered, has essentially reached a diploblastic state. However, such an animal would be a detritus feeder or a grazer and somewhat vagile, and would tend to be polarized anteroposteriorly and develop bilateral symmetry. Nevertheless, a coelenterate ancestor of this general sort has been postulated.

A diploblastic state could also develop via the evolution of a cup-shaped suspension feeder, the cells located centrally within the cup becoming specialized for ingestion and then becoming internalized to promote digestive efficiency. A sessile benthic habitat commonly correlates with a radial symmetry. There is evidence from comparative coelenterate embryology and anatomy, however, that the more primitive

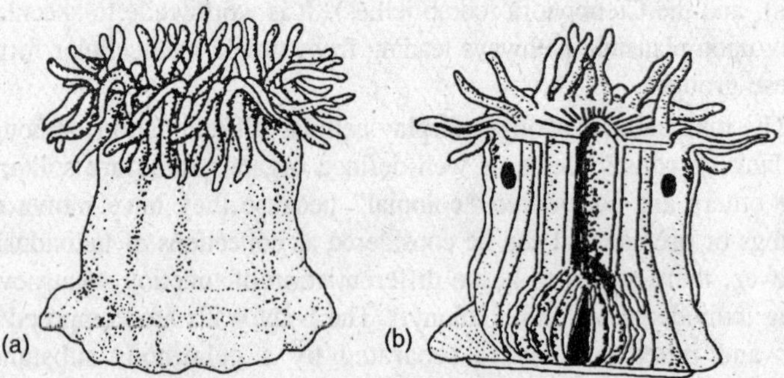

Fig. 7.1. (a) A sea anemone; (b) Vertical cross-section of a sea anemone showing tentacles surrounding mouth, gastric cavity, and an internal mesentary with associated muscles and gonads.

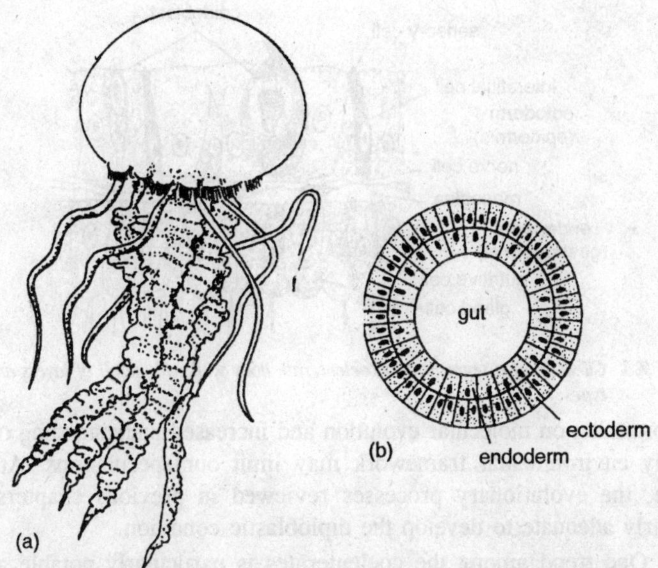

Fig. 7.2. (a) A jellyfish; (b) Horizontal cross-section of simple jellyfish showing two-layered body wall.

form is a medusa, a pelagic animal. Hand (1959, 1963) has argued that the coelenterates arose from a pelagic planula-like organism and that their radial symmetry is adaptive to pelagic life. The mesoglea in medusae is highly functional, serving to antagonize muscles and to provide support, shape, and floatation for the body. Rees (1966) has suggested a simple coelenterate ancestor that has four hollow tentacles and a tetraradiate stomach. He postulates a bentho-pelagic habitat for this form. One can derive such an animal from any of the more generalized multicellular organisms discussed above. Hadzi (1963, and references therein) has sought to derive coelenterates from bilateral flatworms, but this notion has not stood up well to critical examination.

Ctenophora or comb jellies are another diploblastic group that are sometimes treated as a subphylum of Coelenterata or as a separate phylum. However they may be classified, their origin appears to be tied up with that of the coelenterates, with which they share numerous common features. Clearly, the precise adaptive pathway and the precise succession of direct ancestral forms of the coelenterates is not known. However, our problem lies not in imagining a reasonable adaptive route, but in determining which among a wide variety of plausible possibilities is the historical pathway. In the absence of a fossil record we may never be able to decide this question definitively, although

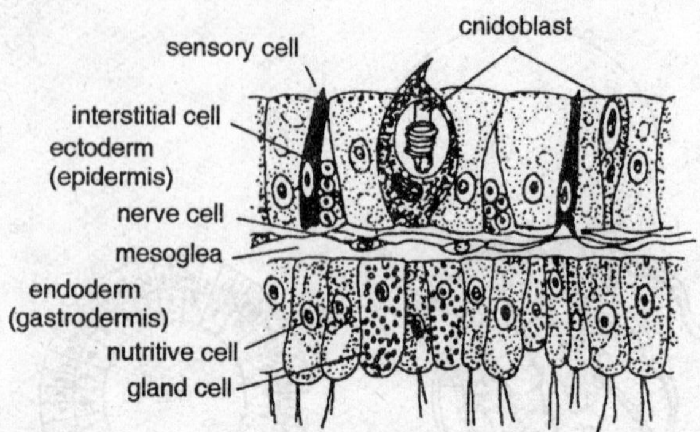

Fig. 7.3. Close-up of a segment of a coelenterate body wall with detail of layers and cell types.

information on molecular evolution and increased understanding of the early environmental framework may limit our speculations. At any rate, the evolutionary processes reviewed in previous chapters are clearly adequate to develop the diploblastic condition.

One trend among the coelenterates is particularly notable as an illustration of the evolutionary tendency toward complexity. This is the development and elaboration of colony formation among individual multicellular organisms. It has occurred a number of times in Coelenterata as well as in other metazoan phyla. The general advantages of coloniality for coelenterates are similar to those that we outlined for primitive multicellular organisms. Noncolonial aggregations of individuals are commonly advantageous, for clumping can lead to useful surface topographies of the aggregates that are advantageous to the individuals, especially in suspension feeding and in reproduction. One disadvantage of aggregation is that it frequently results in overcrowding. The integration of aggregates into colonies can mitigate this problem by controlling the population density; the most advantageous spacing distances and patterns between individuals can be evolved, once these features are under genetic control. Furthermore, coordination among colonial individuals, which are physically attached by tissues, can be achieved through metabolites or common nerve paths. Such coordination can lead to improved efficiency in obtaining food, in eliminating metabolic wastes, or in warning of predatory attack or of the onset of other threatening conditions. The commonness of coloniality can easily be explained by such advantages.

Once colonial organization appears, a further advantage may be gained through the evolution of specialization. Since individuals are so placed within a colony that they encounter food, competitors, or predators more frequently than other cells, and selection for distinctive functions follows. In extreme cases, individuals specializing in feeding, reproduction, defense, or other functions lose nearly all but one major function, employed to the advantage of the colony. This development implies an elaboration of the machinery of genetic regulation. Colonial individuals originate by budding and have identical genomes, yet they become extremely differentiated during ontogeny. In the coelenterate class Scyphozoa, which includes such pelagic animals as the Portuguese man-of-war and the by-the-wind-sailor, differentiation of individuals is so extreme as to produce superorganisms of individuals. One individual forms the float, others serve as feeding organs, digestive organs, or reproductive organs. The same sort of adaptive path that has led from single cells to multicellularity has led a diploblastic line from solitary individuals to a colony that is virtually at a higher grade of organization. However, this particular trend towards complexity ends in the scyphozoans; higher Metazoa arose along other adaptive pathways, although in response to some of the same selective pressures.

Evolution of the Triploblastic Grade

Instead of proceeding from compound superorganisms, higher Metazoa evolved by attaining a truly new major grade of organization at the level of the individual organism. Among the simplest living organisms at this new grade are flatworms, the phylum Platyhelminthes. In the place of the gelatinous mesoglea layer, a meshwork of cells is developed to form a tissue, the *mesenchyme*. This is considered to represent a third tissue layer, or mesoderm, developed between ectoderm and endoderm in all higher Metazoa; the triple layered structure is *triploblastic*.

Flatworms are small elongate animals that employ one of three chief methods of locomotion. First, smaller species move by the activity of cilia packed on the flattened ventral surface. At larger body size the increase in mass is disproportionate to the increase in area and therefore the cilia become too few to provide adequate propulsion. Instead, locomotion is enhanced by the body wall musculature, which can provide great force. The second method is for these muscles to "inch" the worms along. A third method is for these muscles to cause waves (pedal waves) of alternate swellings and hollows in the body. The swellings adhere to the substrate, and the waves are passed backwards by coordinated muscular contractions. New waves form as the anterior portion of the body extends forward beyond the previous leading wave. The elongate, bilaterally symmetrical flatworm is well adapted to directional creeping on reasonably solid substrates. The relatively solid mesenchyme provides the internal rigidity required for locomotion by pedal waves. Flatworms cannot burrow well, although they can penetrate loosely packed sediments, such as sands, making their way within particle interstices.

The only simple living organism that much resembles the imagined flatworm ancestor is the coelenterate planula larva. It has been suggested that flatworms developed from early multicellular organisms resembling planulae. These ancestors may be imagined as metazoan larvae that underwent neoteny, thereby dropping the adult stages out of their life cycles (diploblastic forms, perhaps) and remaining on the substrate as worms.

A benthic diploblastic adult has also been postulated as the flatworm ancestor; some living coelenterates creep, and some ctenophorans, e.g., *Coeloplana*, are highly convergent with flatworms in general body form. However, the ctenophoran and platyhelminth ground plans are very distinct, and this pathway is unlikely. Hadzi.

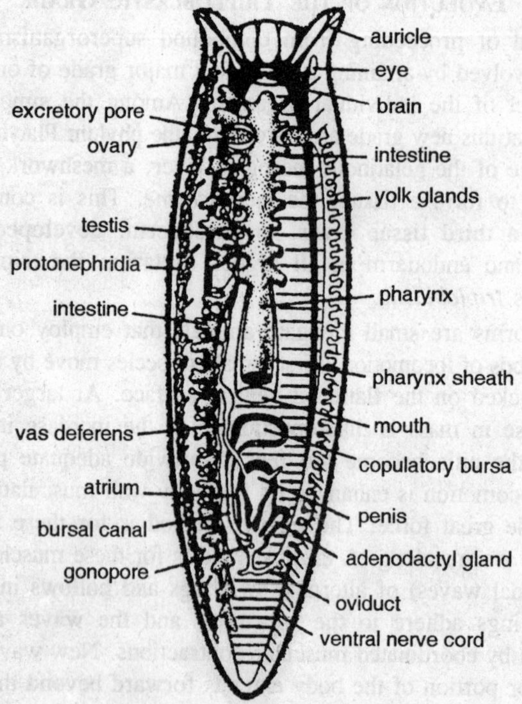

Fig. 7.4. Architecture of a typical flatworm.

(1963) and some previous workers have proposed a phylogeny in which flatworms developed from ciliate synctia, which in turn gave rise to the coelenterates. Some difficulties with this hypothesis were mentioned earlier. At any rate, the rise of the triploblastic grade does not present any conceptual difficulties, although again we, cannot trace a precise phylogenetic route.

The development of the flatworm ground plan involved still other elaborations. Although the flattened form does not require specialized breathing apparatus (since oxygen exchange can occur over the general body surface), the elongate shape and triploblastic construction give rise to problems of nourishment and waste elimination. As body size increases, the interior cells become increasingly remote from food sources in the absence of a circulatory system. The endoderm-lined digestive tract opens ventro-medially and runs both posteriorly and anteriorly to service the length of the organism. In some turbellarian flatworms the endodermal lining of the gut is greatly increased in area by a series of laterally paired diverticulae, which extend the digestive spaces into the lateral body regions. Excretion is provided by protonephridia, and reproduction is provided by special tissues that are concentrated into gonads. These organs are placed in the spaces between gut diverticulae, one on each side in series. This arrangement is of obvious advantage in serving an elongate organism.

The *pseudometamerous* body plan of flatworms, with a bilaterally symmetrical serial arrangement of internal organs, is found in other phyla, such as Nemertea. These are larger triploblastic worms that possess a few circulatory vessels beneath the body-wall musculature but still lack special respiratory organs. Many are far less flattened than the platyhelminths and some lineages have thick cores of mesenchyme. The thickened mesenchyme is associated with a type of locomotion that foreshadows developments in later metazoans: peristalic locomotion. Annular waves of contractions, passing entirely around the body, run consecutively backward along the body length. The dense mesenchymal tissue is not easily compressible, and when squeezed in by say a constricting crevice, it bulges out to each side, so that it forms a hydrostatic skeleton to antagonize the body-wall muscles. The worm flows forward as the waves pass backwards;this process is aided by ciliary activity on the body surface. Nemertines are thus able to burrow, for the peristaltic system permits the pushing aside of enclosing sediment. Nemertine burrowing is probably restricted by the relatively limited hydraulic efficiency of tissues, and by the lack of respiratory and advanced circulatory systems.

Evolution of Coelom

A number of other triploblastic lower metazoan phyla are known, each presenting a fascinating puzzle in adaptation. For our purposes, however, it is best to examine the next evolutionary advance in grade, which freed the invertebrates from many ecological constraints and permitted their diversification into all the extant higher metazoans. This was accomplished by the development of the coelom, an internal, fluid-filled space surrounded by mesoderm and lined with a special tissue, the peritoneum. Unlike the gut, which represents a space that is continuous with the external environment, the coelom is entirely internal. It does communicate with the exterior through special ducts that transmit wastes (renal ducts) or reproductive products (gonoducts).

The origin of the coelomic cavity is uncertain; common theories are that it arose first as gonoducts, which were later expanded, or that it arose as out-pockets of the gut that migrated into the mesoderm. Clark (1964) has argued that the primitive function of the coelom is as a hydrostatic skeleton, and that the coelom arose during the evolution of peristaltic locomotion. Since there are a number of different invertebrate phyla, the coelomates may be polyphyletic.

Four or five basically different coelomate body plans can be recognized among living organisms, but we do not know which, if any, is the most primitive. Although the early coelomates have left fossil traces, these are not distinctive enough to permit the erection of phylogenetic trees. The simplest coelomate architecture consists of a hollow-bodied worm. Among living phyla, Sipunculida may have had a longitudinally unregionated coelom from the beginning. Other worms with annelidan affinities have little segmentation, but this is probably a secondary condition. Unsegmented worms are commonly detritus feeders, living in shallow burrows or nestling in borings or crevices and gathering food by ingestion of sediments or by trapping detritus on mucous nets. These worms are relatively sedentary. Introverts are commonly developed for burrowing and feeding.

A second coelomic architecture is found in forms that have two longitudinal coelomic compartments separated by a septum. This situation occurs among the lophophorate phyla Phoronida, Brachiopoda, and Ectoprocta. The phoronids are vermiform and may be closest to the common ancestral type of lophophorate. These worms are sessiel suspension feeders. Most form a burrow, from which they extend their lophophores above the substrate to feed. The lophophore consists of a set of tentacles surrounding the mouth. Within each tentacle is a coelomic lumen; these spaces are continuous within the tentacular crown, but are separated from the trunk coelom by a septum. The tentacles are manipulated by intrinsic muscles that are antagonized by the tentacular coelom. The cylindrical trunk with its coelomic cavity can antagonize body-wall muscles in burrowing. The hydrostatic systems of the trunk and lophophore function somewhat independently.

The deuterostome invertebrate phyla, the group from which the chordates have arisen, possess a coelomic plan similar to that of the lophophorates. These phyla include Echinodermata, Hemichordata, Urochordata, and some minor groups. The coelom in this alliance is primitively divided into three longitudinal compartments. Presumably this plan arose because three separate hydrostatic functions evolved. One compartment was probably used in burrowing, and another to manipulate tentacles or some similar structure for feeding. It has been suggested that the third compartment may have originally been associated with locomotion, perhaps to move up and down a burrow. Whatever their primitive function, the coelomic architectures of the deuterostomes are much modified during ontogeny.

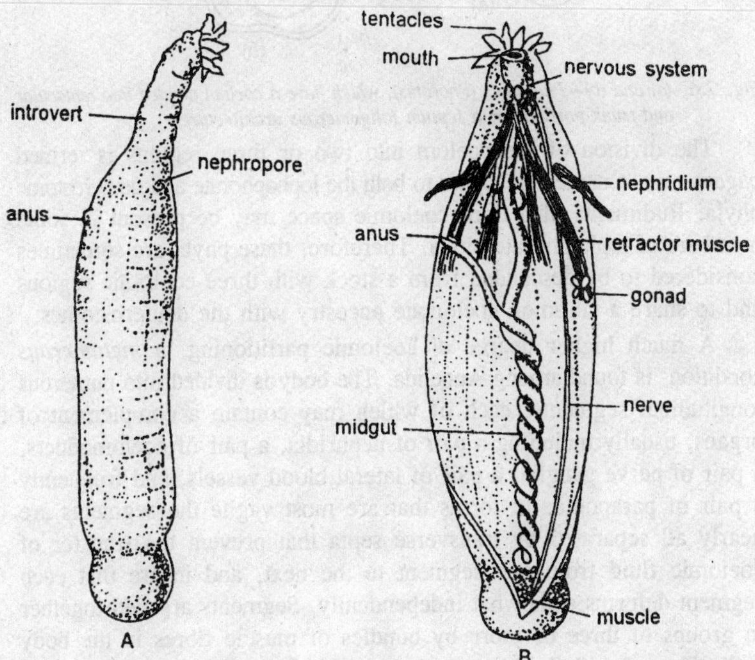

Fig. 7.5. A and B—Sipunculids (Sipunculus), which entirely lack coelomic regionation (amerous architecture).

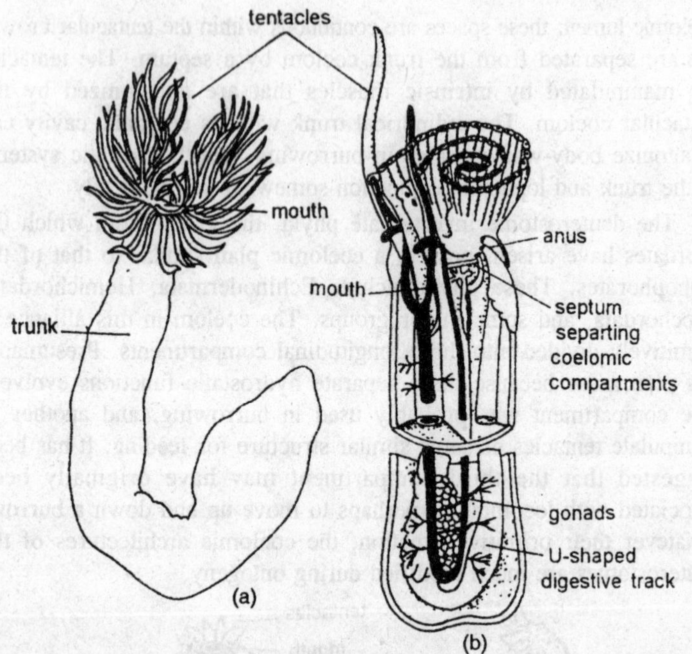

Fig. 7.6. (a) and (b)—Pheronids (Phoronis), which have a coelom divided into tentacular and trunk portions by a septum (oligomerous architecture).

The division of the coelom into two or three regions is termed *oligomerous*, a name that applies to both the lophophorate and deuterostome phyla. Rudiments of a third coelomic space may be present in some members of lophophorate phyla. Therefore, these phyla are sometimes considered to be descended from a stock with three coelomic regions and to share a common coelomate ancestry with the deuterostomes.

A much higher degree of coelomic partitioning, a *metamerous* condition, is found among Annelida. The body is divided into numerous longitudinal segments, each of which may contain a complement of organs, usually including a pair of nephridia, a pair of coelomoducts, a pair of nerve ganglia, a pair of lateral blood vessels, and frequently a pair of parapodia. In forms that are most vagile the segments are nearly all separated by transverse septa that prevent the transfer of coelomic fluid from one segment to the next, and insure that each segment deforms somewhat independently. Segments are tied together in groups of three or more by bundles of muscle fibres in the body wall. The internal fluid pressures resulting from muscular contraction during peristalsis are thus localized within a few segments, and

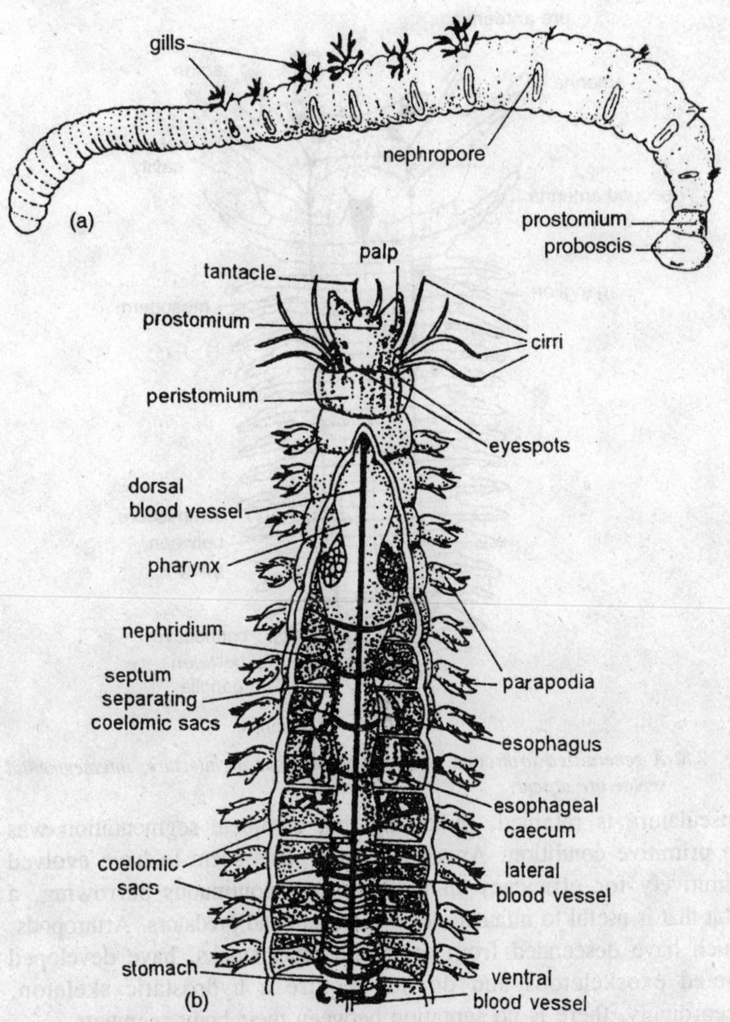

Fig. 7.7. (a) and (b)—Annelids (Arenicola and Nereis), which have a highly segmented coelom and regularized repetition of organs.

succeeding peristaltic waves attain a large measure of mechanical independence, which makes for a relatively high locomotory efficiency.

In annelid lineages that have become largely sedentary, such as the deposit feeding burrowers (*Arenicola*) and tube-dwellers (*serpulids*), many of the septa are reduced or obsolete. These animals pursue a way of life reminiscent of the phoronids; their coelomic architecture is simplified although the regularized repetition of organ systems and

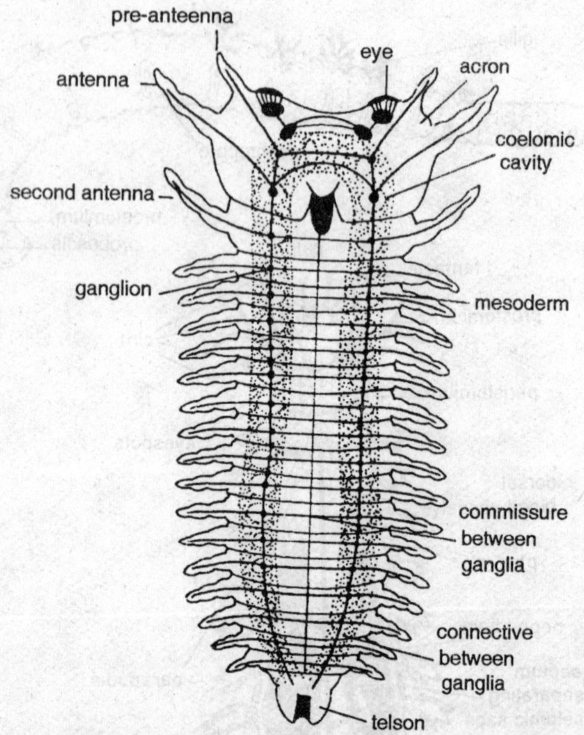

Fig. 7.8. A generalized arthropod, showing metamerous architecture; intersegmental septae are absent.

musculature is retained, indicating that complete segmentation was the primitive condition. Annelid ground plans seem to have evolved primitively for efficiency in more or less continuous burrowing, a habit that is useful to infaunal detritus feeders and predators. Arthropods, which have descended from proto-annelidan stocks, have developed jointed exoskeletons and do not require a hydrostatic skeleton. Accordingly, there is no septation between their body segments.

A final type of coelomic architecture is displayed by the molluscs. Their coelom is neither regionated nor segmented; yet the primitive molluscs (Monoplacophora) display extensive serial repetition of organ systems, including gills (five pair), nephrida (six pair), pedal, musculature (eight pair each of two types), gonads (two pair), and a number of other organs. This sort of pseudometamerism recalls the flatworm condition. The monoplacophorans have rather extensive coelomic spaces. There are no data on monoplacophoran locomotion; presumably they move by pedal waves, like large flatworms and

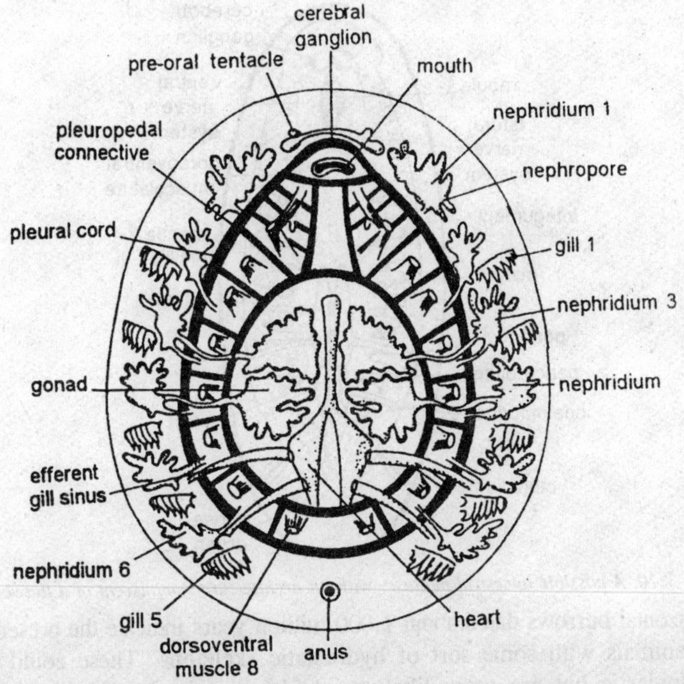

Fig. 7.9. A primitive mollusc (Neopilina, a monoplacophoran) showing serial repetition of some organs.

gastropods. If so, their coelom is not involved. A thorough functional account of the monoplacophorans is greatly needed.

Each of the distinctive coelomic plans can be explained as an adaptation to a distinctive mode of life. Each group has played an important role in the evolution of the biosphere, except perhaps the sipunculids. Lophophorates dominated many marine communities during the Paleozoic, and molluscs dominate these communities today. Arthropods are the most diverse terrestrial phylum today, while deuterostomes, which gave rise to the chordates and hence to man, have always been important marine forms. Even though we cannot yet certify a phylogeny among the early coelomate types, we can erect an adaptive model to explain many features of their anatomy and of the early fossil record.

Evolution of Early Metazoan Radiations

From the adaptive pathways likely followed in the development of metazoan ground plans, and from the data of the fossil record, one can erect a model of the early diversification of the metazoans. The

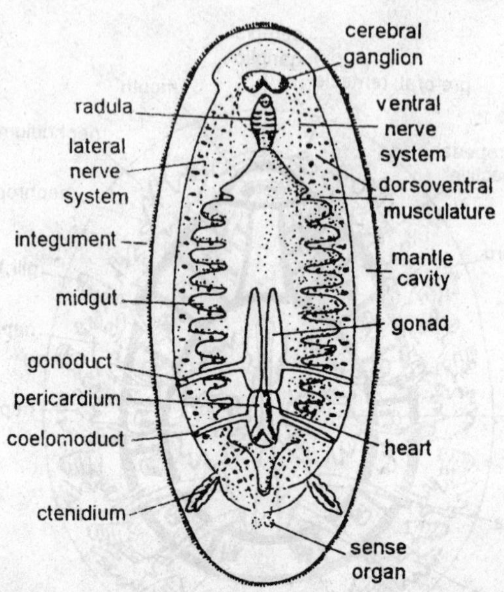

Fig. 7.10. A possible ancestral mollusc, with an architecture reminiscent of a flatworm.

horizontal burrows dated about 1,000 million years indicate the presence of animals with some sort of hydrostatic skeletons. These could be diploblastic but are more likely triploblastic animals with skeletons that are either mesenchymal (tissue), pseudocoelomic (body spaces with fluid but lacking a peritoneal lining), or coelomic (fluid spaces surrounded by a peritoneum). If so, much metazoan evolution had preceded them. Clemmay (1976) suggests that the early burrowers may have become extinct, since traces do not appear again for 300 million years or so. Of course, metazoans at the flatworm grade might have persisted, to give rise eventually to higher grades again. This problem must await further evidence.

From the appearance of the traces near 700 million years ago there is a record of continuous metazoan presence with increasing complexity of type and expansion of the ecospace (the inhabited functional regions of the bisophere). Although it is mechanically possible that the burrows of Ediacaran time were all made by non-coelomates, it is much more likely that primitive coelomate worms are represented, including detritus feeders and perhaps predators and suspension feeders.

Data from cytochrome sequencing tend to support this judgement. According to work by Brown and his colleagues (1972), annelids and deuterostomes separated about 750 million years ago, molluscs and deuterostomes about 720 million years ago. Further work may adjust these dates. It is not certain that the difference between them is real. The dates do suggest that the coelomates were present and that at least three of the major coelomate ground plans had diverged during the late Precambrian. The phylogeny shown is frankly speculative. We reason that the earliest coelomates were not yet segmented or egionated but were seriated, resembling flatworms with coeloms, and that from these forms: (1) the *metamerous annelids* diverged through coelomic segmentation, which promoted improved detritus feeding habits; (2) *oligomerous lineages* diverged as suspension feeders (lophophorates) living within the sediments of the seabed, and perhaps as sessile particle trappers (deuterostomes); (3) the *pseudometamerous molluscan* lineages diverged by development of a dorsal visceral mass, freeing the foot for creeping; and (4) *amerous sipunculids* diverged by suppression of seriation, developing an introvert for feeding. Whatever the actual pathways of descent may be, it is clear that the development of the coelom paved the way for new grades of organization. One immediate effect was to permit efficient borrowing into the sediments of the sea floor—an *infaunal* mode of life—opening up a region of biospace that today is densely populated by a large variety of animals. The diversification of coelomic architecture in the late pre-Cambrian is implied by the diversity of coelomic plans in Cambrian faunas.

With the widespread occurrence of durable skeletons in the Cambrian, the fossil record becomes much better. In all phyla the early skeletonized forms (for which modes of life have been worked out) were essentially *epifaunal*, that is, they lived upon rather than within the seafloor. Although they must have descended from ancestors with one or another of the coelomic architectures described, they were no longer vermiform, but had developed novel body plans.

Peristaltic locomotion is very inefficient on a surface, when only a short arc of a worm's cross section would actually contact the substrate. The energy required to maintain body shape and generate annular peristaltic waves would not be repaid by locomotory progress. One solution, followed by arthropods, was to develop a chitinous jointed exoskeleton with intrinsic musculature to operate appendages for locomotion. Arthropodization may have been the first adaptive pathway leading to a novel epifaunal coelomate. Another solution, favoured by unsegmented lineages, was to employ pedal waves, thereby consolidating body-wall muscles and moving the viscera into a dorsal position to free the locomotory foot. Body shape and muscular purchase is afforded by a consolidated body-wall musculature and a rigid exoskeletal cap. A seriated vermiform ancestor would result in a primitive monoplacophoran, creeping on or at shallow depths within the substrate.

Infaunal worms that were nearly or quite sessile seem also to have given rise to epifaunal descendants. Lophophorate oligomerous worms similar to the phoronids probably developed into solitary epibenthic forms, such as the brachiopods, along several different pathways, and also into the minute colonial epibenthic phylum Ectoprocta. Although the brachiopod skeleton seems to have developed in part for protection, it developed chiefly to enhance suspension feeding, first by framing and supporting the lophophore and later by bringing the feeding currents completely inside the mantel cavity, providing for increased homeostasis.

It is not necessary here to pursue hypothesis as to the rise of each epifaunal lineage. The central point is that the appearance of skeletons nearly 580 million years ago signals the rise of relatively diverse epifaunal communities, and suggests that a number of living phyla had their origin at about that time. As Cloud (1949) has pointed out, the ground plans of many of the durably skeletonized phyla do not make sense in the absence of the skeleton. The anatomical modifications associated with the shifting ecospaces of these lineages were coadapted with skeletonization. The brachiopods, for example, have a soft-part anatomy that is precisely adapted to reside within a bivalved exoskeleton, with pedicle, visceral, and lophophoral spaces and a complicated musculature associated with valve functions. Without a skeleton, a brachiopod would have to be redesigned. The simplest such design would closely resemble a short species of phoronid, which may closely resemble the brachiopod ancestor. We can be reasonably certain that the appearance of the first primitive brachiopod skeleton in the fossil record is very close to the time of evolution of the association of features that constitute the brachiopod ground plan.

A similar line of reasoning holds for many other phyla and classes, although not for groups like corals and bryozoans, in which mineralized skeletons form no part of their anatomical ground plans. These two colonial groups do require rigid skeletons in order to occupy all of the habitats and to perform the array of ecological functions they have achieved, but their ground plans could have been evolved long before their skeletons appear in the record. Indeed, their skeletons first appear well after the bulk of invertebrate groups; perhaps they originated near the early Cambrian but did not become rigidly skeletonized until

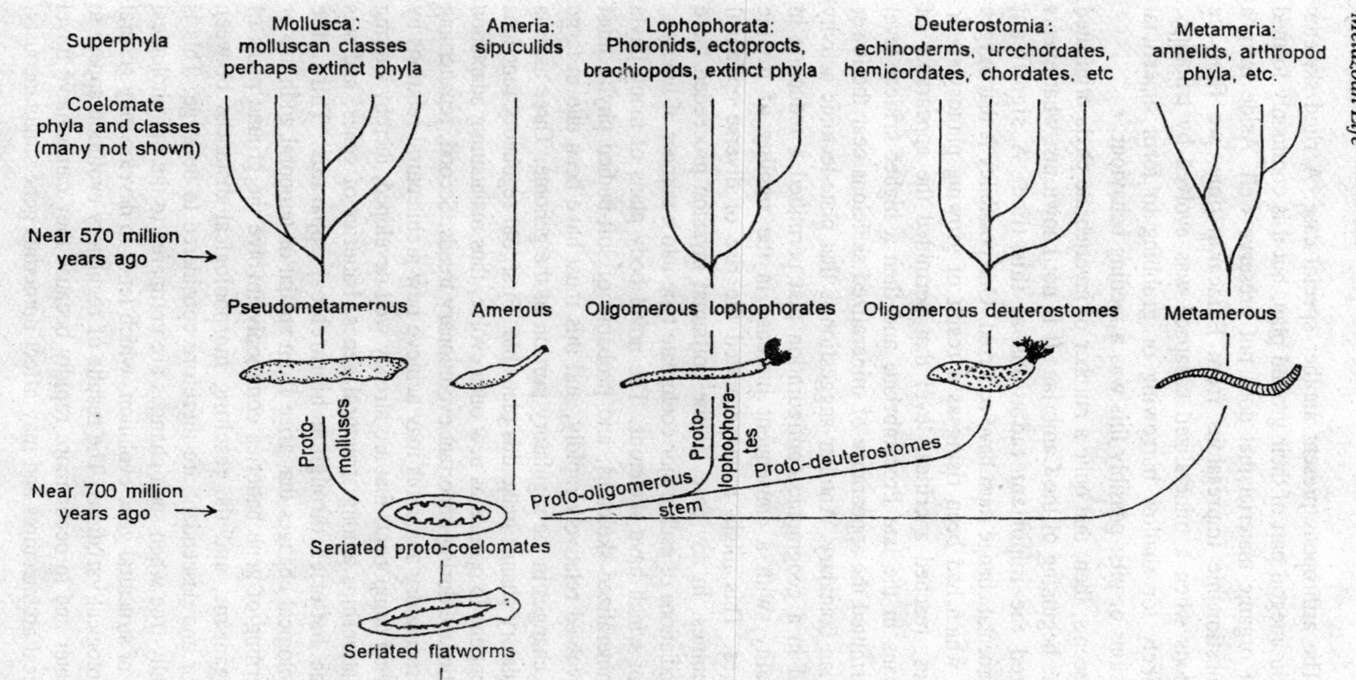

Fig. 7.11. Radiation of the major coelomate stocks from a flatworm ancestor.

later. The arthropods present another special case. A rigid skeleton forms an integral part of their ground plan; but it is commonly formed only of organic material that does not preserve well. Aside from a few questionable impressions from Ediacaran time, we first see arthropods when a mineralized carapace was evolved by trilobites, very likely for shallow burrowing or grubbing to form superficial depressions or pits; possibly this was a feeding behaviour.

It seems, then, that quite a number of invertebrate phyla originated near the beginning of the Cambrian. It is not known just what events localized the important radiations at this time. A significant environmental change must have occurred. One possibility is that oxygen levels, which had been rising as a result of growing photosynthetic activities, reached a critical level that permitted the appearance of metazoans in the late Precambrian, and then a higher critical level that permitted the appearance of mineralized skeletons near the lower Cambrian boundary. Another suggestion is that plate-tectonic activity resulted in a geographic configuration that permitted a reduction in seasonality, with a consequent increase in the stability of trophic resources. This might have permitted the rise of diverse epifaunal communities. In any case, as the epifaunal invasion proceeded there were radiations of each major coelomate type into a number of lineages, many of which have skeletons. The novel body plans of many phyla with mineralized skeletons, and probably of soft-bodied phyla, must have evolved relatively rapidly, and this must have been due in large part to changes in the regulatory portion of the genome. Three aspects of regulatory change make this plausible. First, old regulatory functions may continue to operate as new ones evolve, thus maintaining adaptation of the lineage during important evolutionary trends. Second, repatterning of the regulatory apparatus may achieve new architectural relations by employing components that are already well developed, thereby altering the relationships among structural genes, batteries of genes, and sets of gene batteries through a hierarchical apparatus. Third, the morphological changes that arise from spatial or temporal additions or repatterning of gene batteries commonly involve the ground plans of the organisms, and the resulting morphological distances between ancestor and descendant are therefore considered to be large. This is especially true when such changes are compared to the morphological results of structural gene evolution, which tends to develop only general morphological variations. The results of regulatory evolution appear to be greater and to occur more rapidly because they can achieve novel biological architectures and may affect numerous gene simultaneously.

Evolution of Chordates

For our purposes it is best to restrict the phylum Chordata to the subphyla Cephalochordata and Vertebrata. The most primitive living chordates are probably members of the subphylum Cephalochordata, typified by amphioxus. Amphioxus has a solid dorsal supportive structure, the notochord, which is regarded as a precursor to the vertebral column. The notochord is neither cartilaginous nor bony, but consists of a sheathed series of vacuolated cells that clearly function as a supporting structure. It is present in vertebrate (including mammalian) embryos, and perists into the adult form of some fishes, such as lampreys and even sturgeons. Despite its notochord, amphioxus is a rather sessile filter feeder. It burrows in sand and ingests water through the mouth, expelling it through gill slits—the feeding method inferred for primitive fishes.

A similar feeding system is found in two invertebrate deuterostome phyla: Hemichordata (enteropneusts, pterobranchs), which lack a notochord, and Urochordata (tunicates), which have tadpole-like larvae that possess notochords in their tails. The gill slits of these invertebrates are very similar to those of the primitive vertebtates, and most workers have concluded that they are all inherited from a common ancestor.

The earliest known vertebrate fossils are scraps of bone of early Ordovician age that in all probability represent a group of jawless fishes, the ostracoderms. These primitive fishes have a well-developed dermal armour of bone. They do not have paired fins like more advanced fishes, and have very small mouths. Presumably they were filter feeders like amphioxus, ingesting water through the mouth and expelling it through gill slits. The ostracoderms became quite diverse and evidently gave rise to jawed fishes with paired fins, which appear in Devonian rocks. Thereafter the ostracoderms tended to become more restricted, and most groups died out. Modern lampreys may represent an ostracoderm lineage that has retained many primitive features—they are jawless and lack paired fins, for example—while losing their bony skeletons. It is likely that in lampreys and some other fishes cartilage is an embryonic adaptation retained in the adult stage and is therefore a neotenous feature in such forms.

A plausible deuterostome phylogeny begins with infaunal oligomerous worms that feed by means of a tentacular system resembling a lophophore. From this stock epifaunal forms evolved, probably in the late Precambrian or early Cambrian. These included Echinodermata, a phylum in which the tentacular system was elaborated

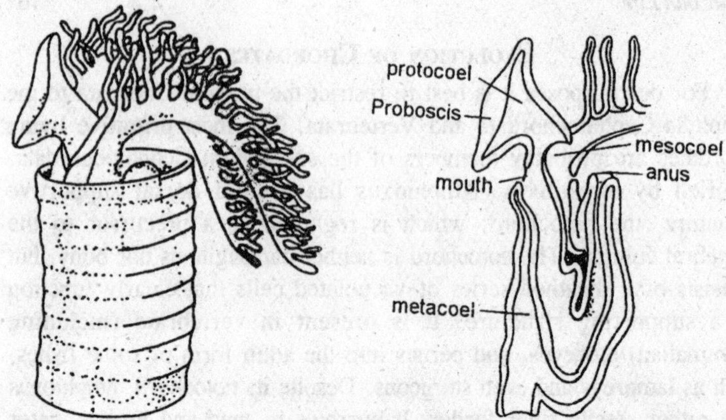

Fig. 7.12. Hemichordate—a pterobranch and a cross-section through a pterobranch.

into an extensive coelomic water-vascular system. This group developed a unique skeletal ground plan based upon numerous small plates or ossicles, and diversified into a large number of classes and orders during the early Paleozoic. Other early epifaunal deuterostome lineages remained without mineralized skeletons, and from these sprang the hemichordates. They developed gill slits, presumably to enhance the feeding system. The slits provided an outlet for feeding currents, which could be directed into the mouth by cilia associated with the tentacular system. Perhaps the tunicates evolved from a hemichordate stock; they have become completely sessile as adults and seem little more than filtering baskets.

The chordates seem to have arisen from an early deuterostome invertebrate. Since the tunicate larvae possess notochords and gill slits and somewhat resemble our expectations for a primitive chordate, it has been suggested that such a larva might have evolved into a chordate by developing the ability to reproduce in the larval stage—a classic case of neoteny, if true. It is also possible that aduti members of an early deuterostome worm lineage became swimmers, and that the notochord was elaborated to enhance this function. The urochordates could be an offshoot of this hypothetical stock, and amphioxus a more direct descendant. Still another phylogenetic hypothesis, argued most forcefully by Jeffries (1967, 1968), derives the vertebrates from carpoids, bilaterally symmetrical organisms with strong calcareous platy skeletons that are usually assigned to Echinodermata. Carpoids appear in the earliest Middle Cambrian and disappear in the Middle Devonian. Jeffries views them as a transitional group between echinoderms and

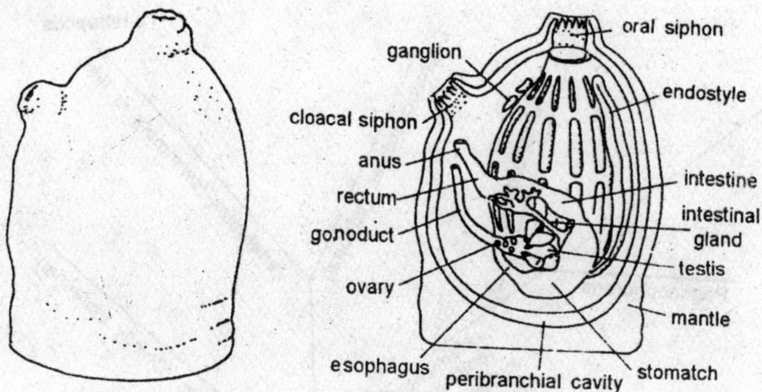

Fig. 7.13. Urochordate—a tunicate and an anatomical scheme of a tunicate.

chordates, with their more advanced members resembling huge armoured tunicate larvae. However, the earliest known chordate is soft-bodied and is also of Middle Cambrian age.

Vertebrate bone is certainly a unique skeletal material, but many of the skeletonized invertebrate phyla also have unique shell structures. For epifaunal invertebates and for the chordates, the appearance of mineralized skeletons seems to have heralded a new level of success, or at least diversification. Skeketonizatons was usually associated with the establishment of a new anatomical architecture that was coadapted with the skeleton to function in a novel environment or novel mode. Many coelomates became skeletonized once the epifaunal ground plans evolved, and commonly the skeleton proved a key to much further elaboration and diversification. The most general reason for the success of mineralized skeletons seems to be that they employ much less energy and are simpler to maintain than skeletal systems based on fluids. Thus, for functions that require support and rigidity, mineralized

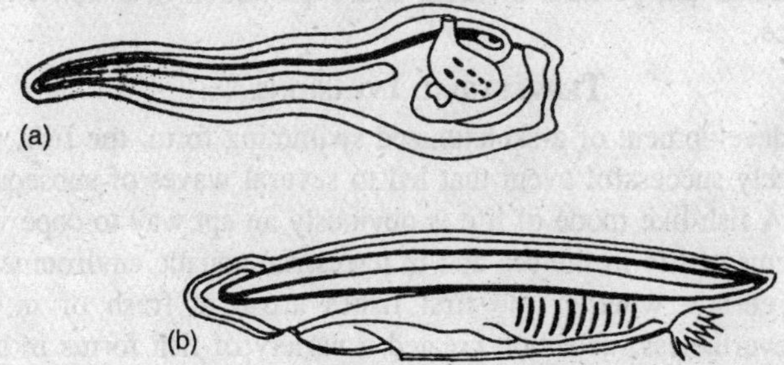

Fig. 7.14. (a) Planktonic larva of a tunicate, resembling a tadpole, with notochord. (b) Cephalochordate—an amphioxus.

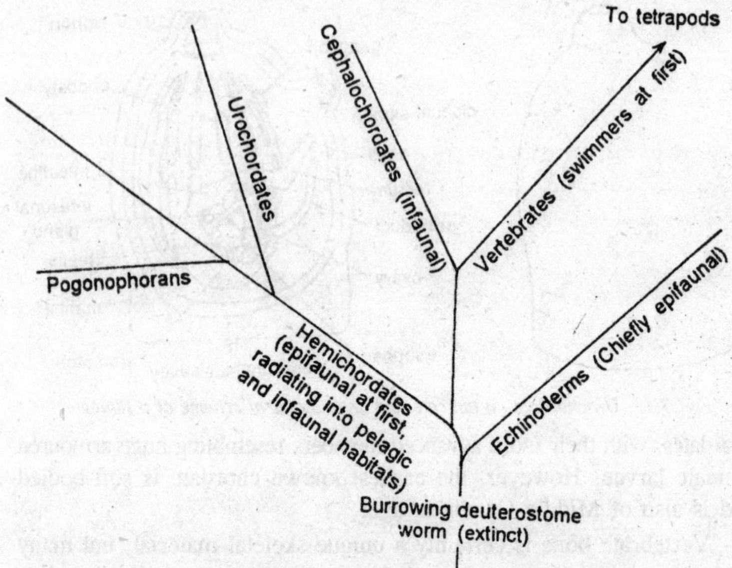

Fig. 7.15. A possible phylogenetic tree of Deuterostomia.

skeletons are superior. Functions that require a variable skeletal system, such as burrowing, are performed efficiently by hydraulic systems; fluid skeletons are often employed for such purposes. Mineralized skeletons can be put to many uses other than support, such as protection and thermoregulation, and so are often subjected to a variety of positive selective pressures. It is therefore not surprising that once a skeleton has appeared within a lineage it should be extensively developed in a short time, and that skeletonized lineages should be potentially versatile and frequently capitalize upon environmental opportunities to become highly diverse. Skeletonization is frequently a significant component of the adaptation that permits invasion and exploitation of a new region of biospace.

Terrestrial Invasions

The development of a skeletonized swimming form, the fish, was an extremely successful event that led to several waves of subsequent radiation. A fish-like mode of life is obviously an apt way to cope with the problems of life in the sea and in terrestrial aquatic environments. It is not certain whether the first fishes arose in fresh or in salt water. Nevertheless, selection created a variety of fish forms in both fresh-water and marine environments, certainly in response to opportunities provided by the mode of life and probably to challenges afforded by environmental changes as well. As selection explored the adaptive potentials of fishes, two morphological innovations proved to have special consequences. These are the development of jaws (from gill arches) and paired fins. Jaws expanded the food sources of fishes, which must have soon become important predators. Paired fins improved swimming balance and orientation, which would promote efficient predation. Jawed fishes quickly diversified and eventually replaced the jawless ostracoderms, both in marine and fresh waters, so that by Middle Devonian time early ray-finned and lobe-finned fishes were well established. The ray-finned forms eventually gave rise to a vast majority of modern fishes. Lobe-fins were represented by lungfishes,

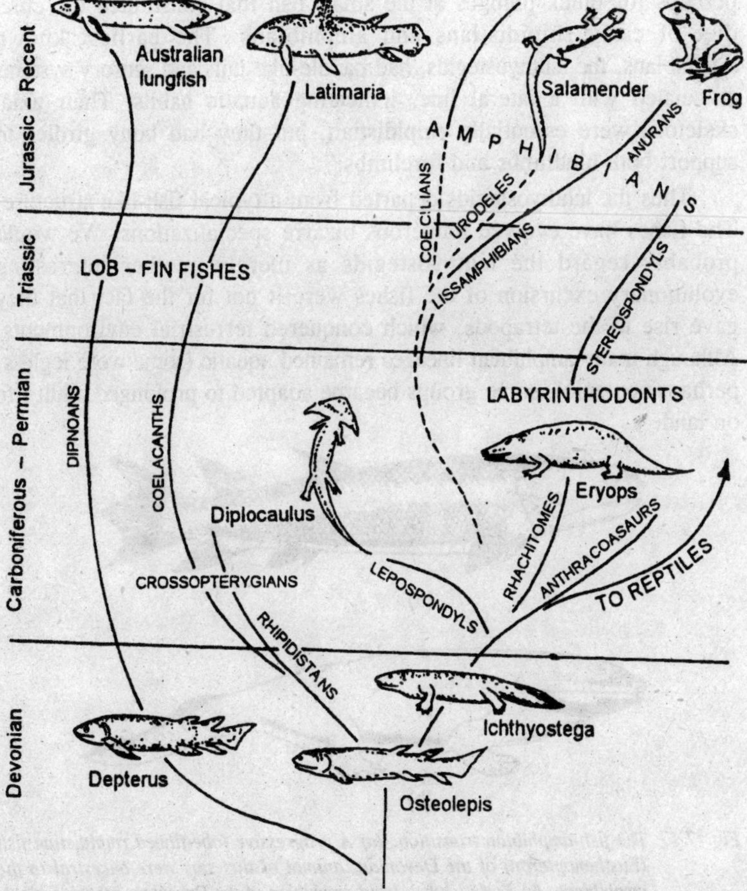

Fig. 7.16. Generalized phylogeny of lobe-finned fishes and amphibians.

coelacanths, and a group called rhipidistians. The rhipidistians became extinct near the end of the Permian, but they have an important place in evolutionary history because the amphibians evolved from early members of this group.

In some rhipidistian lineages the endoskeleton was increasingly ossified, legs developed from fin supports, and air breathing (not at all a unique development among fish) appeared. These features were coadapted with many fish-like features to form components of what must have been a highly adaptive mode of life. It is not certain that legs evolved in response to selection for terrestrial locomotion; they may well have originally served for locomotion in very shallow water, perhaps for quick plunges at the small fish that constituted the chief diet of early rhipidistians and amphibians. The earliest known amphibians, the ichthyostegids, had paddle-like tails and sensory systems associated with a lateral line, indicating aquatic habits. Their axial skeletons were essentially rhipidistian, but they had bony girdles to support both hindlimbs and forelimbs.

Thus the ichthyostegids departed from a typical fish-like structure. The fishes have evolved numerous bizarre specializations. We would probably regard the ichthyostegids as merely another interesting evolutionary excursion of the fishes were it not for the fact that they gave rise to the tetrapods, which conquered terrestrial environments. Although many amphibian lineages remained aquatic (some were legless, perhaps neotenic), some groups became adapted to prolonged adult life on land.

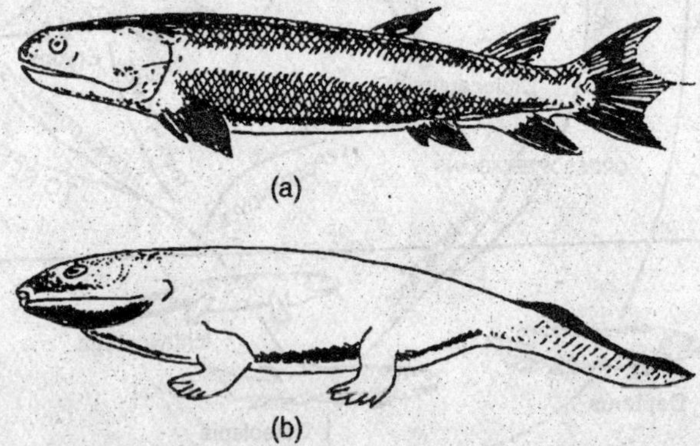

Fig. 7.17. The fish-amphibian transition. (a) A progressive lobe-finned rhipidistian fish (Eusthenopteron) of the Devonian; animal of this sort were ancestral to the amphibians. (b) Early ichthyostegid amphibian of the Devonian, probably highly aquatic descended from a Eusthenopteron-like stock.

As the fossil record of terrestrial habitats is generally poor, the course of the biotic invasions of the land is known only in broad outline. Plants must have been established in fresh water from at least the early Paleozoic, and invertebrates no doubt followed closely. Fish may have originated in fresh water, and if not, were represented there at least by Silurian time. During the middle Paleozoic or earlier, life moved beyond the edges of fresh-water lakes, ponds, and swamps to become established on land. The earliest known vascular (land) plants are Silurian, and probably lived in damp regions near fresh-water bodies. Arthropods must have evolved terrestrial habits soon after, perhaps living at first in subterranean cracks and feeding on the plant detritus then enriching the swampside soils; soon they were feeding on the plants themselves. By Devonian time the green belts were spreading widely into lowland terrains and trees appeared; also at this time flying insects developed. Some rhipidistians followed the burgeoning land biota into terrestrial environments, developing into amphibian tetrapods during the Devonian. Probably their terrestrial foods were chiefly arthropods.

The pressures that cause lineages to evolve in novel habitats are numerous. Populations suffer much internal competition between individuals for food and space, and commonly there is much competition from populations of other species as well. Furthermore, populations are usually heavily preyed upon, and may be subjected to periodic or regular inclement physical conditions. Given the opportunity, selection will act so as to minimize these pressure, even if this requires pioneering in an unusual environment. Scientists have frequently attempted to identify particular pressures as the most important in given cases. For early amphibians, for example, one theory suggests that desiccation of ponds led the ancestral tetrapods to migrate overland to larger pools, and thus to acquire gradually terrestrial adaptations, including limbs. Another theory suggests that overpopulation may have caused rhipidistian—amphibian stocks to evolve some independence from the aquatic habitat and to exploit terrestrial food sources for parts of their life cycles. We are not really sure what the pressures actually were; the complexity of ecological interactions is so great that even if the situation had been studied at the time, an accurate assessment of the contributions of all relevant factors might not have been forthcoming. Clearly, preadaptations commonly permit the successful invasion of a new habitat, especially whether creation of novel modes of life are involved. Partly for this reason, it is tempting to favour the idea that limbs were originally an aquatic adaptation.

The primitive amphibians became increasingly adapted to terrestrial life; improvements were achieved by different lineage at different times. The body could have been raised from the ground as a locomotory improvement and skin layers thickened to impede desiccation. Stahl (1974) gives an account of the functional modifications that occurred in early amphibians. The most diverse amphibian group to appear during the Carboniferous was the labyrinthodonts, which include the primitive ichthyostegids as a stem group and which diversified into a number of distinctive phyletic lineages. As has so commonly happened, it was one of the earlier lineages (the anthracosaurs) that eventually gave rise to the next major advance in grade.

Reptiles, Dinosaurs, and Mammals

Reptiles are known from the late Carboniferous, but may have descended from an amphibian stock that diverged from other anthracoasurs in the early Carboniferous. They possess eggs that can develop out of water, and skins that prevent undesirable water loss to the environment. In principle, these features free the reptile from necessary dependence upon life in water at any stage of the lifecycle. The detailed adaptive pathways that led to the reptiles are not yet understood. Important steps long that pathway seem to have included the development and elaboration of a membrane around the embryo, presumably for respiration at first, the subsequent acquisition of a horny protective covering around the egg to prevent desiccation, and finally the mineralization of the egg shell. Physiological and structural changes required of the embryo itself include the evolution of physiological systems to produce safe metabolites that are relatively nontoxic and can be stored in an enclosed egg. The reptilian egg may have evolved to prevent desiccation in impermanent ponds or drying swamp shallows, or to protect eggs from the numerous aquatic predators, which was accomplished by depositing them on land.

Many of the early reptiles, the captorhinomorphs, lived chiefly or entirely in water, but some seem to have inhabited forests fringing large swamps. They soon diversified, and from their basal stock arose distinctive lineages, including the pelycoasurs. These forms spread into moist terrestrial lowlands, which they eventually shared with advanced captorhinomorphs and other reptiles. Clearly, their integuments were proof against excessive desiccation. Numerous modifications of their skeletal morphology occurred as they radiated into a variety of habitats, developing modes of life that included feeders on plants, on fish, and on each other. Probably by early Permian time a group of active

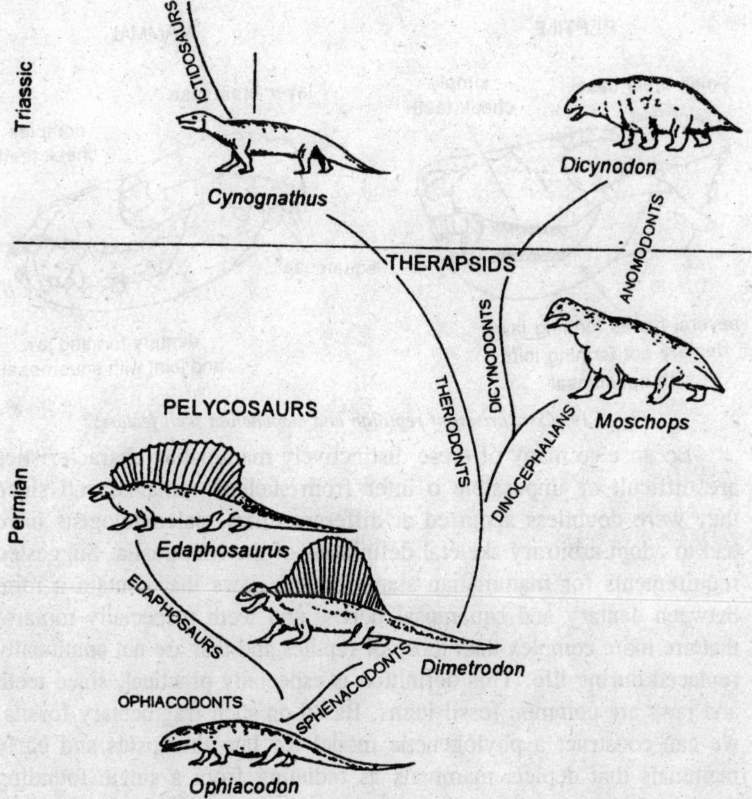

Fig. 7.18. Generalized phylogeny of pelycosaurs and the therapsid reptiles, from which mammals evolved.

predators had split off from the pelycoasurs, in turn diversifying into herbivores as well as carnivores of several types and coming to dominate a wide range of middle and upper Permian terrestrial habitats. These were the therapsids. It is from predatory therapsids that mammals evolved. Developments in axial skeletons, limbs, and skulls in several therapsid groups in the Triasic produced features that were to be emphasized and continued in mammals. It is possible to reconstruct, from clues presented by therapsid skeletons, features of soft-part anatomy that are now possessed by mammals, thought it is not really certain which of these were acquired by therapsids and which by their early mammalian descendants. Such features include the development of a diaphragm for breathing, hair, improved agility, and the facial skin and muscles that permitted suckling of young; even warm-bloodedness may have appeared in therapsids.

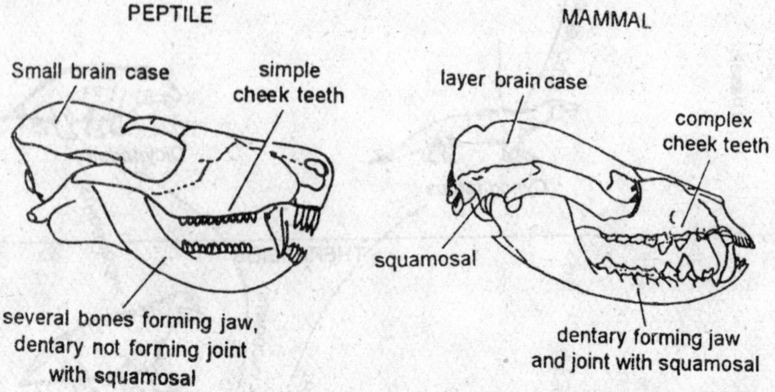

Fig. 7.19. Comparison of reptilian and mammalian skull features.

Because so many of these distinctively mammalian characteristics are difficult or impossible o infer from skeletal remains, and since they were doubtless acquired at different times, paleontologists have had to adopt arbitrary skeletal definitions of the Mammalia. Suggested requirements for mammalian status include jaws that contain a joint between dentary and squamosal bones, and teeth (especially molars) that are more complex than those of reptiles and that are not continually replaced during life. This definition is especially practical, since teeth and jaws are common fossil items. Based on such fragmentary fossils, we can construct a phylogenetic model for late therapsids and early mammals that depicts mammals as radiating from a single founding lineage. It is not certain that this is in fact the historical case. The egg-laying monotremes (platypus and echidna) may represent a lineage descended independently of other mammals.

Mammals diversified and modernized at first only gradually; in some lineages the trend toward a longer fetal period eventually led to the appearance of the placenta, probably during the early Cretaceous and possibly in Asia. By the close of the Cretaceous, placentals were undergoing the diversification that produced the primates; this diversification extended into the early Tertiary and gave rise to modern mammalian groups. Throughout their long Mesozoic history, which embraces fully two-thirds of their existence, mammals were completely overshadowed in abundance, diversity, and size by reptiles, particularly by the dinosaurs. Dinosaurs probably arose from a lineage that diverged from the captorhinomorphs, only a few million years before the mammals appeared. The results of their spectacular radiations are well known, particularly the much depicted large brontosaurs, stegosaurs, ceratopsians, and the great predators. Birds also developed from a dinosaur stock. As mammals have now become far more successful than reptiles, the questions arise why reptiles were so successful in the Mesozoic, and why they have declined to the present status. There are now plausible, though unproven, answers.

These answers have emerged from studies, by Ostrum and others, that have reevaluated such features as posture and thermoregulation in reptiles and mammals. Bakker (1971) has proposed a model in which dinosaurs and early mammals each inherited a sprawling posture from their (separate) reptilian ancestors. By late Triassic time dinosaur lineages had achieved erect posture and probably became quite active, though mammals were still sprawling or semi-erect. Several of the larger dinosaur groups, including the famous predator *Tyrannosaurus*, are envisioned as having fast locomotory gaits. This indicates high body temperatures and suggests homoiothermy. If dinosaurs did attain this physiological grade, it would be easier to explain their numerous adaptive radiations, for we know that the level of activity provided by homoiothermy opens wide arrays of life modes similar to those later developed among larger mammals. It would also account for the apparent ease with which dinosaurs prevented mammals from encroaching upon their biospace; mammal homoiothermy would not have conferred any special competitive advantage. Indeed, visions of charging tyrannosaurs have led some placeontologists to be grateful that dinosaurs have become extinct.

Although more dinosaurs were rather small, down to about the size of a rooster, they were larger than contemporary mammals; even their hatchlings were larger than most of their adult mammalian contemporaries. Large animals have surface—volume ratios favourable to heat retention; and since dinosaurs lacked obvious insulating features, such as feathers or fur, they might have been unable to maintain body heat with small sizes. Early mammals, though small, were furry and could thus achieve homoiothermy despite their small sizes. According to this model, then, dinosaurs and mammals each achieved homoiothermy and high activity levels; the former radiated into biospace appropriate for large animals, the latter for small ones. It was not until dinosaurs declined that mammals radiated into the vacated habitats and took up many of the abandoned modes of life. Although the notion that dinosaurs were warm-blooded has been attacked, it has also been stoutly defended, and remains an attractive hypothesis.

What extinguished the dinosaurs? There is no lack of hypothesis, but none has been adequately tested. Since dinosaurs descended from reptiles, they traditionally have been considered as ecotootherms. Cold-blooded terrestrial animals are restricted in habits and habitats by thermoregulatory problems. Climatic changes to cool whether, to hot weather, or to highly seasonal climates, could in theory easily exceed dinosaur ectothermal potentials, and have often been suggested as a major factor in their extinction. Even if the dinosaurs were homoiotherms, their hatchlings would have been too small to be able to cope with cool temperatures. Climatic change thus remains a likely cause of dinosaur extinction, partly because it would have a wide variety of secondary effects that could increase dinosaur mortality and contribute to breakdowns in their population structure.

PRIMATES

Although the placental mammals originated while dinosaurs still existed, the Mesozoic lineages were chiefly small, generalized insectivore-like forms. However, they began to differentiate somewhat near the end of the Cretaceous and underwent an explosive diversification that established most modern orders during the Paleocene, filling many places in terrestrial ecosystems that had been vacated by extinct dinosaurs. As the post-Pangaean continents became increasingly isolated owing to plate-tectonic processes, distinctive mammalian assemblages developed on different continents; although there were some subsequent interchanges.

The earliest primates appeared among the late Cretaceous orders of placental mammals, differing probably only a little from the insectivores from which they arose. Simons (1972) gives a careful resume of fossil primates. Cretaceous primates are represented only by teeth, so that little is known of their habitats, and Paleocene deposits have yielded few postcranial remains. In the Eocene rather modern, lemur-like prosimians appeared. These Eocene primates had shorter faces and larger eyes than their ancestors, and probably had arboreal habitats. As rodents were differentiating at this time with much apparent success on the ground, the adoption of arboreal habits by primates may have been due in part to their exclusion from forest floors.

Monkeys arose in the alter Eocene or early Oligocene. The living Old World (*catarrhine*) and New World (*platyrrhine*) monkeys seem to have been separated for a long time, and it is sometimes suggested that they arose independently from non-monkey ancestors, thus representing a case of parallel evolution. This is possible, but their common ancestor would have to have been well removed from the prosimian stocks that have living descendants; all monkeys and apes have proven to be rather similar when compared by techniques of molecular biochemistry, and rather different from living prosimians.

Apes are catarrhines; they are not known for certain until the Oligocene. The earliest known ape (*Aegyptopithecus*) had a small brain, a tail, and hands and feet that indicate an arboreal, quadrupedal life. By Miocene time several species assigned to the genus *Dryopithecus* had appeared. Modern chimpanzees, orangutans, and gorillas probably descended from this assemblage. Man seems to have descended from the catarrhines also.

There is nothing very unusual in the evolutionary processes that gave rise to the hominid lineage, although the evolution of the human brain has certainly had spectacular consequences—but so did the evolution of the reptilian egg, the development of the chordate skeleton, the epifaunal radiation of the invertebrates, and the invention of the coelom, the choose only a few of many landmark events in animal evolution. In each case the innovations evolved through selection and solved an immediate adaptive problem that opened the way into vast regions of biospace.

Evolutionary advances have been steadily accompanied by extinctions, which are clearly an integral part of the evolutionary patterns. They seem an inevitable consequence of adaptation to the ambient environment in a changing biosphere. Peak extincting levels probably correspond with reductions in the capacity of the environment to accommodate species. Today, diversity appears to be near a record high. Evolutionists therefore have a rich biosphere to study, furnished with a wide variety of organisms, communities, and provinces. One wonders what the future of this biosphere may be, and whether the potentials are as great as previously.

Chemical Similarities and Differences

All living bodies are astonishingly similar in chemical composition. Not only do they contain atoms of the same elements, but often in similar proportions. The same classes of compounds, particularly nucleic acids and proteins, are found everywhere. Proteins are composed of the same 20 amino acids, and the amino acids, with rare exceptions, are represented only by left optical isomers. Energy carriers (such as adenosine triphosphate, ATP) and enzymes with identical functions (such as cytochrome *c*) are present in most diverse organisms. To an evolutionist, these chemical similarities are meaningful; they affirm that man is kin to all that lives.

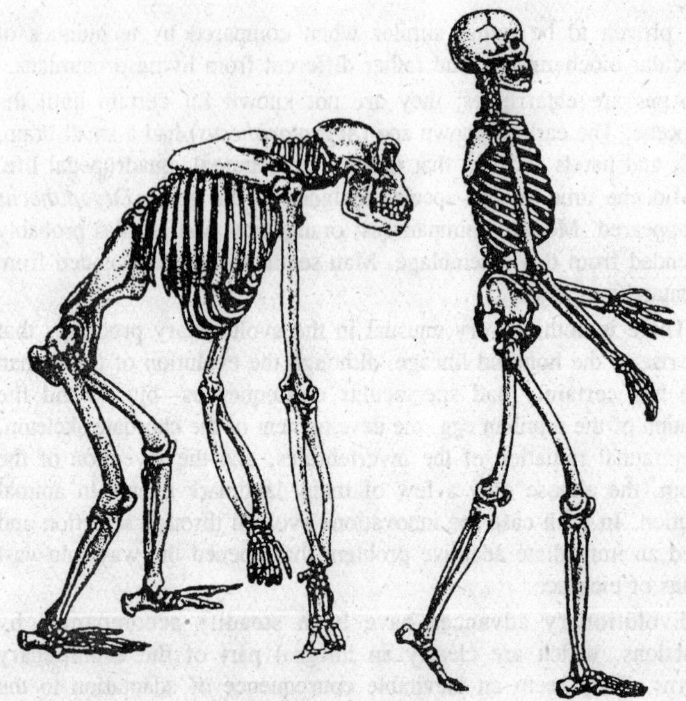

Fig. 7.20. Skeletons of ape and man.

More novel and more interesting are the differences between related compounds in man and other animals. Though some enzymes may play similar physiological functions in man and other organisms, they are often more or less distinct in their chemical composition. The achievements of molecular biology have made possible the detection and quantification of such distinctions. Hemoglobins are oxygen-transporting pigments in the blood of all vertebrate and some invertebrate animals. In adult humans the most abundant form of hemoglobin is A, each molecule of which consists of two alpha and two beta chains, composed respectively of 141 and 146 amino acids and an iron-containing heme group.

The sequences in which the amino acids are arranged in human alpha and beta hemoglobin chains have been worked out by Ingram and many other investigators. Many mutational variants have been discovered, most of them carried in very few or even in single individuals. However, some variants, such as hemoglobin S, are widespread in populations of large geographic regions, where they confer on their carriers a relative immunity to endemic diseases, especially to malarial fevers. A large majority of mutant hemoglobins differ from the prevalent, or "normal," hemoglobin A by substitution of single amino acids in either the alpha or beta chains. Comparison of the amino acid sequences in the hemoglobins of man and other animals is fascinating. The hemoglobin A of the chimpanzee is identical with the human one. The gorilla alpha chain differs from the human one in a single amino acid substitution: glutamic acid in man and aspartic acid in the gorilla at position 23 in the chain. Between man and the gorilla there is also one amino acid difference in the beta chain. The differences between human and gorilla hemoglobins are not much greater than variants in human individuals. One is tempted to ask: could a man live with gorilla hemoglobin, and vice versa? So far no answer to this question is forthcoming.

The chimpanzee and gorilla were judged to be man's closest relatives on the basis of their external and anatomical similarities, as well as their chromosome numbers (46 in man, 48 in the chimpanzee and gorilla). The amino acid sequences in their hemoglobins confirm this judgement. Zuckerkandl, Buettner-Janusch, and their co-workers found that among primates morphologically less similar to man than the chimpanzee and gorilla, the hemoglobins are also less similar. The alpha chain of the baboon differs from human alpha in several amino acids, and the beta chains show even more differences. A comparison of man and *Lemur fulvus* shows six substitution in the alpha chain and 23 substitutions in the beta chain. Studies by many workers disclose the following amino acid substitutions in the hemoglobin chains of various animals compared to man (these figures show-called mutational distances, i.e., the minimum numbers of nucleotide substitutions in the DNA of the genes needed to yield the amino acid changes observed):

Alpha Chains		*Beta Chains*	
Gorilla	1	Gorilla	1
Macaque	5	Macaque	10
Mouse	19	Mouse	31
Sheep	26	Sheep	33
Pig	20	Pig	28
Horse	22	Horse	30
Rabbit	28	Rabbit	16
Chicken	45	Kangaroo	54
Carp	93		
Lamprey	113		

The similarities as well as the differences in the above list are remarkable. As the animals examined become more and more obviously different from man, their hemoglobins become increasingly different. A single mutation could conceivably transform the alpha or beta chains of the gorilla into the respective chains of man; the coincidence of the several mutations needed to transform a human alpha chain into that of a mouse, or even of a macaque, has a probability indistinguishable from zero. It took the accumulation of many separate mutations over millions of years to build up the differences observed. And yet, even the hemoglobins of organisms as different as man and chickens or man and fish, have preserved similarities in amino acid arrangements that are far too great to be explained by chance. The similarities are marks of common ancestry.

Several similarities in amino acid sequence have also been observed in comparisons of other proteins of man and various organisms. The work of Margoliash, Fitch, Jukes, and their colleagues on cytochrome *c*, myoglobins, fibrinopeptides, and other proteins discloses increasing divergence of the amino acid sequences in more and more different organisms.

Man's Early Ancestors and Relatives

When Darwin put forward his conclusion that the human species evolved from ape-like ancestors, the evidence of fossils was almost wholly missing. Of course, in 1856 the skull of the Neanderthal Man had been unearthed in Germany, but the interpretation of this find was at once mired in controversy. Some authorities deemed the Neanderthal to be a pathological specimen of modern man, while others saw it as a "missing link" between *Homo sapiens* and his ape-like ancestors. The brain size of the Neanderthal was equal to or even slightly above the average for modern man, but the shape of the brain capsule, the presence of heavy brow ridges, and a strong and chinless mandible made him distinctive. Numerous Neanderthal skulls and skeletons have since answered to everyone's satisfaction. Some fragmentary remains of australo-pithecines have been found in strata of great antiquity—two to five million years old. Both *Australopithecus robustus* and *A. africanus* lineages seem to be present. On the other hand, some australopithecine remains are much younger, less than one million years. Most unexpected, however, is Leakey's discovery at Lake Rudolf of a more advanced hominid, allegedly a species of the genus *Homo*, dated about 2.6 to 2.9 million years. If this claim is validated, it would seem to follow that *Homo* and *Australopithecus* lived simultaneously and in about the same geographic region. Our ideas of the evolutionary sequence of hominids would then have to be revised.

The transition from *Homo erectus* to *Homo sapiens* also needs clarification. Several fragmentary remains which have been dubbed "pre-sapines," have been dated at the last interglacial period. The Neanderthals who inhabited Europe during the last period of the Ice Age seem to resemble modern *H. sapiens* rather less than the older pre-sapiens did, despite the fact their brain was as large as that of modern man. Some 35-40 thousand years ago the Neanderthals were abruptly replaced in Europe by a full-fledged *H. sapiens*: Cro-Magnon people, whose skeletal parts are indistinguishable from those of modern man.

Homo sapiens (or pre-sapiens) developed from *H. erectus* perhaps during the penultimate glaciation or even earlier, probably intropical and subtropical climates. It subsequently spread to Europe and during the last stage of the Ice Age formed the race *neanderthalensis*, adapted to the severe climatic conditions of that time. This race was finally replaced by a less rugged but culturally more advanced *H. sapiens* people. Thereafter, the success of human populations depended on the possession of superior technologies rather than greater bodily strength. That *neanderthalensis* was a subspecies of *H. sapiens* rather than a separate species has been substantiated by the discovery of intermediates in caves of Mount Carmel in Palestine. Some researchers have interpreted these intermediates as "hybrids," although this interpreation is not a necessary one if by "hybrids" one means the immediate progeny of parents who belong to different subspecies. A more plausible view is that Palestine was then a territory with a population intermediate between *H. sapiens sapiens* and *H. s. neanderthalensis*, just as at present we find territories inhabited by populations transitional between white and Negroid races. The Neanderthals no longer exist at present. It is possible that they were mostly killed off by the invading *H. sapiens* — the earliest suspected case of genocide. It is just as plausible that they interbred with the invaders, and we may carry some of their genes.

8

Diversity of Life

The Protozoa, the most primitive phylum of animals, appears to have been derived from primitive flagellate algae, and within single genera among the euglenoids there may be some species which show predominantly plant characters and others which show predominantly animal characters. Thus there is a high probability that the flagellates are close to the point of separation of the two kingdoms, if, indeed, the separation is complete here, for there are flagellates like *Trypanosoma* which are undoubtedly animals; there are others like *Chlamydomonas* which are undoubtedly plants; and there is the great intermediate group, typified by *Euglena*, which defies any indisputable assignment to the kingdoms.

Diversification of the Protozoa

Once the Protozoa were established, they became greatly diversified, and their relationships are by no means certain. The class Flagellata itself includes a wide range of structural and ecological types even when only indisputably animal types are considered. Feeding may be by engulfing food with pseudopodia, or by a simple mouth. The parasitic forms are generally saprozoic. Several orders of flagellates deserve especial mention. The order Protomonadina includes a wide variety of small, colourless flagellates, typically with two flagella, one of which trails along the side of the animal. Reproduction is always asexual. In the free-living members of the order, this is by simple fission, but among the parasitic species there may be multiple divisions which are difficult to distinguish from spore formation of the Sporozoa. It seems highly probable than the Sporozoa, all of which are parasitic, were evolved from this group. But sexual reproduction was also developed among the ancestors of the Sporozoa. The best known protomonads are the members of the genus *Trypanosoma*, all of which are blood parasites of vertebrates. Quite as important for evolution, however, are the choanoflagellates. These are protomonads which have a protoplasmic collar encircling the base of the flagellum. Its theoretical importance lies in that fact that sponges have similar cells, called choanocytes, and that colonial choanoflagellates, such as *Proterospongia*, resemble simple sponges.

The most highly specialized flagellates are probably the orders Polymastigina and Hypermastigina. The polymastigotes are generally parasitic on the digestive tract of arthropods or vertebrates. The simpler members generally have four flagella, but the more complex members may have large numbers. There may be only a single nucleus, or there may be many. Complicated cytoplasmic organnelles may be present. Reproduction is generally reported to be by simple fission, but Cleveland has reported sexual reproduction in some species, including the unique feature of reduction of the chromosome at a single division. The hypermastigotes are morphologically the most complex flagellates. There are numerous flagella arranged indefinite ways. Cytoplasmic organelles are very complex. All hypermastigotes are parasites or commensals in the digestive tract of termites or cockroaches. They are essential to the nutrition of the hosts for which they digest cellulose, for it has been shown that the host starves if the protozoans are removed experimentally.

The order Rhizomastigina is of especial interest because the members of the order appear to be intermediate between the Flagellata and the Sarcodina, amoeba and its allies. Many flagellates are capable of ameboid movement, but the members of this order, although they do possess a flagellum, are also permanently ameboid, and so they make a nice connecting link between the Flagellata and the Sarcodina. *Mastigamoeba* is a good example.

Ameba and Its Allies

The class Sarcodina is also highly diversified, and it is by no means certain that it is really a unified group. That is different orders of the Sarcodina may have arisen from different flagellate ancestors. By far the best known member of this class, or perhaps of any protozoan class, is *Amoeba proteus*, the familiar study material of every elementary biology laboratory, and the traditional example of a primitive animal. *A. proteus* is quite typical of its order, the Lobosa. But, in addition to such simple, free-living, ameboid organisms, the Lobosa

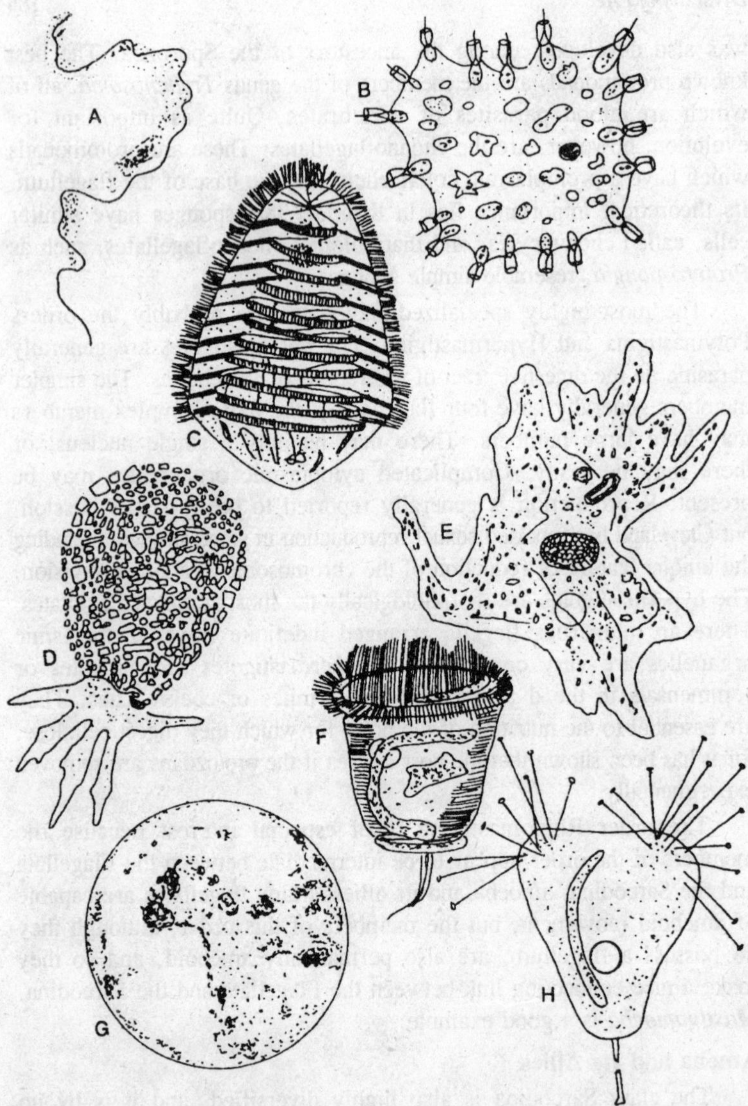

Fig. 8.1. Representative protozoa, A, Trypanosoma; B, Proterospongia; C, Macrospironympha; D, Difflugia; E, Amoeba; F, Vorticella; G, Plasmodium infested red blood cell; H, Acineta, A—C are flagellates; D and E are sarcodinans; F is a ciliate; G is a sporozoan; and H is a suctorian.

includes parasites like *Entamoeba histolytica*, which parasitizes the digestive tract of man; and there are free-living but shelled organisms such as *Diffugia*. The nutrition of all of these, including the parasites, is holozoic. Reproduction by binary fission is the rule, but sexual reproduction has been reported for one species of *Amoeba*, and it is common among the shelled species.

The order Lobosa is generally regarded as a terminal group in evolution, yet the idea that the Myxomycophyta or the Eumycophyta or both may have been derived from this group is gaining prestige among mycologists. And some zoologists treat the Myxomycophyta as protozoans under the name of Mycetozoa. The remaining orders of the Sarcodina are all terminal groups which are characterized by elaborate and often beautiful calcareous or siliceous shells, and by many slender, semipermanent pseudopodia, the characteristics of which are peculiar to each order. These are the Foraminifera and the Radiolaria, both of which are abundantly represented in the fossil record even into pre-Cambrian times, and the Heliozoa, which are less adequately known as fossils.

Ciliata and Suctoria

In contrast to the classes Flagellata, Sprozoa, and Sarcodina, all of which are fairly clearly interrelated, the two remaining classes of Protozoa show no evidence of relationship to the above named classes. These are the classes Ciliata and Suctoria, which some protozoologists would prefer to assign to a separate phylum. The ciliates have a definite shape, maintained by a pellicle, as do some flagellates. There is a definite anteroposterior axis, and the animal is symmetrical, sometimes radially, sometimes bilaterally, and sometimes an irregular deviant from one of these types of symmetry. Nutrition is holozoic, with other minute organisms being ingested by a mouth. Cilia are commonly arranged in specialized tracts which sweep a food-bearing current toward the mouth. It is often said that the Protozoa are the simplest of animals. While this may be debated, there is little doubt that ciliates are the most complicated of cells. The cilia themselves may be arranged as coordinated tracts, or they may be fused in sheets to form undulating membranes, or they may be fused in tufts to form appendage-like cirri. Whatever the arrangement of the cilia, their movements are coordinated by a complicated network of fibrils, the neuromotor system, which exceeds in complexity the simplest nervous systems of Metazoa.

The trichocysts, small bodies which can be discharged to produce filaments, underlie the pellicle. Their function is uncertain, but defense and attachment while feeding have been the principal alternatives suggested. The food vacuoles, formed at the mouth, pass through the body by a regular route and leave at a definite point. In some, it would scarcely be an exaggeration to say that there is a digestive tube and an anus. There are two nuclei, a micronucleus which is concerned with heredity and a macronucleus which is concerned with metabolic functions. This led Kofoid to state that it would be as logical to call a whale unicellular as to call a *Paramecium* unicellular.

Finally, the ciliates have developed a unique type of sexual reproduction called conjugation. The details are quite complicated, but in substance it amounts to this, that the maturation divisions result in a stationary nucleus and a wandering nucleus in each of a pair of conjugants, and the wandering nucleus of each fertilizes the stationary nucleus of the other. The most primitive order of ciliates, the Holotricha, is completely clothed in cilia, and the animals are strong swimmers. But feeding by means of currents seems to lead to a sessile mode of life, for specialization in the ciliates has generally resulted in attachment to the substrate, restriction of the cilia to limited areas usually related to feeding, and specialization of the ciliary tracts to form membranelles, undulating membranes, or similar structures.

The well known *Vorticella* is a good exemplar of such tendencies. In a single order of ciliates, the Hypotricha, specialization has been in the direction of increased efficiency of locomotion. In this order, a band of typical cilia still creates a current for feeding purposes, but locomotion is based upon the use of fused tufts of cilia, the cirri, which function in a leg-like fashion. Finally the class Suctoria is a small group which is undoubtedly related to the ciliates. Adult suctorians have no cilia, nor have they any other locomotor organelles. Nor do they have a mouth, for they use protoplasmic tentacles for the capture and ingestion of food. They are generally permanently attached to their substrate. Up to this point, they show no affinity with the ciliates. However, they do have the two types of nuclei and sexual reproduction by means of conjugation, phenomena which are unknown elsewhere in the world of life except for these two classes. Further, the suctorian zygote becomes a free-swimming, ciliated organism which only later settles down on the substrate and adopts the typical suctorian mode of life. This is interpreted as embryological recapitulation of ancestral history.

Origin of Metazoa

Hyman has said that there is no direct proof of the origin of the Metazoa from the Protozoa, yet the discussion of the origin of the

Metazoa (multi-cellular animals) usually revolves around the question of which protozoan stock would seem to be the most probable progenitor of the Metazoa. Two broad possibilities exist by which the Metazoa could have been formed from the Protozoa. The first is that repeated nuclear division without cytoplasmic division might have led to formation of a plasmodium, like some of the Heliozoa.

Formation of cell membranes would then result in multicellularity, and differentiation might then lead to the true multicellular individual. The second method is the differentiation of cells within a colony or Protozoa, comparable to *Volvox*, leading to interdependence and individuality. A very different, third possibility has been suggested. In protozoan colonies each cell ingests food, but even in simple metazoans, a new method of feeding is used, with a digestive tract feeding for the; whole organism. This transition might be difficult. Hardy suggests that simple plants, like *Volvox*, living in an environment deficient in nitrates and phosphates, may have satisfied the deficiency by capturing smaller organisms. Increasing utilization of this nutritive pathway, together with loss of photosynthesis, would then lead to a simple metazoan.

Insectivorous plants show the feasibility of such a nutritive mechanism, and the fact that unicellular plants seem to have given rise to protozoans more than once lends plausibility to the suggestion that multicellular plants may have achieved animalization at least once. Paleontology is of no help in this problem, for the Metazoa were already well established at the beginning of the Cambrian. It is therefore probable that the origin of the Metazoa will always be speculative. But most zoologists favour the flagellates as the most probable progenitors of the Metazoa. The reasons for this are many. The flagellates are a highly variable group which appear to have given rise to many other groups of plants and to several, perhaps all, other groups of Protozoa. Further, some groups of flagellates show a tendency to form colonies of ever-increasing size and complexity. The evolution of sex has occurred here, and the colonies are definitely divided into somatic and germinal "tissues."

Oogamous reproduction is the rule for such colonial flagellates. The sperm are at once similar to some simpler, noncolonial flagellates and to the typical sperm of Metazoa. Such colonies may also show anteroposterior differentiation. These large, highly specialized colonies occur principally among the plant flagellates, yet colonies of a highly suggestive character also occur among the animal flagellates. None of these reasons is conclusive, yet collectively they carry considerable weight. However much the origin of other Metazoa may be disputed, it seems almost certain that the Porifera (sponges) were derived from choanoflagellates. It is a short step from the structure of the colonial choanoflagellate *Proterospongia* to that of the simplest sponges.

Proterospongia consists of a small mass of gelatinous material in the surface of which are imbedded choanocytes (collared, flagellate cells), and in the interior of which are ameboid cells. The choanocytes can withdraw the collar and flagellum and move into the interior to become ameboid cells. In order to change this to the structure of a simple sponge it would be necessary only to develop a system of channels through the gelatinous mass, to let these channels be lined with choanocytes, and to let the outer surface be covered by a simple epithelium. No organ systems are present, the functions of the sponge being carried on by the component cells individually.

Phylum Porifera

Cooperative activity is at a minimum, and it has often been debated whether a sponge is a true organism or a colony of unicellular organisms. There are enough evidences of cooperative activity to swing the balance of zoological opinion in favour of the former alternative. Thus, while some sponges have no definite shape, many have rather complex shapes. Most sponges produce skeletal elements in the form of spicules which may be quite complex, and so exactly formed in every case that they are among the best taxonomic characters. Further, the beat of the flagella in the canals is not obviously coordinated, yet the water current is unidirectional, and so there must be at least a limited control of the flagellar beat. Sponges can reproduce asexually by formation of groups of cells, the gemmules, but they also reproduce sexually by means of typical eggs and sperm. The zygote develops into a free-swimming larva which is flagellated. After swimming for a brief time, it settles down to become a sponge. Because this larva has typical (not collared) flagella, some zoologists believe that the sponges were derived from some typical flagellates, with choanocytes being a secondary development.

Yet this is not unexpected in view of the recapitulation principles, for the choanoflagellates themselves must have been derived from typical flagellates. Evolution within the Porifera has taken the forms of elaboration of the canal system and of the supporting spicules or fibers. As relationships within the phylum are not at all clear, it may be more profitable to go directly to the problem of the relationships of the Porifera to other animals. The Porifera are a terminal group. Further, they are so different from all other Metazoa in the absence of unified tissues, in the physiological independence of the individual cells, and in their embryology and adult anatomy, that it is generally believed that their origin from the Protozoa must have been independent of that of the rest of the Metazoa. The Profiera are therefore assigned to a separate branch; of the subkingdom Metazoa, the branch Parazoa, in contrast to the Eumetazoa.

Gastrea Theory

Speculation on the origin of the Metazoa has been dominated by the Gastrea theory of Haeckel, a theory based on the literal application of the Biogenetic Law. In its original form, Haeckel interpreted the egg as corresponding to an ameboid ancestor, possibly to *Amoeba* itself. As evidence in favour of this, he pointed to the ameboid eggs of sponges and of some coelenterates. Other types of eggs he assumed to be secondary specializations. The egg, of course, undergoes the cleavage divisions which result first in a solid morula and then in a hollow ball of cells, the blastula. The morula was interpreted as corresponding to a simple hypothetical ameboid colony, the *Synamoeba*, while the blastula was supposed to correspond to a colonial ancestor, the *Blastea*, more or less comparable to *Volvox*, but ameboid rather than flagellate.

The modern exponents of this theory assume flagellate rather than ameboid ancestors for the reasons stated above. There being only a single layer of cells in the *Blastea*, all cell functions were at first shared by pall cells, but then a division of labour occurred, with the posterior cells assuming the nutritive functions. These cells then invaginated, with the result that the organism became a two-layered gastrula, having an outer layer of flagellated cells (ectoderm) and an inner layer of digestive cells (entoderm). Haeckel called this hypothetical organism the *Gastrea*, and he believed it to be ancestral to all Eumetazoa. Some of the coelenterates he regarded as living gastreads. Next, the *Gastrea* developed a third cell layer, the mesoderm, between the first two, and he regarded all of the structures of the higher phyla as derived from these three layers.

The development of the bottom-feeding habit led to elongation of the body and the formation of primitive worms, similar to the living Turbellaria. From these the higher phyla were developed. The Gastrea theory is a beautiful simplification and synthesis of a vast amount of embryological, morphological, and taxonomic data, and it is almost without a serious competitor. As a result, it was, until recently, presented in nearly every elementary textbook of zoology. Unfortunately, however, as Hyman* has pointed out, "it is probably one of those simplification that are too beautiful to be true." Embryology is not a safe basis for construction of pedigrees, especially if comparison is made between embryos of advanced species and adults of their supposed ancestors, as Haeckel did in this case.

At most, embryology should be treated only as one of several corroborative lines of evidence. But there is the further difficulty that even the embryological evidence does not give unequivocal support to the Gastrea theory. For in the coelenterates, the group which is closest to the hypothetical *Gastrea*, gastrulation ordinarily occurs not by simple invagination of the posterior cells, but by the in wandering of many cells from all parts of the blastula. And this does not result in the immediate formation of a typical gastrula, but rather it consists of a ball of ectodermal cells. This type of larva is called a *planula*. Only later does this entodermal core hollow out and a mouth (blastopore) break through to form a typical gastrula. While the type of gastrulation with which Haeckel dealt is known, for example in the starfishes, it is not widespread in the Animal Kingdom, and it appears to be a secondary modification rather than a primitive character.

It is plausible, then, that the ancestor of the Eumetazoa may have been a blastula-like colonial flagellate, in which there occurred a differentiation between somatic and reproductive cells, as in living *Volvox* and then a differentiation between digestive cells and locomotor cells, with the former type moving into the interior of the organism to form either a gastrula or a planula. Yet it is probable that decisive evidence on this basic question will never be obtained. That this primitive metazoan was not identical with any living type is almost certain.

The three most primitive living metazoan phyla are the Mesozoa, the Coelenterata (= Cnidaria), and the Ctenopora. These phyla are generally radially symmetrical (having one differentiated axis), or else biradially symmetrical (having two differentiated axes). Their general grade of organization is more advanced than that of the Porifera, for, while there are no organ systems, there are two well-defined tissues, the ectoderm and entoderm (or epidermis and gastrodermis). In most coelenterates and in the ctenophores, there is between these layers a jelly-like mass, the mesoglea, which also includes some cells. Thus it is not strictly true, as is often stated, that these phyla have only two cellular layers.

Phylum Mesozoa

The correct phylogenetic position of the Mesozoa is a much vexed question. From a structural view point, they are the simplest of metazoans, consisting of an outer, generally ciliated, layer of cells enclosing a core of internal, reproductive cells. They thus resemble a planula, but the internal cells are not digestive cells. Van Beneden named the group in 1877 with the intention of indicating his judgment that the group was extremely primitive, intermediate between the Protozia and the rest of the Metazoa. On the other hand all Mesozoa are parasitic, and their life cycles are somewhat similar to those of the digenetic trematodes.

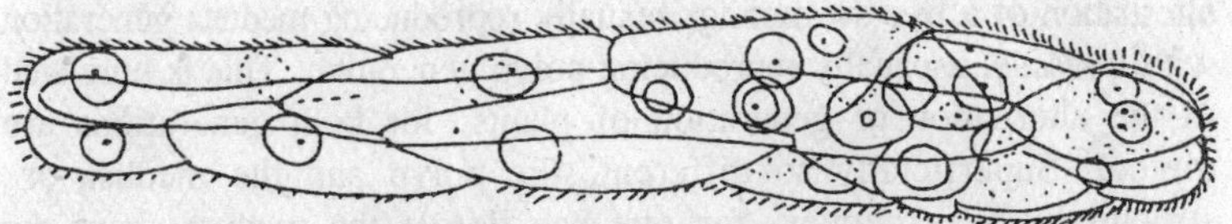

Fig. 8.2. Conocyema deca, a typical mesozoan.

Many zoologists treat the group as a degenerate offshoot from the flatworms. It is could be shown decisively that the characters of the Mesozoa are primitive rather than degenerate, the group would assume great phylogenetic importance, for it could then be reasonably argued that the group must be but little changed from the remote, pre-Cambrian ancestor of the Metazoa. It would prove that the Metazoa were derived from a planula-rather than from a gastrea-type ancestor, and it would leave the Gastrea theory very badly damaged. But the available evidence does not furnish a basis for a final decision on the taxonomic position of the Mesozoa. W. K. Brooks has said that "suspended judgment is the greatest triumph of intellectual discipline," and this appears to be an appropriate place for that achievement. Hyman sets them off as an independent branch of the Metazoa, the branch Mesozoa.

Phylum Coelenterata

The Coelenterata have traditionally been treated as the most primitive of the Eumetazoa. Haeckel regarded them as the source of the flatworms, and hence of all higher phyla. The principal reason for his viewpoint is the obvious resemblance of a hydroid polyp to a gastrula. For the hydroid consists of two simple cell layers, with no organ systems and with only a trace of noncellular mesoglea. The anatomical structure of the polyp could be derived from that of the gastrula simply by the elongation of the body and the drawing out of a circlet of tentacles around the mouth. Food is taken in and waste residues expelled by the mouth, which is simply the blastopore of the balstula. Food is still digeted by the protozoan method, that is, the cells of the gastrodermis engulf food particles, and digestion is carried on intracellularly. But enzymes are also secreted into the gastrovascular cavity, and much digestion occurs there. Muscle tails may be formed in connection with either the epidermis or the gastrodermis. A nerve net is formed from epidermal elements. Thus there is a high degree of tissue differentiation, but no organ systems. Outstanding specializations of the coelenterates include the development of nematocysts, organelles for food procurement and defense; and the alternation of a free-swimming, sexually reproducing medusa generation with a sessile, asexually reproducing polyp generation. This is unrelated to the alternation of generation of plants, for both generations are diploid. Superficially so different, the polyp and the medusa are structurally very similar, for one can derive the medusa from the polyp by simply inverting the latter, greatly increasing the amount of mesoglea and its cellular contents, and drawing the circlet of tentacles away from the mouth.

Haeckel assumed that the anestral coelenterate was a polyp, because of the ease with which a polyp can be derived (in theory) from the Gastrea. But study of what appears to be the most primitive order of hydroids, the Trachylina, has led to the conclusion that the medusa phase is primary and the polyp derived. Once formed, the Coelenterata diverged along three major lines, each comprising a class of the phylum. In the most primitive class, the Hydrozoa, both generations are generally well developed. The class includes hydroids such as the *Hydra* of elementary laboratories and the much more typical

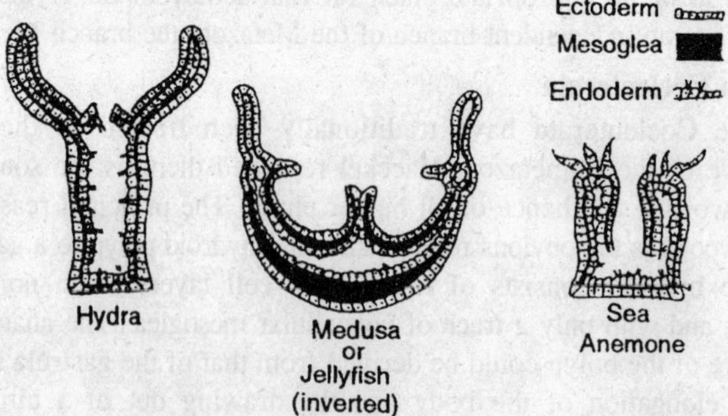

Fig. 8.3. An inverted jellyfish compared to hydra and to a sea anemone, showing their structural similarity.

marine colonial forms, such as *Obelia*. There are also some in which the polyp generation is reduced, as in the order Trachylina. In the class Scyphozoa, including the jelly fishes, the medusa is much the more prominent generation, with the polyp being reduced or absent altogether. The final class, the Anthozoa, includes only the polyp phase, the medusa being suppressed entirely. The class comprises the sea anemones, the corals, and their allies. Many of these, including all of the corals, secrete a calcareous exoskeleton, because of which they have left an excellent fossil record going back to the Ordovician. All three classes are ancient, and it is probable that they diverged in pre-Cambrian times from a primitive hydrozoan type not dissimilar to the Trachylina.

Phylum Ctenophora

The Ctenophora are a small phylum, only about eighty species being known. All are small, marine animals. They are commonly known as "sea walnuts" or "comb jellies" because of the presence of comb-like plates of cilia. They share some important characteristics with the coelenterates. Thus they are radially or biradially symmetrical in contrast to all higher phyla, including the Echinodermata, in which bilateral symmetry is primary, and radial symmetry derived. They are at the tissue grade of construction, with an abundant mesoglea separating the epidermis from the gastrodermis.

In both phylla, the gastrovascular cavity is present, and its branches distribute food through the body. But the ctenophores lack nematocysts, and there is no alternation of generations. Unlike the coelenterats, they are hermaphroditic. There seems to be little doubt that the coelenterates gave rise to the Ctenophora, yet all attempts to relate them to any of the classes of living coelenterates have failed. It seems most probable that the Ctenophora were derived from the same ancient, pre-Cambrian, trachyline stock that gave rise to the three classes of the Coelenterata, and at about the same time.

Primitive Bilateral Phyla

Some of the ctenophores have become elongate and flattened, and it has been suggested that they are related to the ancestors of the flatworms, the phylum Platyhelminthes. But the majority of zoologists believe that the appearance of homology is misleading. Haeckel believed that a primitive hydrozoan was the ancestor of the bilateral phyla, basing his opinion as usual upon the Biogenetic Law. The evidence is insufficient, and it seems at least as probable that flatworms were first derived from the same planula-like stock which gave rise to the coelenterates. But in the flatworms, the middle cell layer (mesoderm) became more highly developed, with organized muscle layers, reproductive system, and an excretory system. Yet no coelom, or body cavity, developed within the mesoderm, as it does in most higher groups. The nervous system is formed from the ectoderm, although it is imbedded in the mesoderm. It is not a diffuse nerve net as in the coelenterates, but rather it is somewhat centralized, being organized about cerebral ganglia at the head end and longitudinal cords.

There are organized sense organs, including eyes. Food is still distributed to the various parts of the body by the branches of the digestive tract, which is now called an intestine rather than a gastrovascular cavity. Thus the flatworms are much advanced beyond the tissue grade of construction which characterizes the radiate phyla, for definite organ systems are present, principally in the mesoderm. Hadzi has urged a radically different phylogeny. He believes that multinuclear ciliates became acoelous flatworms by formation of cell membranes. In evidence, he points out that both are ciliated; hermaphroditism of flatworms he finds homologous with conjugating of ciliates; trichocysts of ciliates he believes are represented by sagitocysts, rod-like inclusions in some of the epidermal cells of acoelous flatworms. He believes that the Anthozoa were derived from flatworms by adoption of the sedentary life, and that the other coelenterates were then derived from anthozoans.

The higher Metazoa were also derived from flatworms, and are grouped into only four phyla, a procedure which unites some highly diverse groups. At this point it may be well to recall the admonition of Hyman that "the exact steps in the evolution of the various grades of invertebrate structure are not and presumably never can be known. Statements about them are inferred from anatomical and embryological evidence and in no case should be regarded as established facts." Although the phylum Platy-helminthes, and especially the most primitive class of this phylum, the Turbellaria, is commonly treated as the stem group from which the higher phyla arose, this is by no means established.

Closely related to the Platyhellminthes, and on the same general level of organization, is the phylum Nemertinea, a small group of marine flatworms. This group is less well known than the former because it is predominantly marine, and because it is a difficult group to study. But the Nemertinea show some characters which qualify them for special consideration as potential forerunners of the higher invertebrates. For the first time, there is an anus present, so that the digestive system is said to be complete. Also, there is a simple blood circulatory system, and the blood contains hemoglobin.

Particularly important according to Kofoid is the fact that the nervous system is based upon cerebral ganglia and eight longitudinal nerve cords, two dorsal, two ventral, and two on each side. He has pointed out that this lends itself to the formation of the principal invertebrate nervous systems by the development of the ventral cords and the suppression of the others; and to the formation of the chordate nervous system by the development of the dorsal cords and the suppression of the others. But the fact the nemerteans capture their food by means of an extensible proboscis, a mechanism found in no other group, suggests that they are a terminal groups. Yet another possibility is that a third group, an unknown phylum of very primitive flatworms derived from the primitive planula by the development of the mesoderm may have been ancestral to both the Platyhelminthes and the Nemertinea. Whatever their origin, once formed, the platyhelminths diverged along three main lines of descent. The first of these is represented by the class Turbellaria, comprising the free-living flatworms, of which *Planaria* is the best known example.

The other two classes have become greatly modified for parasitism, and most of them have developed complicated life cycles to permit their transfer from one host to another. These are the class Trematoda, the flukes, which have retained all of the organ systems of the Turbellaria, and are generally internal parasites of vertebrates; and the class Cestoda, the tapeworms, which; have undergone an extreme degenerative evolution. They are all intestinal parasites of vertebrates. On the same general level of organization as the above discussed acoelomate groups, there are a number of minor phyla, of very uncertain relationships, which are characterized by the possession of a pseudocoel.

A coelom is, by definition, a body cavity formed within mesoderm. A pseudocoel, on the other hand, is a remnant of the cavity of the blastula, and it may be partially filled by large, vacuolated cells. The taxonomy and relationships of these pseudocoelomate organisms are very uncertain. It seems best here simply to acknowledge that the exist, and that they are among the most primitive of bilaterally symmetrical animals, without attempting to unravel their relationships. These are the Rotifera, a phylum of microscopic animals which are known to every student of elementary biology as fascinating contaminants of most protozoan cultures; the phylum Gastrotricha, which are small, worm-like animals of wide distribution; the phylum Kinorhyncha, comprising a group of minute, marine, wormlike animals; the phylum Nematoda, a very important group, the so-called round worms, which includes many little-known free-living worms as well as some of the best known parasites of both plants and animals; the phylum Nematomorpha, or horsehair worms, consists of long, slender worms which are parasitic during larval life, but are generally free-living when mature; and the phylum Priapuloidea, comprising a few species of small, externally segmented marine worms. These six phyla are sometimes grouped together as classes of a single phylum, Aschelminthes.

The phylum Acanthocephala, or spiny-headed worms, are generally small worms the young of which are parasitic on invertebrates, while the adults are parasiti on vertebrates, including man. A final pseudocoelomate phylum is altogether different from these worm-like phyla. This is the phylum Entoprocta, a group of colonial animals, often forming encrusting colonies. Formerly regarded as a class of the phylum Bryozoa, it is now clear that they differ fundamentally from the Bryozoa. The intestine is U-shaped, and both the mouth and the anus are included within a ring of ciliated tentacles, the lophophore, which sets up a current for plankton feeding. This gives them a rather hydroid-like appearance superficially.

Protostomous Phyla

The remaining major phyla of the Animal Kingdom can be arranged in two divergin lines of descent, largely on the basis of embryological criteria. One line culminates in the Annelida, Arthropoda, and Mollusca, while the other culminates in the Echinodermata and the Chordata. Certain minor phyla can be associated with one or the other of these main lines with varying degrees of satisfaction. Haeckel believed that the echinoderm-chordate lien was derived from the Turbellaria, while Kofoid believed that both lines were derived from the Nemertinea. It is at least as likely that both were derived from the unknown, primitive, acoelous flatworm from which both the Playhelminthes and the Nemertinea were probably derived. The cleavage divisions of the annelid-arthropod-molluscan line are both spiral and determinate.

In spiral cleavage, spindles are at right angles to those of the preeding division, so that the cells of each layer alternate with the next layer like bricks in a building. This is not true of the large, yolky eggs of arthropods, yet their position in this series is secure, for their origin from annelids is clear. Determinate cleavage proceeds according to a set pattern, with the part of the body to be formed from each blastomere fixed from the start. Destruction of a blastomere results in a deficient larva. One can designate which blastomere will form ventral surface, which the gut, and so on. The mesoderm is formed from stem cells which multiply to form a pair of ventral bands, growing forward from the posterior end of the larva. The coelom is formed by splitting of these bands, hence these phyla are said to be *schizocoelous*. The blastopore becomes the mouth of the adult, and hence this whole series of phyla is called Protostomia.

Generally, though by no means always, development leads to a trochophore larva. This is a more or less spherical larva with an *apical tuft* of cilia dorsally, and a prototroch, or girdle of cilia around the equator, by which it can swim weakly. There is a digestive system consisting of a mouth, a short forget, an enlarged stomach, a short hindgut, and an anus, a complete digestive system. Some mesoderm is present, and some mesodermal organs, such as a kidney, are formed. This type of larva is characteristic of the annelids and molluscs, but unique larval stages have been developed by the arthropods. Other characters are also held in common by the members of these phyla, but those mentioned are sufficient to indicate the probability of relationship among them.

The actual relationships are, however, largely a matter of conjecture. Origin from some primitive, acoelous flatworm is probable, yet there is little basis for a decision as to which of the possible groups of flatworms was the ancestor, beyond the fact that spiral cleavage is also found in the Nemertinea and in the polyclad Turbellaria. Because of the widespread occurrence of the trochophore larva, it is generally supposed that an independent, trochophore-type animal was an ancestor intermediate between the flatworms and the present-day phyla of the annelid-arthropod-molluscan series.

The only evidence for this is the interpretation of the embryological evidence according to the Biogenetic Law, and it has already been pointed out how insecure this is. Yet, although it may be true, direct evidence is not likely to be obtained, for such an organism would be very poor material for fossilization, and it would have to be sought in pre-Cambrian rocks, for the major invertebrate phyla were all present in the Cambrian.

Phylum Mollusca

Whatever the source, the phylum Mollusca was early established as a group divergent from the others. This is a great phylum, the second largest in the Animal King do, having around 80,000 species encompassing a great variety of forms, and inhabiting almost every type of marine, freshwater, and terrestrial habitat. All of the extant classes of Mollusca were present already in the Cambrian, and literally hundreds of species are known. The classes were just as distinct then as now, and so paleontology is of no help in deciding what the relationship within the phylum may be. In addition to the general characters of all members of the protostome line, the phylum Mollusca is distinguished by many structural characters. In all, the body is divided into four regions: a muscular foot, a head, a visceral hump, and a mantle which generally secretes a calcareous shell. Basically, they are bilaterally symmetrical, but this is obscured in the adults of some classes. The coelom is very much reduced.

Mollusca are traditionally described as unsegmented, yet in 1957 *Neopilina*, a new species from the deep waters of the Gulf of Mexico, was described. This animal has several segmentally arranged organ systems, and its affinities are with a class, Monopalcophoa, which had been thought to be extinct since the Devonian. It indicates that the primitive molluscs may have been segmented. On the basis of anatomical evidence, the class Amphineura is regarded as very primitive, but not necessarily as the progenitor of the more specialized classes. This class comprises the chitons or sea cradles, animals which are familiar parts of the fauna of rocky sea coasts, but there are no freshwater species. The chitons have a broad, flat, elongated foot, over which lies the visceral hump and at the anterior end of which is the head. The mantel coverts the visceral hump and typically secretes a series of eight calcareous plates, the valves, which may be beautifully ornamented.

The class Gasteropoda, which includes the snails, slugs, limpets, and their less well-known allies, has developed a much enlarged visceral hump. In most gasteropods, this hump has grown asymmetrically, with the familiar coiling as a result. Corrrelated with this, there has been a torsion of the visceral hump through 180°, so that structures which were originally posterior have moved to an anterior position, and vice versa.

The class Scaphopoda, or tooth shells, is a small group which is very much specialized for digging. All species are marine, and there are only a few of them. The class Lamellibranchiata includes the bivalves, the familiar clams, mussels, and their allies. They are compressed bilaterally, and enclosed within a two-valved shell, hinged dorsally, and secreted by the two lobes of the mantle. The gills are immensely enlarged to form broad ciliated tracts which produce a current for feeding upon plankton and detritus. The head is very much reduced. The foot is generally wedgeshaped, unlike the broad, flat foot of the more primitive molluscs, but it can be protruded between the valves to serve a locomotor function. In some species, it is specialized for burrowing.

The class Cephalopoda includes the octopi, the squids, the chambered *Nautilus*, and its extinct allies. They are the most complex of mollusca, comparing favourably with the most advanced insects and vertebrates in degree of complexity. The visceral hump, like that of the gasteropods, is much enlarged, but it is still symmetrical. The mouth is surrounded by a ring of tentacles for seizing food. The head and foot are fused so completely that there is no agreement among specialists as to which these tentacles are derived from. There is a siphon by which water may be forcibly ejected from the mantle cavity. The coelom is better developed than in most molluscs. The nervous system is highly centralized, and efficient eyes, superficially quite similar to those of vertebrates, are present.

The more primitive members of the class have a chambered shell of which the animal always lives in the most recently secreted chamber. Each chamber corresponds to a stage in the most recently secreted chamber. Each chamber corresponds to a stage in the growth of the animal, like the successive moults of an arthropod. The shelled cephalopods are today represented only by a single genus, *Nautilus*, but this is the last remnant of a once dominant group in the oceans of the world. The nautiloids appear in the fossil record in the late Cambrian, and rapidly assumed a dominant position in the marine fauna. They reached their peak of development in the Silurian, and then gradually declined. During the Devonian, they gave rise to another suborder, the Ammonoidea, with which they competed unsuccessfully for a long period of the earth's history. The ammonites were the dominant marine molluscs throughout most of the Mesozoic Era, yet they dwindled and became extinct near the end of the Cretaceous.

Meanwhile the nautiloids survived in small numbers, and are known from a single genus at the beginnig of the Cenozoic Era. No longer being in competition with the ammonites, the nautiloids underwent a rapid evolution at the beginning of the Cenozoic, for seven new genera appear during the Paleocene epoch. But only one of these, *Nautilus*, has survived to the present time. Meantime, forms with the shell much reduced and internal, the octopi and squids, have become the principal cephalopods.

Phylum Annelida

The second main branch of the protostome line is the annelid-arthropod branch. The phylum Annelida is best known by the common earthworms, but it includes a wide variety of worms arranged in four classes. All annelids are segmented, that is, the functional and structural units of the body are repeated serially along the length of the body, and the segments may be marked externally. A well-developed coelom separates the digestive tract from the muscular body wall. There is an advanced nervous system based upon a pair of cerebral ganglia and a pair of ventral cords on which ganglia are located in each segment of the body. Whenever there is a larval stage, it is a trochophore.

Most annelids have a thin cuticle, and the typical species have segmentally arranged chitinous bristles called chaetae. The phylum Annelida is best typified by its most primitive class, the Polychaeta. These take their name from their numerous chaetae, which arise from limb-like lobes, the parapodia, on each body segment. There is typically a distinct head which may have appendages. The sexes are separate, and fertilization is external. There is a trochophore larva. The polychaetes are adapted to a wide range of habitats, including pelagic and bottom-dwelling, surface-crawling and burrowing, and permanent tubed welling, and they show very striking morphological adaptations.

In numbers of species, they undoubtedly exceed the remaining classes combined. Yet they are less well-known because they are almost exclusively marine. Better known, but far less typical of the phylum, is the class Oligochaeta, including the earthworms and many small freshwater annelids. The class appears to have been formed from the polychaetes by a process of reduction and simplification. The head region is much reduced and never including appendages. The chaetae are reduced both in size and in numbers, and they are no longer set on parapodia.

Oligochaetes are all hermaphroditic. Development is direct, thee being no larval stages. The embryos develop in a cocoon, an adaptation to terrestrial life. The class is much more uniform than is the Polychaeta.

The class Archiannelia is a small group of extremely simplified marine annelids. These are all small worms which generally lack parapodia and chaetae. They are ciliated, like the young of polychaetes. Their nerve cords retain the primitive connection with the epidermis.

The coelom is only slightly developed. The Archiannelida were named with the intention of inferring that this was the most primitive class of the phylum and the probable progenitor of the other classes. But embryological and morphological studies have led to the conclusion that this class is not truly primitive, but rather has been derived from the Polychaeta by extreme simplification. The class Hirudinea, or leeches, appear to have been derived from oligochaete ancestors, for they share many characters with that class. But all leeches are ectoparasites, and they are extensively modified for that mode of life.

Phylum Arthropoda

The protostome series is climaxed by the great phylum Arthoropoda, the most successful of all groups of animals if numbers of species be an indication of success, for about three fourths of all species of animals are arthropods. This is also the most varied of phyla, for there are arthropods adapted to every imaginable habitat from the abyssal depths of the ocean (crabs and pycnogonids) to aerial heights (insects). Perhaps the key to arthropod evolution is their early development of a thick, chitinous cuticle. This made necessary the joints—thin places in the cuticle—from which the phylum takes its name.

In order to move the hard pieces of the cuticle, the continuous muscular wall, inherited from plychaete ancestors, became broken up into specialized muscles, attached to ingrowths of the cuticle. The limbs, being jointed, are far more adaptable than the parapodia of the polychaetes, and they have become specialized for many functions, including sensation, feeding, and locomotion. This thick cuticle undoubtedly minimizes water loss by evaporation, and it may be that this is a major factor in their successful invasion of the land, for no other invertebrate phylum has so large a portion of terrestrial species. The concentration of the nervous system and the sensory organs in the head region (cephalization), begun in the polychaetes, is carried much further in the arthropods. The coelom is much reduced, and has become largely replaced by a hemococle. Segmentation is prominent externally, but less so internally.

As might be expected with so large a phylum as the Arthropoda, there is no general agreement on the; taxonomic rank to be accorded to the major divisions of the phylum. For present purposes, five major subphyla will be recognized, and no attempt will be made to treat their component classes because of the vastness of the groups. These are the Trilobita, the Crustacea, the Myriapoda, the Insecta, and the Arachnida. These well-defined groups, with the exception of the

Trilobita, are universally familiar, and so need not be defined here. The trilobites are of especial interest because they are the most ancient arthropods known, and because their structure was very generalized and could conceivably have given rise to the other major groups of arthropods.

The trilobites were dominant from the Cambrian through the Silurian. They then declined until the Permian, from which period only a single species is known. With this, the group became extinct. The trilobite body was enclosed in a chitnious skeleton, and was divided into three longitudinal lobes by two longitudinal furrows, hence the name of the group. The body consisted of a head and a segmented trunk. The segments were generally movable, but a variable number of posterior segments were joined together to form a rigid unit, the pygidium. There was a single pair of antennae. The remaining appendages were all simple, undifferentiated, biramous appendages. None were specialized as mouth parts, but all had gnathobases, basal processes which could be used for biting. All species were marine.

Fossil remains of transitional organisms from trilobites to the other major arthropod types are entirely lacking and it may be that the several groups arose independently. Indeed, this would avoid some preplexing inconsistencies. Comparative morphology indicates that the crustaceans, myriapods, and insects from one line of descent, while the arachnids must have diverged very early. In the Crustacea, this

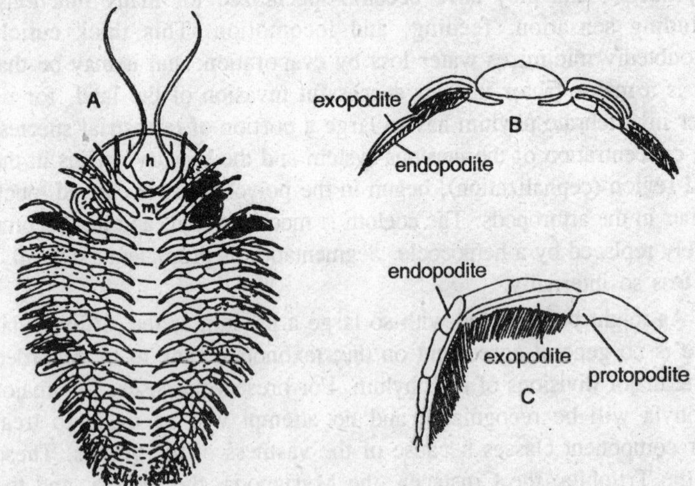

Fig. 8.4. A trilobite, Triarthrus becki. A—Ventral view; B—Section through a thoracic segment.

transformation has involved the division of the body into a highly fused cephalothorax and an abdomen in which the original segmentation is retained. The appendages have become greatly diversified, while still retaining the biramous plan. Almost all crustaceans are aquatic, but there are some terrestrial species, such as the sowbugs. Like the trilobites, the crustaceans are represented in early Cambrian rocks, and it is not improbable that they arose in pre-Cambrian times.

The Myriapoda, including the millipedes, centipedes, and their allies, may have been derived from a crustacean progenitor by the reduction of the exoskeleton and the loss of the gills. The latter were replaced by a system of tracheae, small tubes which carry air to the tissues of the body, thus permitting direct respiration. While not well represented in the fossil record, yet they are known from rocks as old as the Devonian. Insects probably also arose in Devonian times or earlier, but good fossils are first seen in rocks of Pennsylvania age. They may have been derived from myriapods, or they may have come directly from crustacean progenitors. Whichever ancestry they may have had, the tracheal system is always present and is undoubtedly one of the important adaptations which has permitted successful invasion of the land.

The body became divided into three well-defined regions, head, throax, and abdomen. The appendages of the; head segments are all specialized either for sensation or for eating. The thorax bears three pairs of walking legs. Dorsally, it bears two pairs of wings, but one or both pairs may be lacking. The abdominal appendages are all suppressed. The immense variety of insects which has been 660,000, or more than the combined numbers of species of all other living groups. And the numbers of individuals of many of these species are truly immense.

The subphylum Arachnida includes a highly varied array of organisms which have developed along lines quite divergent from the series just discussed. It comprises the horseshoe crabs, scorpions, spiders, harvest men, ticks, mites, and their allies. It is difficult to derive them from the trilobites, but it may be possible. There is a tendency toward suppression of segmentation, with only two body regions, the anterior prosoma and the posterior opisthosoma being marked off. There are six pairs of appendages, of which four pairs are walking legs. Unlike all other arthropods except the trilobites, none of the appendages are specialized as jaws. Gnathobases of the anterior limbs serve this function.

The class Pycnogonida, the sea spiders, is a small and little-known group which is usually placed with the Arachnida, simply because of their superficial resemblance to spiders. Yet their morphology is utterly different. Hedgpeth has studied the group thoroughly, and has concluded that they are undoubtedly arthropods, but that they are so widely divergent that they can not reasonably be grouped with any of the other arthropod types. Even less justification can be found for the common practice of making the Tardigrada a class of archnids. This is a small and little-known group of minute, freshwater organisms which shows some relationship to the Arthropoda, and has usually been treated as a class of the Arachnida. But specialists in the field feel that it should be regarded as an independent phylum of uncertain relationship to the Arthropoda.

Onychophora—A Unique Evolutionary Link

It was mentioned above that the derivation of the Arthropoda from the Annelida is more certain than is the derivation of any other phylum. This depends upon the Onychophora, a group of about eighty species all of which are assigned to a single genus, *Peripatus*. This group of animals shows a peculiar mixture of annelid and arthropod characters. Among the annelid characters may be mentioned the general appearance of the organisms, for they look much like polychaetes in which the parapodia do not bear chaetae. The cuticle is thin like that of annelids, and the muscles of the body wall are continuous. The excretory organs of both annelids and onychophorans are mesodermal tubules segmentally arranged (coelomoducts), while those of the Arthropoda are usually entodermal or ectodermal in origin. The reproductive ducts of the Onychophora are ciliated, but cilia are unknown among the Arthropoda. The eyes of the annelids and onychophorans are simple, whereas those of the arthropods are compound. On the other hand, in contrast to the annelids, the Onychophora and the Arthropoda have jaws derived from appendages. The coelom in each is much reduced and largely replaced by a hemocoel, while the coelom of the annelids is highly developed. The circulatory system of the Onychophora also resembles that of the Arthropoda rather than that of the Annelida.

Finally, the respiratory system of the Onychophora consists of a set of tracheae, a characteristic known nowhere else but in the Arthropoda. Because of this strange mixture of characters, the taxonomic position of the Onychophora has always been a much vexed question. They were orginally treated as a class of the Annelida. But because of the presence of a hemocoel and especially because of the tracheal system, they are now generally treated as a class of the Arthropoda. But these are not the only possibilities. They are sometimes treated as an independent phylum intermediate between the Annelida and the Arthropoda. And Light has urged that, as they are not properly separable from either of the major phyla, the entire annelid-onychophoran-arthropod series ought to be recognized as one great phylum under the name Articulata.

With no other series of phyla are such considerations possible and hence the statement with which this discussion began, that the origin of the Arthropoda from the Anneida is more certain than the origin of any other phylum. It is an unfortunate fact that the Onychophora are scantily known in the fossil record.

Minor Protostomes

A few minor phyla also show the general characters of the protostome line. Two of these, the Sipunculoidea and the Echiuroidea, are worm-like burrowers of the tide flats. They are generally treated as minor annelids, an arrangement which is more convenient than accurate. Yet they probably are more closely related to the Annelida than to any other phylum. The three remaining phyla, the Bryozoa, the Phoronida, and the Brachiopoda, are more difficult to lace. Like the Entoprocta, they feed by means of a lophophor, but unlike the phylum they have a coelom.

The Bryozoa are small, colonial animals which superficially resemble the Entoprocta, but actually differ from them fundamentally. The Phoronida include only two genera. The animals are elongate, wormlike creatures which dwell in tubes on tide flats. When the tide is in, the lophophore projects into the water to feed on plankton and detritus.

The phylum Brachiopoda, or lamp shells, bears a superficial resemblance to the molluscs because of their bivalve shells, but these are dorsal and ventral rather than right and left. Internally, they do nto suggest the molluscs at all. They have a very prominent lophpore. These marine animals are present in the earliest Cambrian deposits, and perhaps the greatest interest in the group drives from the fact that a single genus, *Lingula*, has persisted from the Ordovician to the present time, a span of 400,000,000 years. It may be the oldest genus inexistence. These three phyla were formerly groped together as a single phylum, the Molluscoidea. Yet they have little in common with one another except the lophophor, and there is little indication that any of them have a close relationship to the Mollusca. Hence it seem best to treat them as independent phyla of protostomes, of uncertain relationships to the larger phyla.

DEUTEROSTOMOUS PHYLA

The other major branch of coelomate animals is the Deuterostomia, and it comprises only five phyla: Chaetognatha, Pogonophora, Echinodermata, Hemichordata, and Chordata. This series contrasts with the Protostomia in that group of embryological characters which was used above to characterize the latter. The cleavage divisions are not spiral, and neither are they determinate. The mesoderm is not formed from stem cells, but rather from outpocketings of the entoderm of the gut. This simultaneously establishes the coelom, which is said to be *enterocoelous* (meaning simply that the cavity is established from the gut). The original blastopore becomes the anus, and a new mouth is formed in this series of phyla, hence the name Deuterostomia.

Development does not lead to a uniform larval type. The Chaetognatha have a unique larval type. The Echinodermata have several types of larvae, but all of them first pass through a *Dipleurula* stage to which especial theoretical significance is attached. A typical gastrula is formed, then the entoderm and ectoderm first fuse at one end, then break through to form a mouth. The blastopore is now the anus. The digestive tube now buds off an anterior vesicle which first divides into two lateral compartments, and finally into three segments on each side. These are the coelomic pouches. The cilia which covered the blastula and gastrula evenly now become concentrated in a series of bands arranged around the concave ventral surface of the animal. This is the Dipleurula larva. Because of its universality among the echinoderms, it is generally believed that the echinoderms must be descended from a Dipleurula-like ancestor which; was bilaterally symmetrical and free-swimming.

The Hemochordata have a type of larva, the Tornaria, which is quite similar to the Dipleurula, and even more similar to the Bipinnaria larva of starfishes. When first discovered, the Tornaria was described as a larval starfish, and only much later was the error discovered. This resemblance of the larvae is one of the major arguments for a relationship of the Hemichordata to the Echinodermata. Larvae are not of general occurrence among the Chordata, but the tunicates and amphibians have tadpole larvae, while the cyclostomes have a unique larva, the Ammocoetes.

Phylum Chaetognatha

The Chaetognatha are a small and uniform group of marine worms, the arrow worms, which show little evidence of relationship to any larger group of animals. They are included here because of conformity to the general deuterostome characters, yet they show no more specific affinity to any of the remaining deuterostome phyla, and it seems probable that the phylum branched off very soon after the formation of the deuterostome line. They show a superficial resemblance to Amphioxus, but this is undoubtedly misleading. Aside from a few doubtful specimens, the phylum is not represented in the fossil record.

Phylum Pogonophora

The Pogonophora is a phylum of deep-sea worms, only recently discovered and still but little known. The body consists of small protosoma and mesosoma, and a very elongate metsoma. The protosoma may bear one or more tentacles. It has an unpaired coelom, drained by a pair of nephridial ducts. In the other body segments, the coelom is paired. A nervous mass and ring in the protosoma give rise to a paired dorsal nerve cord. The circulatory system consists of two longitudinal vessels. Musculature is made up of subepidermal longitudinal fibers. There is no digestive system whatever. Larvae are unknown. What little is known of these animals seems to ally them with the Deuterostomia.

Phylum Echinodermata

The Echinodermata are best known by the starfishes, and these perhaps typify the phylum well. All echinoderms have secondarily established radial symmetry in the adult, after beginning life as a free swimming, bilaterally symmetrical larva. The radial symmetry of the adults is generally based upon a pentamerous (five radial segments) plant, or upon a plan derived from such. The development of radial symmetry may have been related to the change from a pelagic to a sessile mode of life by primitive echinoderms, for radial symmetry is a general characteristic of sessile organisms. The fossil record of the echinoderms is one of the best, going clear back to early Cambrian times, and excellent phylogenie scan be constructed within each of the five extant and two extinct classes. But the record does not throw light; upon the origin of the phylum, nor upon its possible relations to other phyla. The questions depend, at present, entirely upon the embryological evidence, with all of its limitations.

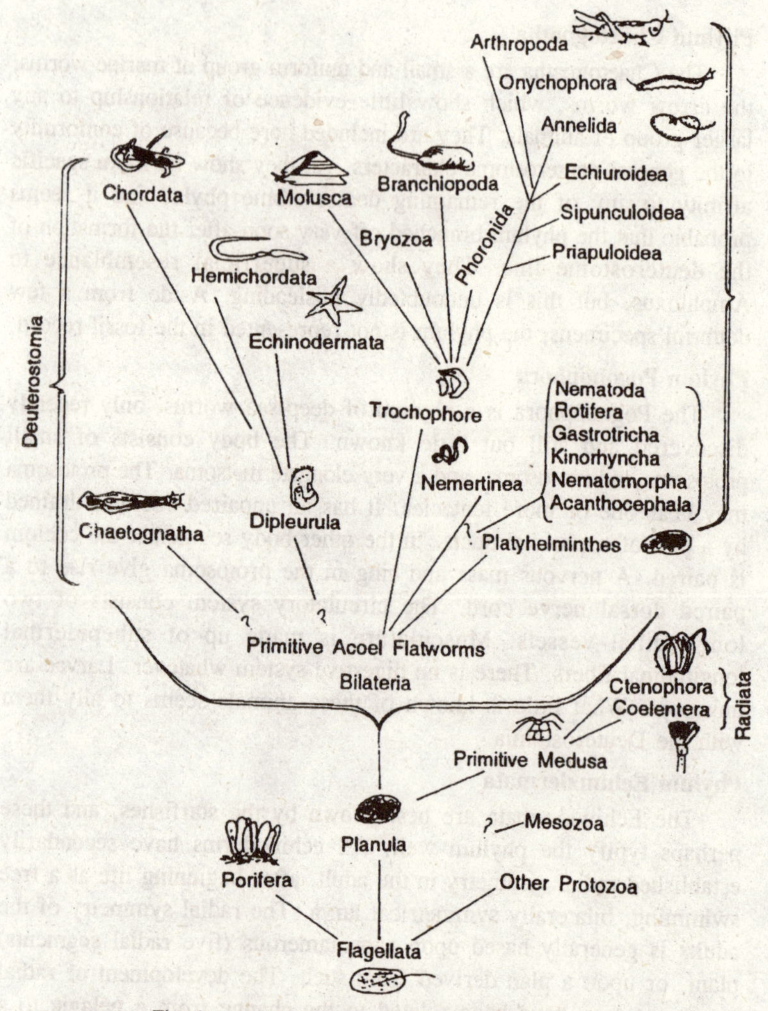

Fig. 8.5. Probable main lines of animal phylogeny.

Phylum Hemichordata and the Origin of the Chordata

The Hemichordata are a small phylum of worm-like marine animals which have been extensively studied because of their supposed relationship to the Chordata. In Haeckel's phylogeny of the vertebrates, the hemichordates were given as the next stage after the primitive flatwrorm. They were originally classed as a subphylum of the Chordata, because they show the three basic diagnostic characters of the Chordata: a dorsal nerve tube, a pharynx modified for respiration, and a notochord. Yet each of these is equivocal. The dorsal nerve tube is confined to the collar region, and the main nervous system is a ventral nerve cord like that of many invertebrates. The pharynx is pierced by numerous gill slits, yet it appears that these function primarily as exit pores for the feeding current rather than as respiratory organs. Yet some respiration probably does occur in the gills. But feeding many have been the primary function in the gills of true chordates also.

Finally, the notochord is a small structure developed as an outgrowth from the digestive tract into the proboscis. It has also been homologized with the pituitary gland, and so its taxonomic value is doubtful. It seems probable that these worms are related to the chordates, but how is not clear. Their relationship to the echinoderms, as indicated by a comparison of the Tornaria and Bipinnaria larvae, has already been discussed. As the phylum Chordata, including as it does the vertebrates and man, will be discussed in the next chapter, it need nto be taken up here beyond the obvious statement that this phylum forms the climax not only of the obscure (from the viewpoint of relationship) Deuterostomia, but also, perhaps, of the entire Animal Kingdom.

Some Evolutionary Generalizations

Now that the greater part of the course of evolution has been sketched, and before taking up that part of the story which is most closely related to those who study it, it may be well to give some consideration to broad tendencies and principles of evolution. At the outset, evolution is not always "upward" or "progressive." Many examples are known of the evolution of simpler or more degenerate types from originally complex types. Thus the fungi may have evolved from algae by the loss of chlorophyll. Grasses have evolved from lily-like ancestors by simplification of parts, especially of flowers. Mistletoe, an angiosperm parasitic upon trees, has undoubtedly evolved from free-living ancestors. Similarly in the Animal Kingdom, many examples of retrogressive or degenerative evolution are known.

The development of sexual reproduction and its great evolutionary importance was emphasized. The Rotifera were undoubtedly evolved from bisexual ancestors, and some species are still bisexual. Nonetheless there are species in which the males are unknown. These still reproduce sexually, for the ova develop parthenogenetically, but the major advantages of sexual reproduction are lost.

The development of the parasitic habit almost always involves degenerative evolution. In the tapeworms, this is extreme. Although derived from free-living flatworms, with well-developed digestive, nervous, reproductive, muscular, and other systems, the tapeworm is reduced substantially to an absorptive sac containing gonads. But, while such degenerative evolution is characteristic of parasites, it is by no means confined to them. As already pointed out, the class Archiannelida was most probably derived from polychaete ancestors by a process of simplification and loss of parts. Thus it is clear that evolution can be retrogressive as well as progressive.

Origin of New Groups from Primitive Ancestors

A second very important generalization relates to the form of the tree of life. It is more properly a shrub than a tree, for new groups do not arise from the most advanced and specialized members of their parent groups, but from the primitive, unspecialized ones. Thus the primitive flagellates have given rise to many additional plant and animal groups, but the more specialized protozoan and algal groups are generally terminal. Again, if indeed the Hemichordata and Chordata were derived from echinoderms, it seems certain that the more advanced phyla must have arisen from the ancestral Dipleurula in a very primitive stage before radial symmetry developed. One more qualification relative to the shape of the phylogenetic tree is necessary for plants. Because of the phenomenon of allopolyploidy—to be discussed in detail below—hybridization of related species may result in new species. Hence branches may form, them fuse again, so that a network results. This phenomenon is of great importance for plants, less so for animals.

Rates of Evolution

The rates of evolution have not always been uniform. In general, periods of major geological change have been periods of rapid evolution, while periods of geological uniformity have been periods of slow evolution. This might be expected *a priori*, for the selective forces themselves should be in a state of flux during geologically unstable periods, while they should be quite stable during geologically uniform periods. Thus in a period of uniformity, organisms generally should tend to become well adapted to their environments. In such a situation, almost any change would be disadvantageous, and would tend to be eliminated by natural selection. But in periods of geological change, only those organisms which also changed could possibly have a selective advantage. Thus there was a rapid burst of evolution in the Siluria and Devonian when plants and animals were colonizing the land. Not only were vast new habitats thus opened up, but this was a time of mountain-building and of great changes in sea level.

In the Mississippian, physical conditions were again quit stable, and evolution was slow. Many such alternations of periods of rapid and slow evolution have occurred. It appears that we are at present in the midst of a period of rapid evolution. Two special situation favour rapid evolution. The first of these is adaptation of a group to a new mode of life, as invasion of land by aquatic plants and animals. During the transition, the organisms must be rather ill-adapted, hence strong selection pressure favours rapid change, and only those that can respond will leave descendants. This will be greatly facilitated if, in the earlier environment, characteristics developed which are *preadaptive* to the later habitat. Thus many Devonian fishes developed lungs, which were adaptive to life in stagnant, drying ponds. This preadaptation late speeded the adaptation to life on land. Simpson has called this "quantum evolution" as it involves sudden change from one"adaptive orbit" to another. He regards the term as unsatisfactory, because quantum events in physics are very small, while quantum evolution is on a large scale. Perhaps *macroadaptive* evolution would be a better term.

Second, when a group has invaved a new major mode of life, many variations of habitat are open to it with a minimum of competition, and hence rapid diversification is favoured until these empty ecological niches are largely filled. This is the adaptive radiation which was discussed above. Entirely apart from changes in the general rate of evolution, the rates in different lines of descent have not been equal. It some, change has been extremely slow for long periods of time. The braciopod *Lingula*, for example, has been substantially unchanged since the Ordovician. In this connection, it should be pointed out that the warm, shallow seas which it inhabits constitute perhaps the most stable environment on earth. On the other hand, the mammals have evolved very rapidly during the Cenozoic Era.

Trends in Size

A very common trend in evolution, sometimes called Cope's Law, is one toward increasing size of individuals. The original studies of the phenomenon were made upon vertebrates, but comparable studies have shown the same tendency in many groups of invertebrates and plants. A review of the paleontology of almost any group shows that its largest representatives are not its earliest ones, though not necessarily its latest ones either. Newell has pointed out that species now living are the largest known representatives of the vertebrates, crustaceans, echinoderms, pelecypods, gastropods, cephalopods, and annelids. Yet the tendency toward size increase has been by no means universal. As already mentioned, the rise of herbs and shrubs is a recent thing, and they have been derived from trees and other large plants. Hooijer has pointed out that progressive size decrease has been characteristic of many vertebrate groups during the Quaternary period, which is now in progress.

Complexity and Efficiently

It is obvious that the general progress of evolution has involved the development of new organ systems and increasing complexity. Yet development of increased efficiency often involves reduction in number and complexity of parts. Vestigial organs in general could perhaps be considered in this light, but it is equally true of actively functional structures. Thus the teeth of fishes are very numerous and usually indefinitely replaceable. They are less numerous in amphibians and reptiles, still less so in mammals, where they reach their maximum degree of specialization and efficiency.

Much the same thing is true of the ververtebrae, and of the bones of the skull. This tendency can also be exemplified by plants, for example by the reduction of numbers of stamens in specialized plants. Increased efficiency is also often obtained by fusion of originally separate parts. Thus the sacrum of mammals is formed by the fusion of three to five originally separate vertebrae, thus making a much stronger attachment of the hind limb to the vertebral column than would otherwise be possible. Another good example is afforded by the pectoralis muscles which, in tetrapods, arise near the midline of the chest region and insert upon the humerus. Muscle slips from many adjacent body segments join to form these muscles. Among plants, the corollas of flowers such as the cucurbits or petunia are formed by the fusion of originally separate petals.

Dollo's Law

Many times during the long history of life, advanced organisms have returned to ancestral habitats and modes of life. This gives selective value to adaptations similar to those of the ancestral species, and raises the question whether evolution might be reversible. Study of such cases shows that always a gross similarity between ancestral and descended structures is achieved without any genuine reversal at all. Thus many reptiles and mammals have reverted to an aquatic mode of life. They have assumed a generally stream-lined, fish-like form, and the limbs have become shortened, webbed, and fin-like. Yet the skeleton of such flippers is always distinctly that of the class to which the animal belongs rather than that of a fish fin. Similarly, many angiosperms have returned to the water and assumed alga-like appearances, but their morphology is still that of flowering, vascular plants.

The evidence indicates that major evolutionary steps, once taken, are never reversed. This is known as Dollo's Law. It even might be expected *a priori*, for major evolutionary steps are compounded of many smaller steps, each preserved by natural selection. That such a sequence, occurring by chance once, should by chance be exactly reversed would be a most extraordinary thing. If not impossible, it is at least most improbable for whole organisms. Attempts to apply Dollo's Law to individual characters have failed, for these are, indeed, reversible by mutation.

Significance of Extinction

Some closing remarks should be made upon the subject of extinction, for this has been the fate of most species since the origin of life. Extinction may have completely different significance in different instances. The dinosaurs were a highly specialized line of Mesozoic reptiles which dominated the earth for a long time. But when conditions became unsuited to their survival, they became extinct and left no descendants. They were succeeded by other unrelated forms. The cynodont reptiles also became extinct, but they were succeeded by their own descendants, the adaptively superior mammals. Thus extinction may mark the end of a line of descent or it may be the accompaniment of the origin of new and superior types. It ought to be added that the major adaptive types—phyla and classes—very rarely become extinct.

9

SMALL MOLECULES IN LIFE

Water spews from the Earth as geysers. It circles the globe as clouds and falls from the sky as the gentle rain or in thundering torrents. Frozen, it covers parts of Antarctica to a depth of 3,000 meters (10,000 feet) or more. Vast oceans of it submerge much of the planet. Water is one of the key ingredients that makes life on Earth possible. It makes up as much as 95 percent of the weight of some living things.

Water is a simple substance containing only three atoms-two of hydrogen and one of oxygen. What does the composition of water tell us about its characteristics, and why is water so important to living systems? We can't answer these questions without more information about atoms in general, and about hydrogen and oxygen atoms in particular. But it is not only the constituents of matter that are important. To understand the Behaviour of something as apparently simple as water, we need to know how its constituent atoms are linked together. The same is true of the other small molecules that are essential to living systems.

The first part of this chapter will address the constituents of matter: atoms—their variety, properties, and capacity to combine with other atoms. Then we'll consider how matter changes. In addition to changes in state (solid to liquid to gas), substances undergo changes that transform both their composition and their characteristic properties. When cells use. oxygen to "burn" glucose, the products are water, carbon dioxide, and energy to power life activities. This transformation is similar to what happens when the fuel propane is burned in a stove, a combustion reaction. By studying simple systems such as the combustion of propane, we can understand better what happens in systems as complicated as living cells. The discussion of general chemical principles and their application to small molecules aids our understanding of the large molecules that form the basis for the life of an organism.

Later in this chapter, we return to a consideration of the structure and properties of water and its relationship to acids and bases. We close with a bridge to the next chapter—a consideration of characteristic groups of atoms that contribute specific properties to larger molecules of which they are part.

ATOMS: CONSTITUENTS OF MATTER

More than a million million (1×10^{12}) atoms could fit in a single layer over the period at the end of this sentence. Each atom consists of a dense, positively charged nucleus, around which one or more negatively charged electrons move. The nucleus contains one or more protons and may contain one or more neutrons. Atoms and their component particles have mass. Mass is a property of all matter. Measuring mass measures the quantity of matter present. The greater the mass, the greater the quantity of matter.

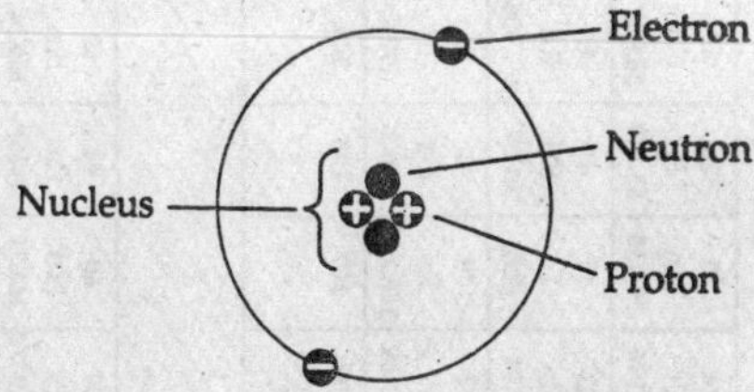

The mass of a proton serves as a standard unit: the atomic mass unit (amu), or dalton (named after the English chemist John Dalton). A single proton or neutron has the mass of 1 dalton, which is 1.7×10^{-24} grams (0.0000000000000000000000017 g). The mass of an electron is 9×10^{-28} g (0.0005 dalton). Because the mass of an electron is so much less than the mass of a proton or a neutron, the contribution of electrons to the mass of an atom can usually be ignored.

The positive electric charge on a proton is defined as a unit of charge. An electron has a charge equal and opposite to that of a proton. Thus the charge of a proton is +1 unit, that of an electron is -1 unit. Unlike charges attract each other; like charges repel. The neutron, as its name suggests, is electrically neutral, so its charge is 0 unit. Because the number of protons in an atom equals the number of electrons, the atom itself is electrically neutral.

An Element is Made Up of Only One Kind of Atom

The element hydrogen consists only of hydrogen atoms; the element iron consists only of iron atoms. An element is a pure substance that

The six elements highlighted in yellow make up 98% of the mass of any living organism.

Elements shown in orange are present in tiny amounts in many organisms.

Vertical columns have elements with similar properties.

Key: Chemical symbol; Atomic number; Atomic mass

1 H 1.0079																	2 He 4.003
3 Li 6.941	4 Be 9.012											5 B 10.81	6 C 12.011	7 N 14.007	8 O 15.999	9 F 18.998	10 Ne 20.179
11 Na 22.990	12 Mg 24.305											13 Al 26.982	14 Si 28.086	15 P 30.974	16 S 32.06	17 Cl 35.453	18 Ar 39.948
19 K 39.098	20 Ca 40.08	21 Sc 44.956	22 Ti 47.88	23 V 50.942	24 Cr 51.996	25 Mn 54.938	26 Fe 55.847	27 Co 58.933	28 Ni 58.69	29 Cu 63.546	30 Zn 65.38	31 Ga 69.72	32 Ge 72.59	33 As 74.922	34 Se 78.96	35 Br 79.909	36 Kr 83.80
37 Rb 85.4778	38 Sr 87.62	39 Y 88.906	40 Zr 91.22	41 Nb 92.906	42 Mo 95.94	43 Tc (99)	44 Ru 101.07	45 Rh 102.906	46 Pd 106.4	47 Ag 107.870	48 Cd 112.41	49 In 114.82	50 Sn 118.69	51 Sb 121.75	52 Te 127.60	53 I 126.904	54 Xe 131.30
55 Cs 132.905	56 Ba 137.34	57–71 La–Lu	72 Hf 178.49	73 Ta 180.948	74 W 183.85	75 Re 186.207	76 Os 190.2	77 Ir 192.2	78 Pt 195.08	79 Au 196.967	80 Hg 200.59	81 Tl 204.37	82 Pb 207.19	83 Bi 208.980	84 Po (209)	85 At (210)	86 Rn (222)
87 Fr (223)	88 Ra 226.025	89–103 Ac–Lr	104	105	106	107	108	109									

Lanthanide series	57 La 138.906	58 Ce 140.12	59 Pr 140.9077	60 Nd 144.24	61 Pm (145)	62 Sm 150.36	63 Eu 151.96	64 Gd 157.25	65 Tb 158.924	66 Dy 162.50	67 Ho 164.930	68 Er 167.26	69 Tm 168.934	70 Yb 173.04	71 Lu 174.97
Actinide series	89 Ac 227.028	90 Th 232.038	91 Pa 231.0359	92 U 238.02	93 Np 237.0482	94 Pu (244)	95 Am (243)	96 Cm (247)	97 Bk (247)	98 Cf (251)	99 Es (252)	100 Fm (257)	101 Md (258)	102 No (259)	103 Lr (260)

Fig. 9.1. The periodic table.

contains only one type of atom. The atoms of each element have certain characteristics or properties that distinguish them from the atoms of other elements. The more than 100 elements found in the universe are arranged in the periodic table. The periodic table arranges elements with similar properties in vertical columns in order of their increasing size. Although there are more than 100 elements in the world, about 98 percent of the mass of a living organism (bacterium, turnip, or human) is composed of just six elements—carbon, hydrogen, nitrogen, oxygen, phosphorus, and sulfur. The chemistry of these elements will be our primary concern.

A substance (such as oxygen gas) that contains only one kind of atom is an elemental substance. A substance that contains more than one kind of atom is a *compound*. Most substances of biological interest are compounds.

Number of Protons Identifies the Element

An atom is distinguished from other atoms by the number of its protons, which does not change. This number is called the atomic number. An atom of hydrogen contains 1 proton, a helium atom has 2 protons, carbon has 6 protons, and plutonium has 94. The atomic numbers of these elements are thus 1, 2, 6, and 94, respectively.

Every atom except hydrogen has one or more neutrons in its nucleus. The *mass number* of an atom equals the total number of protons and neutrons in its nucleus. Because the mass of an electron is infinitesimal compared with that of a neutron or proton, electrons are ignored in calculating the mass number. The nucleus of a helium atom contains 2 protons and 2 neutrons; oxygen has 8 protons and 8 neutrons. Helium, therefore, has a mass number of 4 and oxygen a mass number of 16. The mass number may be thought of as the weight of the atom, in daltons.

Each element has its own one- or two-letter symbol. For example, H stands for hydrogen, He for helium, and O for oxygen. Some symbols come from other languages: Fe (from Latin *ferrum*) stands for iron, Na (Latin *natrium*) for sodium, and W (German *Wolfram*) for tungsten. The periodic table gives the symbols for all of the 92 natural elements, as well as those for 14 elements that do not occur naturally.

Isotopes Differ in Number of Neutrons

We have been speaking of hydrogen and oxygen as if each had only one atomic form. But this is not true. Not all atoms of an element have the same mass number. The different atomic forms of a single element are called *isotopes* of the element. Isotopes of an element differ in the number of neutrons in the atomic nucleus. The common form of hydrogen is ^{1}H, but about one out of every 6,500 hydrogen atoms on Earth has a neutron as well as a proton in its nucleus and is thus ^{2}H, called deuterium. Furthermore, it is possible to create ^{3}H, tritium, which has *two* neutrons and a proton in its nucleus. Because all three types of hydrogen atoms have only one proton, they all have the atomic number 1. Deuterium, tritium, and common hydrogen have virtually identical chemical properties, although ^2H is twice and ^3H three times as heavy as ^1H.

In nature, many elements exist as several isotopes. For example, the natural isotopes of carbon are ^{12}C, ^{13}C, and ^{14}C. Unlike the hydrogen isotopes, the isotopes of most other elements do not have distinct names. Rather they are written in the form shown here and are referred to as carbon-12, carbon-13, and carbon-14, respectively. Most carbon atoms are ^{12}C, about 1.1 percent are ^{13}C, and a tiny fraction are ^{14}C. An element's atomic *mass* (*atomic* weight) is the average of the mass numbers of a representative sample of atoms of the element, with all isotopes in their normal proportions. For example, the atomic mass of carbon is 12.011. In biology, one encounters the terms "weight" and "atomic weight" more frequently than "mass" and "atomic mass"; therefore, we will use "weight" for the remainder of this book. Thus we say that the atomic weight of carbon is 12.011.

Some isotopes, called *radioisotopes*, are unstable and spontaneously give off energy as α (alpha), β (beta), or γ (gamma) radiation from the atomic nucleus. Such radioactive decay transforms the original atom into another type, usually of another element. For example, uranium-238 loses an alpha particle to form thorium-234, and carbon-14 loses a beta particle to form nitrogen-14. Biologists can incorporate radioisotopes into molecules and use the emitted radiation as a tag to identify changes that the molecules undergo in the body or to identify the locations of molecules within the cell. Some radioisotopes commonly used in biological experiments are ^3H (tritium), ^{14}C (carbon-14), and ^{32}P (phosphorus-32).

Although radioisotopes are useful for experiments and medicine, even low doses of radiation from radioisotopes have the potential to damage molecules and cells. Gamma radiation from cobalt-60 (^{60}Co) is used medically to damage or kill rapidly dividing cancer cells. In addition to these applications, radioisotopes can be used to date fossils.

Electron Behaviour Determines Chemical Bonding

In atoms, biologists are concerned primarily with electrons. To understand organisms, biologists study chemical changes that occur in living cells. These changes, called *chemical reactions* or just *reactions*, are changes in the atomic composition of substances. They occur because of the way in which electrons behave. The characteristic number of electrons in each atom of an element determines how the atom reacts with other atoms. All chemical reactions involve changes in the relationships of electrons with each other.

The location of a given electron in an atom at any given time is impossible to determine. We can only describe a volume of space within the atom where the electron is likely to be. The region of space within which the electron is found at least 90 percent of the time is the electron's orbital. In an atom, a given orbital can be occupied by at most two electrons. Thus any atom larger than helium (atomic number 2) must have electrons in two or more orbitals. The different orbitals have characteristic forms and orientations in space.

The orbitals constitute a series of *electron shells*, or energy levels, around the nucleus. The innermost electron shell, called the *K* shell, consists of only one orbital, called an *s* orbital. The *s* orbital fills first, and its electrons have the lowest energy. Hydrogen ($_1$H) has one *K* shell electron; helium ($_2$He) has two. All other atoms have two *K* shell electrons, as well as electrons in other shells. The *L* shell is made up of four orbitals (an *s* orbital and three *p* orbitals) and hence can hold up to eight electrons. The M, N, O, P, and Q shells have different numbers of orbitals, but the outermost orbitals can usually hold only eight electrons.

In any atom, the outermost shell determines how the atom combines with other atoms; that is, these outermost electrons determine how an atom behaves chemically. When an outermost shell consisting of four orbitals contains eight electrons, the atom is stable and will

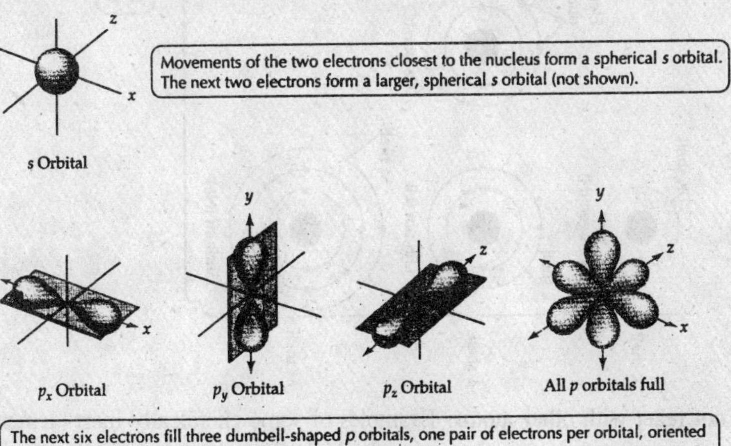

Fig. 9.2. Electron orbitals.

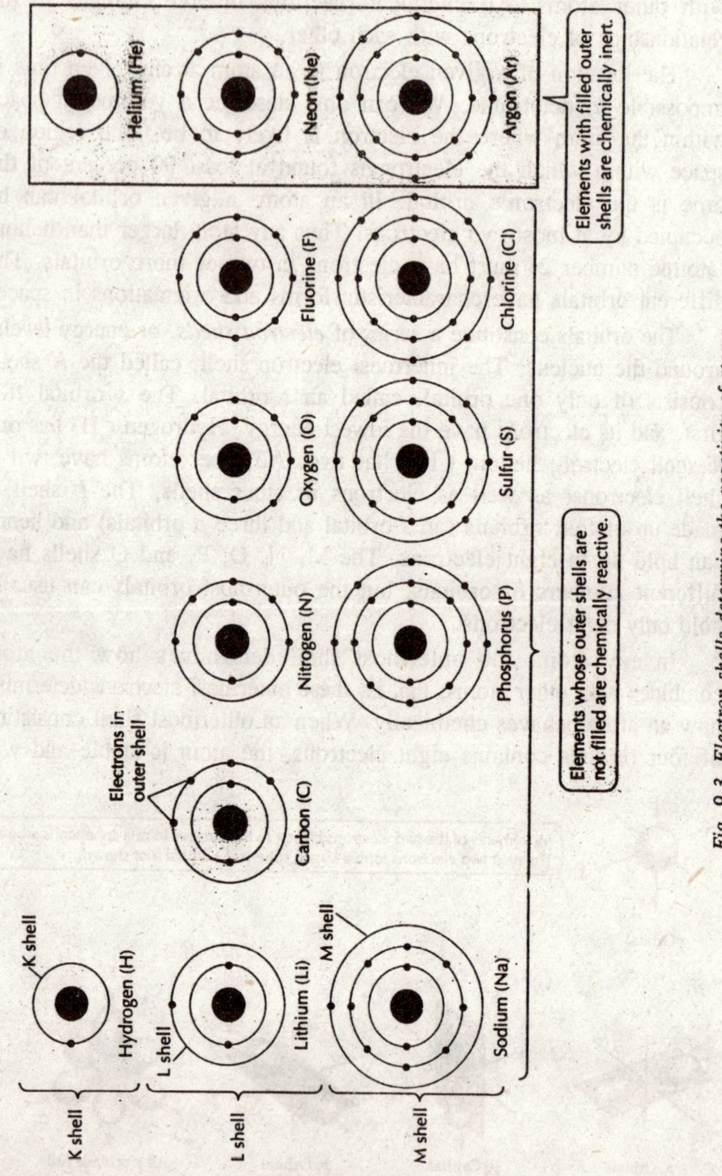

Fig. 9.3. Electron shells determine the reactivity of atoms.

not react with other atoms. Examples of some chemically inert elements are neon and argon, in each of which the outermost shell contains eight electrons. The atoms of other elements seek to attain the stable condition of having filled outer orbitals. They attain this stability by sharing electrons with other atoms or by gaining or losing one or more electrons from their outermost orbitals.

When they share electrons, atoms are bonded together. Such bonds create stable associations of atoms called molecules. A *molecule* can be defined as two or more atoms linked by chemical bonds. The tendency of atoms in stable molecules to have eight electrons in their outermost orbitals is known as the *octet rule*.

Many atoms in biologically important molecules follow the octet rule-for example, carbon (C) and nitrogen (N). However, some biologically important atoms such as hydrogen and phosphorus are exceptions to the rule. Hydrogen (H) attains stability when two electrons occupy its outermost orbital; phosphorus (P) is stable when its outermost orbitals contain ten electrons.

Chemical Bonds: Linking Atoms Together

A *chemical bond* is an attractive force that links two atoms to form a molecule. There are different kinds of chemical bonds, but all strong chemical bonds result from an atom's tendency to attain stability by filling its outermost electron orbitals. Atoms can gain stability in the outermost orbitals by sharing electrons or by losing or gaining one or more electrons. In this section, we will first discuss covalent bonds, the strong bonds that result from sharing of electrons. Then we'll examine hydrogen bonds, which are weaker than covalent bonds but enormously important to biology. Finally, we'll consider ionic bonding, which results when ions form as a consequence of the complete loss or gain of electrons by atoms.

Covalent Bonds Consist of Shared Pairs of Electrons

When two atoms attain stable electron numbers in their outer shells by sharing one or more pairs of electrons, a *covalent bond* forms. A hydrogen atom has one electron in its only shell, but two electrons would be a more stable condition. Imagine two hydrogen atoms, initially far apart but coming closer and closer, until they begin to interact. The negatively charged electron in each hydrogen atom is attracted by the positively charged proton in the nucleus of the other hydrogen atom. When the two atoms are close enough, the two electrons spend time between both nuclei, and the two atoms are covalently bonded together, forming a molecule of hydrogen gas (H_2). The two hydrogen nuclei share the two electrons equally and completely.

The two atoms do not come *too* close together, because their positively charged nuclei strongly repel each other. A certain distance between the coupled atoms gives the most stable arrangement. Pulling the atoms slightly farther apart would require an input of energy because of the "gluing" effect of the shared electrons. Pushing the atoms closer together would require energy because of the mutual repulsion of the protons. So the most stable arrangement of the covalently bonded hydrogens can also be described as an arrangement that has a minimum amount of energy and that is less reactive than are the individual atoms alone, each of which has an incompletely filled orbital in the K shell.

A carbon atom has a total of six electrons; two electrons fill its inner shell and four are in its outer L shell. Because the L shell can hold up to eight electrons, this atom can share electrons with up to four other atoms. Thus it can form four covalent bonds. When an atom of carbon reacts with four hydrogen atoms, a substance called methane (CH_4) forms, resulting from the overlapping of electron orbitals. Thanks to electron sharing, the outer shell of methane's carbon atom is filled with eight electrons, and the outer shell of each hydrogen atom is also filled. Thus four covalent bonds—each bond consisting of a shared pair of electrons—hold methane together.

Orientation of Bonds in Space

Not only the number, but also the spatial orientation, of bonds is important. The four filled orbitals around the carbon nucleus of methane distribute themselves in space so that the bonded hydrogens are directed to the corners of a regular tetrahedron with carbon in the center. Although the orientation of orbitals and shapes of molecules differ depending on the kinds of atoms and how they are linked together, it is essential to remember that all molecules occupy space and have three-dimensional shapes. The shapes of molecules contribute to their biological functions.

Multiple Covalent Bonds

A covalent bond is represented by a line between the chemical symbols for the atoms. Bonds in which a single pair of electrons is shared are called *single bonds* (for example, H-H, C-H). When four electrons (two pairs) are shared, the link is a *double bond* (C=C). In the gas ethylene ($H_2C=CH_2$), two carbon atoms share two pairs of electrons. *Triple bonds* (six shared electrons) are rare, but there is one in nitrogen gas (N-N), the chief component of the air we breathe. In the covalent bonds in these five examples, the electrons are shared more or less equally between the nuclei; consequently all regions of

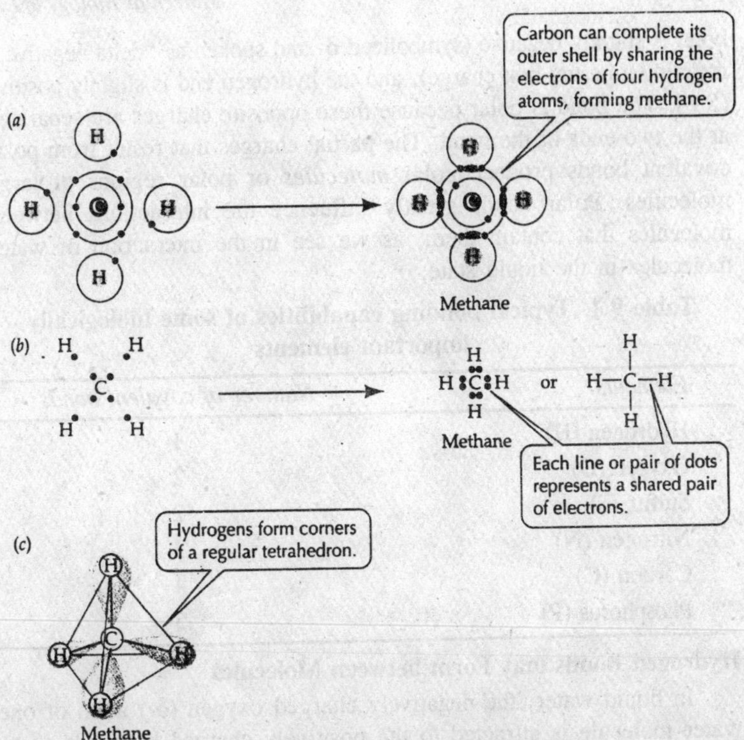

Fig. 9.4. Electrons are shared in covalent bonds.

the bonds are identical. However, when electrons are shared unequally in a covalent bond, regions of partial electric charge exist.

Uniqual Sharing of Electrons

So far we have discussed the covalent bonds that result from the equal sharing of electrons between two nuclei. Now we want to consider the kind of covalent bond that results from unequal sharing of the electrons.

Some atoms hold electrons to themselves more firmly than other atoms do. This characteristic is called *electronegativity*. Highly electronegative atoms that form covalent bonds include oxygen and nitrogen. When these atoms are covalently bonded to atoms with weaker electronegativity, such as carbon and hydrogen, the bonding pair of electrons is unequally shared between the two atoms, and the result is a *polar covalent bond*. For example, when oxygen is bonded to hydrogen, the bonding electrons spend much more time near the oxygen nucleus than near the hydrogen nucleus. Consequently, the oxygen end of the bond is slightly negative (symbolized δ- and spoken as "delta negative," meaning a partial unit charge), and the hydrogen end is slightly positive (S+). The bond is polar because these opposite charges are separated at the two ends of the bond. The partial charges that result from polar covalent bonds produce polar *molecules* or polar regions of large molecules. Polar bonds greatly influence the interactions between molecules that contain them, as we see in the interaction of water molecules in the liquid state.

Table 9.1. Typical bonding capabilities of some biologically important elements

Elements	*Number of covalent bonds*
Hydrogen (H)	1
Oxygen (O)	2
Sulfur (S)	2
Nitrogen (N)	3
Carbon (C)	4
Phosphorus (P)	5

Hydrogen Bonds may Form between Molecules

In liquid water, the negatively charged oxygen (δ-) atom of one water molecule is attracted to the positively charged hydrogen (δ+) of another water molecule. (Remember, negative charges attract positive charges.) The bond resulting from this attraction is called a *hydrogen bond* and is usually symbolized by a series of dots. Hydrogen bonds are not restricted to water molecules. They may form between any covalently bonded hydrogen and an electronegative atom, usually oxygen or nitrogen: -H···O- or -H···N-. Hydrogen bonds form between small molecules or between different parts of large molecules. Covalent bonds and polar covalent bonds, on the other hand, are always found *within* molecules.

A hydrogen bond is a weak bond; it has about onetwentieth (5 percent) of the strength of a covalent bond between a hydrogen atom and an oxygen atom. However, where many hydrogen bonds form, they have considerable strength and greatly influence the structure and properties of substances. Later in this chapter we'll discuss further how hydrogen bonding in water contributes to many of the properties of water that are significant for living systems. Hydrogen bonds also play important roles in determining and maintaining the three-dimensional shapes of giant molecules such as DNA and protein.

Ions Form Bonds by Electrical Attraction

When one interacting atom is much more electronegative than the other, a complete transfer of one or more electrons may take place. For example, a sodium atom has only one electron in its outermost shell; this condition is unstable. A chlorine atom has seven electrons in its outer shell, another unstable condition. The reaction between sodium and chlorine makes both atoms more stable. When the two atoms meet, the highly *electronegative chlorine atom takes the single unstable* electron from the sodium atom. The result is two electrically charged particles, called *ions*. Ions are electrically charged particles that form when atoms gain or lose one or more electrons.

The sodium ion (Na^+) has a +1 unit charge because it has one less electron than it has protons. The outermost electron shell of the sodium ion is full, with eight electrons, so the ion is stable. The chloride ion (Cl^-) has a –1 unit charge because it has one more electron than it has protons. This additional electron gives Clan outer shell with a stable load of eight electrons. Negatively charged ions are called *anions*; positively charged ions are called *cations*.

Some elements form ions with multiple charges by losing or gaining more than one electron to achieve a stable electron configuration in their outer shell. Examples are Ca^{2+} (the calcium ion, created from a calcium atom that has lost two electrons), Mg^+ (magnesium ion), and Al^{3+} (aluminum ion). Two biologically important elements each yield more than one stable ion: Iron yields Fee^+ (ferrous ion) and Fe^{3+} (ferric ion), and copper yields Cu^+ (cuprous ion) and Cue^+ (cupric ion). Groups of covalently bonded atoms that carry an electric charge are called *complex ions*; examples include NH_4^+ (ammonium ion), SO (sulfate ion), and PO_4^{2-} (phosphate ion).

Once they form, ions 'are usually stable, and no more electrons are lost or gained. As stable entities, ions can enter into stable associations through ionic bonding. Thus stable solids such as sodium chloride (NaCl) and potassium phosphate (K_3P0_4) are formed: Although a very complex solid, bone has as one of its major components the simple ionic compound $Ca_3(PO_4)_2$.

Ionic Bonds: Electrical Attraction

Ionic bonds are the bonds formed by electrical attractions between ions bearing opposite charges. In solids such as table salt (NaCl), the cations and anions are held together by ionic bonds. In solids, the ionic bonds are strong because the ions are close together. However, when ions are dispersed in water, the distance between them can be

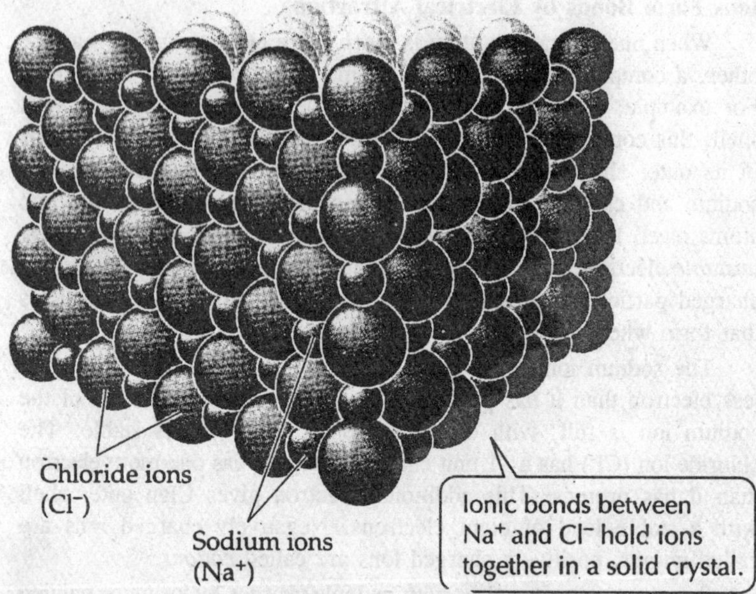

Fig. 9.5. Ionic binding in a solid.

large; the strength of their attraction is thus greatly reduced. Under the conditions that exist in the cell, an ionic bond is less than one-tenth as strong as a covalent bond that shares electrons equally, so an ionic bond can be broken much more readily than a covalent bond.

Not surprisingly, ions with one or more unit charges can interact with polar substances as well as with other ions. Such interaction results when table salt or any other ionic solid dissolves in water. The hydrogen bond that we described earlier is a weak type of ionic bond, because it is formed by electrical attractions. However, it is weaker than most ionic bonds because the hydrogen bond is formed by partial charges (δ- and δ+) rather than by whole unit charges (+1 unit, -1 unit).

Nonpolar Substances have no Attraction for Polar Substances

We have been discussing the bonds that result from electrical attractions between positive and negative charges (ionic bonds and hydrogen bonds). Now let's return to a brief consideration of substances that have "pure" covalent bonds.

These bonds form between atoms that have equal or nearly equal electronegativities-such as carbon and hydrogenwhich share the bonding electrons equally. Such bonds are abundant in the compounds of hydrogen

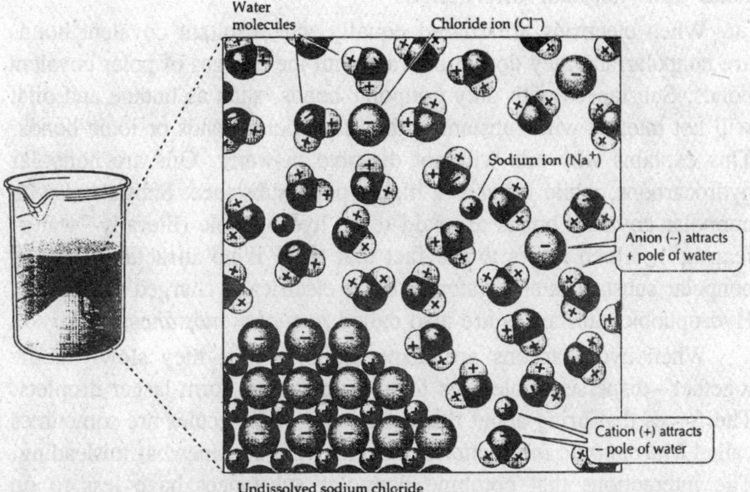

Fig. 9.6. Water molecules surrounded ions.

and carbon—the hydrocarbons. Molecules such as ethane (CH_3—CH_3) and butane (CH_3—CH_2—CH_2—CH_3) are small hydrocarbons, but in living systems, molecules exist with hydrocarbon chains consisting of 16 or more carbon atoms.

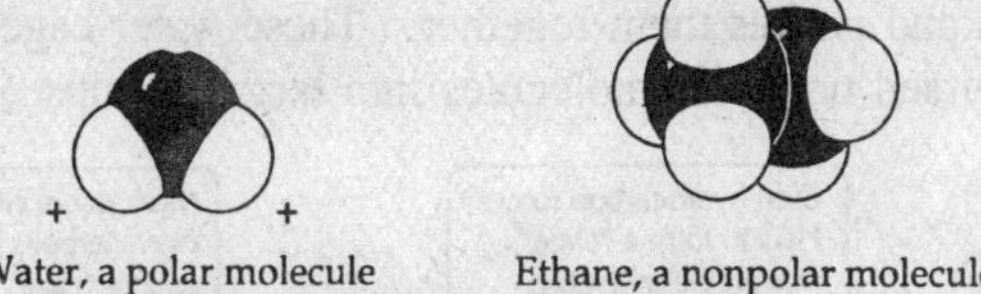

Fig. 9.7. Nonpolar molecules.

Attraction between Nonpolar Molecules

Nonpolar substances such as oils and fats show *van der Waals attractions* between molecules. These attractive forces operate only when nonpolar substances come very close to each other. The random variations in the electron distribution in one molecule create an opposite charge distribution in the adjacent molecule, and the result is a brief, weak attraction. Although each such interaction is brief and weak at any one site, the summation of many such interactions over the entire span of a nonpolar molecule can produce substantial attraction. Thus van der Waals interactions are important in holding together the long hydrocarbon chains that make up the inner portion of biological membranes. They also stabilize portions of the DNA double helix and the intricate folded structure of proteins.

Polar and Nonpolar Interactions

When electrons are shared equally, the resultant covalent bonds are nonpolar and they do not interact with the charges of polar covalent bonds. Substances with only nonpolar bonds, such as butane and oils, will not interact with substances that have polar bonds or ionic bonds. This explains why oils will not dissolve in water. Oils are nonpolar hydrocarbons, while water is a highly polar substance. Substances with nonpolar covalent bonds are said to be hydrophobic (literally "water-fearing"), which refers to the fact that there is no attraction between nonpolar substances and water or other electrically charged substances. Hydrophobic substances are also called *nonpolar substances*.

When hydrocarbons are dispersed in water, they slowly come together—dispersed molecules form droplets that form larger droplets. The forces that bring about this combining of molecules are sometimes called *hydrophobic interactions*, but this term is somewhat misleading. The interactions that combine nonpolar substances have less to do with forces between the nonpolar molecules than with the hydrogen bonding of the water that surrounds the molecules.

When nonpolar substances such as hydrocarbons are introduced into water, they cause a disruption in the usual hydrogen bonding between water molecules. In the vicinity of the hydrocarbon, the water molecules form a hydrogen-bonded "cage" that surrounds the nonpolar hydrocarbons and pushes them together. These water cages can bring together dispersed nonpolar molecules into larger groups.

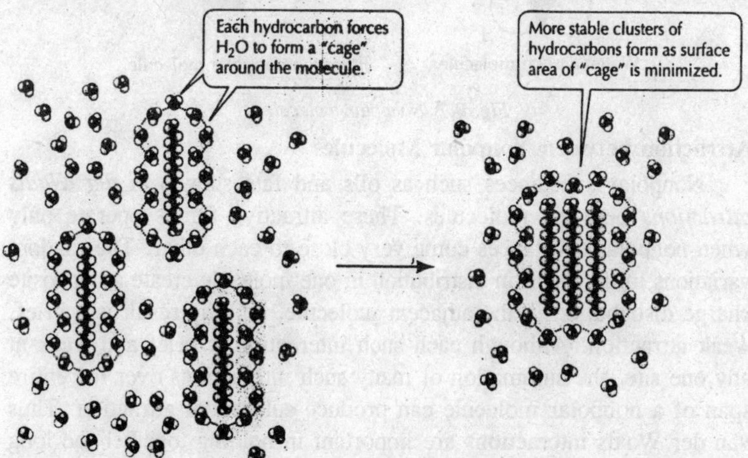

Fig. 9.8. Interaction of water and nonpolar substances.

Eggs by the Dozen, Molecules by the Mole

In modern biology, as in chemistry, the question "How much?" is as important as "What kind?" In this section, we will examine briefly how chemists and biologists deal quantitatively with atoms and molecules.

The molecular formula uses chemical symbols to identify different atoms, and subscript numbers to show how many atoms are present. For example, the molecular formula for methane is CH_4 (each molecule contains one carbon atom and four hydrogen atoms), that for oxygen gas is O_2, and that for sucrose (table sugar) is $C_{12}H_{22}O_{11.}$ The hormone insulin is represented by the molecular formula $C_25_4H_{377}N_{65}O_{76}S_6$! Although molecular formulas tell us what kinds of atoms and how many of each kind are present in the molecule, they tell us nothing about which atoms are linked to which. *Structural formulas* give us this information.

Each compound has a *molecular weight* (molecular mass): the sum of the atomic weights of the atoms in the molecule. The atomic weights of hydrogen, carbon, and oxygen are 1.008, 12.011, and 16.000, respectively. Thus the molecular weight of water (H_2O) is $(2 \times 1.008) + 16.000 = 18.016$, or about 18. What is the molecular weight of sucrose ($C_{12}H_{22}O_{11}$)? Your calculations should tell you that the answer is approximately 342. If you remember the molecular weights of a few representative biological compounds, you will be able to picture the relative sizes of molecules that interact with one another. Experiments require quantitative information, Suppose we want to compare how sodium chloride (NaCJ), potassium chloride (KCI), and lithium chloride (LiCI) affect a biological process. At first you might think we could simply give, say, 2 grams (g) of NaCl to one set of subjects, 2 g of KCI to another, and 2 g of LiCI to the third. But because the molecular weights of NaCl, KCI, and LiCI are different, 2-gram samples of each of these substances contain different numbers of molecules. The comparison would thus not be legitimate. Instead, we want to give *equal numbers of molecules* of each substance so that we can compare the activity of one molecule of one substance with that of one molecule of another. How can we measure out equal numbers of molecules?

We can Calculate Numbers of Molecules by Weighing

Measuring the number of molecules is essential to studying how many molecules take part in chemical reactions. Consider two large barrels, one filled with bolts and the other with pennies. How can we determine the number of units in each barrel? We could count them, but that would be tedious (and it is impossible to count individual molecules directly). However, if we know the weight of one bolt and one penny, we can calculate the number of units by dividing the total weight by the weight of each unit. The same principles apply to determining the number of molecules in a weighed sample.

To determine the number of molecules in a sample of a pure substance, we first determine the weight of the substance in grams, then we divide the grams by the relative weight of one molecule (the molecular weight defined earlier). Therefore, to measure out quantities of substances containing equal numbers of molecules, we weigh out in grams a quantity of each substance equivalent to its molecular weight. The molecular weight of NaCl is 58.45. Therefore 58.45 g of NaCl will have the same number of molecules as 74.55 g of KCI, whose molecular weight is 74.55.

If we divide the 74.55 g by the molecular weight of 74.55, the answer is 1. But one what? Surely not one molecule! The calculated numbers here are not the exact number, because we did not divide by the actual weight in grams of a molecule of potassium chloride (which would be about 1×10^{-22} g). Instead we divided by the relative weight. Therefore, instead of knowing the exact number of molecules, we know the relative number, which is called a *mole*. *One mole of a substance is an amount whose weight in grams is numerically equal to the molecular weight of the substance.* Potassium chloride (KCI) has a molecular weight of 74.55, so 1 mole of KCI weighs 74.55 g. Likewise, 1 mole of NaCl weighs 58.45 g, and 1 mole of LiCl, 42.40 g.

A mole of one substance contains the same number of molecules as does a mole of any other substance. This number, known as *Avogadro's number*, is 6.023×10^{23} molecules per mole. The concept of the mole is important for biology because it enables us to work easily with known numbers of molecules. Since we can neither weigh nor count individual molecules, we work with moles. The mole concept is analogous to the concept of a dozen or a gross. We buy some things, such as eggs, by the dozen (or another similar unit, such as the gross) because the individual items are inconvenient to count. Just as a dozen of anything contains twelve of that thing, a mole of any substance contains Avogadro's number of molecules of that substance.

Reactions Take Place in Solutions

In cells and in the laboratory, reactions take place in solutions formed when substances (solutes) are dissolved in water (the solvent).

The amount of substance dissolved in a given amount of water is the concentration of the solution. Knowing the concentrations of solutions is essential in performing experiments and interpreting their results.

A solution that contains 1 mole of solute per liter of solution has a one-molar concentration, abbreviated 1 *M*. A solution containing a half mole per liter is referred to as 0.5 *M*, or half-molar. How would you make 100 milliliters (ml) of a 0.5 *M* sucrose solution? The molecular weight of sucrose is 342, so 1 liter (1,000 ml) of a *1 M* sucrose solution contains 1 mole, or 342 g, of sucrose. You were asked to make just 100 ml of 0.5 *M* solution. Since 34.2 g of sucrose would make 100 ml of *1 M* sucrose, to make 100 ml of a 0.5 *M* sucrose solution, you would use 0.5 × 34.2 g = 17.1 g of sucrose.

Chemical Reactions: Atoms Change Partners

When atoms combine or change bonding partners, a chemical reaction is occurring. Consider the combustion reaction that takes place in the flame of a propane stove. When propane (C_3H_8) reacts with oxygen gas (O_2), the carbon atoms become bonded to oxygen atoms instead of to hydrogen atoms, and the hydrogen atoms become bonded to oxygen instead of to carbon. As the covalently bonded atoms change bonding partners, the composition of the matter changes, and propane and oxygen gas become carbon dioxide and water. This chemical reaction can be represented by the balanced equation

$$C_3H_8 + 5O_2 \rightarrow 3CO_2 + 4H_2O$$

In this equation, the propane and oxygen are the *reactants*, and the carbon dioxide and water are the *products*. The arrow symbolizes the chemical reaction. The numbers preceding the molecular formulas balance the equation and indicate how many molecules react or are produced. In this and all other chemical reactions, matter is neither created or destroyed. The total number of carbons on the left equals the total number on the right. However, there is another product of this reaction: energy.

The heat of the stove's flame and its blue light reveal that the reaction of propane and oxygen releases a great deal of energy. *Energy* is defined as the capacity to do work, but on a more intuitive level, it can be thought of as the capacity to change. Chemical reactions do not create or destroy energy, but *changes* in energy usually accompany chemical reactions. The energy released as heat and light was present in the reactants in another form, called *potential chemical energy*. In some chemical reactions, energy must be supplied from the environment (for example, some substances will react only after being heated), and some of this supplied energy becomes stored as potential chemical energy in the bonds formed in the reactants.

We can measure the energy associated with chemical reactions using the unit called a calorie (cal). A calorie is the amount of heat energy needed to raise the temperature of 1 g of pure water from 14.5°C to 15.5°C. (The nutritionist's Calorie, with a capital C, is what biologists call a kilocalorie (kcal) and is equal to 1,000 heat-energy calories.) Another unit of energy that is increasingly used is the joule (J). When you compare data on energy, always compare joules to joules and calories to calories. The two units can be interconverted: 1 J = 0.239 cal, and 1 cal = 4.184 J. Thus, for example, 486 cal = 2033 J, or 2.033 kJ. Although defined in terms of heat, the calorie and the joule are measures of any form of energy-mechanical, electric, or chemical.

Within living cells, chemical reactions called oxidation-reduction reactions take place that have much in common with this combustion of propane. The fuel for these biological reactions is different (the sugar glucose, rather than propane), and the reactions proceed by many intermediate steps that permit the energy released from the glucose to be harvested and put to use by the cell. But the products are the same: carbon dioxide and water. We will present and discuss oxidation-reduction reactions and several other types of chemical reactions that are prevalent in living systems in the chapters that follow.

WATER: STRUCTURE AND PROPERTIES

Water, like all other matter, can exist in three states: solid (ice), liquid, and gas (vapour). Liquid water is the medium in which life originated on Earth more than 3.5 billion years ago, and it is in water that life evolved for about a billion years. Today water covers three-fourths of Earth's surface, and all active organisms contain between 45 and 95 percent water. No organism can remain biologically active without water, which is important in both the outer and inner environments. Within cells, water participates directly in many chemical reactions, and it is the medium (or solvent) in which most reactions take place. In this section we will consider the structure and interactions of water molecules, exploring how these generate properties essential to life.

Water has a Unique Structure and Special Properties

The shape of a water molecule is determined by the distribution in space of the four pairs of electrons in the outer shell of the oxygen atom. Each pair of electrons is confined to an orbital. Two of these pairs are bonded to hydrogens, but the other two pairs (nonbonding pairs) also influence shape. Because of their negatively charged electrons, the four orbitals, each containing a pair of electrons, repel one another. In seeking to be as far apart as possible, they give the water molecule a nearly tetrahedral shape.

The shape of the water molecule, its polar nature, and the formation of hydrogen bonds between water molecules or with other substances give water its unusual properties. For example, ice floats, and compared to other liquids, water is an excellent solvent, making it an ideal medium for biochemical reactions. Water is both cohesive (sticking to itself) and adhesive (sticking to other things). And the energy changes that

Ice Floats

In its solid state (ice), water is held by its hydrogen bonds in a rigid, crystalline structure in which each water molecule is hydrogen-bonded to four other molecules. Although these molecules are held firmly in place, they are not as tightly packed as they are in liquid water. In other words, *solid water is less dense than liquid water,* which is why ice floats in water. If ice sank in water, as almost all other solids do in their corresponding liquids, ponds and lakes would freeze from the bottom up, becoming solid blocks of ice in winter and killing most of the organisms living in them. Once the whole pond had frozen, its temperature could drop well below the freezing point of water. However, because ice floats, it forms a protective insulating layer on the top of the pond, reducing heat flow to the cold air above. Thus fish, plants, and other organisms in the pond can survive the winter at temperatures no lower than 0°C, the freezing point of pure water.

Melting and Freezing

Compared to other nonmetallic substances of the same size, ice requires a great deal of heat energy to melt. Melting 1 mole of water molecules requires the addition of 5.9 kJ of energy. This value is high because more than a mole of hydrogen bonds must be broken for 1 mole of water to change from solid to liquid. In the opposite process, freezing, a great deal of energy must be lost for water to transform from liquid to solid. These properties help make water a moderator of temperature changes.

Another property of water that moderates temperature is the high heat capacity of liquid water. *Heat capacity* is the amount of heat energy that is required to raise the temperature of a substance 1°C.

Raising the temperature of liquid water takes a relatively large amount of heat. The temperature of a given quantity of water is raised only 1°C by an amount of heat that would increase the temperature of the same quantity of ethyl alcohol by 2°C, or of chloroform by 4°C. This phenomenon contributes to the surprising constancy of the temperature of the oceans and other large bodies of water through the seasons of the year. The temperature changes in coastal land masses are also moderated by large bodies of water. Indeed, water helps minimize variations in atmospheric temperature throughout the planet.

Cohesion and Surface Tension

In liquid water, the molecules are free to move about. The hydrogen bonds between the water molecules continually form and break. In other words, liquid water has a dynamic structure. On the average, every water molecule forms 3.4 hydrogen bonds with other water molecules. This number represents fewer bonds than exist in ice, but it is still a high number.

These hydrogen bonds explain the cohesive strength of water. The *cohesive strength* of water is what permits narrow columns of water to stretch from the roots to the leaves of trees more than 100 meters high. When water evaporates from leaves, the entire column moves upward in response to the pull of the molecules at the top.

Water also has a high *surface tension*, which means that the surface of water exposed to the air is difficult to puncture. The water molecules in this surface layer are hydrogen-bonded to other water molecules below. The surface tension of water permits a container to be filled slightly above its rim without overflowing, and it permits small animals to walk on the surface of water.

Evaporation and Cooling

Water has a high *heat of vaporization*; which means a lot of heat is required to change water from its liquid state to its gaseous state (the process of evaporation). This heat is absorbed from the environment in contact with the water. Evaporation thus has a cooling effect on the environment-whether a leaf, a forest, or an entire land mass. This effect explains why sweating cools the human body: As the sweat evaporates off the skin, it takes with it some of the adjacent body heat.

Water Molecules Sometimes Form Ions

The water molecule has a slight but significant tendency to come apart into a hydroxide ion (OH^-) and a hydrogen ion (a proton, H^+). Actually, *two* water molecules participate in ionization.

The hydronium ion is in effect a. hydrogen ion bound to a water molecule. For simplicity, biochemists tend to use a modified representation of the ionization of water:

$$H_2O \rightarrow H^+ + OH^-$$

Even though only about one water molecule in 500 million is ionized at any given time, the transformation is significant because H^+ and OH^- ions participate in many important biochemical reactions.

Acids, Bases, and the pH Scale

The ionization of water is very important for all living creatures. This fact may seem surprising, since the ionization is so slightonly one ionization out of 5×10^8 water molecules. But we are less surprised if we focus on the abundance of water in living systems and the reactive nature of H^+ produced by that ionization. Remember that H^+ is a proton, a tiny bit of charged matter-smaller than any atom or molecule in the cell.

Usually it attaches to another water molecule, but in the cell it can attach to other molecules and substantially change their properties. Because acids and bases donate and accept H^+, they can profoundly alter the water environment in which the chemical reactions and processes of life take place.

In this section we'll examine acids, bases, and their measurement using the pH scale. We'll close by looking at how buffers limit the changes in pH.

Acids Donate H^+, Bases Accept H^+

In pure water, the concentration of hydrogen ions exactly equals that of hydroxide ions (OH^-), and this "solution" is said to be *neutral*. Now suppose we add some HCl (hydrochloric acid). As it dissolves, the HCl ionizes, releasing H^+ and Cl^- ions:

$$HCl \rightarrow H^+ + Cl^-$$

Now there are more H^+ than OH^- ions. Such a solution is *acidic*. A *basic*, or alkaline, solution is one in which there are more OH^- than H^+ ions. A basic solution can be made from water by adding, for example, sodium hydroxide (NaOH), which ionizes to yield OH^- and Na^+ ions, thus making the concentration of OH^- ions greater than that of H^+ ions.

$$NaOH \rightarrow Na^+ + OH$$

An acid is any compound that can *release* H^+ ions in solution. HCl is an acid, as is H_2SO_4 (sulfuric acid). One molecule of sulfuric acid may ionize to yield two H+ ions and one $S04^-$ ion. Biological compounds such as acetic acid and pyruvic acid, which contain -COOH (the carboxyl group) are also acids, because —COOH → -COO⁻ + H⁺.

Bases are compounds that can *accept* H^+ ions. These include the bicarbonate ion (HCO_3^-), which can accept a H^+ ion and become carbonic acid (H_2CO_3); ammonia (NH_3), which can accept a H+ ion and become an ammonium ion (NH_4^+); and many others.

Note that although —COOH is an acid, $—COO^-$ is a base, because $-COO^- + H^+ \rightarrow —COOH$. Acids and bases exist as pairs, such as —COOH and $—COO^-$, because any acid becomes a base when it releases a proton, and any base becomes an acid when it gains a proton.

You may have noticed that the two reactions just discussed are the opposites of each other. The reaction that yields $—COO^-$ and H^+ is reversible and may be expressed as

$$—COOH \rightleftharpoons —COO^- + H^+$$

A *reversible reaction* is one that can proceed in either direction-left to right or right to left-depending on the relative starting concentrations of reacting substances and products. In principle, all chemical reactions are reversible.

pH is the Measure of Hydrogen Ion Concentration

The terms "acidic" and "basic" refer only to *solutions*. How acidic or basic a solution is depends on the relative concentrations of H^+ and OH^- ions in it. "Acid" and "base" refer to *compounds* and *ions*. A compound or ion that is an acid can donate H^+ ions; one that is a base can accept H^+ ions.

How do we specify how acidic or basic a solution is? First, let's look at the H^+ ion concentrations of a few contrasting solutions. In pure water, the H+ concentration is 10^{-7} *M*. In 1 M hydrochloric acid, the H^+ concentration is 1 *M;* and in 1 *M* sodium hydroxide, the H^+ concentration is 10^{-14} *M*. Because its values range so widely-from more than 1.0 *M* to less than 10^{-14} M the H^+ concentration itself is an inconvenient quantity. It is easier to work with the logarithm of the concentration, because logarithms compress this range. We indicate how acidic or basic a solution is by its pH (a term derived from "potential of Hydrogen"). The pH value is defined as the negative logarithm of the hydrogen ion concentration in moles per liter (molar concentration). In chemical notation, molar concentration is often indicated by putting brackets around the symbol for a substance; thus $[H^+]$ stands for the molar concentration of H^+. The equation for pH is

$$pH = -\log_{10}[H^+]$$

Since the H^+ concentration of pure water is 10^{-1} *M,* its pH is -log(10^{-7}) = –(–7), or 7. A smaller negative logarithm means a larger number. In practical terms, a lower pH means a higher H^+ concentration, or greater acidity. In 1 *M* HCl, the H^+ concentration is 1 *M,* so the pH is the negative logarithm of 1 (–log 10^0), or 0. The pH of 1 *M* NaOH is the negative logarithm of 10^{-14}, or 14. A solution with a pH of less than 7 is acidic: It contains more H^+ ions than OH^- ions. A solution with a pH of 7 is neutral. And a solution with a pH value greater than 7 is basic. Because the pH scale is logarithmic, the values are exponential: A solution with a pH of 5 is 10 times more acidic than one with a pH of 6 (it has ten times as great a concentration of H^+); a solution with a pH of 4 is 100 times more acidic than one with a pH of 6.

Buffers Minimize pH Changes

An organism must control the chemistry of its cellsin particular, the pH of the separate compartments within cells. Animals must also control the pH of their blood. The normal pH of human blood is 7.4, and deviations of even a few tenths of a pH unit can be fatal. The control of pH is made possible in part by *buffers*, systems that maintain a relatively constant pH even when substantial amounts of acid or base are added. A buffer is a mixture of an acid that does not ionize completely in water and its corresponding base-for example, carbonic acid (H_2CO_3) and bicarbonate ions (HCO_3^-). If acid is added to this buffer, not all the H+ ions from the acid stay in solution. Instead, many of the added H^+ ions combine with bicarbonate ions to produce more carbonic acid, thus using up some of the H+ ions in the solution and decreasing the acidifying effect of the added acid:

$$HCO_3 + H^+ \rightleftharpoons H_2CO_3$$

If base is added, the reaction reverses. Some of the carbonic acid ionizes to produce bicarbonate ions and more H^+, which counteracts some of the added base.

In this way, the buffer minimizes the effects of added acid or base on pH. A given amount of acid or base causes a smaller change in pH in a buffered solution than in an unbuffered one. Buffers illustrate the reversibility of chemical reactions: Addition of acid drives the reaction in one direction; addition of base drives it in the other.

Properties of Molecules

Molecules vary in size. Some are small, such as H_2 and CH_4. Others are larger, such as a molecule of table sugar (sucrose), which has 45 atoms. Still other molecules, such as proteins, are *gigantic*, sometimes containing tens of thousands of *atoms bonded* together in specific ways. Whether large, *medium*, or small, most of the molecules in living systems contain carbon atoms and are thus referred to as *organic molecules*. Most organic molecules include hydrogen and oxygen atoms, and many also include nitrogen and phosphorus.

All molecules have a specific three-dimensional shape. For example, the orientation of the bonding orbitals around the carbon atom gives the molecule of methane (CH_4) the shape of a regular tetrahedron, while in carbon dioxide (CO_2) the three atoms are in line with one another. Larger molecules have specific complex shapes that result from the number and kinds of atoms present and the ways in which these atoms are linked together. Some large molecules have compact ball-like shapes. Others are long, thin, ropelike structures. The shapes relate to the roles these molecules play in living cells.

In addition to size and shape, molecules have certain properties that characterize them and determine their biological roles. Chemists can use the characteristics of composition, structure (three-dimensional shape), reactivity, and solubility to distinguish a sample of one pure compound from another. For example, a compound such as sugar is soluble in water, which means that the solid disperses to form a uniform homogeneous mixture, but the same compound is insoluble in oil (the solid sugar remains a solid), and the resulting mixture is heterogeneous. Another substance, such as butane, is insoluble in water but highly soluble in oil. These solubility differences are due to the different atoms present and to their arrangement in these two kinds of molecules.

The presence of polar or charged sites on a molecule plays important roles in determining the molecule's solubility in water. Such sites can also determine the kinds of chemical reactions in which the molecule participates. The sizes, shapes, solubilities, and reactivities of molecules are significant to understanding the structures and operations of cells and organisms. Certain groups of atoms found together in a variety of molecules simplify our understanding of the reactions that molecules undergo.

Functional Groups give Specific Properties to Molecules

On the basis of atomic composition, structure, and reactivity, we can distinguish families of molecules from other families. Each member within a family of compounds has some characteristics in common with the other members of that family but differs from them in other characteristics. For example, organic acids are a family of carbon compounds that are all acidic but differ in other ways. They all contain a characteristic group of atoms called the *carboxyl group*,

$$\begin{array}{l}\quad O \\ \quad \| \\ -C-OH, \text{ or } COOH\end{array}$$

that is the source of the H^+ that defines an acid:

$$\begin{array}{ccc}\quad O & & \quad O \\ \quad \| & & \quad \| \\ -C-OH & \rightarrow & -C-O^- \; + H^+\end{array}$$

The carboxyl group is a functional group. Functional groups are groups of atoms that are part of a larger molecule and have particular reactive characteristics. The same functional group may be part of very different molecules. In addition to the carboxyl group, you will encounter several other functional groups in your study of biology.

Several classes of biologically important compounds are defined by the functional groups they contain. When the functional group is a hydroxyl group (—OH), the product is an alcohol. Perhaps the most familiar alcohol is ethanol (also called ethyl alcohol, CH_3CH_2OH). Small alcohols like ethanol are soluble in water because of the hydrogen bonding possible between the polar hydroxyl group and water molecules, but larger alcohols are not soluble in water because of their long hydrocarbon chains.

Sugars contain both hydroxyl and carbonyl groups. The carbonyl *group* has a central carbon atom with a double bond to an oxygen atom (C=O). If one of the other two bonds of the carbon atom in a carbonyl group is attached to a hydrogen atom, the compound is an *aldehyde*.

Carboxyl groups are found in organic acids. Some organic bases, called amines, possess an *amino group* ($—NH_2$), which has a tendency to react with H^+ to produce the positively charged ammonium group ($—NH_3^+$). This H^+-accepting characteristic accounts for the classification of amines as bases.

Amino acids are important compounds that possess both a carboxyl group and an amino group attached to the same carbon atom, the a (alpha) carbon. Also attached to the a carbon atom are a hydrogen atom and a side chain designated by the letter R. Different side chains have different chemical compositions, structures, and properties. Each of the 20 amino acids found in proteins has a different side chain that gives it its distinctive chemical properties. Because they possess both carboxyl and amino groups, amino acids are simultaneously acids and bases. At the pH values commonly found in cells, both the carboxyl and the amino groups are ionized: The carboxyl group has lost α proton, and the amino group has gained one.

Two other functional groups deserve our attention: the sulfhydryl group and the phosphate group. The *sulfhydryl group* (—SH) is an important constituent of the side chains of two amino acids. When two of these sulfhydryl groups react, the two hydrogen atoms are lost and a covalent bond, called a disulfide bond, is formed between the sulfur atoms (—S—S—). As we'll see in the next chapter, disulfide bonds help maintain the three-dimensional structure of proteins that is required for normal protein functioning.

The *phosphate group* ($—OPO_3^-$) is found on many different kinds of molecules. It plays important structural roles in DNA and RNA and participates in reactions that transfer energy. The removal (hydrolysis) of phosphate groups from some molecules releases energy that can be used to fuel other energy-requiring reactions

Isomers have Different Arrangements of the Same Atoms

Isomers are compounds that have the same chemical formula but different arrangements of the atoms. (The prefix "iso-" means "same and is encountered in many technical terms.) Of the different kinds of isomers, we will consider two: structural isomers and optical isomers.

Structural isomers are isomers that differ in how the atoms are joined together. Consider two simple molecules, each composed of four carbon and ten hydrogen atoms bonded covalently, with the formula C_4H_{10}. These atoms can be linked in two alternative ways in which carbon can form four bonds and hydrogen one bond. These two forms are called butane and isobutane. Their different bonding relationships are distinguished in structural formulas, and they have different chemical properties.

```
      H  H                     CH3
      |  |                      |
H3C—C—C—CH3               H3C—C—CH3
      |  |                      |
      H  H                      H
     Butane                 Isobutane
```

Many molecules of biological importance, particularly the sugars and amino acids, have optical *isomers*. Optical isomers (also called enantiomers) are related to each other in the way an object is related to its mirror image. Optical isomers occur whenever a carbon atom has four *different* atoms or groups attached to it. These instances allow two different ways of making the attachments, each the mirror image of the other. Such a carbon atom is an asymmetric carbon, and the pair of compounds are optical isomers of each other. Your right and left hands are optical isomers. Just as a glove is specific for a particular hand, so some biochemical molecules can interact with a specific optical isomer of a compound but are unable to "fit" the other.

The a carbon in an amino acid is an asymmetric carbon because it is bonded to four different groups. Therefore, amino acids exist in two isomeric forms, called D-amino acids and 1.-amino acids. "D" and "L" are abbreviations for right and left, respectively. Only L-amino acids are commonly found in most proteins of living things.

The compounds discussed in this chapter include some of the more common ones found in organisms.

Between these small molecules and the world of the living stands another level, that of the giant macromolecules. These huge molecules—the proteins, lipids, carbohydrates, and nucleic acids—are the subject of the next chapter.

10

Large Molecules in Life

The lives of cells are dances and dramas with tens of thousands of different kinds of molecules. Their dramatic choreography is what scientists reveal as they investigate the molecules of living systems, their physical properties, and their chemical reactions. What molecules are present? What are their chemical structures and properties? And what biological functions do these molecules perform? In different ways, we will be concerned with these questions throughout much of this book.

In this chapter, we'll look at gigantic molecules called macromolecules (*macro-*, "large") and the subunits from which they are constructed. Macromolecules may contain hundreds or thousands of atoms and have very large molecular weights. For example, human hemoglobin has a molecular weight of 64,500 and contains more than 6,000 atoms. The molecular complexity of such a structure can be understood in terms of the structures and properties of its subunits, which are fewer in number and easier to understand. Macromolecules are chains of small, individual units called monomers (*mono-*, "one"; *-mer,* "unit") covalently bonded together to form a polymer (*poly-*, "many").

In living systems there are three types of macromolecules: polysaccharides, proteins, and nucleic acids. *Polysaccharides* are constructed from sugar monomers such as glucose; *proteins* are constructed from amino acids; and *nucleic acids* are composed of nucleotides. Another important group of molecules that can form large structures is the *lipids*. But because the individual components (the lipids) do not covalently bond together, the resulting large structures are not true macromolecules.

The monomers that form the basis of polysaccharides, proteins, and nucleic acids are identical in different species. For example, a molecule of glucose in human blood is identical to a molecule of glucose from a reptile or a cabbage plant. The same group of 20 different amino acids is used to construct proteins for all living things—from bacteria to birds, bats, and humans. However, this cannot always be said of the larger molecular forms built from these monomers.

Proteins and nucleic acids in particular are informational macromolecules that are different in different species. For example, human hemoglobin is different from the hemoglobin found in fish or birds. No animal acquires its macromolecules directly from its food. Instead, it uses the subcomponents of its food to construct new macromolecules suited to its unique needs. For example, we eat proteins constructed by other animals and plants, but we break these proteins down into their amino acids and then reassemble the amino acids into the chemically different proteins of our own bodies. This process is like picking up Lego toys that somebody else has made, taking them apart, and putting the parts together again to make the toys we want.

Before we turn to the main focus of this chapter—the macromolecules: carbohydrates, proteins, and nucleic acids—let's first take a look at lipids and the large macromolecule-like structures that they form.

Lipids: Water-Insoluble Molecules

Lipids are a chemically diverse group of hydrocarbons. The property they all share is an insolubility in water that is due to the presence of many nonpolar covalent bonds. Nonpolar molecules can associate together and form massive structures, but these structures are not considered macromolecules, because the bonds between the separate molecules are not covalent bonds. These nonpolar hydrocarbon molecules are literally pushed together by surrounding water molecules, which are not attracted to the nonpolar substance. When the nonpolar molecules are sufficiently close together, weak but additive van der Waals forces hold them together. Although insoluble in water, lipids are soluble in other lipids or in nonpolar solvents such as ether or benzene.

In addition to their diverse chemical structures, lipids have many different biological roles. Some of them store energy (the fats and oils). Others play important structural roles in cell membranes (phospholipids). The carotenoids help plants capture light energy, and

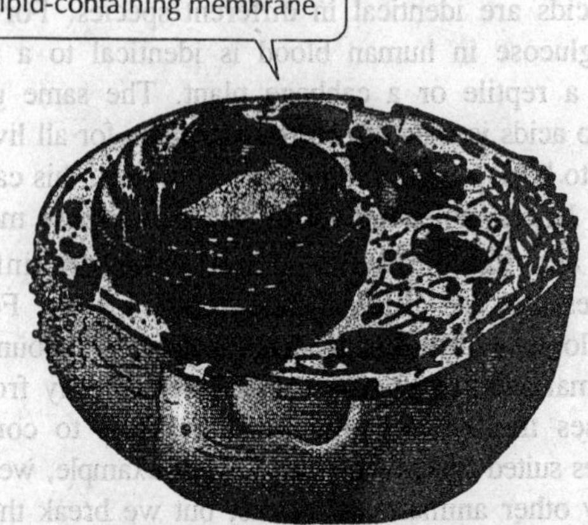

Fig. 10.1. Lipids from cellular membranes.

the steroids and some lipids play regulatory roles. We'll begin our discussion by considering the structures and properties of the fats and oils.

Fats and Oils Store Energy

Chemically, fats and oils are triglycerides, also known as *simple lipids*. Triglycerides that are solid at room temperature (20°C) are called fats; those that are liquid at room temperature are called *oils*.

Triglycerides are composed of two types of building blocks: fatty acids and glycerol. *Glycerol* is a small molecule with three hydroxyl (—OH) groups. Fatty *acids* have long nonpolar hydrocarbon tails and a polar carboxyl functional group (—COOH). Fatty acids can be of different lengths (usually 16 to 20 carbon atoms) and may be either Theturated or unsaturated.

In *saturated fatty acids*, all the bonds between the carbon atoms in the hydrocarbon chain are single bonds-there are no double bonds. That is, all the bonds are *saturated* with hydrogen atoms. Saturated fatty acids include palmitic acid, which has 16 carbon atoms, and stearic acid, which has 18. These molecules are relatively rigid and straight, and they pack together tightly, like pencils in a box.

In *unsaturated fatty acids*, the hydrocarbon chain contains one or more double bonds. Oleic acid is a monounsaturated fatty acid that has 18 carbon atoms, and one double bond near the middle of the hydrocarbon chain causes a kink in the molecule. Fatty acids, such as

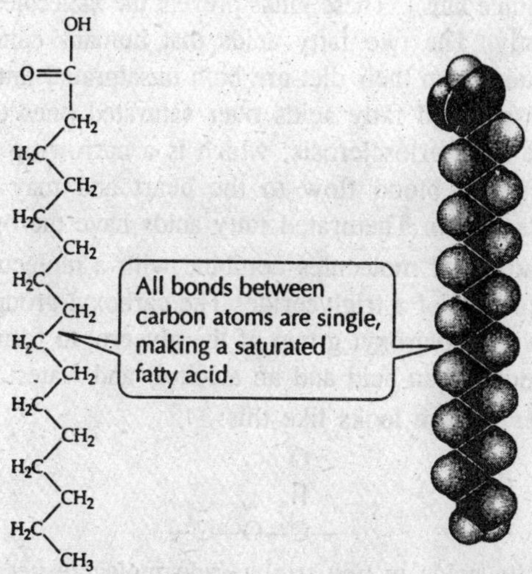

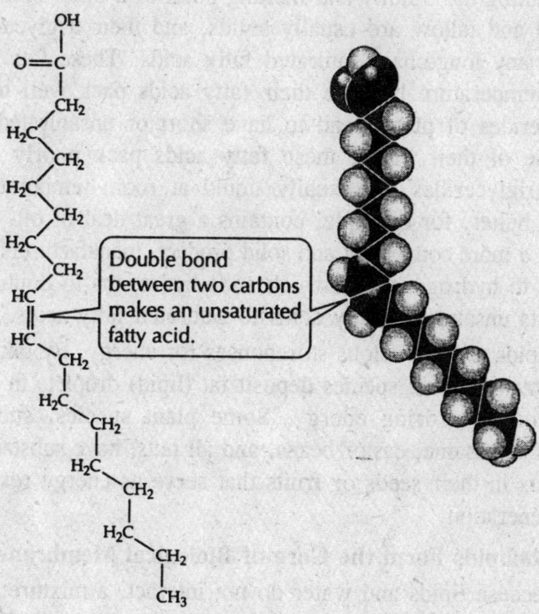

Fig. 10.2. Fatty acids.

linoleic acid, that have more than one double bond are polyunsaturated and have multiple kinks. These kinks prevent the molecules from packing together tightly. The two fatty acids that humans cannot synthesize and must obtain from their diet are both unsaturated fatty acids. Diets favouring unsaturated fatty acids over saturated ones tend to reduce the incidence of arteriosclerosis, which is a narrowing of the arterial wall that restricts blood flow to the heart and may lead to heart attacks. Diets rich in Theturated fatty acids have the opposite effect.

Three fatty acid molecules combine with a molecule of glycerol to form a molecule of a triglyceride. The carboxyl group of each fatty acid reacts with a hydroxyl group of the glycerol to form an *ester* (the reaction product of an acid and an alcohol) and water.

The ester linkage looks like this:

$$\begin{array}{c} O \\ \| \\ -C-O-C- \end{array}$$

The three fatty acids in one triglyceride molecule need not all have the same length, nor do they all have to be either saturated or unsaturated. The kinks associated with double bonds are important in determining the fluidity and melting point of a lipid. Animal fats such as lard and tallow are usually solids, and their triglycerides tend to have many long-chain saturated fatty acids. These fats are solids at room temperature because their fatty acids pack well together. The triglycerides of plants tend to have short or unsaturated fatty acids. Because of their kinks, these fatty acids pack poorly together and these triglycerides are usually liquid at room temperature. Natural peanut butter, for example, contains a great deal of oil. However, to market a more convenient and solid product, manufacturers often subject the oil to hydrogenation, which adds hydrogens to double bonds and converts unsaturated fatty acids to saturated fatty acids.

Lipids are marvelous storehouses for energy. By taking in excess food, many animal species deposit fat (lipid) droplets in their cells as a means for storing energy. Some plant species, such as olives, avocados, sesame, castor beans, and all nuts, have substantial amounts of lipids in their seeds or fruits that serve as energy reserves for the next generation.

Phospholipids Form the Core of Biological Membranes

Because lipids and water do not interact, a mixture of water and lipids forms two distinct phases. Many biologically important substances—such as ions, sugars, and free amino acids—that are soluble in water are insoluble in lipids. These two properties—water solubility and lipid insolubility—have proved essential to distinguishing the interior of cells from their external environment.

Suppose that you must design water-filled compartments, separated from each other and from their environment by barriers that limit the passage of materials. Given the properties of lipids, a seemingly effective way to accomplish this task is to use membranes that contain a special class of lipid called the phospholipid. This is the system that has evolved in nature. Molecular traffic within an organism or into and out of its compartments is constrained by the properties of the lipid portion of the surrounding membrane. Compounds that dissolve readily in lipids can move rapidly through biological membranes, but compounds that are insoluble in lipids are prevented from passing through the membrane or must be transported across the membrane by specific proteins.

Like triglycerides, phospholipids have fatty acids bound to glycerol by ester linkages. In phospholipids, however, any one of several phosphate-containing compounds may replace one of the fatty acids. Many phospholipids are important constituents of biological membranes. If you think carefully about the structure and properties of phospholipids, you will find it easy to understand how they are oriented in membranes. The phosphate functional group has negative electric charges, so this portion is hydrophilic, attracting polar water molecules. But the two fatty acids are hydrophobic, so they are pushed together by water.

In a biological membrane, phospholipids line up in such a way that the nonpolar, hydrophobic "tails" pack tightly together to form the interior of the membrane, and the phosphate-containing "heads" face outward (some to one side of the membrane and some to the other), where they interact with water, which is excluded from the interior of the membrane. The phospholipids thus form a bilayer, a sheet two molecules thick. Because membranes are so important.

Because the word "lipid" defines compounds in terms of their solubility rather than their structural similarity, a great variety of different chemical structures are included as lipids. The next two lipid classes we'll discuss—the carotenoids and the steroids—have chemical structures very different from the structures of triglycerides and phospholipids and from the structures of each other.

Carotenoids Trap Light Energy

The *carotenoids* are a family of light-absorbing pigments found in plants and animals. Betacarotene (β-carotene) is one of the pigments that traps light energy in leaves during photosynthesis. It is the β-carotene in plants that senses light and causes their parts to grow toward or away from the light. In humans, a molecule of β-carotene can be broken down into two vitamin A molecules, from which we make the pigment rhodopsin, which is required for vision. Carotenoids are responsible for the colour of carrots, tomatoes, pumpkins, egg yolks, and butter.

Steroids are Signal Molecules

The *steroids* are a family of organic compounds whose multiple rings share carbons. Some steroids are important constituents of membranes. Others are hormones, chemical signals that carry messages from one part of the body to another. *Testosterone* and the *estrogens* are steroid hormones that regulate sexual development in vertebrates. *Cortisol* and related hormones play many regulatory roles in the digestion of carbohydrates and proteins, in the maintenance of salt balance and water balance, and in sexual development.

Cholesterol is synthesized in the liver and contributes to the structure of some cellular membranes. It is the starting material for making testosterone and other steroid hormones, as well as the bile salts that help break down dietary fats so that they can be digested.

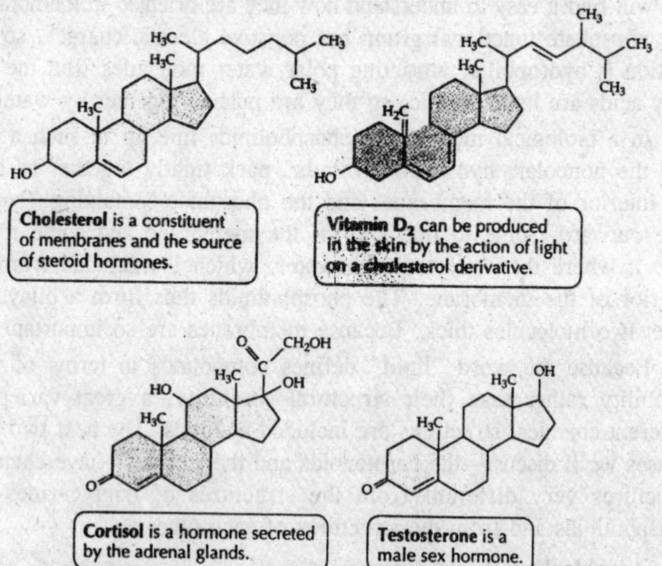

Fig. 10.3. All steroids have the same ring structures.

We absorb cholesterol from foods such as milk, butter, and animal fats. When we have too much cholesterol in our blood, it is deposited in our arteries (along with other substances), a condition that may lead to arteriosclerosis and heart attack.

Some Lipids are Vitamins

A large group of fat-soluble substances, including the carotenoids and steroids, are synthesized by covalent linking and chemical modification of isoprene to form a series of isoprene units:

$$CH_2 = \underset{\displaystyle |}{\overset{\displaystyle CH_3}{C}}\!-\!CH = CH_2$$

The fat-soluble vitamins (A, D, E, and K) are formed in this manner by plants and bacteria. These substances are not synthesized by humans and must be acquired from dietary sources. *Vitamin A* is formed from β-carotene found in green and yellow vegetables. Among other roles, vitamin A is directly involved in the reception of light by our eyes. Deficiency in vitamin A leads to dry skin, eyes, and internal body surfaces; retarded growth and development; and night blindness, which is a diagnostic symptom for the deficiency. *Vitamin D* regulates the absorption of calcium from the intestines. It is necessary for the proper deposition of calcium in bones; a deficiency of vitamin D can lead to rickets, a bonesoftening disease.

Vitamin E is not a single vitamin, but a group of related lipids that seem to protect cells from damaging effects of oxidation-reduction reactions. These lipids appear to have an important role in preventing unhealthy changes in the double bonds in the unsaturated fatty acids of membrane phospholipids. Commercially, vitamin E is added to some foods to slow spoilage. *Vitamin K* is found in green leafy plants and is also synthesized by bacteria normally present in the human intestine. This vitamin is essential to the formation of blood clots. Predictably, a deficiency of vitamin K leads to slower clot formation and potentially fatal bleeding from a wound.

Because of their insolubility in water, in the body some lipids may require water-soluble carrier proteins in order to be transported in the blood, which is mostly water. Among the lipids, only triglycerides and phospholipids assemble into large structures in cells: triglycerides form droplets of stored fat, and phospholpids form the bilayers of membranes. But in spite of their size, these droplets and bilayers are not usually considered macromolecules.

Now that we have seen the macromolecule-like structures that lipids form, let's turn to the large structures that constitute true macromolecules.

Macromolecules: Giant Polymers

As noted earlier, *macromolecules* are giant *polymers* constructed by the covalent linking of smaller molecules called *monomers*. These monomers may or may not be identical, but they always have similar chemical structures. Molecules with molecular weights exceeding 1,000 are usually considered macromolecules, and the polysaccharides, proteins, and nucleic acids of living systems certainly fall into this category.

In the cell, each type of macromolecule performs some combination of a diversity of functions: energy storage, structural support, protection, catalysis, transport, defense, regulation, movement, and heredity. These roles are not necessarily exclusive. For example, both carbohydrates and proteins can play structural roles-supporting and protecting cells and organisms. However, only nucleic acids specialize in information and function as hereditary material, carrying both species and individual traits from generation to generation.

The functions of macromolecules are directly related to their shapes and the chemical properties of their monomers. Some macromolecules, such as catalytic and defensive proteins, fold into compact spherical forms with surface features that make them watersoluble and capable of intimate interaction with other molecules. Other proteins and carbohydrates form long, fibrous systems that provide strength and rigidity to cells and organisms. Still other long, thin assemblies of proteins can contract and cause movement.

Because macromolecules are so large, they contain many different functional groups. For example, a large protein may contain hydrophobic, polar, and charged groups. These groups give specific properties to local sites on a macromolecule. As we will see, this diversity of properties determines the shapes of macromolecules and their interactions with both other macromolecules and smaller molecules.

Macromolecules Form by Condensation Reactions

The polymers of living things are constructed by a series of reactions called *condensation reactions*, or dehydration hydration reactions (both words refer to the loss of water). Condensation reactions covalently bond monomers. Water is lost in the reaction:

$$A\text{—}H + B\text{—}OH \rightarrow A\text{—}B + H_2O$$

A—H is a molecule with a reactive hydrogen; B—OH is a molecule with a hydroxyl group. The products of the reaction are A—B and H_2O. The atoms that make up the water molecule are derived from the reactants: one hydrogen atom from one reactant, and an oxygen atom and the other hydrogen atom from the other reactant. Condensation reactions are not limited to polymer formation; for example, we encountered them in the linkage (esterification) of fatty acids to glycerol.

The condensation reactions that produce the different kinds of macromolecules differ in detail, but in all cases polymers will form only if energy is added to the system. In living systems, specific energy-rich molecules supply the energy, and there are additional steps to the reaction. The reverse of a condensation reaction is a *hydrolysis reaction* (*hydro-*, "water"; *-lysis,* "breakage"). These reactions digest polymers and produce monomers. Water reacts with the bonds that link the monomers together, and the products are free monomers. The elements of the reactant H_2O become part of the products. Hydrolysis reactions of all sorts are very important in cellular functioning and are not limited to the digestion of polymers.

Carbohydrates: Sugars and Sugar Polymers

Carbohydrates are a diverse group of compounds based on the general formula CH_2O. Some are relatively small, with molecular weights less than 100. Others are true macromolecules, with molecular weights of hundreds of thousands. There are four categories of biologically important carbohydrates. *Monosaccharides* (*mono-*, "one"; *saccharide,* "sugar") such as glucose or fructose—are the monomers out of which the larger forms are constructed. *Disaccharides* (*di-*, "two") consist of two monosaccharides. *Oligosaccharides* (*oligo-*, "several") have several monosaccharides (3 to 20). *Polysaccharides* (*poly-*, "many")-such as starch, glycogen, and cellulose—are composed of hundreds of thousands of glucose units.

The relative proportions of carbon, hydrogen, and oxygen indicated by the general formula for carbohydrates, CH_2O, is true for monosaccharides. However, for disaccharides, oligosaccharides, and polysaccharides, these proportions differ slightly from this general formula because two hydrogens and an oxygen are lost during the condensation reactions.

Monosaccharides are Simple Sugars

All living cells contain the monosaccharide glucose, whose formula is $C_6H_{12}O_6$. Green plants produce glucose by photosynthesis, and other

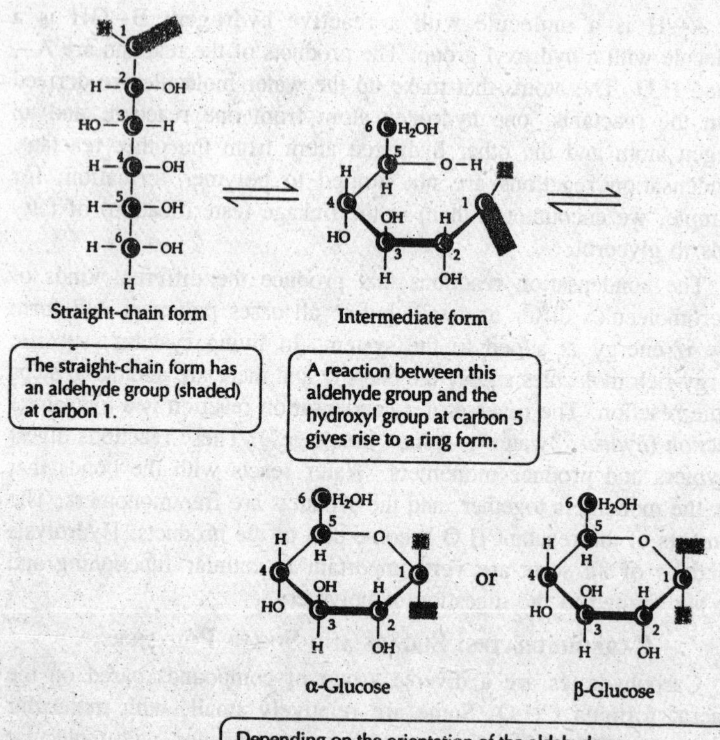

Fig. 10.4. Glucose: from one form to the other.

organisms acquire it directly or indirectly from plants. Cells use glucose as an energy source, changing it through a series of reactions that release stored energy and produce water and carbon dioxide.

Two forms of glucose, the straight chain and the ring, exist in equilibrium with each other when dissolved in water, but the ring form predominates (>99%). The two distinct ring forms (a and b-glucose) differ only in the placement of the —H and —OH attached to carbon 1. Most of the monosaccharides found in living systems belong to the D series of optical isomers. But there are structural isomers-composed of the same kinds and numbers of atoms, but with the atoms combined differently in each. All *hexoses* (*hex-*, "six"), a group of structural isomers, have the formula $C_6H_{12}0_6$. Included among the hexoses are fructose (so named because it was first found in fruits), mannose, and galactose.

Three-carbon sugar

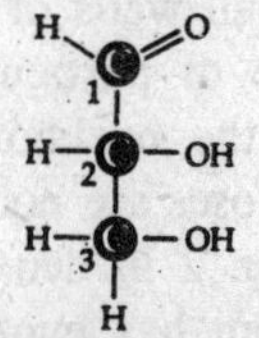

Glyceraldehyde is the smallest sugar and exists only as the straight-chain form.

Five-carbon sugars

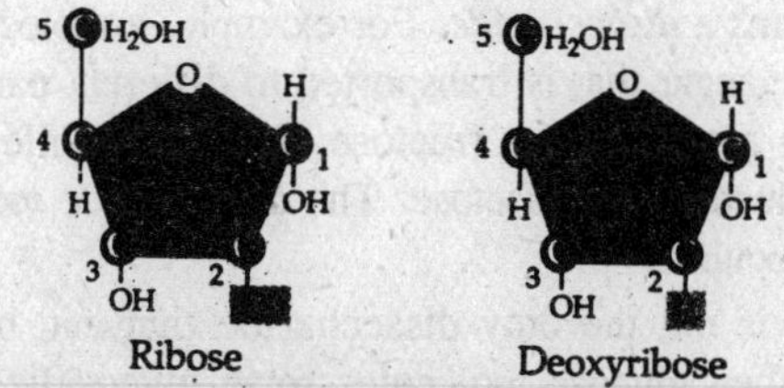

Ribose and deoxyribose each have five carbons, but very different chemical properties and biological roles.

Six-carbon sugars

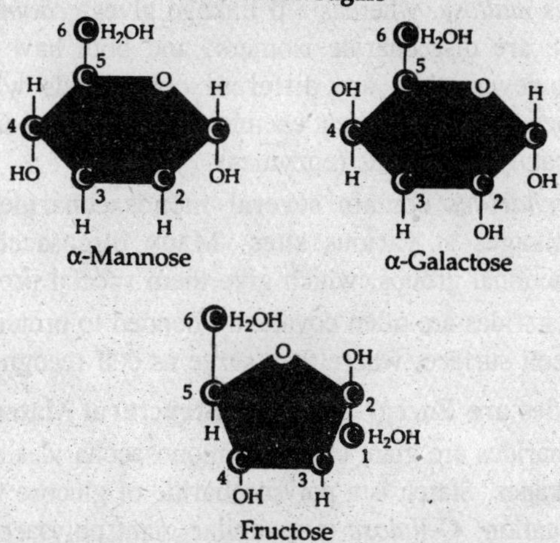

These hexoses are isomers. All have the formula $C_6H_{12}O_6$–but each has distinct chemical properties and biological roles.

Fig. 10.5. Monosaccharides are simple sugars.

Pentoses (*pent-*, "five") are five-carbon sugars. Some pentoses are found primarily in the cell walls of plants, as are several of the hexoses. Two pentoses are of particular importance: *Ribose* and *deoxyribose* form part of the backbones of RNA and of DNA, respectively. These two pentoses are not isomers; rather, one oxygen atom is missing from carbon 2 in deoxyribose (de, "absent"). The absence of this oxygen atom has enormous consequences for the functional distinction of RNA and DNA.

Glycosidic Linkages bond Monosaccharides Together

Monosaccharides are covalently bonded together by condensation reactions that form *glycosidic linkages*. Such a linkage between two simple sugars forms a *disaccharide*. For example, a molecule of *sucrose* (table sugar)—the sugar that is transported to different parts of plants—is formed from a glucose and a fructose molecule, while *lactose* (milk sugar) contains glucose and galactose. The disaccharide *maltose* contains two glucose molecules.

But maltose is not the only disaccharide that can be made from two glucoses. When glucose molecules form glycosidic linkages, the disaccharide product must be one of two types: α-linked or β-linked, depending on whether the molecule that bonds by its 1 carbon is α-glucose or β-glucose. An α linkage with carbon 4 of a second glucose molecule gives *maltose*, whereas a β linkage gives *cellobiose*. Maltose and cellobiose are disaccharide isomers, and both have the formula $C_{12}H_{22}O_{11}$. However, they are different compounds with different properties. They undergo different chemical reactions and are recognized by different catalytic proteins (enzymes).

Oligosaccharides contain several monosaccharides linked by glycosidic linkages at various sites. Many oligosaccharides gain additional functional groups, which give them special properties.

Oligosaccharides are often covalently bonded to proteins and lipids on the outer cell surface, where they serve as cell recognition signals.

Polysaccharides are Energy Stores or Structural Materials

Polysaccharides are giant chains of monosaccharides connected by glycosidic linkages. Starch is a polysaccharide of glucose with linkages in the a orientation. *Cellulose* is a similar giant polysaccharide made up solely of glucose, but its individual units are connected by β linkages instead of a linkages. Cellulose is the predominant component of plant cell walls and is by far the most abundant organic compound on this planet. Both starch and cellulose are composed of nothing but glucose, but their biological functions and chemical and physical properties are entirely different.

Starch can be more or less easily degraded by the actions of chemicals or catalytic proteins (enzymes). Cellulose, however, is very stable because of its β glycosidic linkages. Thus starch is a good storage form that can be easily degraded to supply glucose for energy producting reactions, while cellulose is an excellent structural component that can withstand harsh environmental conditions without changing.

Starch is not just one chemical substance, but large family of giant molecules of broa structure. All starches are large polymers of glucose with a linkages, but different starches can be distinguished by the amount of branching between carbons 1 and 6. Some starches are highly branched; others are not. Plant starches, called *amylose*, are not highly branched. The polysaccharide glycogen, which stores glucose in animal livers and muscles, is highly branched.

What do we mean when we say that starch and glycogen are storage compounds for energy? Very simply, these compounds can readily be hydrolyzed to yield glucose monomers. Glucose, in turn, can enter into a series of reactions that liberate its stored energy in forms that can be used for cellular activities. However, glucose can also serve the cell's need for raw materials—that is, carbon atoms. Glucose can undergo other chemical reactions that disassemble parts of the molecule and rearrange its carbon atoms to form the skeletons of other compounds needed by cells. Glycogen and starch are thus storage depots for carbon atoms as well as for energy.

Derivative Carbohydrates Contain other Elements

Sometimes carbohydrates are modified by chemical changes in their structure or by the addition of functional groups such as phosphate and amino groups, thus becoming *derivative carbohydrates*. For example, carbon 6 in glucose may be oxidized from —CH_2OH to a carboxyl group (—COOH), producing glucuronic acid. Or a phosphate group may be added to one or more of the —OH sites. Some of these *sugar phosphates*, such as fructose 1,6-bisphosphate, are important intermediates in cellular energy reactions. When an amino group is substituted for an —OH group, *amino sugars* such as glucosamine and galactosamine are produced. Galactosamine is a major component of cartilage, the material that forms caps on the ends of bones and stiffens the protruding parts of the ears and nose. A derivative of glucosamine produces the polymer *chitin*, which is the principal structural polysaccharide in the skeletons of insects, crabs, and lobsters, as well as in the cell walls of fungi. Fungi and insects (and their relatives) constitute more than 80 percent of the species ever described, and chitin is one of the most abundant substances on Earth.

Proteins: Amazing Polymers of Amino Acids

Proteins are an extraordinary group of macromolecules. Constructed of amino acids, proteins are fascinating because they have such intricate and diverse structures and because they perform so many different functions for cells. Among the cellular functions of macromolecules listed earlier, only energy storage and heredity are not functions of proteins. Proteins are involved in structural support, protection, catalysis, transport, defense, regulation, and movement. Of particular importance are the catalytic proteins, called *enzymes*, that increase the rates of chemical reactions in cells. In general, each chemical reaction requires a different enzyme, because proteins show great specificity for the smaller molecules with which they will interact.

Proteins range in size from the small RNA-digesting enzyme *ribonuclease A*, which has a molecular weight of 5,733 and 51 amino acid residues, to gigantic molecules such as the cholesterol transport protein *apolipoprotein B*, which has a molecular weight of 513,000 and 4,636 amino acid residues. (The word "residue" is used for a monomer when it is part of a polymer.) Each of these proteins consists of one chain of amino acids folded into a specific three-dimensional shape that is required for protein function. Some proteins have more than one polymer chain. For example, the oxygen-carrying protein *hemoglobin* has four chains that are folded separately and associate together to make the functional protein. The largest protein complex known has more than 40 separate chains of amino acids.

All of these different proteins have a characteristic amino acid composition, but every protein contains neither all 20 amino acids nor an equal number of different amino acids. Nor is there a simple regular sequence in which the amino acids are linked. The diversity in amino acid content and sequence is the source of the diversity in protein structures and functions. In some proteins, different kinds of chemical substances, called *prosthetic groups*, may be attached to the protein. These prosthetic groups include carbohydrates, lipids, phosphate groups, the iron-containing heme group, and metal ions such as copper and zinc. Whether they are small or large, and whether or not they have prosthetic groups, there is nothing "casual" about the structures of proteins. Proteins have specific three-dimensional shapes that are necessary for their specific functions.

To understand this stunning variety of functions, we must explore protein structure. First, we will examine the properties of the 20 amino acids and the characteristics of how they link to form proteins. Then we will systematically examine the four levels of protein structure and look at how a linear chain of amino acids is consistently folded into a compact three-dimensional shape.

Proteins are Composed of Amino Acids

The 20 different amino acids commonly found in proteins show a wide variety of properties. We considered the structure of amino acids and identified four different groups attached to a central carbon atom: a hydrogen atom, an amino group, a carboxyl group, and a side chain, or R group. Since amino acids are linked together by reactions between their amino and carboxyl groups, these groups are not exposed and do not give the protein its distinguishing properties and functional specificity-the side chains fill this role.

The *side chains* of amino acids are a protein's reactive groups and show a wide variety of chemical properties. Side chains control the function of a protein and contribute to its structure. The sequence of side chains determines how a protein folds into a three-dimensional shape (which we will discuss shortly). Although side chains are very important, they are often omitted from diagrams of proteins when the focus is on other aspects of structure. In these cases, the side chains are identified by the letter R (for "residue") and are sometimes called R groups.

One useful classification of amino acids is based on whether their side chains are electrically charged (+1, -1), polar (δ+, δ-), or nonpolar and hydrophobic. The five amino acids that have electrically charged side chains attract water and oppositely charged ions of all sorts. The four amino acids that have polar side chains tend to form weak hydrogen bonds with water and with other polar or charged substances. Eight amino acids have side chains that are nonpolar hydrocarbons or very slightly modified hydrocarbons. In the watery environment of the cell, the hydrophobic side chains may cluster together.

Three amino acids—cysteine, glycine, and proline—are special cases, although their side chains are generally hydrophobic. Two *cysteine* side chains, which have terminal —SH groups, can react to form a covalent bond in a *disulfide bridge* (—S—S—). Hydrogen bonds and disulfide bridges help determine how a protein chain folds. When cysteine is not part of a disulfide bridge, its side chain is very hydrophobic. The *glycine* side chain consists of a single hydrogen atom; thus glycines may fit into tight corners in the interior of a protein molecule, where a larger side chain could not fit. *Proline* differs from other amino acids because it possesses a modified amino group.

Peptide Linkages Covalently Bond Amino Acids Together

When amino acids polymerize, the carboxyl group of one amino acid reacts with the amino group of another, undergoing a condensation reaction that forms a peptide linkage. Gives a simplified description of the reaction. (In living cells, other molecules must activate the reactants in order for this reaction to proceed, and there are intermediate steps.) A linear polymer of amino acids connected by peptide linkages is a *polypeptide*. A *protein* is made up of one or more polypeptides.

At one end of the polypeptide is a free amino group. This end is the N terminus, named for the nitrogen atom in the amino group. At the other end of the polypeptide—the C terminus—is a free carboxyl group. The other amino and carboxyl groups are bound in peptide linkages. Thus a protein has direction. For example, the dipeptide glycine-alanine, in which glycine has the free amino group, differs from alanine-glycine, in which alanine has the free amino group. In cells, the synthesis of polypeptide chains begins with the N terminus.

In the peptide linkage, the C=O oxygen carries a slight negative charge (δ-), whereas the N—H hydrogen is slightly positive (δ+). This asymmetry of charge favours hydrogen bonding within the protein molecule itself and with other molecules, contributing to both the structure and the function of many proteins.

Primary Structure of a Protein is its Amino Acid Sequence

Protein structure is elegant and complex—so complex that it is described as consisting of four different levels: *primary*, *secondary*, *tertiary*, and *quaternary*. However, proteins are always linear chains; there are no branches. The precise sequence of amino acids in a polypeptide constitutes the primary *structure* of a protein. The *peptide backbone* of this primary structure consists of a repeating sequence of three atoms (—N—C—C—) from the amino group, the central carbon, and the carboxyl group of each amino acid.

In cells, the primary structure of a protein is dictated by the precise sequence of nucleotides in a linear segment of a DNA molecule. The elucidation of this relationship between DNA primary structure and protein primary structure was one of the triumphs of molecular biology.

The theoretical number of different proteins is enormous. Since there are 20 different amino acids, there are $20 \times 20 = 400$ distinct dipeptides, and $20 \times 20 \times 20 = 8{,}000$ different tripeptides. Imagine this process of multiplying by 20 extended to a protein made up of 100 amino acids (which is considered a small protein): There could be 20^{100} of these small proteins, each with its,own distinctive primary structure. How large is the number 20^{100}? In the entire universe there aren't that many electrons!

At the higher levels of protein structure, local coiling and folding give the final functional shape of the molecule, but all of these levels derive from the primary structure. The different properties associated with a precise sequence of amino acids determine how the protein can twist and fold. By twisting and folding, each protein adopts a specific stable structure that distinguishes it from every other protein.

Secondary Structure of a Protein Requires Hydrogen Bonding

Although the primary structure of each protein is unique, the secondary structure of many different proteins may be the same. A protein's *secondary structure* consists of regular, repeated patterns in different regions of a polypeptide chain. One type of secondary structure, the *α helix* (alpha helix), is a right-handed coil that is "threaded" in the same direction as a standard wood screw. The amino acid side chains extend outward from the peptide backbone of the helix.

This helical structure of a polypeptide chain results from hydrogen bonds between elements of the peptide bonds that are distributed along the back bone of the chain. Hydrogen bonds form between the slightly positive hydrogen of the N—H of one peptide bond and the slightly negative oxygen of the C=O of another peptide bond. When this pattern of hydrogen bonding is established repeatedly over a segment of the protein, it stabilizes the twisted form, resulting in an a helix. However, the ability of a protein to form an a helix depends on its primary structure. Amino acids with larger side chains that distort the coil or otherwise prevent the formation of the necessary hydrogen bonds will keep the a helix from forming.

Alpha helical secondary structure is particularly evident in the fibrous structural proteins called *keratins*. Keratins constitute many of the protective materials found in mammals and birds, such as fingernails and claws, skin, hair, wool, and feathers. Hair can be stretched because this stretching requires that only hydrogen bonds in an a helix, and not covalent bonds, be broken; when the tension on the hair is released, both the helix and the hydrogen bonds reform.

Another type of secondary structure, the β pleated sheet, is found in the protein silk. In silk, rather than being coiled, the protein chains are almost completely extended and lie next to one another, stabilized by hydrogen bonds between the elements of the peptide linkages. The R pleated sheet may be found between separate polypeptide chains, as in silk, or between different regions of the same polypeptide that is bent back on itself. Many enzymes contain regions of a helix and of β pleated sheet in the same polypeptide chain.

Tertiary Structure of a Protein is Formed by Bending and Folding

In most proteins, to establish the compact structure the polypeptide chain must be bent at specific sites and folded back and forth. The resulting overall shape of a protein is its tertiary structure. Although the a helices and β pleated sheets contribute to the tertiary structure, more frequently only limited portions of the molecule have these secondary structures, and large regions consist of structures unique to a particular protein.

A complete description of the tertiary structure specifies the location of every atom in the molecule in three-dimensional space, in relation to all the other atoms. Bear in mind that this tertiary structure and the secondary structure derive from the protein's primary structure. If lysozyme is heated carefully, causing only the tertiary structure to break down, the protein will return to its normal tertiary structure when it cools. The only information needed to specify the unique shape of the lysozyme molecule is the information contained in its primary structure.

Hemoglobin is the protein that transports oxygen (O_2) from the lungs to the tissues. In muscle tissue, hemoglobin delivers O_2 to myoglobin, a smaller but similar protein, for storage. The hemoglobin molecule consists of four similar polypeptide chains that have tertiary structures made up almost entirely of a helices but that lack any disulfide bridges. The helical segments bend and fold against each other, forming a pocket that encloses a *heme group*, an ironcontaining prosthetic group that binds O_2. The stabilizing interaction of hydrophobic side chains on the inner sides of the a helices helps ensure that the helices fold against one another correctly as each polypeptidc assumes its tertiary structure.

Quaternary Structure of a Protein Consists of Subunits

As we mentioned earlier, some proteins have two or more polypeptide chains folded into their own unique tertiary structures.

Each of these folded polypeptides is considered a protein subunit. *Quaternary structure* results from the ways in which these protein subunits fit together and interact; it can be illustrated by hemoglobin. Hydrophobic interactions, hydrogen bonds, and ionic bonds all help hold the four subunits together to form the functional hemoglobin molecule. As the hemoglobin molecule takes up one O_2 molecule, the four subunits shift their relative positions slightly, changing the quaternary structure. Ionic bonds are broken, exposing buried side chains that enhance the binding of additional O_2 molecules.

Each subunit of hemoglobin is folded like a myoglobin molecule, suggesting that both hemoglobin and myoglobin are evolutionary descendants of the same oxygen-binding ancestral protein. But on the surfaces where its subunits come in contact with each otherregions that on myoglobin are exposed to aqueous surroundings and are hydrophilic-hemoglobin has hydrophobic side chains. Again, the chemical nature of side chains on individual amino acids determines how the molecule folds and packs in three dimensions.

Molecular Chaperones Help Shape Proteins

The primary structure of a protein constrains the secondary, tertiary, and quaternary structures (if subunits exist). By determining the primary structure, DNA also determines the higher levels of structure. However, other factors also affect the tertiary structure that is required for proper protein function.

Elevated temperatures, pH changes, or altered salt concentrations can cause a protein to adopt a different, biologically inactive tertiary structure. Biological function depends on a specific three-dimensional structure. Increased temperature causes more rapid molecular movement and thus can break weak hydrogen bonds and hydrophobic interactions. Altered pH can change the pattern of ionization of carboxyl and amino groups in the side chains of amino acids, thus disrupting the pattern of ionic attractions and repulsions that contribute to normal tertiary structure.

The loss of appropriate tertiary structure is called *denaturation*, and it is always accompanied by a loss of the normal biological function of the protein. Denaturation can be caused by heat or high concentrations of polar substances such as urea that disrupt the hydrogen bonding that is crucial to protein structure. Nonpolar solvents may also disrupt normal structure. Usually denaturation is irreversible, particularly if many denatured proteins interact in random nonspecific ways. However, in some cases denaturation is reversible, and upon return to normal environmental conditions, the protein may return to its active form, as does lysozyme. This return is called *renaturation*.

How can such a complicated molecule spontaneously fold into its normal tertiary structure? Biologists are just beginning to be able to predict how a protein will fold, given its primary structure. The study of protein folding has a long way to go. The situation in the living cell is complex-even more complex than we first imagined. The protein doesn't form all at once. In a human cell it may take minutes from the beginning of synthesis until the entire protein squeezes into the compartment where it folds-and things are very crowded in that compartment. There, other proteins present inappropriate potential partners for interaction (in the form of amino acid side chains).

A special group of proteins, called the *chaperone proteins*, at least in part help prevent newly formed proteins from reacting inappropriately. The chaperones attach to a new protein while it is forming and protect it from interactions with other proteins until the new molecule has its correct shape. Most chaperone proteins can act on many different forming proteins. Some act on only a few, and others may act on only a single, specific protein.

What if the chaperone proteins fail to do their job, or if for other reasons a protein folds abnormally? Evidence suggests that abnormal protein folding underlies certain infectious diseases, including mad cow disease and, in humans, Creutzfeldt-Jakob disease. Alzheimer's disease may also result from protein misfolding.

Nucleic Acids: Informational Macromolecules

The nucleic acids are linear polymers specialized for the storage, transmission, and use of information. There are two types of nucleic acids: DNA (deoxyribonucleic acid) and RNA (ribonucleic acid). DNA molecules are giant polymers that encode hereditary information and pass it from generation to generation (through reproduction). The information encoded in DNA is also used to make specific proteins through the intermediate RNA. RNA molecules of various types copy the information in segments of DNA to specify the sequence of amino acids in proteins. Information flows from DNA to DNA in reproduction. But for nonreproductive activities of the cell, information flows from DNA to RNA to proteins, which ultimately carry out these functions. What compositions, structures, and properties of nucleic acids permit them to play these fundamental roles in living systems?

Nucleic Acids have Characteristic Structures and Properties

Nucleic acids are composed of monomers called *nucleotides*, each of which consists of a pentose sugar, a phosphate group, and a nitrogen-containing base-either a pyrimidine or a purine. Molecules consisting of a pentose sugar and a nitrogenous base, but no phosphate group, are called nucleosides. In DNA, the pentose sugar is deoxyribose, which differs from the ribose found in RNA by only one oxygen atom.

In both RNA and DNA, the backbone of the molecule consists of alternating sugars and phosphates (—sugar—phosphate—sugar—phosphate—). Bases are attached to the sugars and project from the chain. The nucleotides are joined by covalent bonds in what are called *phosphodiester* linkages between the sugar of one nucleotide and the phosphate of the next ("-diester" refers to the two bonds formed by reacting —OH groups with acidic phosphate groups). The phosphate groups link carbon 3' in one ribose to carbon 5' in the adjacent ribose.

Most RNA molecules are *single-stranded*, consisting of only one polynucleotide chain. DNA, however, is usually *double-stranded*; it has two polynucleotide chains held together by hydrogen bonding between their complementary nitrogenous bases. The two strands of DNA run in opposite directions. You can see what this means by drawing an arrow through the phosphate group from carbon 3' to carbon 5' in the next ribose. If you do this for both strands, the arrows point in opposite directions. This antiparallel orientation is necessary for the strands to fit together.

Uniqueness of a Nucleic Acid Resides in its Base Sequence

Only four nitrogenous bases—and thus only four nucleotides—are found in DNA. The DNA bases and their abbreviations are: adenine (A), cytosine (C), guanine (G), and thymine (T). A key to understanding the structures and functions of nucleic acids is the principle of *complementary base pairing* through hydrogen bond formation. In double-stranded DNA, adenine and thymine always pair (AT), and cytosine and guanine always pair (CG).

Base pairing is complementary because of two factors: the corresponding sites for hydrogen bonding and the molecular sizes of the paired bases. Adenine and guanine are both purines consisting of two fused rings. Thymine and cytosine are both pyrimidines consisting of only one ring. The pairing of a large purine with a small pyrimidine ensures a stable and consistent dimension to the double-stranded molecule of DNA. We'll discuss in more detail how complementary strands of DNA separate and are faithfully copied.

Ribonucleic acids also have four different monomers, but the nucleotides differ from those of DNA. In RNA the nucleotides are termed ribonucleotides. They contain ribose rather than deoxyribose, and instead of the base thymine, RNA uses the base uracil. The other three bases are the same as in DNA.

Although RNA is generally single-stranded, complementary hydrogen bonding between ribonucleotides can take place. These bonds play important roles in determining the shapes of some RNA molecules and in the associations between RNA molecules during protein synthesis. During the DNA-directed synthesis of RNA, complementary base pairing also takes place between ribonucleotides and the bases of DNA. In RNA, guanine and cytosine pair as in DNA, but adenine pairs with uracil. Adenine in an RNA strand can pair either with uracil (in another RNA strand) or with thymine (in a DNA strand).

The three-dimensional appearance of DNA is strikingly regular. Through hydrogen bonding, the two complementary polynucleotide strands pair and twist to form a double helix. Compared to the complex and varied tertiary structures of different proteins, this formation seems amazingly regular. But this structural contrast makes sense in terms of the functions of these two classes of macromolecules.

DNA is a purely informational molecule. *The information in DNA is encoded in the sequence of bases carried in its chains.* This sequence is, in a sense, like the tape of a tape recorder. The message must be read easily and reliably, in a specific order. A uniform molecule like DNA can be interpreted by standard molecular machinery, and any cell's machinery can read any molecule of DNA-just as a tape player can play any tape of the right size.

Table 10.1. Distinguishing RNA from DNA

Nucleic acid	*Sugar*	*Bases*
RNA	Ribose	Adenine Cytosine Guanine Uracil
DNA	Deoxyribose	Adenine Cytosine Guanine Thymine

Proteins, on the other hand, have good reason to be so varied. In particular, different enzymes must recognize their own specific "target" molecules, They do this by having a unique three-dimensional form that can match at least a portion of the surface of their target molecules. In other words, structural diversity in the molecules with which enzymes react requires corresponding diversity in the structure of the enzymes themselves. *In DNA the information is in the sequence of bases; in proteins the information is in the shape.*

DNA is a Guide to Evolutionary Relationships

Because DNA carries hereditary information between generations, a series of DNA molecules with changes in base sequences stretches back- through time. Of course, we cannot study all of these DNA molecules, because many of their organisms have become extinct. However, we can study the DNA of living organisms, some of which are judged to belong to lineages that are more ancient than those of other living forms. Comparisons and contrasts of these DNAs add to the evolutionary record.

Closely related living species should have more similar base sequences than species judged by other criteria to be more distantly related. Indeed this is the case. The examination of base sequences confirms many of the evolutionary relationships that have been inferred from the study of microscopic or macroscopic structures or studies of biochemistry and physiology. For example, the closest living relative of humans (*Homo sapiens*) is the chimpanzee (genus *Pan*), which shares more than 98 percent of its DNA base sequence with human DNA.

Confirmation of well-established evolutionary relationships gives credibility to the use of DNA to elucidate relationships when studies of structure are not possible or are not conclusive. For example, DNA studies revealed a close evolutionary relationship between starlings and mocking-birds that was not expected on the basis of studies of anatomy or behaviour. DNA studies support the division of the prokaryotes into two kingdoms (Eubacteria and Archaebacteria). Each of these two groups of prokaryotes is as distinct from the other as either is from the other four kingdoms into which living things are classified. In addition, DNA comparisons support the hypothesis that certain subcellular compartments of eukaryotes (the organelles called *mitochondria* and *chloroplasts*) evolved from early bacteria that established a stable and mutually beneficial way of life inside larger cells.

Interactions of Macromolecules

We have been treating the classes of macromolecules as if each were completely separate from the others. In cells, however, certain macromolecules of different classes may be covalently bonded to one another. Proteins with attached oligosaccharides are called *glycoproteins* (*glyco-*, "sugar"). The specific oligosaccharide chain determines the placement of a glycoprotein within the cell.

Other carbohydrate chains bind to lipids, resulting in *glycolipids*, which reside in the cell surface membrane, with the carbohydrate chain extending out into the cell's environment. In humans, the substances that determine the ABO blood types are carbohydrates attached to either proteins or lipids. When a human body cell becomes cancerous, the glycolipids and glycoproteins on the cell surface are modified, and these changes serve as a recognition signal for the body defenses to destroy the abnormal cell.

Not all the associations among macromolecules or between a macromolecule and a smaller molecule are covalent. Weaker linkages such as ionic bonds, hydrogen bonds, and even van der Waals forces may also be involved. Enzymes bind to their reactants using a variety of weak bonds. Ionic bonds link proteins to DNA, forming *nucleoproteins* that regulate the activities of DNA. Still other proteins, in combination with cholesterol and other lipids, form *lipoproteins*. The lipoproteins make it possible to move very hydrophobic lipids through water-rich environments such as the blood and tissue fluid and deliver cholesterol to all the body's cells.

11

Role of pH in Life

pH: is measured when a potential difference (emf) develops within a circuit consisting of a reference electrode (generally calomel half-cell) and an "ion-selective" electrode (glass plus a silver-silver chloride half-cell) connected by a combination of a salt bridge and the solution of unknown hydrogen ion concentration.

The readout mechanism (digital of galvanoemetric) is a millivoltmeter with a scale graduated in pH units.

pCO_2: is measured as the change in pH of a weak bicarbonate buffer after it has reached equilibrium by dialysis across a semipermeable membrane with the CO_2 physically dissolved in an unknown solution, usually blood.

pO_2: is measured as the current flowing between a platinum and a silver-silver chloride electrode immersed in a buffer (phosphate and potassium chloride) which has been equilibrated with dissolved oxygen in an unknown solution by dialysis across a semipermeable membrane. The current results from oxygen uptake of electrons at the platinum electrode's surface after the electrode is polarized using a standard dc voltage, usually a mercury battery.

Principles

A metal rod, inserted into a solution of its own metallic ions, will develop an electrical potential (voltage) between the metal and the solution. This voltage results from the production of electrons as some of the metal atoms become dissolved in the solution. Free electrons are produced by this reaction:

$$M^{\circ} \rightarrow M^{+} + e$$

where

M° = metal atom

M^+ = metallic ion

e = electron

This system is a *half-cell* and the metal rod can be considered the electrode. Because oxidation is occurring at this electrode, it is also the anode.

The ions of another metal in solutions and in contact with their metal atom may favour information of metal atoms with acceptance of a electron.

$$M^+ + e \rightarrow M°$$

This second electrode would the be the cathode because reduction is taking place. This, too, is a half-cell.

The two half-cells can be connected by means of a conducting solution (salt bridge). This establishes the circuit and enables a current to flow. This a complete cell is formed.

A comparison of the voltage of one half-cell to another half-cell can be made. The International Union of Pure and Applied Chemistry in 1953 adopted the method of comparing all half-cell potential to he standard hydrogen electrode, which potential they arbitrarily assigned as 0.0000 V. The standard electrode potential (E_0) for each half-cell is determined by a comparison to the standard hydrogen electrode. This standard electrode potential is defined as the emf of the half-cell when all activities are unity and is usually given at 25°C. A few examples follow:

Half-cell	*Reaction*	*emf*
½ H_2	$H^+ + e$	0.0000 V
Ca	$Ca^{2+} + 2e$	+2.87 V
Fe	$Fe^{2+} + 2e$	+0.441 V
Cu	$Cu^{2+} + 2e$	-0.337 V

If the ions in the cells are not at unit activity, the potential *E* is given by the Nernst equation.

$$E = E_0 + \frac{RT}{nF} \ln a$$

where

E = observed potential (emf)

E_0 = standard electrode potential (emf)

R = molar gas constant (8.314 J/mole per °K)

T = absolute *T* in °K (25°C = 298°K)

n = number of electrons transferred

F = Faraday's constant (96,500 C/Eq weight)

ln = log to the base *e* (natural log)

a = activity of the ion

At 25°C, with conversion of the natural $\log_e$ to $\log_{10}$ insertion of the values for *R* and *F*, this equation simplifies to

$$E = E_0 + \frac{0.0591}{n} \log_{10}[a]$$

From this equation it can be seen that voltage is dependent on the activity of the ion. Since a change in ion activity will change the measurable voltage, pH can be determined by a difference in voltage. A pH measurement is an indirect measurement of the hydrogen ion activity. The hydrogen ion activity coefficient approaches unity in very dilute solutions. Therefore, this activity measurement can become a workable approximation for concentration.

ELECTRODES

Calomel Reference Electrode

This electrode consists of a combination of mercury (Hg_2), mercurous chloride (Hg_2Cl_2), and KCl paste in a saturated potassium chloride solution. The chloride ions become saturated with mercury at the surface of the mercury. Under these conditions, a typical calomel electrode will produce a *constant* reference potential of 244 mV to which other electrodes may be compared. For proper operation, the surface of the electrode tip must be completely immersed in the electrolyte.

It is best to have the electrode's electrolyte and the bridge salt solution identical; in this case, saturated potassium chloride is used. Other bridge solutions are less satisfactory. With dilute bridging salt solutions, liquid junction potential are much less predictable. The calomel electrode should be separated from the KCl salt bridge by a porous ceramic plug which will reduce the chance of back diffusion and contamination.

Potential contaminating factors for the calomel electrode include:

Bromide contamination

Concentration of bromide in KCl should be less than 0.005%.

Thermal instability

The calomel reference electrode system contains at least nine thermal sensitive equilibrium sites with possible junction potential. Thermal stability within $\pm 0.05°C$ is necessary.

Calomel potential

The calomel electrode provides a potential independent of pH when connected to a glass electrode by a KCl bridge whenever the pH is greater than 1.5 and less than 12. For best operation, the calomel electrode should be intact, appear silver grey, and have a open air vent during the time of use.

Silver—Silver Chloride Electrode

These usually consist of a platinum wire on which silver and then silver chloride are deposited. The potential developed at the electrode surface can be described by the following equation:

$$AgCl + e^- \rightleftharpoons Ag° + Cl^-$$

These may be used as reference electrodes. However, KCl may not then be used as a electrolyte or salt bridge as silver chloride is soluble in KCl. Both HCl and NaCl may be used.

Hydrogen Electrode

This so-called ultimate standard electrode is created by bubbling pure hydrogen gas through a solution in which a platinum electrode is immersed and establishing an equilibrium between the hydrogen ion and the hydrogen gas in that solution.

$$H^+ + e^- = \frac{1}{2} \; H_2(g)$$

The platinum electrode is covered by a surface catalyst such as platinum or palladium black, which reduces the energy barrier, increases the equilibrium pressure of hydrogen, and makes the electrode reversible in response to hydrogen ion. The electrode has disadvantages of working with hydrogen gas, having a slow equilibrium, and requiring large volumes of solution.

Antimony Electrode

Antimony in solution is sensitive to hydrogen ion concentration.

$$Sb_2O_3 + 6H^+ + 6e^- \rightleftharpoons 2Sb° + 3H_2O$$

The antimony electrode is not in general use because it is not reproducible, not always reversible, and is sensitive to oxidation and reduction agents.

Quinhydrone Electrode

The quinhydrone electrode consists of a slightly soluble, equimolar solution of benzoquinone ($O{=}C_6H_4{=}O$) and hydroquinone ($HO{-}C_6H_4{-}OH$) which shows a reversible response to hydrogen ion as per the following equation:

$$Q + 2H^+ + 2e^- \rightleftharpoons H_2Q$$

where Q equals benzoquinone and H_2Q represents hydroquinone. Such a compound in solution and in contact with a gold or platinum electrode may develop a potential in response to a change in hydrogen ion concentration, but it use is limited to solutions of pH 8.0 or less. It is seriously handicapped by the development of errors in response of proteins, contamination by oxidizing agents, high salt concentrations, and instability above 30°C.

Glass Electrode

Certain thin glass membranes will develop a potential difference when ionic solutions of differing pH are preset on the membrane's two sides. The thin glass membrane separates a unknown solution from a reference electrode (AgCl electrode) immersed in a solution of a know pH. The electrode will reversibly respond to changes in pH within this solution. The electrode is usually HCl. These ion-selective electrodes measure hydrogen ion activity, not hydrogen ion concentration.

Each surface of the glass membrane (of the electrode) develops a hydrated glass lattice consisting of a network of oxygen atoms held together in an irregular chain by silicon atoms. This lattice contains anionic sites capable of attracting cations of appropriate size-to-charge ratio.

The mineral composition of the glass is critical. Minor change in composition will produce major changes in the electrode sensitivity. Three types of glass electrodes (the pH type, the cation type, and sodium type) are available. The general formula for all includes approximately 72% SiO_2, 6% CaO, and 21% Na_2O. One per cent Al_2O_3 produces a good pH-sensitive glass membrane with little metal response. The concentration of oxides of sodium and aluminium are varied to make all three electrode types.

At the hydrated glass surface, exchange of alkaline metal ions in the lattice for hydrogen ions in solution plays a major part in the development of the pH response. Hydration generally occurs to the extent of 50-100 mg of water per cm^3 of glass, producing an average membrane with a dry glass thickness of 50 μ and an overlying hydrated layer on each surface of 50-100 Å. The hydration process causes swelling of the glass, but constant dissolution of the surface of the hydrated area maintains the glass membrane's total thickness at a steady state. The glass needs to be of relatively high chemical durability, or alkaline earths within the glass will dissolve too rapidly and, in solution, will erode glass and electrode alike over a period of months. The rate of dissolution of glass is one major factor determining the electrode's life.

For best performance, pH-sensitive glass electrodes should be:

1. stored in a buffer solution *near* the pH at which the electrode will be used
2. never scracthed
3. never cleaned with chromic acid or a dehydrating agent
4. washed, if necessary, with 6M hydrochloric acid or only with solutions recommended by the manufacturer.

Precipitate Electrode

These anionic-sensitive electrodes exchange anions through a membrane containing cationic sites. One type uses polymerized silicon rubber diffusely permeated with small grains (5-10 μ) of a silver halide such as AgI. Each electrode exchanges best for the anion in common with its precipitate; that is, I^- for AgI impregnated electrodes, Cl^- for AgCl, and Br^- for AgBr. The electrodes are relatively insensitive to redox interferences and surface poisonings. They are relatively insensitive to cation effect.

Solid-State Electrode

These electrodes, similar to precipitate electrodes, separate an external (unknown) from an internal (standard solution containing) electrode by an inorganic single crystal "doped" with a rate earth. In the fluoride-sensitive electrode, this thin membrane consists of crystalline lanthanum fluoride doped with europium (II), added to reduce electrical resistance and facilitate ionic charge transport. These solid-state electrodes show remarkable specificity for fluoride ion. The hydroxyl ion offers the only major interference.

Liquid-Liquid Electrode

In these electrodes, custom-made liquid ionic exchange solutions of high specificity are separated from the test solution by synthetic membrane or porous glass permeable to the ion in question. Transport of such ions across the membrane, to or from the liquid ion exchanger, causes development of measurable potential difference. The ionic

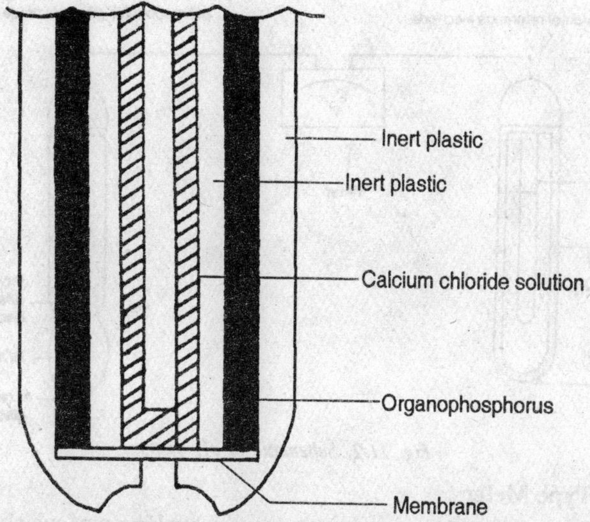

Fig. 11.1. Schematic of calcium ion-specific electrode.

exchange solution must be in contact with the sample solution (via membrane), yet mixing must be minimal. The exchanger must be relatively insoluble in the sample to reduce contamination and must exhibit photochemical and thermal stability.

In the calcium electrode, calcium organophosphorus is used as the ionic exchange compound. It is active between pH 5.5 and 11. Higher pH levels cause interference by calcium hydroxide formation. Reagents that complex calcium and prevent ionization, such as phosphate buffers, and so forth, cannot be used. A calomel reference electrode is used to develop a circuit.

A standard of ionic calcium is used (which has sodium, potassium, and magnesium ions added at levels approxi-mately those of the unknown).

pH Meter

The pH meter[3] is an instrument for the potentiometric comparison of a change in emf (voltage) between a glass electrode and a reference electrode created by the relative hydrogen ion concentration in the unknown solution completing the circuit.

The meter must be capable of measuring emf to 0.1 mV (equivalent to 0.002 pHu). At this millivoltage an extremely small current will flow, in the neighborhood of 10^{-8} A, because of the high internal resistance of the glass electrode.

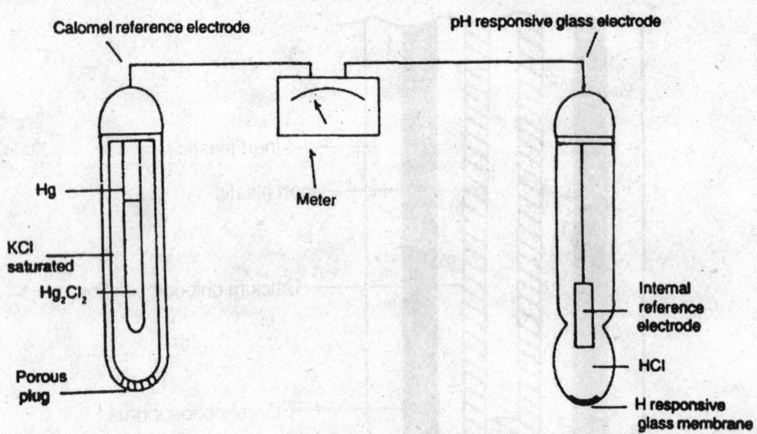

Fig. 11.2. Schematic of pH system.

Null-Type Meter

The null-type instrument introduces a bucking emf which balances the circuit and returns the meter to its original position after it is changed by introduction of a solution with hydrogen ion concentration different from that of the calibrating solution. The null principle is similar to that used in spectrophotometers with a galvanometer measuring system.

Direct Reading-Type Meter

In a direct reading instrument, the actual emf produced by a change in pH is amplified and displayed digitally or on a meter.

At any given temperature, the change in voltage resulting from a change in $[H^+]$ will be directly proportional to the change in log $[H^+]$. Thus, the change in voltage is *linear* with respect to change in pH.

The instrument converts a change in to a change in pH units at the various electrode temperatures. The instrument's temperature control knob alters the constant proportionally between *voltage* and *scale reading*. For example, an electrode which would reflect a change of 59.1 mV for a change of 1.00 pH units at 25°C would show a change of some 74 mV for the same change in pH, 1.00, at 100°C. In other words, the instrument's temperature control knob will accommodate the temperature coefficient. Temperature compensation therefore represents a slope adjustment relating the changes in pH to changes in millivolts.

The pH meter is calibrated by standardizing it against buffers of known pH at two or three different values, above and below the value of the unknown. By using buffers of pH and ionic strength similar to those of the unknown and standards, and carefully reproducing the physical and chemical characteristics of the liquid junction, the variables can be effectively limited to the hydrogen ion concentration in the unknown solution.

Glass electrodes operate best at pH between 2 and 12 and ionic strength between 0.01 and 0.1.

Proper pH measurements require the following:

1. The sample and standard buffers must be treated identically.
2. The electrode must reflect the correct voltage to pH ratio at all times.
3. The pH meter must translate millivolts to pH units correctly regardless of the temperature at which the measurements are taken.
4. Meter standardization is best with buffers covering the entire useful rage. Utilizing a buffer with temperature, pH, and ionic strength similar to that of the unknown will reduce the variables created by liquid junction potential difference and thermal effects.

Location of Trouble

pH meter problems are generally due to the following:

1. scratched or cracked glass electrodes
2. dirt on electrodes
3. contaminated KCl
4. clogged ceramic aperture
5. inaccurate buffer
6. unusable or improperly collected sample.

Most of these should be avoided or detected prior to faulty use by an appropriate system of preventive maintenance and quality control.

Functional characteristics of a pH meter such as reproducibility, drift, noise, response time, and linearity should be periodically determined and can be utilized with advantage in any quality control program. A through review of check systems for pH meters should be consulted for guidance in setting up a quality control program.

Partial Pressure of Carbon Dioxide

The pCO_2 electrode takes advantage of the fact that pH has a linear relationship to the log pCO_2 over the range of 10-90 mm Hg (1.4-11.4%).

The following equilibrium exists in blood as well as in bicarbonate buffer:

$$PCO_2 \overset{k}{\rightleftharpoons} H_2CO_3 \rightleftharpoons HCO_3^- + H^+ \quad ...(1)$$

The dissociation equation can be written as follows:

$$K = \frac{[H^+][HCO_3^-]}{[H_2CO_3]} \text{ or } K = \frac{[H^+][HCO_3^-]}{[k \bullet pCO_2]} \quad ...(2)$$

and, substituting into eq. (2),

$$\log [H^+] + \log [HCO_3^-] = \log K + \log k + \log pCO_2 \quad ...(3)$$

Because pH equals the negative log of the hydrogen ion concentration and pK equals the negative log of the dissociation constant k,

$$pH - \log [HCO_3^-] = pK + pK - \log pCO_2 \quad ...(4)$$

pK and log k are constants. Therefore,

$$pH \cong \log [HCO_3^-] - \log pCO_2 \quad ...(5)$$

and pH has a linear dependence on log pCO_2 when equilibrium has been established.

The pCO_2 Electrode

The pCO_2 electrode interposes a semipermeable (to CO_2 dissolved as a gas only) membrane between the unknown solution (usually blood) and a weak bicarbonate buffer into which is immersed both a pH-sensitive glass electrode and a reference electrode (usually silver-silver chloride).

After the CO_2 physically dissolved in the sample or as a calibrating gas or solution reaches equilibrium with the bicarbonate buffer across the semipermeable membrane, some hydrogen ion concentration change within the buffer may occur.

Types of Glass Electrodes

Several different types of glass electrodes are commercially available. Reasonably rugged glass electrodes which operate well from 0 to 80°C and up to a pH of about 10 are not very expensive. Several different reference electrodes are also stocked by most scientific supply houses. In more recent years *combination electrodes*, in which both the reference electrode and the glass electrode are incorporated into one slender glass tube, have come into general use. These electrodes may be used to determine the pH of small volumes of solutions; however, they have the disadvantage that, if one element of the electrode misfunctions, the entire electrode must be discarded.

Use and Care of pH Meters

Before using a pH meter the electrodes must be washed thoroughly with distilled water and blotted gently with tissue. The temperature compensation knob should be adjusted to the temperature of the test solution and the standard buffers to be used to calibrate the meter. Be aware of the fact that the pH values of the standard buffers are temperature-dependent. This temperature-dependence is a natural consequence of the temperature-dependence of the dissociation constant of the buffers. For example, a common standard buffer has a pH of 7.0 ± 0.02 at 25°C, but at 0°C the pH is 7.12 ± 0.02. To use the meter, immerse the electrodes into the buffer solution and adjust the meter with the calibration knob to the known pH of the buffer at that temperature. Rinse the electrodes and check the operation of the pH meter with another standard buffer, for example a buffer of pH 4 or pH 10. Ideally, the pH of each of the pair of buffers selected to calibrate the meter should overlap the expected pH of the test solution. If the pH of the second buffer is within 0.05 pH of the expected value, the pH meter is working properly. The use of two standard buffers of different pH is the only reliable way of checking a pH meter. Since the characteristics of the glass electrode change with time, the calibration of the meter must be checked at regular intervals.

Once the meter is properly calibrated, the pH of the test solution may be determined. After use, rise the electrodes with distilled water. Since the glass electrode is very sensitive to dehydration, it is important not to let it dry out excessively. Keep the glass electrode immersed in distilled water when not in use.

Errors in pH Measurement

The Sodium Error

Most pH-responsive glasses are somewhat sensitive to Na^+. In fact, glass electrodes have been prepared which are more sensitive to alkaline earth cations than they are to H^+ (the so-called *cation electrodes*). At high pH as little as 0.1 *N* Na^+ can cause an apparent pH decrease of 0.4 unit. A curve for the correction of the sodium error is provided by the manufacturers of glass electrodes. This problem may be alleviated by the use of potassium salts rather than sodium salts or by the use of the ewer glass electrodes with reduced sensitivity to Na^+.

Temperature and Ionic Strength

The dissociation of acids and bases is temperature-dependent. Thus, the pH of a buffer solution can change dramatically with temperature. For example, tris (hydroxy methyl) aminomethane (tris buffer), a very commonly used buffer in biochemical research, has a dissociation constant which is strongly temperature-dependent. The $\Delta pK_a/\Delta T$ is — 0.031 per degree. In practice this means that a solution of Tris with a pH of 8.4 at 0°C would have a pH of 7.4 at 37°C. It is, therefore, imperative that the pH of a buffer solution be adjusted at the temperature at which it will be used.

Activity coefficients of the charged species of a buffer are affected by the ionic strength. Thus, addition of a salt at concentrations exceeding 0.025 *M* to a buffer solution can have profound effects on the pH. Further, dilution of a buffer will also generally have an effect on the pH. The importance of temperature and ionic strength cannot be overemphasized. The pH of a buffer solution should be determined directly under the conditions that the buffer is to be used.

Fouling of the Electrodes by Biological Samples

A pronounced drift in the pH of a solution may sometimes be noticed. In some cases, this drift may be caused by a small crack in the electrode; in other cases, however, it may be caused by the deposition of a thin coating of protein on the glass membrane. This coating may be removed by rinsing the electrode in a strong detergent solution.

If the pH of samples containing suspended materials is determined, the fiber junction of the calomel electrode can readily become plugged. Since the fiber junction forms the salt bridge with the solution, it is imperative that it be clean. A plugged junction may be cleaned by briefly immersing the tip of the electrode into a solution of warm, concentrated nitric acid.

12

Activator of Life

A living organism is a magnificent assembly of chemical reactions, an entity rather than chaos because most of its components make an orderly contribution to existence of the whole. Many of these reactions occur at significant rates only because specific proteins—the enzymes—are present to catalyze them. The enzymes are constructed to control both the kinds of reaction and the rates at which they occur, and it is the existence of these proteins that creates a complex harmony of chemical function.

Discovery of Biological Catalysts

In the early nineteenth century the Swedish chemist Jon Berzelius discovered that an extract of potatoes is more effective than concentrated sulphuric acid in promoting the breakdown of starch. Berzelius concluded that this effect is due to the presence of biological catalysts and with remarkable insight went on to predict that all materials found in living organisms are synthesized under the influence of such catalysts. This observation stimulated interest in the biochemical identity of these catalysts, and Louis Pasteur soon postulated that the catalytic effect is intimately associated with cell structure and so cannot be separated from living cells. If true, this meant that the catalysts could not be isolated and studied in purified form. It was therefore a great milestone when, in 1897, Edward Buchner demonstrated that an extract of yeast, from which all intact cells had been removed, is capable of catalyzing the breakdown of glucose. This was an especially significant observation because glucose degradation is a major metabolic pathway in cells. This finding firmly established that biological catalysts, in the absence of any cell organization, are capable of enhancing the rates of metabolic reactions.

The increasing importance of this class of molecules soon led to suggestions for a uniform nomenclature system to facilitate communication among biochemists. The term *enzyme* was adopted as a general designation for biological catalysts, and it was decided that individual enzymes would be named by adding the suffix "ase" to the name of the *substrate* (the substance upon which the enzyme acts). For example, the enzyme catalyzing the degradation of urea was termed urease, enzymes catalyzing hydrolysis of nucleic acids were designated nucleases, the enzyme catalyzing glucose oxidation was called glucose oxidase, and so forth. This basic approach to enzyme nomenclature is still in use today, although some older names introduced prior to the adoption of this system (e.g., trypsin pepsin) have been retained.

In the years immediately following the initial demonstration of cell-free enzyme activity by Buchner, a large number of other enzymes were identified by biochemists. But little progress was made in understanding the structure or mechanism of these cellular catalysts. An example of the difficulties experienced during this period is the classic story of Richard Willstatter's encounters with the enzyme *peroxidase*, one of the first enzymes to be obtained in a highly purified form. When Willstatter chemically analyzed his purified enzyme preparation, he found no lipid, no sugar (and therefore no nucleic acid), and no protein. These results led to the prevailing opinion in the early twentieth century that enzymes belong to none of the commonly recognized classes of biological molecules.

In 1926, however, James Summer reported the first crystallization of an enzyme, in this case urease. He found that the crystals consisted of pure protein, even after repeated cycles of recrystallization. Since crystallization is generally recognized as the ultimate criterion of purity, these results suggested to Summer that the enzyme urease is a protein. Although this conclusion initially met with skepticism, similar experiments by John Northrop and his colleagues on other enzymes soon led to general acceptance of the idea that enzymes are made of protein. Willstatter, it turns out, had been misled by the fact that enzymes are active in extremely small quantities. Thus his peroxidase preparation exhibited measurable enzyme activity even though the protein content of the sample was too low to be detected by the analytical techniques then available.

In the years since the pioneering work of Summer and Northrop, over a thousand enzymes have been purified and identified as proteins. Though the number of different reactions catalyzed by these enzymes is very large, one can subdivide this group into a relatively small number of categories. All cellular reactions are catalyzed by enzymes that fit into one of these categories.

As Berzelius predicted many years ago, enzymes act as extraordinarily efficient chemical catalysts. Enzymes therefore exhibit the four defining characteristics shared by all catalysts: (1) They modify the rates of chemical reactions. (2) They are not consumed in the process. (3) They are effective in extremely small quantities. (4) They do not alter the equilibrium of the reaction being catalyzed.

Enzymes are Highly Specific

Enzymes are highly specific both in the reaction catalyzed and in their choice of reactants, which are called *substrates*. An enzyme usually catalyzes a single chemical reaction or a set of closely related reactions. The degree of specificity for substrate is usually high and sometimes virtually absolute.

Let us consider *proteolytic enzymes* as an example. The reaction catalyzed by those enzymes is the hydrolysis of a peptide bond.

$$\underset{\text{Peptide}}{\ldots\text{N(H)–C(H)(R}_1\text{)–C(=O)–N(H)–C(H)(R}_2\text{)–C(=O)}\ldots} + H_2O = \underset{\text{Carboxyl component}}{\ldots\text{N(H)–C(H)(R}_1\text{)–C(=O)O}^-}$$

$$+ \underset{\text{Amino component}}{{}^+H_3N\text{–C(H)(R}_2\text{)–C(=O)}\ldots}$$

Most proteolytic enzymes also catalyze a different but related reaction, namely the hydrolysis of an ester bond.

$$\underset{\text{Ester}}{R_1\text{–C(=O)–O–}R_2} + H_2O = \underset{\text{Acid}}{R_1\text{–C(=O)O}^-} + \underset{\text{Alcohol}}{HO\text{–}R_2} + H^+$$

Proteolytic enzymes vary markedly in their degree of substrate specificity. Subtilisin, which comes from certain bacteria, is quite undiscriminating about the nature of the side chains adjacent to the peptide bond to be cleaved. Trypsin, as was mentioned in previously is quite specific in that it splits peptides bond on the carboxyl side of lysine and arginine residues only. Thrombin, an enzyme participating in blood clotting, is even more specific than trypsin. The side chain on the carboxyl side of the susceptible peptide bond must be arginine, whereas the one on the amino side must be glycine.

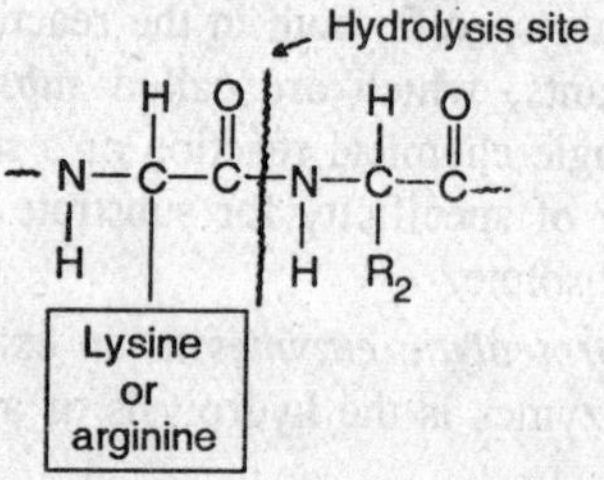

Hydrolysis site

Lysine or arginine

Fig. 12.1. Specificity of trypsin.

Another example of the high degree of specificity of enzymes is provided by DNA polymerase I. This enzyme synthesizes DNA by linking together four kinds of nucleotide building blocks. The sequence of nucleotides in the DNA strand that is being synthesized is determined by the sequence of nucleotides in another DNA strand that serves as a template DNA polymerase I is remarkably precise in carrying out the instructions given by the template. The wrong nucleotide is inserted into a new DNA strand less than once in a million times.

Some enzymes are synthesized in an *inactive precursor form* and are activated at a physiologically appropriate time and place. The digestive enzymes exemplify this kind of control. For example, trypsinogen is synthesized in the pancreas and is activated by peptide-bond cleavage in the small intestine to form the active enzyme trypsin.

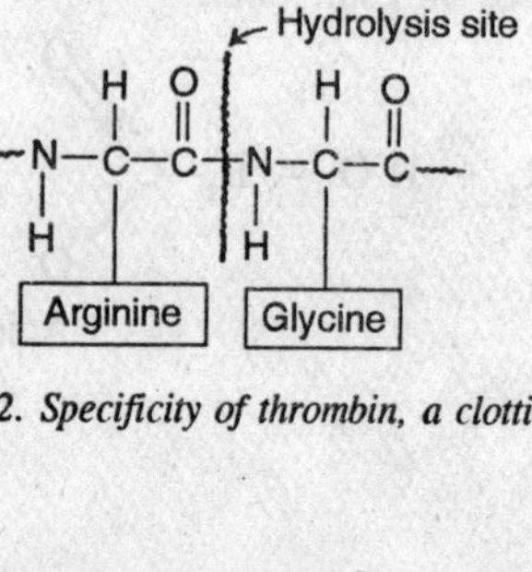

Hydrolysis site

Arginine Glycine

Fig. 12.2. Specificity of thrombin, a clotting factor.

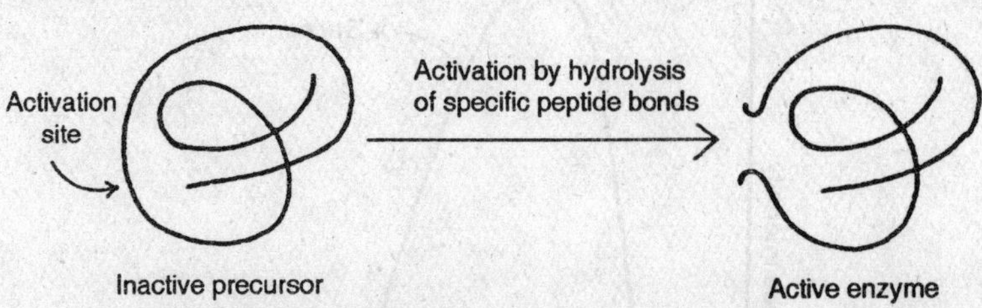

Fig. 12.3. Zymogen activation by hydrolysis of specific peptide bonds.

This type of control is also repeatedly used in the sequence of enzymatic reactions leading to the clotting of blood. The enzymatically inactive precursors of proteolytic enzymes are called *zymogens*.

Another mechanism that controls activity is the covalent insertion of a small group on an enzyme. The control mechanism is called *covalent modification*. For example, the activities of the enzymes that synthesize and degrade glycogen are regulated by the attachment of a phosphoryl group to a specific serine residue on these enzymes. This modification can be reversed by hydrolysis. Specific enzymes catalyze the insertion and removal of phosphoryl and other modifying groups.

A different kind of regulatory mechanism affects many reaction sequences resulting in the synthesis of small molecules such as amino acids. The enzyme that catalyzes the first step in such a biosynthetic pathway is inhibited by the ultimate product. The biosynthesis of isoleucine in bacteria illustrates this type of control, which is called *feedback inhibition*. Threonine is converted into isoleucine in five steps, the first of which is catalyzed by threonine deaminase. This enzyme is inhibited when the concentration of isoleucine reaches a sufficiently high level. Isoleucine binds to a regulatory site on the enzyme, which is distinct from its catalytic site. The inhibition of threonine deaminase is mediated by an *allosteric interaction*, which is reversible. When the level of isoleucine drops sufficiently, threonine deaminase becomes active again, and consequently isoleucine is again synthesized.

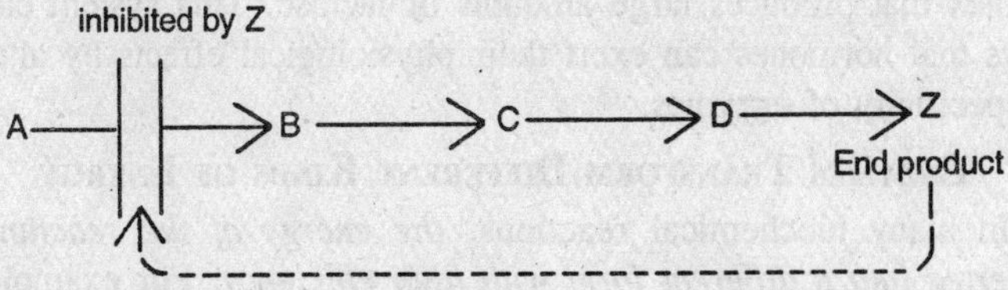

Fig. 12.4. Feedback inhibition of the first enzyme in a pathway by reversible binding of the final product.

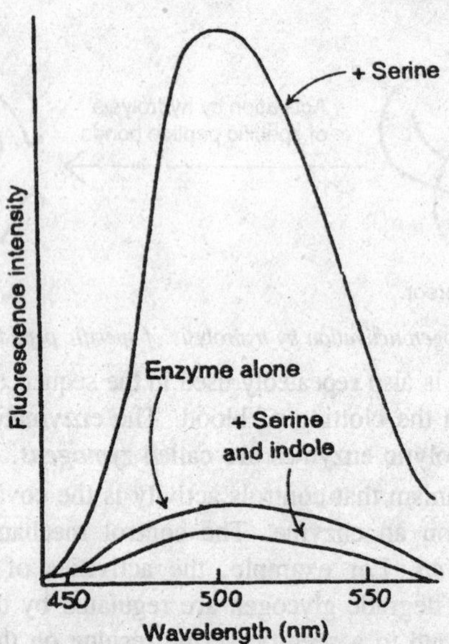

Fig. 12.5. Florescence intensity of the pyridoxal phosphate group at the active site of tryptophan synthetase changes upon addition of serine an dindole, the substrates.

The specificity of some enzymes is under physiological control. The synthesis of lactose by the mammary glands is a particularly striking example. Lactose, synthetase, the enzyme that catalyzes the synthesis of lactose, consists of a catalytic subunit and a modifier subunit. The catalytic subunit by itself cannot synthesize lactose. It has a different role, which is to catalyze the attachment of galactose to a protein that contains a covalently linked carbohydrate chain. *The modifier subunit alters the specificity of the catalytic subunit* so that it links galactose to glucose to form lactose. The level of the modifier subunit under hormonal control. During pregnancy, the catalytic subunit is formed in the mammary gland but little modifier subunit is formed. At the time of birth, hormonal levels change drastically, and the modifier subunit is synthesized in large amounts. The modifier subunit then binds to the catalytic subunits to form an active lactose synthetase complex that produces large amounts of lactose. This system clearly shows that hormones can exert their physiological effects by altering the specificity of enzymes.

Enzymes Transform Different Kinds of Energy

In many biochemical reactions, *the energy of the reactants is converted into a different form with high efficiency*. For example, in photosynthesis, light energy is converted into chemical-bond energy. In mitochondria, the free energy contained in small molecules derived from foods is converted into a different currency, that of adenosine triphosphate (ATP). The chemical-bond energy of ATP is then utilized in many different ways. In muscular contraction, the energy of ATP is converted into mechanical energy. Cells and organelles have pumps that utilize ATP to transport molecules and ions against chemical and electrical gradients. These transformations of energy are carried out by enzyme molecules that are integral parts of highly organized assemblies.

Enzymes do not alter Reaction Equilibria

An enzyme is a catalyst and consequently is cannot alter the equilibrium of a chemical reaction. This means that an enzyme accelerates the forward and reverse reaction by precisely the same factor.

Consider the interconversion of A and B. Suppose that in the absence of enzyme the forward rate (k_F) is 10^{-4} sec^{-1} and the reverse rate (k_R) is 10^{-6} sec^{-1}. The equilibrium constant K is given by the ratio of these rates :

$$A \underset{10^{-6}\ sec^{-1}}{\overset{10^{-4} sec^{-1}}{\rightleftharpoons}} B$$

$$K = \frac{[B]}{[A]} = \frac{k_F}{k_R} = \frac{10^{-4}}{10^{-6}} = 100$$

The equilibrium concentration of B is 100 times that of A, whether or not enzyme is present. However, it would take several hours to approach this equilibrium without enzyme, whereas equilibrium would be attained within a second when enzyme is present. Thus, enzymes accelerate the attainment of equilibria but do not shift their positions.

Enzymes Decrease the Activation Energies of Reactions Catalyzed by Them

A chemical reaction, A=B, goes through a *transition state* that has a higher energy than either A or B. The rate of the forward reaction depends on the temperature and on the difference in free energy between that of A and the transition state, which is called *Gibbs free energy of activation* and symbolized by $\Delta G^{\ddagger}$.

$$\Delta G^{\ddagger} = G_{\text{transition state}} - G_{\text{substrate}}$$

The reaction rate is proportional to the fraction of molecules that have a free energy equal to or greater than $\Delta G^{\ddagger}$. The proportion of

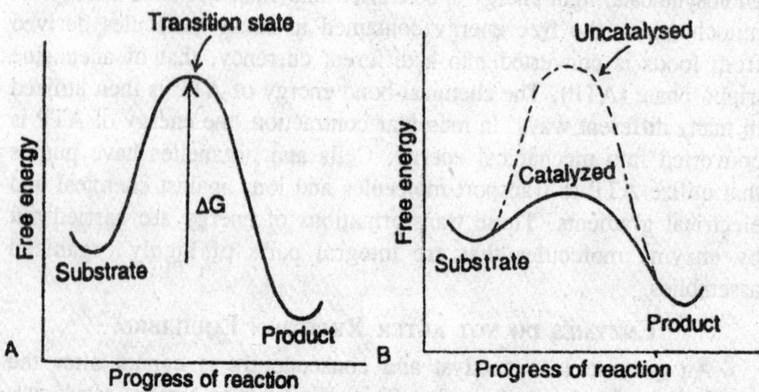

Fig. 12.6. A—Definition of $\Delta G^{\ddagger}$, the frec energy of activation. B—Enzymes accelerate catalysis by reducing $\Delta G^{\ddagger}$.

molecules that have an energy equal to or greater than $\Delta G^{\ddagger}$ increases with temperature.

Enzymes accelerate reactions by decreasing $\Delta G^{\ddagger}$, the activation barrier. The combination of substrate and enzyme creates a new reaction pathway whose transition-state energy is lower than it would be if the reaction were taking place in the absence of enzyme.

Formation of an Enzyme-Substrate Complex is the First Step in Enzymatic Catalysis

The making and breaking of chemical bonds by an enzyme are preceded by the formation of an *enzyme-substrate* (ES) complex. The substrate is bound to a specific region of the enzyme called the *active site*. Most enzymes are highly selective in their binding of substrate. Indeed, the catalytic specificity of enzymes depends in large part on the specificity of the binding process. Furthermore, the control of enzymatic activity may also take place at this stage.

The existence of ES complexes has been shown in a variety of ways :

1. ES complexes have been *directly visualized by electron microscopy and X-ray crystallography*. Complexes of nucleic acids and their polymerase enzymes are evident in electron micrographs. Detailed information concerning the location and interactions of glycyl-L-tyrosine, a substrate of carboxypeptidase A, has been obtained from X-ray studies of that ES complex.
2. The *physical properties* of an enzyme, such as its solubility or heat stability, frequently change upon formation of an ES complex.

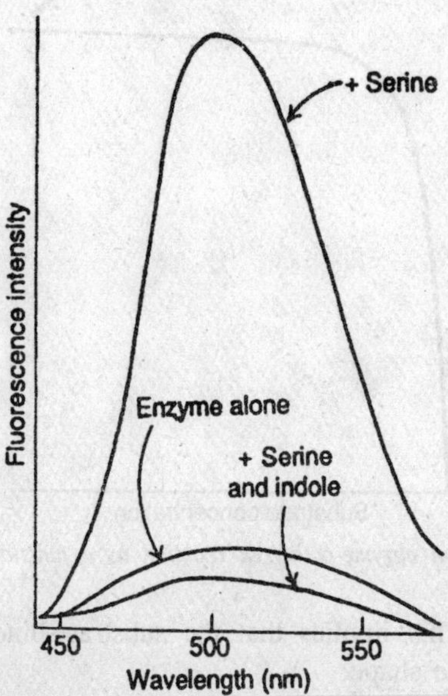

Fig. 12.7. Fluorescence intensity of the pyridoxal phosphate group at the active site of tryptophan synthetase changes upon addition of serine and indole, the substrates.

3. The *spectroscopic characteristics* of many enzymes and substrates change upon formation of an ES complex just as the absorption spectrum of deoxyhemoglobin changes markedly when it binds oxygen or when it is oxidized to the ferric state, as described previously. These changes are particularly striking if the enzyme contains a coloured prosthetic group. Tryptophan synthetase, a bacterial enzyme that contains a pyridoxal phosphate prosthetic group, affords a nine illustration. This enzyme catalyzes the synthesis of L-tryptophans from L-serine and indole. The addition of L-serine to the enzyme produces a marked increase in the fluorescence of the pyridoxal phosphate group. The subsequent addition of indole, the second substrate, quenches this fluorescence to level lower than that of the enzyme alone. Thus, fluorescence spectroscopy reveals the existence of an enzyme-serine complex and of an enzyme-serine indole complex. Other spectroscopic techniques, such as nuclear and electron magnetic resonance, also are highly information about ES interactions.
4. A high degree of *stereospecificity* is displayed in the formation of ES complexes. For example, D-serine is *not* a substrate of tryptophan synthetase. Indeed, the D-isomer does not even bind to

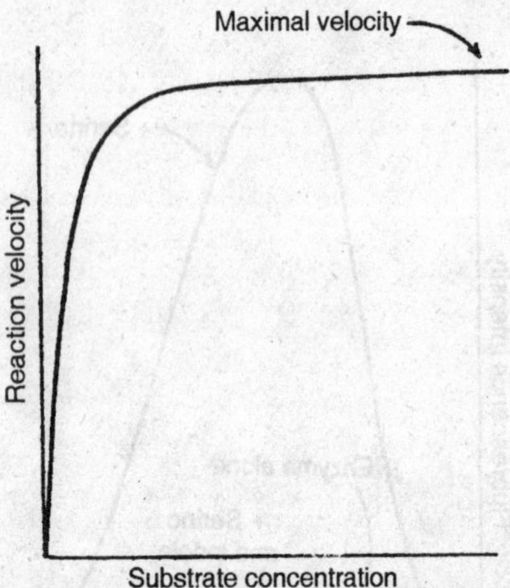

Fig. 12.8. Velocity of an enzyme-catalyzed reaction as a function of the substrate concentration.

the enzyme. This implies that the substrate-binding site has a very well define shape.

5. ES complexes can sometimes be *isolated in pure form*. For an enzyme that catalyzes the reaction A + B = C, it is sometimes possible to isolate an EA complex. This can be done if the enzyme has a sufficiently high affinity for A and B is absent from the mixture.
6. At a constant concentration of enzyme, the reaction rate increases with increasing substrate concentration until a maximal velocity is reached. In contrast, uncatalyzed reactions do not show this saturation effect. In 1913, Leonor Michaelis interpreted the *maximal velocity of an enzyme-catalyzed reaction* in terms of the formation of a discrete ES complex. At a sufficiently high substrate concentration, the catalytic sites are filled and so the reaction rate reaches a maximum. This is the oldest and most general evidence for the existence of ES complexes.

Some Features of Active Sites

The active site of an enzyme is the region that binds the substrates (and the prosthetic group, if any) and contributes the residues that directly participate in the making and breaking of bonds. These residues are called the *catalytic groups*. Although enzymes differ widely in structure, specificity, and mode of catalysis, a number of generalizations concerning their active sites can be stated :

1. *The active site takes up a relatively small part of the total volume of an enzyme.* Most of the amino acid residues in an enzyme are not in contact with the substrate. This raises the intriguing question of why enzymes are so big. Nearly all enzymes are made up of more than 100 amino acid residues, which gives them a mass greater than 10 kdal and a diameter of more than 25Å.

2. *The active site is a three-dimensional entity.* The active site of an enzyme is not a point, a line, or even a plane. It is an intricate three-dimensional form made up of groups that come from different parts of the linear amino acid sequence-indeed, residues far apart in the linear sequence may interact more strongly than adjacent residues in the amino acid sequence, as has already been seen for myoglobin and hemoglobin. In lysozyme, the important groups in the active site are contributed by residues numbered 35, 52, 62, 63 and 101 in the linear sequence of 129 amino acids.

3. *Substrates are bound to enzymes by relatively weak forces.* ES complexes usually have equilibrium constants that range from 10^{-2} to 10^{-8}M, corresponding to free energies of interaction ranging from –3 to –12 kcal/mol. These values should be compared with the strengths of covalent bonds, which are between –50 and 110 kcal/mol.

4. *Active sites are clefts or crevices.* In all enzymes of known structure, substrate molecules are bound to a cleft or crevice from which water is usually excluded unless it is a reactant. The cleft also contains several polar residues that are essential for binding and catalysis. The nonpolar character of the cleft enhances the binding of substrate. In addition, the cleft creates a micro-environment in which certain polar residues acquire special properties essential for then catalytic role.

5. *The specificity of binding depends on the precisely defined arrangement of atoms in on active site.* A substrate must have a matching shape to fit into the site. Emil Fischer's metaphor of the lock and key, stated in 1890, has proved to be an essentially correct and highly fruitful way of looking at the stereospecificity of catalysis. However, recent work suggests that the active sites of some enzymes are not rigid. In such an enzyme, the shape of the active site is modified by the binding of substrate. The active has a shape complementary to that of the substrate only *after* the substrate is bound. This process of dynamic recognition is called *induced fit*. Further, some enzymes preferentially bind a strained form of the substrate corresponding to the transition state.

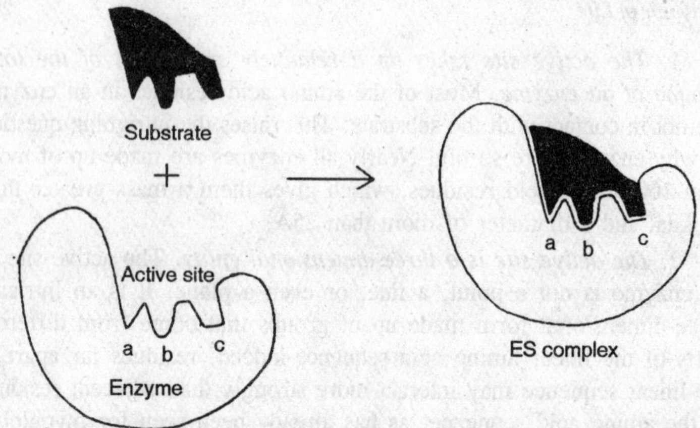

Fig. 12.9. Lock-and-key model of the interaction of substrates and enzymes.

Mechanism of Enzyme Actions

There is a very close structural relationship between the molecular surface of an enzyme and its substrates. Enzymes are protein molecules with definite surface geometry. The functional groups of the enzyme are exactly complementary to those of the substrate. Only particular types of substrate molecules will fit with a given enzyme molecule. For example, substrates A and B will fit into the enzyme E but not substrate C. This is referred to as a 'lock-and-key' mechanism. Thus reactions involving A and B will be speeded up but, not reactions involving C.

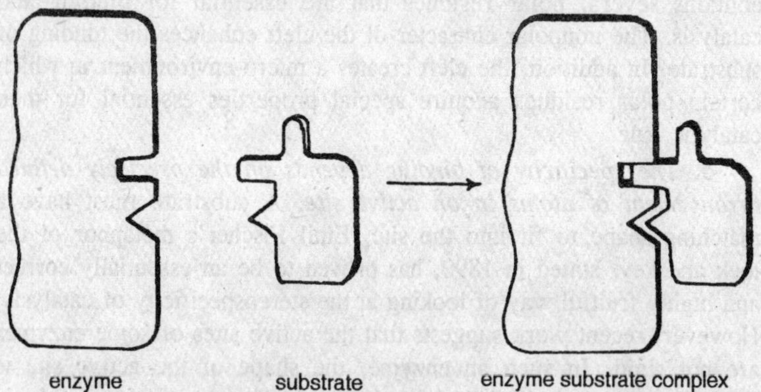

Fig. 12.10. Diagram illustrating how only particular substrates react with an enzyme.

The *enzyme* enters into a chemical combination with the *substrate* to form an *enzyme-substrate complex* (Michaelis-Menten hypothesis).

$$E + S \rightarrow ES$$

The enzyme substrate complex then breaks down to give the products of reaction. The enzyme is released and can be used over and over again.

$$ES \rightarrow E + Products$$

Active Site

An enzyme has a distinct cavity or cleft in which the substrate is bound. The cleft contains an *active centre* in which the amino acids are grouped together in such a way as to enable them to combine with the substrate. The reactive amino acids may lie widely separated in the polypeptide chain. The chain, however, undergoes folding in such a manner that the reactive amino acids come together in the active site.

It is believed that when the substrate molecule binds to the active site, its parts are hold together in such a way as to cause distortion of the chemical bonds, i.e., the bonds are weakened. This distortion of the chemical bonds of the substrate increases its reactivity, and thus speeds up the rate of the reaction. The products of the reaction are released because they are less firmly bound. The mechanism suggested above has been called the *strain model* of enzyme catalysis.

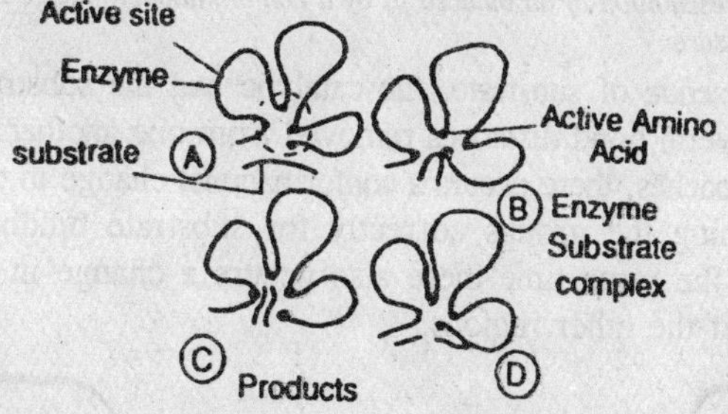

Fig. 12.11. Diagrammatic representation of an enzyme and its active site. A—Enzyme with its reactive site consisting of four reactive amino acid. B—Substrate binds to enzyme to form enzyme-substrate complex. C—Reaction occurs in which the substrate is broken down into its products. D—Products removed from active site.

Another model, called the *rack model*, suppose that the conformational change in the substrate after binding leads to *increased* distortion of the substrate molecule. This bending after binding makes the substrate more reactive.

Another way in which the reactive site is believed to speed up a reaction is by *excluding water molecules* of the solvent from the site of reaction. The tight fit of the substrate in the active site of the enzyme molecule does not permit water molecules at the site of reaction. It is known that in the case of certain compounds exclusion of water molecules greatly affects the rate of reaction.

Induced Fit Model

This model was given by Koshland (1966). In the Fischer model, i.e., lock-and-key mechanism, the catalytic site is presumed to be preshaped to fit the substrate. In the induced fit theory, the substrate induces a conformational change in the enzyme. This aligns amino acid residues or the other groups on the enzyme in the correct spatial orientation for substrate binding, catalysis or both. At the same time, the other amino acid residues may get buried in the interior of the enzyme.

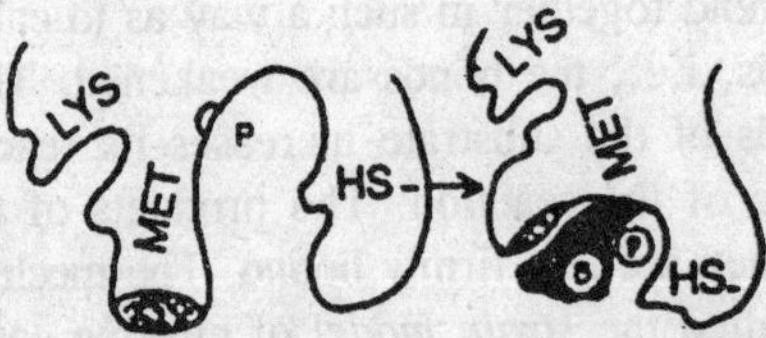

Fig. 12.12. Representation of an induced fit by a conformational change in the protein structure.

In the absence of substrate, the catalytic and the substrate-binding groups are several bond distances removed from one another. When the substrate approaches, there occurs a conformational change in the enzyme protein, aligning the groups correctly for substrate binding and for catalysis. At the same time there also occurs a change in the spatial orientations of the other regions.

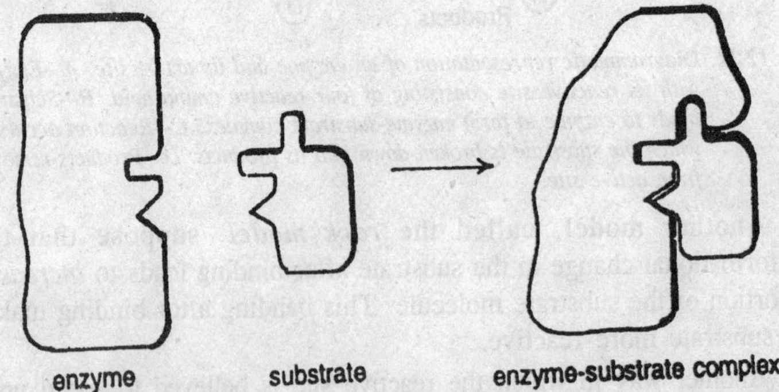

Fig. 12.13. Representation of alternative reaction paths for a substrate induced conformational change.

The main evidence in favour of induced fit model comes from demonstration of conformational changes during substrate binding and catalysis with creative kinase, phosphoglucomutase, and several other enzymes. Upon this time, it could not be established about the exact sequence of events in a substrate-induced conformational change. There may be several possibilities which are depicted.

Even if one knows the complete primary structure of an enzyme, it is not very easy to decide exactly which residues exactly constitute the catalytic site.

FACTORS AFFECTING ENZYME ACTIVITY

Substrate Concentration

Enzyme activity has been found to first increase with increase in concentration of the substrate to a maximum and then it levels off. At this concentration of substrate, amount of enzyme may probably become limiting.

Two types of graphs are obtained from the plotting of substrate concentration and enzyme activity. Straight line or Michaelis type of curve is obtained with the enzymes if they are made up of one unit.

If the enzyme are made up of the several subunits and they exhibit interaction amongst subunits (allosteric enzymes) then sigmoidal type of curve is obtained.

Table 12.1. Amino acid sequence of active sites of various enzymes

Enzyme	*Sequence of Active Site*
Trypsin	Gly—Asp—Ser—Gly—Gly—Pro
Chymotrypsin	Gly—Asp—Ser—Gly—Gly—Pro
Elastase	Gly—Asp—Ser—Gly—Gly—Pro
Thrombin	Asp—Ser—Gly
Alkaline phosphate	Asp—Ser—Ala
Futyrylcholine esterase	Gly—Glu—Ser—Ala—Gly
Acetylcholine esterase	Glu—Ser—Ala
Aliesterase	Glu—Ser—Ala

Concentration of Enzymes

Increase in enzyme concentration has been found to increase enzyme activity. When an excess of substrate is present, increasing the enzyme concentration by two times generally doubles the rate of formation of the product. With the given concentration of the enzyme also, a point

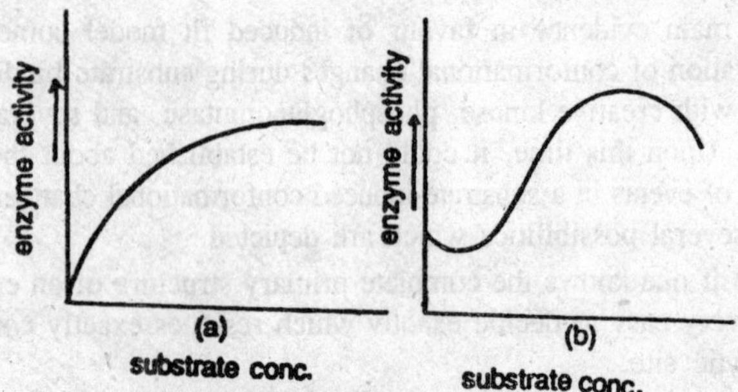

Fig. 12.14. Effect of substrate concentration on enzyme activity, (a)—Michaelis-Menten type curve, (b)—Sigmoidal curve.

may be reached where all the substrate molecules get bound to the enzyme and further increase in enzyme does not have any effect on the formation of the product.

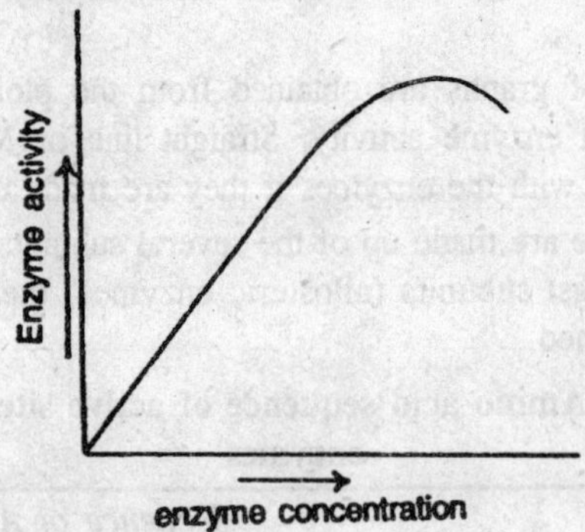

Fig. 12.15. Effect of enzyme concentration on enzyme activity.

Effect of pH

As enzymes are made up of proteins, they are very sensitive to change in pH. A graph of enzyme activity against pH is generally bell shaped. There is a optimum pH, which depends upon the nature of the enzyme. However, some enzymes may be having two optima of pH or no sharp pH but a range of pH.

At optimum pH the activity of enzymes is maximum. For most enzymes the effective pH range is 4-9, Beyond these limits denaturation of enzymes takes place. The optimum pH for pepsin is 2.0 and for trypsin 8.0.

Effect of ions

Many enzymes only become active in the presence of a cation such as Mg^{2+}, Ca^{2+}, Mn^{2+}, Zn^{2+}, Na^{+} or K^{+}. In some cases, the

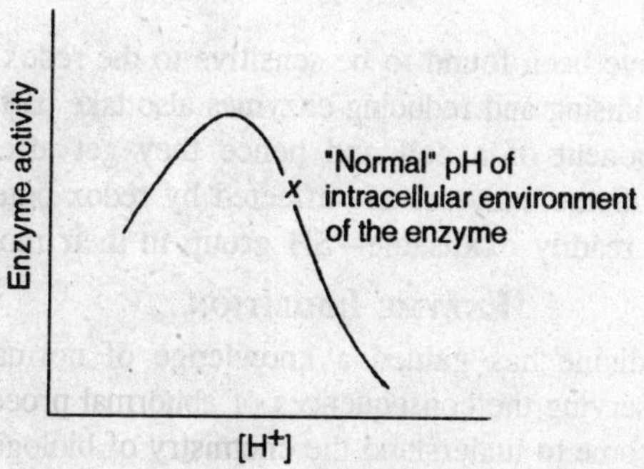

Fig. 12.16. Effect of pH.

cations get loosely bound to the enzyme, while in other cases they get bound to the substrate. Anions have also been found to increase the enzyme activity. For example, chloride enhances the activity of salivary amylase. Concentration of ions also influences the enzyme activity.

Effect of Temperature

Enzyme action is greatly affected by temperature. If the temperature is increased by 10°C the rate of most chemical reactions is doubled. However, at 40°C—60°C there is loss of enzyme activity because denaturation of proteins occurs at this temperature. There are few enzymes which continue to be active above 60°C also. The enzymes of dry tissues such as seeds and spores, are more resistant to high temperature.

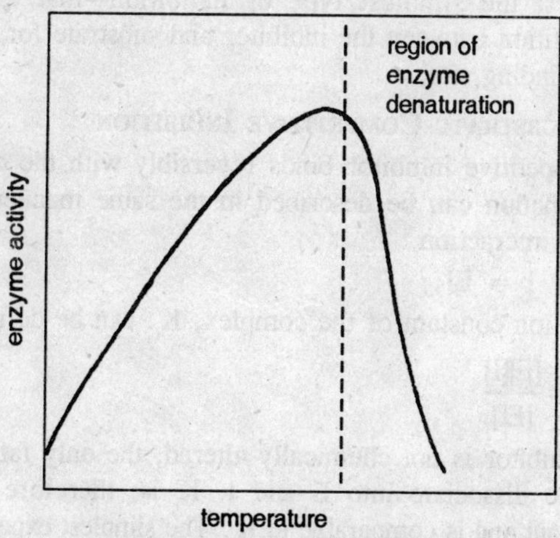

Fig. 12.17. Effect of increasing temperature on the rate of an enzyme-catalyzed reaction.

Redox Potential

Enzymes have been found to be sensitive to the redox potential of the cell also. Oxidising and reducing enzymes also take part in changing the redox component of a cell and hence they get affected by the redox potential. Some enzymes are affected by redox potential due to the presence of readily oxidisable—SH group in their molecule.

Enzyme Inhibition

Just as medicine has gained a knowledge of normal metabolic pathways by observing the consequences of abnormal processes, so the biochemist has come to understand the chemistry of biological catalysis through analyses of enzyme inhibition. Directly or indirectly, all inhibitors affect the union of enzyme and substrate and/or the catalytic consequences thereof. The resulting inhibition of catalytic function can be categorized as competitive, uncompetitive, or noncompetitive. In all cases the effects of the inhibitor will be reflected as changes in the kinetics of enzyme-substrate interactions:

$$E + S = ES$$

To characterize the mechanism by which a particular inhibitor is adversely affecting an enzyme reaction, questions such as the following must be answered. Is the degree of inhibition at a fixed enzyme concentration and fixed inhibitor concentration dependent on [S]? Is the degree of inhibition time dependent, suggesting that the inhibitor may be combining irreversibly with the enzyme?

Consider first the simplest type of inhibition—that due to a reversible competition between the inhibitor and substrate for binding at the substrate-binding site.

Substrate-Competitive Inhibition

When a competitive inhibitor binds reversibly with the enzyme, the complex formation can be described in the same manner as the enzyme-substrate interaction :

$$E + I = EI$$

The dissociation constant of the complex, K_i, can be defined as

$$K_i = \frac{[E][I]}{[EI]}$$

Since the inhibitor is not chemically altered, the only fate of the EI complex is to dissociate into E and I. K_i is, therefore, a true dissociation constant and is comparable to K_s. The simplest experimental test of substrate-competitive inhibition is to determine whether the degree of inhibition at a fixed inhibitor concentration can be decreased by increasing the substrate concentration. Referring again to the relationship between v_0 and [S], the presence of a fixed amount of inhibitor with varying amounts of S will give the following plots:

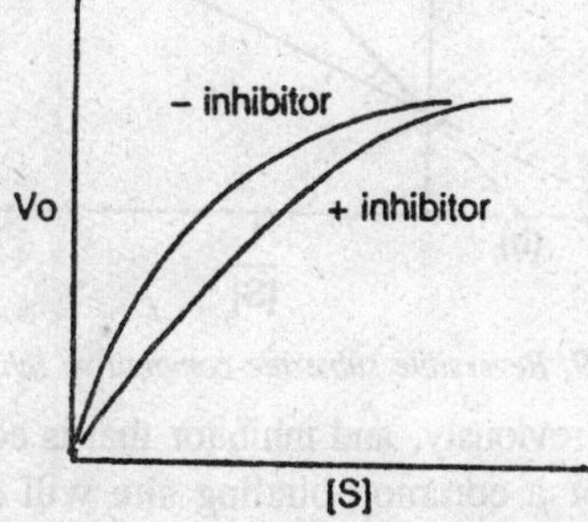

As would be predicted from the Michaelis-Menten model, a higher [S] is required to reach half-maximal velocity for the inhibited reaction than for the uninhibited reaction. Theoretically, the inhibition can be completely reversed by making the [S] [I] ratio very high; the two curves will merge asymptotically as V_{max} is approached.

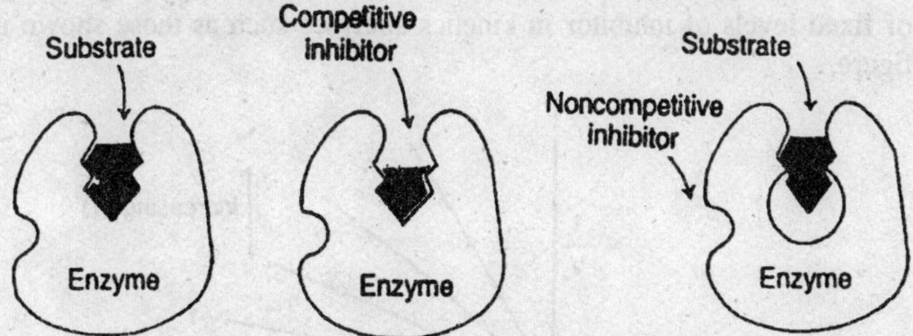

Fig. 12.18. Distinction between a competitive inhibitor and a noncompetitive inhibitor: (left) enzyme-substrate complex; (middle) a competitive inhibitor prevents the substrate from binding; (right) a noncompetitive inhibitor does not prevent the substrate from binding.

Just as the kinetic parameters of the Michaelis-Menten relationship are conveniently evaluated by a double-reciprocal plot the kinetics of an inhibited enzyme reaction can also be analyzed by plots of 1/[S] versus $1/v_0$. Thus, when the [I] is kept constant and the [S] is varied, the uninhibited and competitively inhibited reactions will be related. The extrapolated intercepts will be

$$\text{(a)} -\frac{1}{K_M} \text{ and } \text{(b)} -\frac{1}{K_M(1+[I]/K_I)}$$

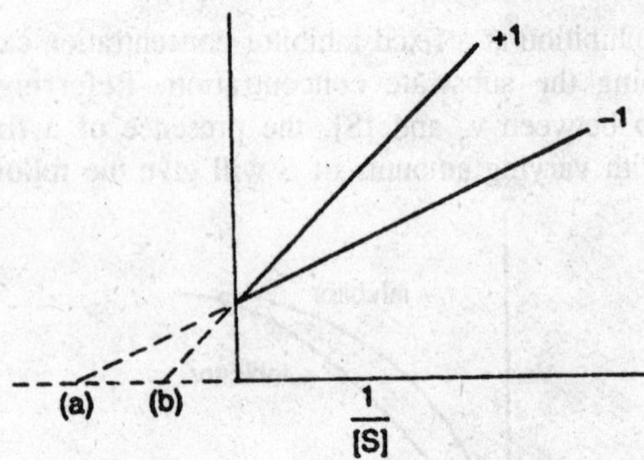

Fig. 12.19. Reversible substrate-competitive inhibition.

As emphasized previously, and inhibitor that is competing reversibly with the substrate for a common binding site will not affect the V_{max} of the system. This type of inhibitor does not affect the rate of breakdown of the enzyme-substrate complex to E+S or to E+P. Accordingly, one criterion of substrate-competitive inhibition is the common intercept. $1/V_{max}$, in plots of $1/v_0$ versus 1/[S]. Corroborative evidence for concluding that an inhibition is of the single substrate-competitive type and also reversible is obtained by employing a range of fixed levels of inhibitor in kinetics analyses such as those shown in figure.

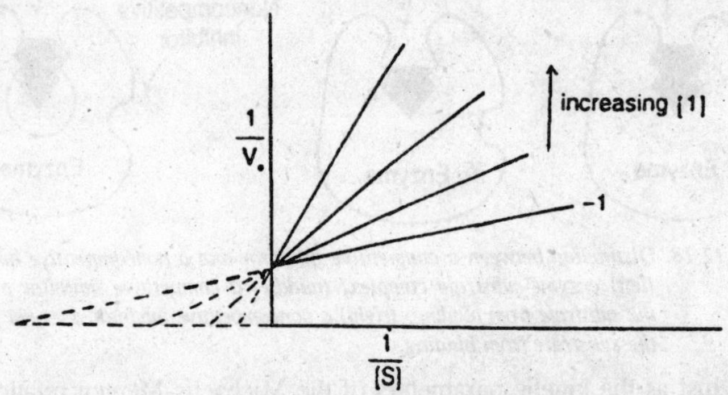

The mathematical formulation describing the reversible competitive between inhibitor and substrate must account for the following processes :

$$E + S = ES = E + P$$

$$E + I = EI \text{ (inactive)}$$

The final form of the Michaelis-Menten relationship in this case is

$$\frac{1}{v_0} = \frac{K_M}{V_{max}}\left(1 + \frac{[I]}{K_i}\right)\frac{1}{[S]} + \frac{1}{V_{max}}$$

In the reciprocal plot of v_0 versus [S] of the inhibited reaction, the intercept on the ordinate, I/V_{max}, remains the same as for the uninhibited reaction as pointed out above. The slope of the line now becomes

$$\frac{K_M}{V_{max}}\left(1 + \frac{[I]}{K_I}\right)$$

Compared to the uninhibited reaction the slope has been increased by a factor of 1 + $[I]/K_I$. Equation can also be used to calculate K_I, the dissociation constant of the inhibitor-enzyme complex. As will be discussed in later section, a knowledge of the values of K_M and K_I is not only relevant to our understanding of enzyme mechanisms but also has application in the design and clinical use of chemotherapeutic agents.

It was noted in comparing the hyperbolic plots of v_0 versus [S] in the substrate-competitive inhibitor system that to attain the same v_0 in both the uninhibited and inhibited reactions, the [S] in the inhibited reaction must be higher than in the uninhibited system. This is a reasonable consequence of the two competing reactions:

$$E + S = ES = E + P$$

and $$E + I = EI$$

At any given moment some of the enzyme, being in the form of EI, is unavailable for ES formation. It is important to realize, however, that the presence of I affects neither the association of E and S nor the dissociation of the ES complex. The value of K_M is the same in the presence of I as in its absence. Just as the I does not alter the K_M, neither does it affect V_{max}. As seen in the plot on previous page and in Figure. When the [S]/I ratio is very large, the V_{max} of the inhibited reaction is the same as the V_{max} of the uninhibited system.

The biochemical literature is replete with examples of substrate-competitive inhibition of enzyme systems and their application to medicine. Almost without exception, the inhibitors and substrates have striking similarities with regard to functional groups and/or molecular conformation.

Noncompetitive Inhibition

Alteration of the conformation of an enzyme can be expected to affect its catalytic activity. An reagent that can combine reversibly with the enzyme in some region or with some functional group outside the substrate-binding site may cause a structural change in the protein if the affected region or site is needed to maintain the native shape of the molecule. Such a conformational change in turn results in a modified substrate-binding site. The kinetic consequences of such structural changes can be both a decreased rate of ES formation as well as decreased rate of ES breakdown to E and reaction product(s). The inhibitor can combine with either E or ES :

$$E + I = EI$$

and

$$ES + I = ESI$$

The dissociation constants of these inhibitor complexes are

$$K_{EI} = \frac{[E][I]}{[EI]}$$

and

$$K_{ESI} = \frac{[ES][I]}{[ESI]}$$

It may also be assumed that EI can react with S :

$$EI + S = ESI$$

The K for this reaction, K'_{ESI}, may not have the same value as the K_{ESI} for reaction (22). Thus, the "composite kinetics" of a noncompetitively inhibited enzyme reaction is complicated. Assuming that the three K values cited above are identical and that EI and ESI are inactive, the reciprocal form of the Michaelis-Menten relationship is

$$\frac{1}{v_0} = \frac{K_M}{V_{max}}\left(1+\frac{[I]}{K_I}\right)\frac{1}{[S]} + \frac{1}{V_{max}}\left(1+\frac{[I]}{K_I}\right)$$

The plot of this equation is shown in figure. The common intercept obtained by extrapolation of the inhibited and noninhibited plots is — $1/K_M$. Although the K, for the two reactions

$$E + S = ES \text{ and } EI + S = ESI$$

can be expected to be different because the active site has been deformed by the noncompetitive inhibitor, the K_M for the two system can be the same, Why" Assuming that both the EI and ESI complexes are inactive, it is as if the noncompetitive inhibitor has merely removed some of the enzyme. That is, the v_0 and V_{max} for the inhibited enzyme are the velocities that would be obtained if the kinetic analysis of the uninhibited control had been repeated using less enzyme.

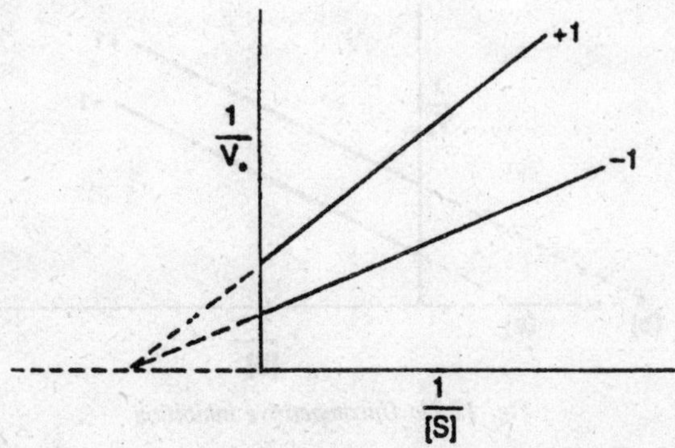

Fig. 12.20. Noncompetitive inhibition.

One test for reversible, noncompetitive inhibition is that the effects of the inhibitor cannot be reversed no matter how much the [S] is increased. The value of V_{max} cannot be increased to that of the uninhibited reaction. The inhibition can only be reversed by removing the inhibitor.

UNCOMPETITIVE INHIBITION

Uncompetitive is the term used to describe an inhibition of enzyme activity resulting from the reversible combination of the inhibitor with the ES complex

$$ES + I = ESI$$

The ESI is inactive and the dissociation constant is :

$$K_I = \frac{[ES][I]}{[ESI]}$$

In the reciprocal form, the relationship between the v_0, [S], and dissociation constants for this type of inhibition is :

$$\frac{1}{v_0} = \frac{K_M}{V_{max}} \frac{1}{[S]} + \frac{1}{V_{max}}\left(1 + \frac{[I]}{K_I}\right)$$

The plot of initial velocities versus [S] in a system containing a fixed concentration of uncompetitive inhibitor is shown in Figure. Although the V_{max} of the inhibited reaction will be less than for the uninhibited system, the K_M of the substrate-enzyme complex itself is not affected by an uncompetitive inhibitor. It will be seen that the degree of inhibition by a fixed concentration of uncompetitive inhibitor can actually be increased by increasing the [S].

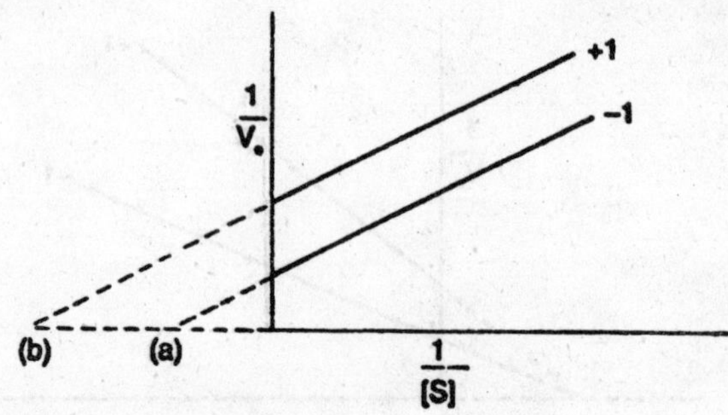

Fig. 12.21. Uncompetitive inhibition.

Irreversible Inhibition

It should be emphasized again that with all three types of inhibition discussed in the previous sections, the interaction between inhibitor and enzyme or between inhibitor and the transient enzyme—substrate complex is reversible and rapidly established. Enzymes can also be inhibited in an irreversible manner. As in the case of all other proteins, enzymes can be modified by agents that can react covalently with functional groups in the protein or which can in some other fashion alter structure. Such derivatization includes oxidation of —S—H groups to the —S—S—form, alkylation, and phosphorylation. The irreversible inhibition of an enzyme elicited by such agents is most often slow and time related, usually increasing with time. The kinetics of this type of inhibition do not fit the Michaelis-Menten model.

13

Energy in Life

Respiration is one of the most fundamental phenomenon of life and indeed life itself is inseparable from respiration in some forms. Respiration involves the oxidative breakdown of certain organic substrates present in the cells. Respiration, therefore, is not the concern of special organs or tissues, it is an integral part of the life of every cell. Cellular respiration is the aerobic oxidation of carbohydrates in the cell with the help of atmospheric oxygen, in the course of which energy is liberated and carbon dioxide is given out. In glycolysis and fermentation, glucose is oxidized in the absence of oxygen, i.e. *anaerobically*. Organisms that live in the presence of oxygen are aerobes, those that live in its absence are anaerobes.

The first series of reactions in oxidative break-down of glucose is called the *Embden-Meyerhof (E-M) pathway*. This was described by two German biochemists *Embden* and *Mayerhof*. The term *glycolysis* is sometimes used as a synonym for the E-M pathway, but biochemists usually apply the term glycolysis to the E-M pathway plus a form of fermentation, which together lead from glycolysis to lactic acid. The E-M pathway is a several-step process that occurs in the cytoplasmic matrix of cell. It can proceed in either the presence or absence of oxygen but makes available only a small portion of the potential energy that could be expected from glucose.

I. The E-M Pathway

From six-carbon glucose molecule the E-M pathway starts and the formation of 2-three-carbon molecules of pyruvic acid is the final step of this pathway. Following steps take place in the entire pathway.

1. *Phosphorylation.* In the first step glucose reacts with adenosine triphosphate (ATP) to form glucose-6-phosphate in the presence of *hexokinase* and Mg^{++} ions.

$$\text{Glucose} + \text{ATP} \xrightarrow[\text{Mg}^{++}]{\text{hexokinase}} \text{Glucose-6-phosphate} + \text{ADP}.$$

This reaction is exergonic in nature so that it is irreversible.

2. *Isomerization.* Glucose-6-PO_4, in the presence of the enzyme *phospho-hexose-isomerase*, is isomerized into fructose-6-PO_4.

$$\text{Glucose-6-phosphate} \xrightarrow{\text{Phosphohexose -isomerase}} \text{Fructose-6-phosphate.}$$

3. *Phosphorylation.* Fructose-6-PO_4 reacts with one molecule of ATP in the presence of Mg^{++} ions and *phosphofructokinase* to form fructose-1, 6-diphosphate. This reaction is also irreversible due to its strongly exergonic nature.

$$\text{Fructose-6-phosphate} + \text{ATP} \xrightarrow[\text{Mg}^{++}]{\text{Phospho fructokinase}} \text{Fructose-1-6-disphosphate.}$$

4. *Cleavage.* Fructose-1, 6-diphosphosphate is cleaved under the influence of an enzyme *aldolase* with the formation of dihydroxy acetone phosphate or phosphodihydroxy acetone (DHAP) and 3-phosphoglyceraldehyde or glyceraldehyde-3-phosphate (PGAL). Both the products are triose phosphates. The two trioses can be reversibly changed into one another by enzyme *Triosphosphate isomerase.*

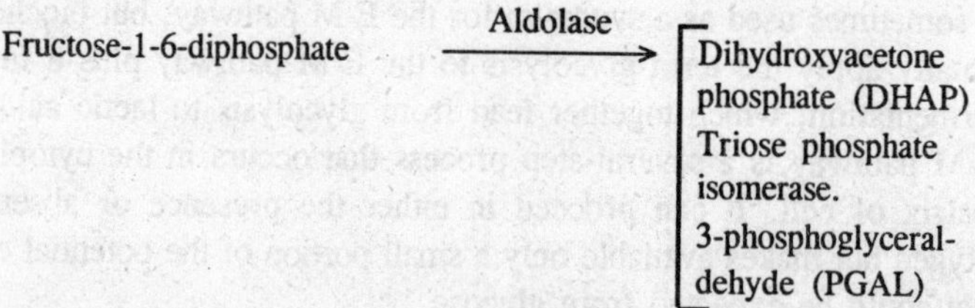

5. *Phosphorylation and oxidative dehydrogenation.* This is the first reaction in the energy-yielding phase of the E-M pathway. 3-phosphoglyceraldehyde (PGAL) reacts with phosphoric acid (H_3PO_4) to form 1-3-diphosphoglyceraldehyde in the presence of enzyme *glyceraldehyde-3-phosphate dehydrogenase.*

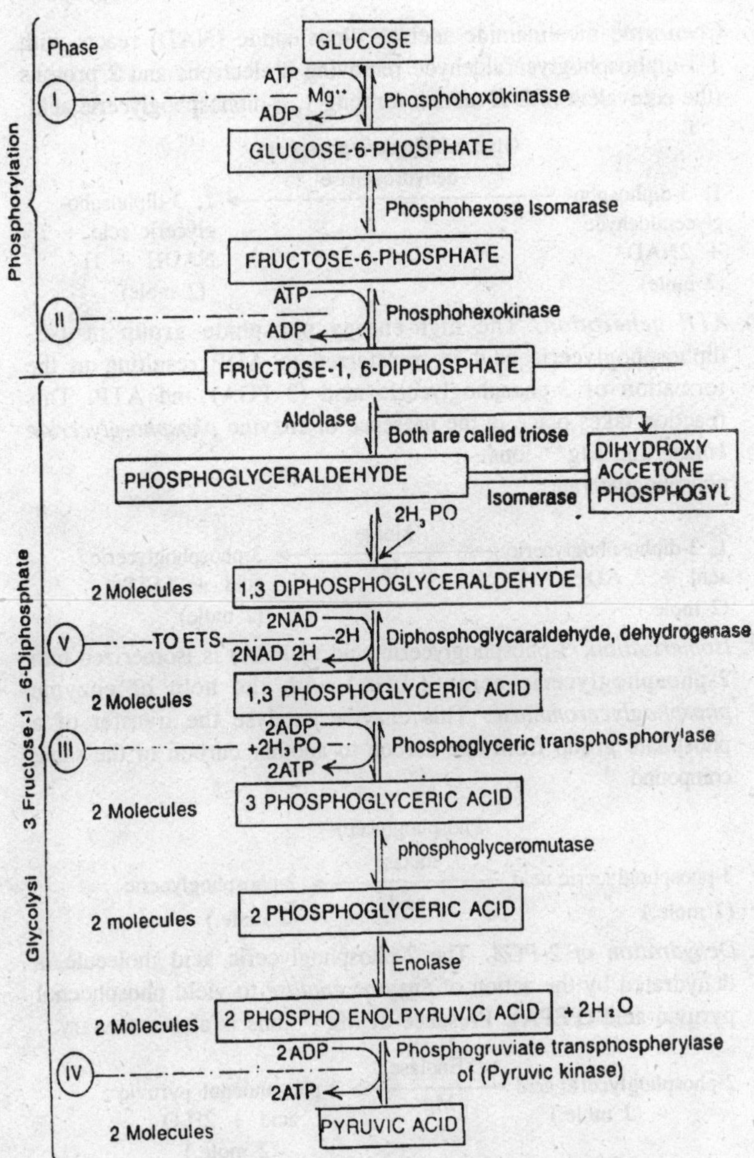

Fig. 13.1. Glycolysis of Embeden-Meyerhof pathway.

$$\underset{\text{(2 molecule)}}{\text{3-phosphoglyceraldehyde}} + \underset{\text{(2 mole.)}}{H_3PO_4} \rightarrow \underset{\text{(2 mole.)}}{\text{1, 3-diphosphoglyceraldehyde.}}$$

Coenzyme nicotinamide adenine dinucleotide (NAD) reacts with 1-3-diphosphoglyceraldehyde removing 2 electrons and 2 protons (the equivalent of 2 H atoms) forming 1, 3-diphosphoglyceric acid.

Glyceraldehyde-3-phosphate dehydrogenase

1, 3-diphospho-glyceraldehyde + $2NAD^+$ (2 mole) ⟶ 1, 3-diphospho-glyceric acid + 2 NADH + H (2 mole)

6. *ATP generation.* The high-energy phosphate group in 1-3-diphosphoglyceric acid is transferred to ADP resulting in the formation of 3-phosphoglyceric acid (3 PGA) and ATP. This reaction takes place in the presence of enzyme *phospho-glycerate kinase* and Mg^{++} ions.

phosphoglycerate kinase, Mg^{++}

1, 3-diphosphoglyceric acid + 2 ADP (2 mole.) ⟶ 3-phosphoglyceric acid + 2ATP (2 mole)

7. *Isomerization.* 3-phosphoglyceric acid (3-PGA) is isomerized into 2-phosphoglyceric acid (2-PGA) with the help of enzyme *phosphoglyceromutase*. This enzyme catalyse the transfer of a phosphate group from one carbon to another carbon of the same compound.

Phosphoglycero-mutase

3-phosphoglyceric acid (2 mole.) ⟶ 2-phosphoglyceric acid (2 mole.)

8. *Dehydration of 2-PGA.* The 2-phosphoglyceric acid molecule is dehydrated by the action of enzyme *enolase* to yield phosphoenol pyruvic acid (PEPA). Presence of Mg^{++} ions is also necessary.

Enolase, Mg^{++}

2-phosphoglyceric acid (2 mole.) ⟶ 2-phosphoenol pyruvic acid + $2H_2O$ (2 mole.)

9. *ATP generation.* In the final step of E-M pathway the high energy phosphate group in 2-phosphoenol-pyruvic acid (PEPA) is transferred to ADP in the presence of *pyruvate kinase* and Mg^{++} ions with the formation of ATP and pyurvic acid.

$$\begin{array}{c}\text{2-phosphoenolypyruvic} \\ \text{acid + 2ADP} \\ \text{(2 mole.)}\end{array} \xrightarrow[\text{Mg}^{++}]{\text{Phruvative kinase}} \begin{array}{c}\text{Pyruvic acid + 2ATP} \\ \text{(2 mole.)}\end{array}$$

Energy Yield of the E-M Reactions

How much energy or ATP has the cells gained after all these reactions? That depends upon whether the reactions have taken place under anaerobic or aerobic conditions. In general, the steps of energy production in E-M pathway are (1) during conversion of 1,3-diphosphoglyceric acid into 3-phosphoglyceric acid and (2) during conversion of 2-phosphoenol pyruvic acid into pyruvic acid. If ADP is present in these steps, energy of energyrich compounds is captured with the formation of ATP. One molecule of ATP at each of the above two steps in synthesized per molecule of 1, 3-diphosphoglyceric acid or 2-phosphoenolpyruvic acid utilized for forward reaction. Thus, in all, 4 molecules of ATP are formed at these two steps per molecules of glucose used under anaerobic conditions.

Besides these two steps, these exists one more site of energy production which comes into play under aerobic conditions. This step takes place during conversion of 3-phosphoglyceraldehyde into 1, 3-diphosphoglyceric acid when one molecule of reduced NAD, is oxidized with the formation of 1, 3-diphosphoglyceric acid.

In order to evaluate the net gain of energy the energy expenditure has also got to be taken into account and must be substracted from total energy production. Under anaerobic conditions 2ATP molecules are used up during reactions and there is a net gain of only 2ATP molecules.

II. Oxidative Decarboxylation

Pyruvic acid, when enters the mitochondria does not itself enter Kreb's cycle reactions. Entry into the Kreb's cycle requires preparation of a 2-carbon fragment called *acetyl group*. This takes place in several steps. In the first step there is a decarboxylation (a carbon dioxide molecule is released), that shortens the carbon chain by one carbon. A molecule of NAD is also reduced. The remaining 2-carbon fragment i.e., acetyl group is attached to a carrier molecule, *coenzyme A*. This complex, made up of two carbon fragment and coenzyme A is called *acetyl CoA*. The acetyl CoA is in an active form ready to be fed into Kreb's cycle.

$$\underset{\text{(Pyruvic acid)}}{CH_3\ COCOOH} + \underset{\text{(CoA)}}{CoASH} + NAD \xrightarrow[\substack{\text{Thymine pyrophosphate} \\ \text{Lipoic acid amide}}]{\substack{\text{Pyruvic dehydrogenase} \\ \text{multienzyme complex}}}$$

$$\underset{\text{(acetyl-S-CoA)}}{CH_3COS\ CoA} + NADH + H^+ + CO_2$$

III. Krebs cycle or Tricarboxylic Acid Cycle (TCA)

The krebs cycle completes the oxidation of carbohydrates and supplies electrons to be passed through electron transport system. Enzymes catalyzing Kreb's cycle reactions are located in the mitochondrial matrix.

The cyclical nature of the reactions by which further oxidations beyond pyruvic acid occur was established after a series of experimental analyses by *Albert Szent-Gyorgyi*, *Hans Krebs* and others. *Krebs* was awarded Nobel Prize in 1953 for this work.

The Kreb's cycle is a *cycle* because it is series of reaction events, each catalyzed by a specific enzyme, that ends up back at its original point. The reactions involve a series of organic acids with different number of carbon atoms.

(i) *Formation of citric acid.* Acetyl Co-A, formed from pyruvic acid condenses with oxaloacetic acid forming citric acid. This reaction is catalyzed by enzyme *citrate synthatase* or condensing enzyme.

$$CH_3CO\text{-}S\text{-}CoA + \text{Oxalicacid} + H_2O \xrightarrow[\text{CoASH}]{\substack{\text{Citrate} \\ \text{synthetase}}} \text{Citric acid} + CoASH$$

(ii) *Dehydration.* Citric acid is converted into *cis*-aconitic acid by the enzyme *aconitase* with release of one molecule of water.

$$\text{Citric acid} \xrightarrow{\text{Aconitase}} \textit{Cis}\text{-aconitic acid} + H_2O$$

(iii) *Hydration I.* By hydration of *Cis*-aconitic acid, isocitric acid is formed. This reaction is catalyzed by enzyme *aconitase*.

$$\textit{Cis}\text{-aconitic acid} + H_2O \xrightarrow{\text{Aconitase}} \text{Isocitric acid.}$$

(iv) *Dehydrogenation I.* Isocitric acid is dehydrogenated to form oxalo succinic acid in the presence of *isocitric dehydrogenase* and one pair of hydrogen atoms removed is accepted by NAD+ or NADP+ which forms NADH + H^+ or NADPH + H^+ (Mitochondria have been found to contain two types of isocitric dehydrogenases, one is NAD specific and other i NADP specific).

$$\text{Isocitric acid} + NAD+ \text{ or } NADP \xrightarrow[\text{}]{\text{Isocitric dehydrogenase}} \text{Oxalosuccinic acid} + NADH + H^+ \text{ or } NADPH + H^+$$

(v) *Decarboxylation I.* In the presence of enzyme *carboxylase*, the oxalosuccinic acid undergoes decarboxylation to form a-Ketoglutaric acid. One CO_2 is removed. Thus a-Ketoglutaric acid is 5-carbon molecule.

$$\text{Oxalosuccinic acid} \xrightarrow{\text{Carboxylase}} \alpha\text{-Ketoglutaric acid} + CO_2$$

(vi) *Dehydration II and decarboxylation II.* a-Ketoglutaric acid undergoes simultaneous decarboxylation and dehydration and joins with coenzyme A to form succinyl Co-A, a 4-carbon atom derivative of coenzyme A. This reaction is catalyzed by enzyme a-*Ketoglutaric*

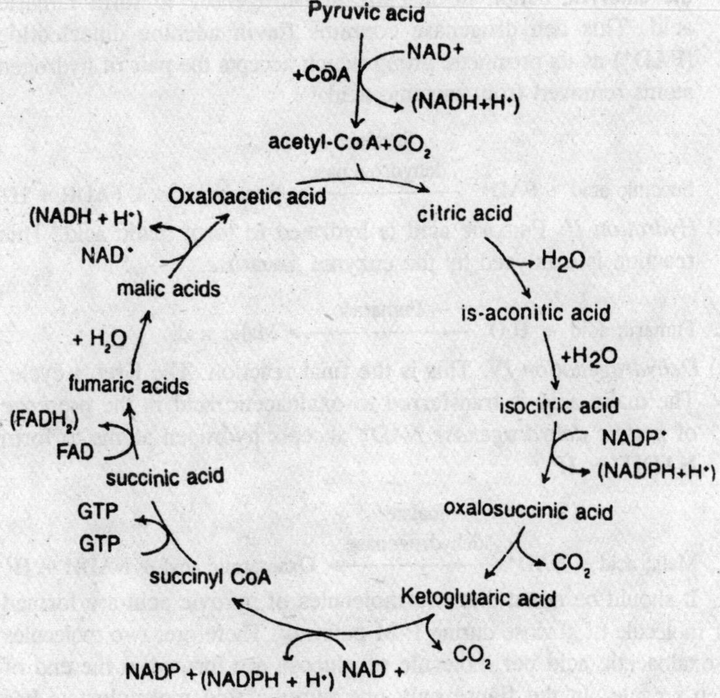

Fig. 13.2. The Krebs cycle.

dehydrogenase in the presence of thiamine pyrophosphate (TPP) and lopoic acid. The hydrogen atoms released are accepted by NAD^+ to form NADH + H^+

$$\alpha\text{-Ketoglutaric acid} + \text{Co-ASH} + NAD^+ \xrightarrow[\text{dehydrogenase}]{\alpha\text{-Ketoglutaric}} \text{Succinyl CoA} + CO_2 + NADH + H^+$$

(vii) *Phosphorylation.* Succinyl CoA on hydrolysis yields to the formation of succinic acid and CoA. Succinyl CoA possesses a high-energy thiol group. In the presence of guanosine diphosphate (GDP) and inorganic phosphate the energy of thiol group is transferred to GDP with the formation of guanosine triphosphate (GTP). This reaction is catalyzed by enzyme *succinic thiokinase*.

$$\text{Succinyl- CoA} + GDP + H_3PO_4 \xrightarrow{\text{Succinic thiokinase}} \text{Succinic acid} + GTP + CoASH.$$

(vii) *Dehydrogenation III.* Succinic acid undergoes dehydrogenation under the catalytic action of *succinic dehydrogenase* to form fumaric acid. This dehydrogenase contains flavin adenine dinucleotide (FAD^+) as its prosthetic group which accepts the pair of hydrogen atoms removed from succinic acid.

$$\text{Succinic acid} + FAD^+ \xrightarrow{\text{Succinic dehydrogenase}} \text{Fumaric acid} + FADH + H^+$$

(ix) *Hydration II.* Fumaric acid is hydrated to form malic acid. This reaction is catalyzed by the enzyme *fumarase*.

$$\text{Fumaric acid} + H_2O \xrightarrow{\text{Fumarase}} \text{Malic acid.}$$

(x) *Dehydrogenation IV.* This is the final reaction. The Kreb's cycle. The malic acid is transferred to oxaloacetic acid in the presence of *malate dehydrogenase NAD*$^+$ accepts hydrogen atoms to form NADH + H^+.

$$\text{Malic acid} + NAD^+ \xrightarrow{\text{malate dehydrogenase}} \text{Oxaloacctic acid} + NADH + H^+$$

It should be noted that two molecules of pyruvic acid are formed per molecule of glucose during E-M pathway. Therefore, two molecules of oxaloacetic acid per molecule of glucose are formed at the end of Kreb's cycle. In the figure only one pyruvic acid molecule fate has been traced.

During transfer of hydrogen atoms from one enzyme carries in electron transport system, large amount of energy is released (as shown in Table). The enzymes which participate in the process are called respiratory chain enzymes or electron transport system enzymes. All such enzymes are found in the inner membrane of mitochondria. These are arranged in the following sequence—succinic dehydrogenase which remains associated with coenzyme NAD, flavoprotein (FAD), non-haem ion protein or iron-sulphur protein, flavoprotein (FAD or FP), cytochrome b, ubiquinone or Co Q, cytochrome c_1, cytochrome c, cytochrome a cytochrome a_3 and three coupling sites where phosphorlylation coupled with oxidation leads to the production of ATP (*Hall et. al.* 1974).

The respiratory chain system manufactures the ATP molecules during oxidation of reduced coenzymes in the following steps:

1. *Dehyderogenase* enzymes accept hydrogen from different substrates of Krebs' cycle. The hydrogen atom becomes ionized into protons ($2H^+$) and electrons (2e).

   Protons or hydrogen ions reduce the oxidized NAD

   $NAD^+ + 2H^+ \longrightarrow NADH + H^+$

   The NADH molecule passes through M-side of inner mitochondrial membrane and enters the respiratory chain system.

2. The NADH is oxidized into NAD^+ by transferring its proton (H^+) to flavoprotein FAD. The FAD acts as a hydrogen carrier. From the flavoproteins each proton is discharged into perimitochondrial space through the C-side of inner mitochondrial membrane and electrons are passed onto non-haem iron protein.

3. In normal course, reduced FADH is reoxidized by ubiquinone (Co Q) which is the next member of ETS. The ubiquinone is reduced to dihydroubiquinone which is oxidized by ferric (Fe^{++}) cytochrome—C—(Fe^{+++}) which is reduced to ferrocytochrome. At this level of oxidation, hydrogen atoms undergo ionization to form $2H^+$ and $2e^-$. The protons cannot be further carried by ETS, however, the electrons can be carried easily by the cytochromes. One molecule of each cytochrome is capable of carrying only one electron at a time. Thus, for each molecule of hydrogen which releases two electrons, two molecules of each cytochrome are required. At the level of cytochrome-b, two protons per molecule of hydrogen are released in the mitochondrial system. Ferrocytochrome b is oxidized in the presence of ferric cytochrome

c_1. From ferrocytochrome c_1 these electrons are taken by cytochrome a and than to a_3 (Both cyto-a and a_3 together called cytochrome oxidase.

4. These electrons from cytochrome oxidase react with one atom of oxygen making it reactive.
5. This activated oxygen react with $2H^+$ (protons) forming molecule of water.

$$O + 2e \longrightarrow O^- \xrightarrow{2H^+} H_2O$$

With the release of 2 electrons and 2 H atoms from every molecule of NADH + H^+ and their reaction.with oxygen, 3ATP molecules are synthesized. The position of ATP molecules. Energy released at these positions reach and stored in ATP molecules. The process is known as oxidative phosphorylation because it occurs in respiration in the presence of oxygen hence it is called oxidative phosphorylation.

Now we can summarize the energy retum from the complete oxidation of a glucose molecule under aerobic conditions. Under aerobic conditions, the NADH produced in the E-M pathway can transfer electons to the ETS inside mitochondria, these by salvaging more usable energy than is possible under anaerobic conditions.

However, reduced NAD cannot enter a mitochondrion directly, so it cannot pass its electrons directly to an ETS in the inner mitochondrial membrane. Instead, the electrons are passed to a shuttal carrier that takes them into the mitochondrion and donates them to an electron transport system.

These are several such electron shuttles for example malate-asperate shuttle, glycerol phosphate shuttle etc., but the common one

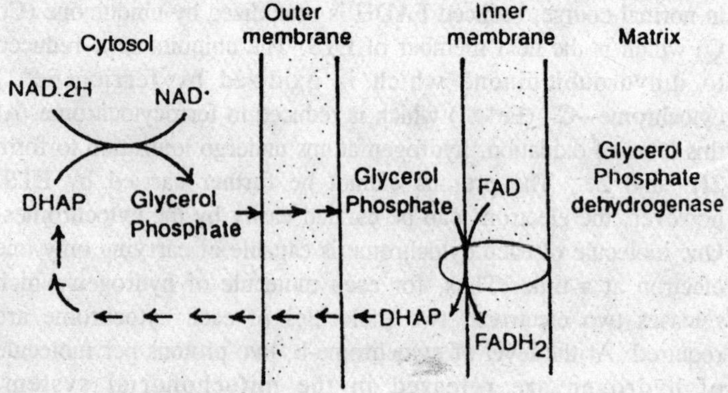

Fig. 13.3. Glycerol-phosphate shuttle.

is *glycerol-phosphate shuttle*, which picks up electrons from reduced NAD, carries them into mitochondrion and donates them not to NAD, but to ubiquinone. In this case the electrons do not start at the 'top' of the ETS, so the possibility for one phosphorylation is bypassed. If mitochondria use this particular mechanism, a reduced NAD in the cytoplasm ends up being worth only 2ATP molecules. In this case complete oxidation of glucose will result in a gain of 36 ATP molecules.

In *malate-asperate shuttle*, the electrons are transferred from NADH in cytoplasm to malate which traverses the inner mitochondrial membrane and reoxidized to form NADH + H^+ thus resulting in the formation of oxaloacetate. Oxaloacetate does not readily cross the inner mitochondrial membrane 20 a transmission reaction is needed to form asperate which does traverse this barrier. 3ATP molecules are generated for each pair of electrons. In this case complete oxidation of glucose will result in a gain of 38 ATP molecules.

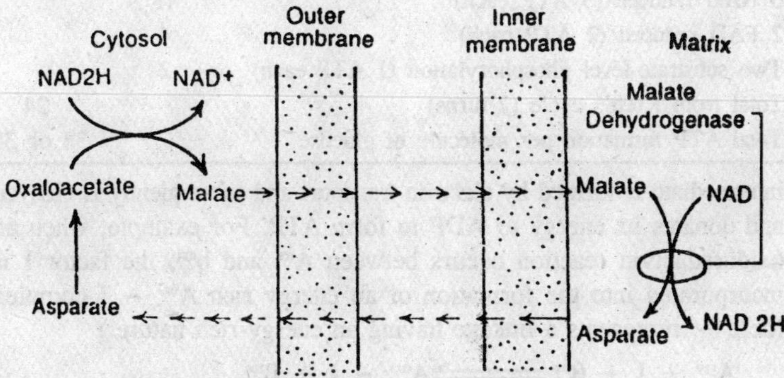

Fig. 13.4. Malate-aspartate shuttle.

Oxidative Phosphorylation Mechanism

Oxidative phosphorylation was discovered in 1939 but the mechanism of coupling between ETS and ATP synthesis is not yet known. However, three principal hypothesis have been proposed for the coupling of oxidation and phosphorylation. These are:

1. Chemical Coupling hypothesis.
2. Cheiniosomotic Coupling hypothesis.
3. Conformational hypothesis.

Chemical Coupling Hypothesis

According to this hypothesis the ET is coupled to ATP synthesis by a sequence of consecutive reactions in which a high-energy covalent

Table 13.1. Showing ATP Formation During Oxidation of Glucose.

Embden-Meyerhof Pathway	*ATP calculations*	*Sum*
ATP expanded to phosphorylate sugar molecule	– 2	
Substrate-level phosphorylation	4	
ATP formed as a result of electron transport from two NADS reduced during E-M pathway reactions (2 or 3 ATP each)	4 or 6	
Net gain from E-M pathway aerobic conditions		6 or 8
Two pyruvic acid to two acetyl CoA	6	
Two NAD reduced (3 ATP each)		6
Krebs cycle (two turns)		
6 NAD reduced (3 ATP each)	18	
2 FAD reduced (2 ATP each)	4	
Two substrate-level phosphorylation (1 ATP each)	2	
Total from Kreb's cycle (2 turns)		24
Total ATP formation per molecule of glucose		36 or 38

intermediate is formed by electron transport and subsequently is cleaved and donates its energy to ADP to form ATP. For example, when an oxidoreduction reaction occurs between A^{red} and b^{oxi}, the factor I is incorporated into the formation of an energy rich $A^{oxi} \sim I$ complex (here ~ represents a linkage having an energy-rich nature.)

$$A^{red} + I + B^{oxi} \rightleftharpoons A^{oxi} \sim I + B^{red}$$

In the following reaction an enzyme (E) replaces A^{oxi} from the above complex forming an energy rich $E \sim I$ complex. Later on inorganic phosphate (P_i) reacts with $E \sim I$ complex, forming phosphoenzyme complex ($E \sim P$). The phosphoenzyme complex reacts with ADP to form ATP liberating enzyme free.

$$A^{oxi} \sim I + E \rightleftharpoons A^{oxi} + E \sim I$$

$$E \sim I + P_i \rightleftharpoons E \sim P + I$$

$$E \sim P + ADP \rightleftharpoons ATP + E$$

An extensive search has been made for the nature of $E \sim I$ and $E \sim P$, but non have been found in mitochondria. The chemical coupling hypothesis is only of historical importance today.

Table 3.2. Free-Energy Change at Different Steps in the Electron Transport Chain.

Steps of Transport			*Difference in Redox-potential (ΔE^0), (Volt)*	*Free energy change (Kcals)*
NAD	$\xrightarrow[2e^-]{2H^+}$	FAD	+0.22	–10.1
FAD	$\xrightarrow[2e^-]{}$	cyt.b	+0.15	–6.9
2Cyt.b	$\xrightarrow[2e^-]{}$	2 cyt.c	+0.19	–8.77
2 Cyt.c	$\xrightarrow[2e^-]{}$	2 cyt.a	+0.05	–2.30
2 Cyt.a	$\longrightarrow$	Oxygen	+0.52	–24.00
Total			1.13 Volts	52.07 Kcals.

Chemiosmotic Coupling Hypothesis

This hypothesis was proposed by *Peter Mitchell* in 1961. He proposed that electron transport and ATP synthesis are coupled by a proton gradient, rather, than by a covalent high-energy intermediate or an activated protein. This theory suggests that a transmembrane electrochemical gradient is critically important to the phosphorylation process.

Chemiosmotic coupling hypothesis suggests that during electron transport, a gradient of H^+ ions is produced across the inner mitochondrial membrane. The mobile electron carrier that gains H^+ ion is reduced one side of the membrane and loses proton as it is oxidized near the other side. These reactions lead to the transport of a pair of H^+ across the inner mitochondrial membrane from the matrix to the space between the mitochondrial membrane. Thus there is a H^+ ion gradient across the membrane. There is a strong tendency for these H^+ ions to move back toward the matrix because of the excess of +ve charge is the space between the two membrane.

Mitchell further suggests that there are special channels through which the H^+ ions can move in response to the forces acting on them (electrical as well as concentration differences). In same way, the movement of H^+ ions back through the membrane is coupled in ATP synthetase complex to the phosphorylation of ADP to form ATP.

Oxidation-reduction loop

Mitchell has proposed that reducing equivalents are transferred as H atoms by some of the electron carriers (ubiquinone) and as electrons by other (Fe-S center and cytochromes). According to him hydrogen carrying and electron carrying proteins alternate in the respiratory chain to form three functional loops called *oxidation-reduction loops* (O/γ loops).

A single loop consists of a hydrogen carrier and an electron carrier. In each loop two H^+ are carried outward through the inner membrane and delivers two H^+ to the C side, the corresponding pair of electron is then carried back from outer to the inner surface of the membrane. Thus, each pair of reducing equivalents passing through such loop carries $2H^+$ from the matrix to outside. Each loop is thought to provide the osmotic energy to form one molecule of ATP.

Mitchell's hypothesis is supported by a wealth of evidences.

Conformation Coupling Hypothesis

According to this proposed by *Boyer* (1965), the electron transport and phosphorylating systems are in molecular contact and that information is transmitted by short-arange interactions which may, in

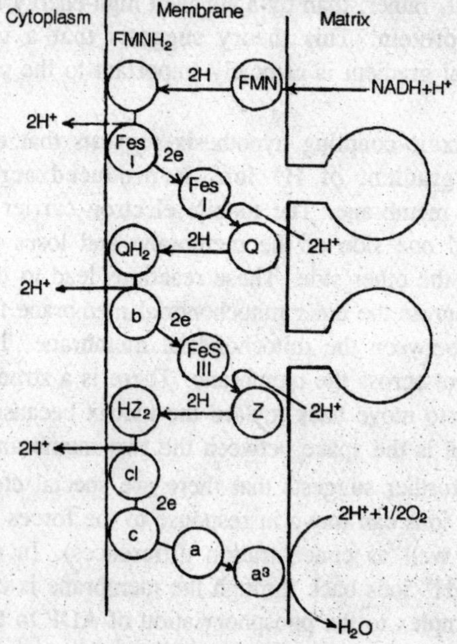

Fig. 13.5. Showing Michell's chemiosmotic coupling hypothesis.

part, be electrostatic in nature. It is postulated that upon acceptance of the two electrons, each electron transfer protein undergoes a conformational change which is then transmitted to the ATPase by short range interactions. However, the mode of conformational changes that take place in the inner mitochondrial membrane is not yet clearly understood.

According to some recent views (*Ernster,* 1977) the conformational and chemiosmotic mechanisms of electron transport linkage to phosphorylation do not represent mutually exclusive alternative but may indeed be two aspects of the same mechanism.

INDEX

INTERNATIONAL ENCYCLOPAEDIA OF LEARNING TO LIVE TOGETHER - 2

GLOBALIZATION AND LIVING TOGETHER

Editor

Dr. Digumarti Bhaskara Rao
M.Sc., M.A., M.A., M.Ed., Ph.D.
Dean, Faculty of Education
Member, Academic Senate
Ex-Chairman, Board of Studies in Education
Member, Research Advisory Committee
Acharya Nagarjuna University
D-43, S.V.N. Colony
Guntur - 522 006 (India)

DISCOVERY PUBLISHING HOUSE PVT. LTD.
NEW DELHI-110 002

First Published-2005

Reprinted - 2015

ISBN: 978-81-7141-848-0 (Set)

ISBN: 978-81-7141-990-6

Globalization and Living Together

Published by:
DISCOVERY PUBLISHING HOUSE PVT. LTD.
4383/4B, Ansari Road, Darya Ganj
New Delhi-110 002 (India)
Phone: +91-11-23279245, 43596064-65
Fax: +91-11-23253475
E-mail: discoverypublishinghouse@gmail.com
sales@discoverypublishinggroup.com
web: www.discoverypublishinggroup.com

Printed at:
Infinity Imaging Systems
Delhi

FOREWORD

The development and reform of school curricula is an ongoing preoccupation for educational authorities in all countries. The approach of the new millennium has given new urgency to efforts by governments to provide all citizens with access to quality education, at least for the basic level, while improving and widening access to secondary education. More than ever before, governments are being called upon to equip children and young people through education with the capacity to lead meaningful and productive lives in a world of bewilderingly rapid and complex change. Existing curriculum content and pedagogical methods are increasingly being called into account, as pupils leave schools ill-prepared for the world of work and adulthood, unready and unmotivated to carry on learning throughout their lives. The meaning and role of education, of teaching and learning, are constantly being redefined in an effort to meet the real needs and demands of individual and society.

The implication of globalization for societies around the world is at the heart of present concerns to improve and upgrade education systems. While globalization is often defined primarily in terms of its economic dimensions, the Report to UNESCO of the International Commission on Education for the Twenty-first Century—the *Delors report*—sees the most important consequence of this complex phenomenon to be its socio-cultural and ethical dimensions. It draws attention to the growing interdependence and interrelationships between peoples and cultures the world over: 'the far-reaching changes in the traditional patterns of life require of us a better understanding of other people and the world at large today; they demand mutual understanding, peaceful interchange and indeed harmony'.[1] However, the report

1. Delors, J., et al. *Learning: the treasure within*. Paris, UNESCO, 1996, p. 22. (Report to UNESCO of the International Commission on Education for the Twenty-first Century.

stresses that 'learning to live together'—one of the pillars of education—will only occur through the possession of self-knowledge and understanding, and appreciation of one's own origins and culture.

It is now widely recognized that designing a curriculum is mainly a national concern, normally shared between educational protagonists at the central, local and school level. The principle of subsidiarity suggests that curriculum issues should not be addressed at the supranational level. Indeed, experience shows that, until quite recently, international exchanges and co-operation in curriculum development were limited, being restricted to professional associations of curriculum specialists. However, two recent trends have contributed to bringing international attention to bear on curriculum matters:

1. The globalization of economies and societies raises a new challenge, requiring the adaptation of educational content to meet both national demand and international concerns;
2. The diversification of actors, both national and international, involved in the delivery of education (in particular with the growing use of Information and Communication Technologies—ICTs), as illustrated by the significant share of non-formal education, have resulted in the emergence of new concepts and norms for educational content; as indicated by such terms as the 'common core', 'universal values', 'basic life skills', etc. Once again, this means sharing responsibility for educational content, as well as presenting us with new opportunities for international co-operation.

It is against this background and in an effort to respond to the numerous contemporary concerns about the content of education, that the International Bureau of Education (IBE), designated as the UNESCO institute responsible for strengthening the capacity of Member States in curriculum development, is focusing its new programme activity on the adaptation of the content of education to the challenges of the twenty-first century. The IBE's programme is divided into two components: (i) integrating the concern of living together into the content of education; and (ii) adapting the content of education in order to cope with some of the challenges raised by a globalized world.

The IBE programme of co-operation in 'research and studies', 'training and capacity building'; and 'the exchange of information and expertise' is based upon two major assumptions:

(a) Although different Member States of UNESCO have very uneven and heterogeneous experiences in the design and adaptation of their educational content, there is room for beneficial exchanges between countries;

(b) Although there are some common views on how to address the demand that content should be modified, a great deal remains to be done to improve the process of adaptation.

The approach to implementing the programme followed by the IBE is based on the assumption that the programme proposes to include: (a) an international platform of information on educational content; and (b) a number of regional and sub-regional co-operation projects. The preparation of such programmes is carried out through regional workshops.

This sub-regional workshop took place in 1999 in New Delhi to cater to the concerns of South and South-East Asia. The purpose of the workshop was:

(a) To collect information about the status of educational content in the region—a 'fact-finding' exercise;

(b) To encourage the assessment of experiences, both among Asian countries and a selected number of countries from other parts of the world; and

(c) To explore and map out some areas for potential co-operation among the participating Member States.

Two key phrases of the IBE's new programmes were incorporated into the umbrella theme of the course, which was The adaptation of content to the demands of the globalization and the need for learning to live together.

This publication is a compilation of the various presentations made during the workshop. Both through the diversity of topics covered and the list of issues raised, it offers a rich and varied picture of the challenges facing Asian countries seeking to adapt educational content to meet the social demand of the coming decade.

The IBE would like to thank all of the contributors, both from the participating countries and from other parts of the world, who have shared their information and experience during the workshop. Indeed, the distinction between 'faculty' and 'participants' proved to be of little relevance in this workshop, where all contributors learned from each other.

On behalf of the International Bureau of Education, I would like to express our deep appreciation to the Government of India—and in particular to the Central Board of Secondary Education—for its generous support of the workshop. India both financially supported and hosted the meeting and contributed significantly to the debates and to enriching the information shared among the participants. Finally, I should mention with thanks the financial and intellectual contribution of the Deutsche Stiftung fur internationale Entwicklung, which ensured the success of the workshop.

Jacques Hallak
Assistant Director-General,
Director of the IBE.

PREFACE

Education is gearing towards preparing individuals to live in together since ancient times. However, the notion of 'Learning to Live Together' is clearly shaped only very recently by the UNESCO's International Commission on Education for the Twenty-first Century. The Commission emphasized that education had a treasure within and developed a vision of education for the 21st century based on four pillars, namely Learning to know, Learning to do, Learning to be, and Learning to live together. The first three pillars are essential for the sound development of persons, communities or individual nations, but the fourth pillar, learning to live together, is of a different, more global nature: its omission may result in the annihilation of all educational, cultural, health and developmental efforts through war, terrorism, deterioration of resources, pandemics, etc.

The theme for which the UNESCO convened from 5 to 8 September 2001 in Geneva the 46th session of the International Conference on Education (ICE), organized by the UNESCO's International Bureau of Education, was "Education for All for Learning to Live Together: Contents and Learning Strategies—Problems and Solutions".

The ICE brought together over 600 participants from 127 countries, including in particular 80 ministers and 10 vice-ministers of education, as well as representatives of inter-governmental and non-governmental organizations. The themes of ICE are very relevant all over the world with regard to the necessity and complexity of living together as well as to the role and limitations of education in this respect.

This book, Learning to Live Together, is in four volumes, namely, *Volume 1:* International Conference on Learning to Live Together, *Volume 2:* Globalization and Living Together, *Volume 3:* Curriculum for Learning to Live Together, and *Volume 4:* Science

Education for the Contemporary Society. This multi-volume book will make all the partners of education aware of the entire spectrum of learning to live together and involve them in making the 'Learning to Live Together' a reality.

I am very much grateful to Ms. Cecilia Braslavsky, Director of the UNESCO's International Bureau of Education and Secretary-General of the 46th Session of the International Conference on Education for her kind co-operation in this academic endeavour. I am thankful to Mr. John Fox, Head of the Publications Unit of UNESCO-IBE for according me the permission to reproduce the material from various documents that appeared in this book. I am thankful to all the contributors and editors of various documents incorporated in this book. I am also thankful to the UNESCO's International Bureau of Education, UNESCO, Indian Ministry of Human Resource Development, Indian Central Board of Secondary Education, Ministry of Education of Cuba, Caribbean Network of Educational Innovation for Development, Organization of Inero-American States, The Oman National Commission for UNESCO, UNESCO—Principal Regional Office for Asia and the Pacific, Thailand Ministry of Education, and The Chinese National Commission for UNESCO.

Sai Soudha **Dr. D.B. Rao**
D-43, S.V.N. Colony
Guntur-522 006
Andhra Pradesh
India

CONTENTS

Part I

The Impact of Globalization on Curriculum Development

1

Introduction

This publication is divided into five parts. In Part One, entitled *The impact of globalization on curriculum development*, the keynote lecture by J. Hallak and M. Poisson outlines the implications for education of the phenomenon of globalization, and stresses the renewed urgency for curricula to promote peaceful coexistence and co-operation among pupils.

The subsequent round-table debate examines the relevance of new trends in the teaching of science, social science and the humanities to the challenge of globalization and the principle of learning together. The presenter for the humanities focuses on the primordial role of language instruction in the curriculum. The science presenter identifies the opening up of schools to society and a growing recognition of the need to develop scientific attitudes and behaviour in students as principal trends in science teaching. The final presenter stresses how the social sciences are an ideal medium for transmitting the values and attitudes necessary for harmonious existence within society and the natural environment.

In the following working group discussions, the science group addresses the following four issues: the adequate coverage within the curriculum of both basic scientific knowledge and ongoing scientific developments; the organization of experimental activities; the use of, and respect for, the environment; and bridging the gap between scientific and traditional knowledge. The humanities and social science groups discuss how both subject areas may support general education by defining new opportunities for curriculum design, teaching/learning methods, pupil assessment and implementation strategies.

Part Two, entitled *Some challenges for the adaptation of content raised by the principle of learning to live together,* is comprised of a number of presentations on various key current concerns in curriculum reform. R.H. Dave and J.S. Rajput stress the need for comprehensive reforms to existing approaches to teacher education, with the emphasis to be put on career-long education and training, and the development of a professional ethic among all teachers. J.M. Sani and M.M. Pant outline the possibilities which Information and Communication Technologies (ICTs) offer for innovating the educational process, underlining the need in this context for a redefinition of the role of the teacher while stressing that this role remains key to ensuring the effective exploitation of ICTs as tools for learning. The need for vocational education to meet the demands of globalization is stressed by Arun K. Mishra, who points out not only the economic, but also the social goals, of this essential area of education. His presentation indicates the possibilities which vocational education offers for promoting values of co-operation and respect for others.

Part Three, *Interdisciplinarity, school-based management and non-school science education: a few topics for reflection by curriculum developers,* includes three case studies from countries outside of the Asian region. Two cases examine major curriculum reforms, one at the national, and the other at the provincial level. E.M. Skaflestad discusses how the new national curriculum in Norway was designed as a connective model establishing clear links between all educational levels and emphasizing an interdisciplinary approach in subject teaching, with core values being taught across the curriculum. Though centrally prescribed, the curriculum provides for local and individual adaptation. Kenneth Ross describes the introduction of the most decentralized system of schooling in the history of public education in Australia, in the state of Victoria. Under this system, in which each school in collaboration with the community draws up its own charter outlining its particular vision and aims, institutions design their own educational programmes based on a core curriculum and standards framework established by the Department of Education. The final country study from France describes the valuable role played by one of the world's major science and technology museums, La Cité des sciences et de l'industrie (La Villette), in making available a wealth of educational resources to the country's schools. This study illustrates the importance of collaboration between non-school institutions providing learning opportunities and schools, stressing the need for schools to emerge from their traditional isolation and to become more involved in society.

In Part Four, *Current trends in the adaptation of educational content in South and South-East Asia,* U. Bude's lecture focuses on the need for curriculum development to be a continuous holistic process aimed at real and meaningful change in the classroom. The importance of flexibility and openness in curriculum policy and design, permitting regional and local adaptation of a core document, based on the needs of diverse socio-cultural groups within a country, is emphasized. The advisability of providing for broad-based participatory approaches to the curriculum development process is stressed, with the fundamental importance of the teacher in successful curriculum design and implementation underlined.

The report concludes with an overview of the country reports on curriculum development presented in Part Five.

2

Education and Globalization: Learning to Live Together

Jacques Hallak and Muriel Poisson

This text will first define the phenomenon of globalization, considered as the most widespread trend on the eve of the twenty-first century, and will present some of its implications for knowledge, employment and poverty. Second, it will identify some of the challenges raised by globalization in the specific sphere of education. Third, it will focus on the principle of teaching/learning to live together. This is considered a key issue in helping young generations to cope with the challenges raised by globalization. This section will also cover, more precisely, the different means of introducing teaching to live together in educational content and the problems that this may entail. Fourthly, it will pay careful attention to the role of teachers in this context. They are undoubtedly the key actors in any reforms that may take place. In particular, we describe how to help them adapt their teaching methods and develop a 'sense of ownership' towards the curriculum so that change can actually be guaranteed at the school level. This section will conclude with the main topics that were selected for debate during the course.

GLOBALIZATION, A NEW PHENOMENON

The phenomenon of globalization[1] has resulted from the worldwide integration of economic and financial sectors. It has existed historically since the development of international trade.

1 For more information on globalization, one can consult the following booklet: J. Hallak, *Education and globalization.* Paris, UNESCO: International Institute for Educational Planning, 1998. (IIEP contributions, no. 26.)

However, it can be considered, to a certain extent, as a more recent phenomenon, since over the past few years it has experienced a high acceleration, due to the following factors:

- ***Geo-political changes:*** The erosion of the power of nation states, in a context of transfer of sovereignty from governments to regional entities (ASEAN, CIS, EU, MERCOSUR, SADC, etc.), and the development of multinational corporations have contributed to a dramatic increase in trans-border exchanges.
- ***A dominant ideology of regulation by market forces:*** The end of the regulation of monetary, financial and economic markets, following, in particular, the setting-up of the Bretton Woods system, the GATT agreements and, finally, the expansion of the free-market ideology after the collapse of the Soviet Bloc, has greatly contributed to the interpenetration of national economies
- ***Fast and significant technological progress***: Technical progress in the communications field has permitted users to access and exchange information at any time and from any place in the world, which has largely facilitated the speeding-up of production, as well as the sharing of goods, services, capital flows—and also ideas.
- **The aim of *increasing return on capital investment,*** added to the possibility of locating the units of production of goods and services almost anywhere in the world, have also contributed to uniting or globalizing our planet.

A few figures can illustrate the importance of the phenomenon of globalization in Asia. *Table 2.1* shows that the volume of exchanges among the Asian countries has

Table 2.1: Developing Asia and Pacific region: external trade—total value (millions of US$)

	1986	1990	1995	1986-95
Imports	230 993	456 132	945 701	+309%
Exports	175 693	416 441	889 524*	+406%

Exports to Asia, 52%; North America, 21%; Western Europe, 17%.

Source: United Nations, 1997.

mushroomed during the last decade. Thus, in developing Asia and the Pacific, the total value of imports increased three-fold from 1986 to 1995 and the total value of exports more than quadrupled during the same period. About half of these exports took place within the Asia region (52% for 1995), which proves that the phenomenon of regionalization is linked to globalization.

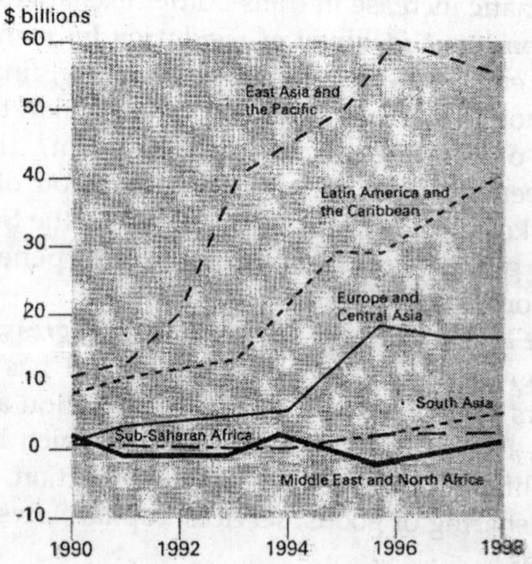

Source: World Bank, 1998.

Figure 2.1: Foreign direct investment in decline in East Asia

Figure 2.1 illustrates the fact that the development of the economies of different countries in Asia is linked to foreign investment, and that, consequently, countries of the region are integrated in the global economy and dependent on outside partners. Direct foreign investment has thus increased from US$ 10 billion to US$ 60 billion between 1990 and 1996 in East Asia and the Pacific, whereas in other regions, such as Europe and Central Asia, the Middle East and North Africa or Sub-Saharan Africa, it has stayed at the same level, or has only slightly increased.

Among the main implications of the phenomenon of globalization, one can mention:

- the *emergence of learning societies* due to the multiplication of sources of information and communication;

- the *transformation of the nature of work* with, in particular, the need for more flexibility and mobility, the importance of communication skills, the necessity for teamwork, the increasing use of new technologies, etc.,
- the *progression of social exclusion*—a large part of the world's population does not participate in this process.

In this context of a changing world, the central role played by education to favour social and professional integration appears to be hugely reinforced.

SOME OF THE EDUCATIONAL CHALLENGES RAISED BY GLOBALIZATION

The educational challenges raised by globalization are a multitude. We may mention the following as examples:

- ***The need to rethink the delivery of content, integrating new sources of information.*** The multiplication of sources of information (newspapers, radio, television, the web, CD-ROMs and other multimedia materials, etc.) and the fast evolution of knowledge, in particular, in the fields of science and technology, imply not only the need to update educational content regularly, but also to review the design of curricula and the teaching of subjects in an interdisciplinary manner. This also raises the need for better co-operation with 'information brokers'—such as science museums—that can play the role of intermediaries between 'knowledge producers' (universities) and 'knowledge consumers' (pupils/students).
- ***The need to favour the development of skills alongside knowledge.*** The fast evolution of today's societies—and, more specifically, of labour markets—requires new skills from individuals. These were summarized in a report published by UNESCO in 1996 as follows: learning to know, learning to do, learning to live together and learning to be *(see Box 1)*. They call for the redefinition of educational programmes on a competency-based approach: this implies that from now on, curricula should focus on students attaining a stated number of clearly defined skills or competencies at the end of each stage and level of

school education. This, of course, should have consequences on teaching/learning methods, student assessment procedures and certification.

Box 1: The four pillars of education, as defined by the Delors Commission

- **Learning to know,** by combining a sufficiently broad general knowledge with the opportunity to work in depth on a small number of subjects.
- **Learning to do,** in order to acquire not only an occupational skill, but also the competence to deal with many situations and to work in teams.
- **Learning to live together,** by developing an understanding of other people and an appreciation of interdependence in a spirit of respect for the values of pluralism, mutual understanding and peace.
- **Learning to be,** so as better to develop one's personality and be able to act with ever greater autonomy, judgement and learning responsibility (including memory, reasoning, and communication skills).

Source: J. Delors, et al. *Learning: the treasure within,* Paris, UNESCO, 1996. (Report to UNESCO of the International Commission on Education for the Twenty-first Century.)

- *The need to adapt curricula to the needs of different socio-cultural groups, and to maintain the national and social cohesion of the country.* The phenomenon of globalization has helped to widen the gap between those who globalize, and those who are globalized—or left out—of the process at the local, national, regional and international levels. This raises the problem for public authorities of how to maintain the cohesion of a country. As a result, there are implications for education, since the situation and the socio-cultural background of learners are seldom taken into consideration when conceiving and implementing curricula. Trying to cope with this situation means not only having to refer to the national identity or promote the concept of citizenship, but also having to include the teaching of peaceful co-existence in educational content.

HOW TO INTRODUCE TEACHING ABOUT LIVING TOGETHER INTO EDUCATIONAL CONTENT

If we refer to the definition given by the Delors Report mentioned in Box 1, teaching to live together is synonymous with developing an understanding of other people and an appreciation of interdependence in a spirit of respect for the values of pluralism, mutual understanding and peace'. There are several ways of introducing this concept into educational content. The three main strategies are as follows:

- *By defining new scopes for 'old disciplines' (history and geography, foreign languages, etc.)*: A study led by the University of Geneva, under the leadership of the IBE, has shown that traditional subjects, such as history and geography, can contribute, to a great extent, to teaching about living together. It thus categorized three different ways of teaching geography *(see the schemes included in Box 2)*. In the case of the 'closed model', a country is presented as a homogenous space, delimited by its borders. The 'opposing model' takes into account the existence of several overlapping entities (geographical, administrative, economic, linguistic, etc.) in the national territory. Finally, the 'comprehensive model' puts forward the existence of various entities, both inside and outside the national territory, and their mutual interactions. The only model favouring learning to live together is obviously the comprehensive one, which brings together different geographical scales to permit an understanding of the continuity from the local to the

Box 2: The role played by geography in learning to live together

Closed Model:

Opposing Model:

Comprehensive Model:

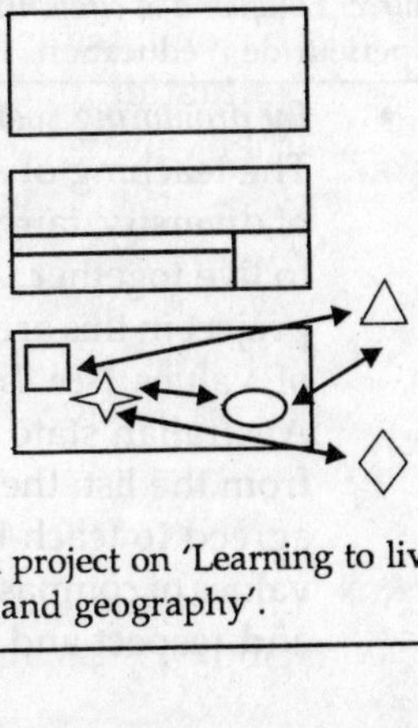

Source: IBE-University of Geneva research project on 'Learning to live together through the teaching of history and geography'.

global, and which allows not only the concept of identity to develop, but also those of otherness, interdependency and universality.

- *By introducing new subjects (i.e. education towards active citizenship, environmental issues, etc,) in curricula*: The widespread concern for global issues, such as population control, urbanization, environment, consumption, citizenship, etc., favours their integration in educational programmes. The example of the project entitled 'Educating towards active citizenship'. Launched by Quebec (Canada) in 1998 *(see Box 3)*, shows that the introduction of these new concepts demands a review of curriculum design by adopting a comprehensive approach, developing more participatory teaching/learning methods and working in partnership with out-of-school actors so as to enable students to put what they have learned into practice.

Box 3: Educating towards active citizenship—the experience of Quebec (Canada)

To respond to its linguistic, cultural and ethnic diversity, in 1998 Quebec launched an innovative programme on education towards active citizenship. It was based on a comprehensive approach, involving all school subjects and the different levels of education. It emphasizes the use of participatory pedagogical approaches, aimed at encouraging debate, decision-making, mutual understanding, etc. It also favoured opportunities to put citizenship into practice out of school, thanks to the co-operation of parliament, associations, etc.

Source: Eduquer à la citoyenneté, Quebec. (Rapport annuel du Conseil supérieur de l' éducation, 1997-98.)

- *By promoting social consensus on a common core of values:* The teaching of specific values, such as the acceptance of diversity, fairness, tolerance, etc., can help in learning to live together. Australia has developed an innovative project in this area. A 'charter of values', including a list of values *(see Table 2.2)*, has thus been set up. Each Australian state has the possibility of selecting values from the list: the state of Queensland, for instance, has agreed to teach in its schools, among other things, the values of compassion, equality, respect for human rights and respect and the search for truth.

Table 2.2: Towards charters of values in Australian states

Value	States/years				
	WA 1985	SA 1989	Qld 1991	NSW 1991	Qld 1994
Adaptability	•	•	•	•	•
Authority, respect for legitimate				•	
Compassion	•				
Concern for the welfare of others		•	•	•	•
Co-operation	•	•		•	•
Co-operation, international				•	•
Creativity					•
Critical thinking					•
Diversity, cultural sub-groups, acceptance			•		•
Ecological sustainability					•
Equality		•	•	•	
Ethical standards				•	
Excellence	•			•	
Family, place of, in society				•	
Freedom, individual liberties		•	•	•	
Honesty	•			•	•
Initiative, enterprise					•
Integrity	•				•
Intellectual enquiry					•
Justice, fairness	•	•	•	•	•
Lifelong learning					•
Life, respect for	•				•
Partnership: school/home/community					•
Peaceful resolution of conflict	•	•		•	
Rationality, respect for reasoning	•	•	•	•	
Responsibilities, social	•		•	•	•
Rights, human, respect for	•		•	•	
Self-respect					•
Sensitivity: physical, aesthetic, emotional, spiritual	•				
Stewardship of the earth and its resources					•
Tolerance	•			•	
Truth, respect and search for		•	•		•

Source: Education Western Australia (WA), 1985; Education South Australia (SA), 1991; Education Queensland (Qld), 1989; Education New South Wales (NSW), 1991; Queensland Curriculum Review, 1994.

Nevertheless, a few warnings can be issued about the introduction of teaching to live together in curricula. They are summarized in *Table* 2.3.

Table 2.3: Risks and remedies of teaching about living together

Risks	Remedies
The 'Panic approach'	Rigorous planning of the development of curricula
Over-loading of curricula	Focus on the basics Review the balance between subjects. Inter-disciplinarity
Accumulation of a fragmented knowledge	Bridges between general education and the world of Work
Limited impact on learners' behaviour	Modern teaching/learning methods

The development of new teaching/learning methods appears to be fundamental in obtaining a significant change in learners' behaviours. An article published in the *Journal of vocational education and training* gives a good illustration of the importance of the pedagogy used *(see Box 4):* students from the Hong Kong Polytechnic University were failing to integrate into the food and management sector, although they had been trained to do so. This was partly due to the irrelevancy of the teaching methods used, which did not help them feel sufficiently confident to communicate in English with foreigners, who were to be their main customers.

Box 4: Food and beverage management teaching in Hong Kong

The Hong Kong Polytechnic University offers training in food and management. But as much of the terminology used in this sector is derived from Western cultures, Chinese students had difficulty in communicating. To help students fit better with the needs of their future employers, it was proposed:

- to improve their listening skills in English;
- to develop their confidence in speaking, thanks to a student-centred learning methodology.

In other words, learning how to interact with the Western system.

Source: Journal of vocational education and training, vol. 49, no. 3, 1997.

HOW TO HELP TEACHERS ADAPT THEIR TEACHING METHODS

Teachers are the key actors if any actual reform of educational programmes is to take place. Several policies can be followed in order to help them adapt their teaching methods, three of which are presented hereafter:

- *To help them develop new skills through initial and in-service training (involvement in teamwork, etc.).*

The most obvious strategy consists of renewing their initial training through in-service courses designed to integrate new subjects that will be taught in an interdisciplinary way, at the same time familiarizing them with participatory and student-centred approaches. Regular in-service training should help teachers to update their knowledge and share their experience with others.

- *To modify the design of textbooks to allow them to be used differently.* In order to modify their teaching methods, teachers should be given the opportunity to use newly-designed educational materials. A report published by the French Ministry of National Education contains recommendations to review the design and use of textbooks *(see Box 5).* It suggests, in particular, that textbooks should be defined as 'a reference tool', summarizing the main concepts which children should acquire. They should not to be too long, nor contain too many pictures, quotations, etc. This would enable teachers to organize their course autonomously, referring to textbooks, but using other materials as well—in particular, multi-media sources—to illustrate their presentation as they want.

Box 5: A few recommendations to review the design and use of textbooks

Prepare textbooks for each educational level with a view to ensuring the coherence of the teaching of a discipline over several years.

Encourage school principals and teachers to define an educational project, so that the selection of books should be made according to a global—and not a disciplinary—approach.

Make the textbook, defined as a reference tool, separate from other sources of information (documents images, multimedia), so that teachers may create their own pedagogy.

Source: Inspection générale de l' éducation nationale, *Le manuel scolaire,* Paris, Ministére de l' éducation nationale, 1998.

- *To make new evaluation means available to teachers, defined together with different actors.* In order to be ready to adopt 'competency-based' curricula, teachers should be provided with new evaluation tools. The example of the United States' General Educational Development Testing (called GED Testing) can give a few ideas on this point *(see Box 6)*. In its 2000 version, it includes a series of tests aimed at evaluating the capacity of students on information processing and at assessing their communication skills. A cross-disciplinary test also makes it possible to see if students are able to solve a global problem, by making use of what they have learned in different subjects.

Box 6: General Educational Development Testing in the United States

GED Testing is designed to measure the academic outcomes traditionally acquired by the end of a typical high school study programme. The GED 2000 series tests should take into account:
Information processing, including knowing how to determine what information is needed, how to conduct a search for that information, how to synthesize information from diverse materials and in different media, and how to organize and present that information,
Communication skills, including knowing how to develop a message for a variety of audiences and purposes.
A **cross-disciplinary test** will allow examinees to use information processing skills from a variety of academic disciplines to interpret the material.
Source: Connections, Philadelphia, PA, NCAL, University of Pennsylvania, 1998.

- *By giving them the possibility of resorting to non-formal educational resources (science museums, etc.):* The creation of innovative science museums—such as the Exploratorium in San Francisco, USA, or La Cité des Sciences de la Villette, in Paris, France—should be regarded as key partners for teachers to access the most recent knowledge and to experience other ways of teaching (development of projects, group work, etc.).

But above all, favouring a 'sense of ownership' towards the curriculum on the part of teachers should be looked at as a key factor for change. Two main strategies can be explored in this area:

- *To involve teachers in the development of curricula:* Innovative experiments have been conducted in Sub-Saharan Africa in order to enable teachers to take part actively in the setting-up of educational programmes *(see Box 7).* Curriculum conferences have thus been organized to allow teachers to decide how prescribed curriculum guidelines should be put into practice. These conferences have proved to be successful, helping teachers to feel involved and, as a result, responsible for what they teach.

Box 7. Curriculum conferences in Sub-Saharan Africa
Curriculum conferences offer opportunities for active participation for all those involved in reforming primary education (practising teachers, subject specialists, teacher trainers, etc.). The main emphasis of such conferences is the transformation of nationally prescribed curriculum guidelines into practical lesson units that reflect local conditions and concentrate on prevailing cultural trends. By favouring 'a sense of ownership' of the reform among those involved in transforming curricula into action at school level, curriculum conferences facilitate the implementation of the recommended core curriculum.

- *To favour school-based curricula:* An example from Scotland illustrates another way of associating teachers with the development of educational programmes *(see Box 8).* For a while, Scottish teachers have been authorized to select, adapt and develop, at school level, the limited guidelines provided to them by the government. They have had, for instance, no textbooks available to help them. This experience was said not only to favour a sense of ownership of curricula among teachers, but also to help improve the relevance of what was taught, by adapting it both to local and school needs.

To conclude, from all the topics presented above, a selected number were chosen to be discussed during the sub-regional course on curriculum development, namely:

Box 8. School-based curriculum development in Scotland
In Scotland, schools are free to develop their own curriculum within the limits of a broad, nationally prescribed framework. From the national curriculum framework, each school selects, adapts and develops courses and modules. The central educational authorities assist schools by making guidelines, course materials, nationally approved full courses, short courses and modules available to them. It permits the development of curricula that are more relevant to local and school needs.
Source: A. Lewy, *National and school-based development,* Paris, IIEP, 1991.

- The relevance of new trends in the teaching of science, social science and humanities to the challenges of globalization and the principle of living together.
- Decentralization of curriculum development to local or school level.
- Integrated and interdisciplinary approaches in curriculum reform.
- The adaptation of teacher training to curriculum change.
- The potentials and challenges of information and communication technologies in the adaptation of curricula.
- Non-school science resources and collaboration between formal and non-formal educational institutions.
- The need for vocationalizing curricula.

3

Round-Table: The Relevance of New Trends in the Teaching of Science, Social Science and Humanities to the Challenge of Globalization and the Principle of Learning to Live Together

C.J. Daswani

Four resource persons participated in the round-table debate, which was then followed by group discussions on issues relating to the three broad subject areas examined: humanities, science and social science. The full text of the presentation on the Humanities (Language and the School Curriculum by C.J. Daswani) is included below, followed by summaries of the other presentations.

THE HUMANITIES: LANGUAGE AND THE SCHOOL CURRICULUM

The human child is born to learn. The learning process begins at birth and continues throughout life. The human capacity to learn is best demonstrated by the child's ability to learn a first language. Bombarded with ungrammatical, incomplete and mutilated baby talk, nevertheless the child learns to speak his/her mother-tongue perfectly by the age of 4 or 5. In this way, the child is able to control his or her total environment—social, emotional, and psychological—through language. A child is even able to learn more than one language—two, sometimes more—if the social environment so requires.

Very few of us realize or appreciate that 'normal' children perform a miracle in mastering their mother-tongue so quickly, and can learn more than one language if they are exposed to more languages. We can appreciate this miracle when we see the problems that adults have in trying to learn a 'new' language.

By age 4 or so, every child has learned to speak the mother-tongue or home language perfectly; understands what others say, and makes himself/herself understood. By this age, the child is able to use the language creatively and may be likened to a 'born' poet. The only difference between the child and the adult is in the range of vocabulary that a child controls. And, of course, the child has yet to learn to read and write.

The Child and the School

When children enter school they are already accomplished users of the mother-tongue or the home language. But the school often ignores this linguistic competence of the child, since a uniform standard form of language is 'taught' to all children. Often this standard language is so different from the child's home language that the child has problems comprehending it—as well as the school itself.

It is bad enough when the child's home language is simply a social or regional dialect of the standard school language. The problem is compounded when the school language is totally different from the home language of the child.

In many bi- or multi-lingual countries, children have to start school using a language that is entirely different from their mother tongue. When this happens the child not only feels lost, but is traumatized. In such a situation, linguistically the child becomes an infant again, while still actually being an accomplished user of the mother-tongue.

The situation is even more complex, because the schoolteacher believes that the child has to be taught basic language skills—listening and speaking, followed by reading and writing—in a language that the child does not actually understand in the first place. Very soon, the teacher becomes convinced that the child will not be able to learn the school language efficiently and writes him/her off as a poor learner. In the absence of basic vocabulary, grammatical intuition and communication skills, the child will not able to take on the other school subjects either.

In some school systems this complex problem is by-passed by a simple strategy whereby the teacher trains the child to memorize facts and standard responses which can be reproduced during examinations. And this is passed off as formal learning. The child learns to don two personalities: one of limited competence in the school, and another of fluent communication at home.

Unfortunately, in such bi- or multi-lingual settings the school language, which can be regional, pan-national or international, often enjoys social prestige and leads to economic advantage. Consequently, the child gradually learns to favour the prestigious 'other' tongue in preference to the mother-tongue. As a result, by the time they complete their formal education children end up as linguistic cripples, having poor competence in the other language, and having allowed their mother tongue to atrophy through disuse.

Place of Language in School

The place and function of language in school ought to be very different from this situation, especially in bi- and multi-lingual school systems. Language education in school must primarily aim at sharpening the linguistic skills that children bring with them. Primary education should essentially be focused on language education, because at this stage children need to build on the linguistic abilities that they already acquired.

If the child is functioning in a monolingual setting, he/she has to transit gradually from the oral mode of communication to the written mode, which can take four or five years of formal primary school. During this period, children should not only acquire basic literacy skills, but should be allowed to build on their intuition in the mother-tongue. Children should be made aware of the immense power of language and the potential that they already possess to use that language creatively. Children have to build on their lexical resources, experiment with descriptive devices, and learn to reflect on their environment—physical, social, psychological and emotional.

On the other hand, if the child has to acquire a new language in school, the school should build on the child's existing skills in the mother-tongue when learning the new language. Although languages appear to be different from each other, there are crucial underlying similarities in the manner in which human languages are organized. The structures of the mother-tongue can be exploited to teach a second and even a third language. All young children have a natural ability to learn other languages quickly, provided they perceive them as essential for day-to-day communication.

What young children do not understand is why they should learn a new language, which is obviously not needed for communication. If left to themselves, children would learn a new language only if it hastened their integration into a milieu. The motivation to learn a language for instrumental purposes comes at a later stage. However, the school curriculum can be designed so that young children can become motivated to learn a second language that will be useful to them later in life.

The Language Curriculum

In order to plan and implement a creative language curriculum, it is necessary to understand how children learn languages. It is necessary to map the communicative strategies that children naturally employ and to build on these so that they internalize their expanding experiences in school.

The first step in the language curriculum should be to enlarge the child's vocabulary in the mother-tongue and link it to the language used in school. If the school language is different from the mother-tongue, the curriculum should allow the child to use the mother-tongue vocabulary as a basis for learning the lexicon of the language used in school. The enriching of the child's vocabulary can be achieved in various creative ways.

The second step is to ensure that children perceive the linguistic/cultural categories that their mother tongue employs for looking at the real world. This is significant, especially in bilingual and multilingual settings. Different cultures categorize reality in different ways. Colour terms, relational terms and spatio-temporal relationships are perceived differently by different cultures, and these differences are reflected in the languages. For children to acquire intuition in a second or third language, it is crucial that they should apprehend the linguistic/cultural categories of a language.

The third step is to equip children with adequate descriptive devices so that they can describe their experiences and their environment. In the current school curricula, children are seldom encouraged to give expression to their thoughts, perceptions, feelings or emotions. Instead, they are given samples of excellent writing by well-known authors that they are required to emulate. This results in a reluctance on the part of the children to experiment with language. In bilingual and multilingual settings, children can be guided to discover how different languages employ descriptive devices. It is possible for a child to enrich his/her linguistic repertoire by personally comparing two or more languages.

The fourth step is to enable the child to reflect on his/her reality in the mother-tongue and the other tongue(s). Children learn to ask 'why?' at a very early age. Unfortunately, the school often drills out this natural inquisitiveness in the child. This happens when children are expected to conform to the standard norm, and are not allowed to be different from each other.

Language education is inadequate if children do not learn to reflect and work out, on their own, equations such as 'X... therefore Y', 'if not X... then what?', and many more. Reflection is the ultimate goal of language education.

A language curriculum that enables and supports the child's natural capacity to learn can be designed, provided the school system sees the child as the focus of education. Education is clearly unsatisfactory if it consists merely of learning information and superficial facts.

The Language Teacher

No curriculum will suffice unless the teacher is sensitive and innovative. A language teacher has a major role to play in enabling the child to become an excellent communicator. In fact, all teachers are language teachers, no matter what subject(s) they may teach. Each discipline uses the resources of language to impart knowledge. Therefore, it is incumbent on every teacher to become aware of the vast potential of a language.

Language teachers need to be trained to transact the language curriculum creatively. They must understand that the four fundamental skills of listening, speaking, reading and writing are merely teaching devices, and not ends in themselves. The goal of language education is to enlarge the natural resources of the mother-tongue that the child brings to school.

The teacher's job is to ensure that children are provided with sufficient opportunities to practice and sharpen their natural skills. One way of ensuring that this happens is to encourage children to read on their own. Reading stimulates the imagination. Reading enables each child to create magnificent images in the mind, an ability that can be severely impaired by the more modern practice of watching television images created by some other person.

In order to ensure all this, the teacher must also be a learner and observe what the child is capable of doing. The teacher must continuously enlarge his/her capacity to enthuse children to use their natural language skills in order to understand the world.

Language Globalization

In the present information age, the vast reservoir of human knowledge is available in a dominant international language. It is not surprising, therefore, that many school systems find it both necessary and profitable to teach an international language. Currently, the most profitable global language is English.

However, it is important to be alive to the dangers of adopting an international language at the cost of indigenous languages. While English may assist in the process of globalization, neglect of local and national languages may result in their marginalization. And, of course, the impact of such neglect on human cultures the world over cannot even be imagined or calculated.

SCIENCE

Two principal new trends in the teaching of science are the opining up of schools to society and an increasing recognition of the importance of developing scientific methodology and ways of thinking among students.

The Opening up of Schools to Society

For the past twenty years the school has grown to recognize the need to keep up with changes taking place in society. This trend can be observed through the evolution of the subjects in the curriculum, the participation of different actors from the wider community in the classroom, and the increasing use in the classroom of technical tools used in the world of work, in particular information and communication technologies. Seven characteristics of this new trend may be identified:

- **The end of isolation between school and the rest of the society.** If the school is to prepare a larger number of pupils for adult life more effectively, the education system must be less isolated from the rest of the society. This end of isolation is characterized not only by the evolution of the teaching tools available in school, but by all the following points.
- **The increasing incorporation of global and inter-disciplinary topics into the curriculum.** To understand the effects of science and technology on society, it is necessary to teach these subjects in an interdisciplinary manner, relating specific themes to the broader subject and showing their relevance to other disciplines in the curriculum.
- **Vocational education and lifelong learning.** One of the new demands on the school is adequate preparation of pupils for the world of work. The right balance has to be found between general education and the vocational preparation for working life. There needs to be better knowledge of the job market, its structure and organization, and the training required to enter it.
- **Citizenship education.** Another new role for schools is to prepare children to assume their responsibilities as citizens. This should involve not only learning about the organization of society and the role of its institutions, but should also provide models of democratic behaviour, foster rigorous moral attitudes and generate debates on subjects of general interest. All of these concerns are relevant to the effective teaching of science and technology.
- **The new function of non-school organizations.** To help schools assume their new role, diverse non-school organizations can be solicited: the media, museums, libraries, trade unions, businesses and industries can all help schools in this new mission.
- **Basic knowledge versus 'up-to-date' contents.** The vast amount of available knowledge creates the problem of what to include in the science and technology curriculum and what to leave out. Fundamental knowledge is essential, but it is also

necessary to be familiar with the latest developments in science in order to understand the society we live in. These developments often differ from basic knowledge in that they are evolving and our understanding of them is constantly being modified. Therefore we need an increasing quantity of knowledge. One of the solutions (but not the only one) would be to determine what basic knowledge is required, as well as the principal new developments to be taught in the classroom. In addition, students must be taught the skills of research, inquiry and interpretation so that they may continue learning and expanding their knowledge beyond the classroom and beyond the school.

The Increasing Importance of Scientific Methodology and Ways of Thinking

The second main trend for teaching science is the importance of training in scientific attitudes and the ability to relate these to the real world around us. Today, it is vital to acquire a general and scientific education at school, but also crucial to have the skills and rigorous attitudes required for research and lifelong learning. Seven aspects of this scientific behaviour can be identified:

- **Skills of observation, analysis, comparison and classification**. These intellectual skills are as important as acquiring and owning knowledge. Developing them demands rigour and objectivity.
- **Carrying out experiments to understand and respect reality**. Experimental activity is one of the most important approaches in science. Starting from a question, students have to be taught to state a hypothesis, to establish the experimental protocol to confirm or invalidate the hypothesis and, armed with the outcome, to progress from this new temporary understanding of reality to another question.
- **Documentary research**. Developing information research skills is essential to learning science. Students must be taught universal research methods, including the selection of appropriate keywords and the

identification of suitable sources and resources. This method can draw on books, documents and newspapers, but also on the huge quantity of information available on the Internet.

- **Developing critical thinking skills**. One important attitude to be developed in conjunction with documentary research is the training of a critical mind. Pupils must be taught to check and assess all the information they encounter, compare it with other sources, and weigh up the probability of truth in each. This behaviour is not only useful in documentary research, but also during any activity in which information is being transmitted: debates, meetings, radio, television, etc.
- **Active learning and collaborative learning**. For all the types of activities described above, the pupils have to be active learners and participants in the process. The advantage in working in small groups is that it makes possible the sharing of intellectual resources and supports each other's learning.
- **Developing a pedagogy of the project**. The global dimension of a project, with the involvement of all pupils throughout its development is a vital means of giving pupils a sense of ownership, of meaningful participation in the learning process.
- **Starting from the pupils' representations/initial knowledge as the starting point for differentiated learning**. It is essential to take into account the differences among pupils' initial representations of science and technology and the diversity of their learning styles. It is very important to adapt the learning process to the initial conceptions of individual pupils and assess the evolution of their representations at the end of the lesson. This presentation clearly underlines the urgent need to change traditional approaches to teaching, and to pay attention to the selection and use of instructional materials, and to the organization of the class. For these changes to be carried out in schools, innovation and reform in teacher education are vital.

SOCIAL SCIENCES AND RELIGIOUS/MORAL EDUCATION

Focusing on the recent curriculum reform in Norway, the speaker discussed how subject syllabi in the social sciences, humanities and religious/moral education can support general education especially in the light of the principle of learning to live together.

General Concerns

A major concern of curriculum development should be how the various subject areas can support general education for lifelong learning. The curriculum should ideally be seen as a document not just for schools and teachers, but for, and belonging to, society. The ultimate aim of curriculum development should be the achievement of education for all in a unified system, which at the same time acknowledges and caters to difference and diversity.

A major preoccupation is how to develop a balanced curriculum while avoiding overload. How does one decide what to exclude? One way of coping with the excessive amount of knowledge is to develop and foster a spirit of inquiry in the student which will permit him/her to go on learning throughout life. The best approach is to first assist pupils to discover the reality around them and then to expand their inquiry to the wider, global environment. The development of research skills should be an integral part of the core curriculum.

The curriculum should focus on the all-round development of the individual. Students should be taught to see themselves as both citizens of their own country and of the world.

Social Studies in Compulsory School

It is a prerequisite for a truly democratic society that its members are familiar with, and support, certain democratic values. Each new generation needs to learn the value of participation, and to uphold the democratic rules that govern various sectors of the society. Social studies are intended to prepare pupils for different tasks in society by equipping them with knowledge, confidence in their own values, and the desire to undertake tasks for the common good. These aims must be achieved in close co-operation with the wider society outside school.

The social studies disciplines show how people, through their interaction with nature and with each other, have developed different kinds of social life and forms of society. Developing considered attitudes to society past and present is an important aim of social studies.

The methods used in social studies must provide the pupils with an opportunity to gain insight into the subject matter, relevant skills and the ability to co-operate with others, as well as preparing them for action. Their work must stimulate their curiosity and ability to ask questions. Pupils must be given scope to learn to conduct arguments, to exercise critical analysis, to work independently, and to experience how to organize work in co-operation with others.

Throughout schooling, social studies are based on close contacts and co-operation between schools and their local communities. Active participation in school affairs through pupil council and other co-operative bodies makes school work in a wider sense of the term an important feature of training in social studies.

The Structure of the Subject

In the social studies syllabus, objectives and areas of study are arranged and presented under the disciplines history, geography and society.

'History' covers the chronological dimension, discusses different views of people and different forms of society through the years, and considers the driving forces behind social development.

'Geography' deals with space and changes in space, and the localization and distribution on the Earth's surface of natural and man-made phenomena and processes. Geography discusses and explains society as it appears at the intersection between people and nature, and surveys the world's living conditions, ways of life and standards of living.

'Society' deals with politics and socialization and shows why society needs laws and rules, organization and government in order to function. Society studies human interplay and conflict, and discusses the rights and obligations, the influences and challenges that individuals encounter as participants in various communities.

At the primary stage, social studies to a large extent form integrated parts of broad inter-disciplinary areas. Later on, these three disciplines emerge more clearly as separate units, but social studies should aim throughout to combine history, geography and society into an overall picture.

Religious and Moral Education

In Norway, the study of this subject is intended to give the pupils a thorough insight into the State religion and what the Christian view of life implies, as well as familiarity with other world religions and philosophies. The subject is to be approached openly and contribute to insight, respect and dialogue across the boundaries between faiths and philosophies, and promote understanding and tolerance in religious and moral questions. The subject mediates knowledge of a faith—the classroom is not intended to be a place for the preaching of any particular faith. It must respect each pupil's identity in terms of his or her faith, while at the same time furthering dialogue within a shared culture.

In order to meet other faiths and views of life with understanding, one needs to be able to place them in a context that is already familiar. The subject thus has various functions in basic schooling: to transmit a tradition, to maintain a sense of identity, and to build bridges which advance insight and dialogue.

It is natural to have links between the religious and aesthetic heritage, for instance by giving emphasis to graphic arts, architecture, music, drama and literary texts. At the lower secondary stage, the pupils are better able to adopt a critical approach to both the teaching material and divergent opinions, and understand both metaphors and symbols. Discussions, lectures, and excursions all offer relevant approaches to the subject. More weight is given at this stage to comparing various features of religions and philosophies.

In view of its aesthetic dimension, this subject invites co-operation with art and crafts, music and drama, adding variety and raising pupils' levels of activity. What pupils learn about religion and philosophy also brings them into contact with social circumstances and literature, so that an interdisciplinary approach with social studies and mother-tongue instruction is relevant. The human view of nature, and developments in cosmology, research and science, have also depended on religious outlook; co-operation can also be envisaged with natural sciences and technology.

4

Summary of the Round-Table Working-Group Discussions

GROUP ONE: SCIENCE

This group addressed four issues related to the improvement of science education:

1. How to cover both basic scientific knowledge and on-going scientific developments in the curriculum in an adequate manner.
2. The role of experimental activities and how they should be arranged.
3. How to teach about the concepts of exploiting and respecting the environment as necessary complements of each other.
4. How to bridge the gap between scientific knowledge and aspects of traditional knowledge, attitudes and beliefs.

Question One: Scientific Knowledge and On-going Scientific Developments

First, it was necessary to define 'basic scientific knowledge' or what was meant by a core curriculum. It was immediately recognized that decisions about the core curriculum are not neutral and objective but raise fundamental value-laden issues. One suggestion was for this kind of knowledge to consist of matters that had continued relevance over time and would likely remain relevant for the future. Another suggestion was that 'core' should be defined as 'processes' and 'competencies' rather than as 'content'.

The basic science curriculum should be realistic and pragmatic in its aims. Rather than seeking to prepare all students for a career as scientists, it should focus on teaching students the fundamentals of science in order to be able to interpret their environment and effectively function within society. Students leaving school should be able to make informed decisions about issues related to science which affect their daily lives, whether or not they go on to study science at a higher level.

Core contents are essential if students are to learn to apply scientific processes. At the same time, they are fundamental to the understanding of new ideas and developments in science. Core elements of the curriculum and current developments in science should therefore not be considered as separate issues but should be seen as complementary, in the sense that the core can be taught within a contemporary framework. The curriculum should seek to teach basic knowledge with reference to its relevance in contemporary society.

The key concerns that should inform decisions about the broader curriculum include an awareness of the cross-curricular nature of many everyday science issues and the universal implications of science.

An essential component of the regular inclusion of on-going scientific developments into the curriculum is the in-service training of teachers. Countries described the problems of inadequately trained teachers and of limited facilities for science teaching. It was agreed that teachers need to be trained in the management of information if they are to stay abreast of developments in science.

It was suggested that one way of teaching about new developments in science is through project work in which students work on both individual and group projects. The provision of extra-curricular activities, such as science clubs, also provides opportunities for discussion and debate on current issues.

Question Two: The Organization of Experimental Activities

Related to this first issue was a fundamental question: Are schools providing adequate opportunities for learning science? Experimentation should be considered as not simply the 'hands-on' approach to learning science but should be taught as a whole process. From this perspective, the availability of expensive state-of-the-art equipment is not the priority. It is rather a matter of providing adequate opportunities for experimentation.

Experimental activities must be adapted to the available resources and may take place in a number of settings, not simply in the traditional laboratory environment; for example they can take place through play and outdoor activities.

The overall objectives of any experimental activities should be: (a) to help students understand concepts; and (b) to teach them to check theories through the results they obtain themselves. It is essential for the teacher to define the specific objectives of the experimental activity when designing it and for these to be made clear to the student.

The role of the imagination in the development of scientific thought was discussed. From this perspective science should not be taught simply as a precise, rational subject, but should provide room for hypothesis and creativity through experimentation. The need for science instruction to be made enjoyable is vital to improved learning. Experimental activities, if properly designed, can aid students' enjoyment of science, while leading to meaningful learning. From this persepctive, pedagogy needs to be more child-centred.

Three types of 'doing' activity were identified, namely: (a) simple manipulation or the 'hands-on' technique; (b) investigation and enquiry through research projects; and (c) experimental activities providing elaborate processes of confronting students with reality. All of these may be combined in the experimental process. Experimentation clearly calls for varied teaching methods, once again pointing to the need for the adequate training of science teachers. Teachers should ideally be both specialists in their subject and trained in up-to-date pedagogical methods. The lack of a suitably trained teaching force is one of the major obstacles to the improvement of educational quality in developing countries.

Question Three: Use of, and Respect for, the Environment

The concepts of use of, and respect for, the environment should be taught as complementary to each other and should be integrated into the whole curriculum, becoming part of the 'school ideology'. Environmental education is integral to the issues of globalization and of learning to live together.

Environmental education should focus on relevant issues found in the locality, relating them in turn to the global context and showing the interconnectedness of the natural world and of human activity. In defining the environmental education curriculum, a distinction should be made between the science of ecology (nature study) and the much broader spectrum of environmental issues and concerns which enter all aspects of daily life and should be integrated across the curriculum. Environmental education should thus be taught as a systemic discipline. Some participants indicated that his approach is already used in their countries. Integrating the environmental education curriculum also assists in reducing curriculum load.

The evaluation of the effect of environmental education on individual attitudes and behaviour is difficult to measure. It is easier to evaluate the students' grasp of the concepts taught, but measuring application of this learning on daily behaviour would require another type of evaluation over a long-term period. It was suggested that teaching aspects of environmental law within the local context would assist in developing environmental awareness and changing behaviour.

It is very important for students to be aware of the difference between scientific knowledge and social opinion and choice. The evolution of science does not occur in a vacuum but is influenced by a range of political and social issues. Science needs to be taught with the aim of helping students to become informed decision-makers with respect to these complex issues and their impact on people' lives. They need to be aware of the many different actors involved in these issues and the roles they play. Scientific knowledge is essential for true social participation and thus should be considered as part of education for democratic citizenship.

***Question Four:* Scientific/Traditional Knowledge**

While it should be recognized that not all traditional knowledge is opposed to scientific knowledge, this question was considered to be a sensitive and difficult area in the curriculum, because of the potential conflict between scientific belief and socio-cultural (particularly religious) values.

While the universal application and relevance of scientific findings should be stressed, students should be made aware that science itself is not exact but is constantly changing and being revised—uncertainty and hypothesis being integral parts of scientific thought. It should be recognized that scientific though and traditional knowledge and beliefs involve differing ways of looking at the world. Teachers should be trained to acknowledge both viewpoints and to exercise objectivity with respect to both in the classroom.

It is valuable to deal with these conflicting points of view with respect to traditions, some of which may in fact be harmful to individual or collective survival in areas such as health and the environment. An example was given in a project called 'science across the world' project where opportunities were provided for both scientific and traditional beliefs and knowledge to be explored in an objective manner. When teaching these issues, the level of maturity of the students must be taken into account.

The media can play an important role in changing traditional beliefs that may stand in the way of development. However, if the curriculum is to foster democratic attitudes, care should be taken for ideas and beliefs to be discussed and not to be imposed.

One effect of globalization on science is that scientific knowledge is penetrating traditional societies much more rapidly. Care should be taken to examine both the beneficial and the harmful effects of science.

GROUPS TWO AND THREE—SOCIAL SCIENCES AND HUMANITIES

Social Studies—From Core Values to Classroom Practice

Based on national experiences, the group discussed how social sciences (history, geography, society) can support general education (values education, socio-civics, citizenship education) by defining new opportunities for curriculum design, teaching/ learning methods, pupil assessment and implementation strategies.

It was recognized that the aims and content of general education need to be clearly defined. General education was seen not simply as the set of compulsory subjects to be studied by all students, but also as the core values to be taught through the curriculum. The issue of curriculum overload was a common concern.

The group made the following recommendations with regard to *the curriculum:*

- The existing curriculum should undergo review in order to define minimum learning competencies (MLCs) for general secondary education.
- Social studies are an ideal area for teaching about learning to live together and should inculcate a concept of global citizenship.
- The teaching of the social sciences, particularly at primary level, should follow an integrated approach. A key example of where an integrated approach may be adopted is in the broad issue of the interdependence among countries, and the need for respect towards other cultures. This should be taught across subjects, with the key thematic areas already identified during curriculum design.
- Teaching of the social sciences should progress from the immediate to the more distant environment.
- Human rights education should be a component of the social sciences curriculum.
- Citizenship/civic education should include the teaching of values.
- Religious education should perhaps be taught separately from social science subjects, although values found within religion can also be integrated across the curriculum.

The group made the following recommendations with regard to *teaching/learning methods:*

- There is a need for new methodologies, with the focus being on child/student-centred learning. Strategies should include activity-based learning, co-operative learning and individualized learning. Role-play and simulation should also be used. Process skills and higher-order thinking skills should be taught.
- In implementing the strategies, the teacher's role must become a more flexible one ranging from a lecturer to a motivator, guide and facilitator.

The group recommended that *pupil assessment* should aim to move towards a formative mode of assessment through *continuous assessment* using:

- regular feedback;
- possibilities for self-evaluation;
- portfolios;
- formal marking of work;
- project reports.

In the same way, *summative assessment* should use:

- paper and pencil tests;
- formal marking.

The group made the following recommendations with regard to *implementation strategies:*

- new/supplementary learning materials should be prepared, including parents' books (guides to the curriculum), teacher's guides, textbooks and pupils' manuals;
- teacher orientation/training on new teaching methods should be provided, both in terms of pre-service and in-service teacher education (INSET);
- alternative assessment processes should be introduced;
- monitoring/follow-up in schools should be guaranteed.

A phase should be foreseen during which a widely applied trial scheme would be implemented, allowing for adequate review and revision. Heads of schools should be given curriculum frameworks so that they would be in a position to monitor implementation. The process should take place over a period of five to ten years.

Humanities—from core values to classroom practice

The group was given the objective of discussing how the teaching of languages can be adapted to the needs of different socio-cultural groups for developing communicative competence and participation. This includes the following questions: What are the implications for curriculum design, teaching/learning methods, pupil assessment and implementation strategies?

In the field of curriculum design, the principal concerns were:

- The complex situation in multilingual societies leading to a diversity of socio-cultural groups in the classroom;
- The need to develop adequate communication skills among the students.

The group made the following recommendations with regard to *the humanities:*

- when designing curricula, the local environment (natural, social, and cultural) should be borne in mind;
- within the perspective of language as a tool for learning, the mother-tongue should be the medium of instruction at the primary level;
- set priorities with regard to the choice of languages to be taught;
- develop a language curriculum that will seek to promote communicative competence among all learners;
- develop, select and use stimulating instructional materials.

The group made the following recommendations with regard to *methods of teaching and learning:*

- modify teaching/learning methodologies;
- strengthen mother-tongue competence.

Teacher Competence—From Core Values to Teaching Practice

The group was given the objective of discussing the challenge of developing teacher competence in relation to general education. It was decided that there was a need for teachers to use different methodologies and it was expected that teachers should be competent communicators. The group made the following recommendations:

- teachers should be properly prepared to implement a revised language education curriculum;
- the curriculum of pre-service teacher education programmes should be reformed/adapted to accommodate the concerns of general education;
- there should be adequate development of in-service teacher education programmes to equip teachers to handle the concerns of general education more effectively.

Part II

Some Challenges for the Adaptation of Content Raised by the Principle of Learning to Live Together

5

Adaptation of Content to Address the Principle of Learning to Live Together: The Challenge for Teacher Training

R.H. Dave and J.S. Rajput

EFFECTIVE TEACHER EDUCATION: A CORE CONDITION FOR THE QUALITY OF SCHOOL EDUCATION

Education is a process of human enlightenment and empowerment for the achievement of a better and a higher quality of life. A sound and effective system of education results in the continuous development of learners' potentialities, the strengthening of their skills and the fostering of positive interests, attitudes and values.

All progressive societies have recognized the enormous potential of education and have committed themselves to the universalization of elementary education with the explicit aim of providing 'quality education for all'. They have also recognized the need to expand secondary education to as near universal levels as possible, while simultaneously improving its quality for the increased empowerment and advancement of an ever-increasing percentage of the population. While higher education also has great potential in this respect, it is often accessible to only a small section of society. However, at present, primary education can be provided to practically all members of society and, therefore, its quality and efficiency attain a special significance within the personal, social and national context.

Thus, effective teacher education acquires an even more crucial importance, becoming a key factor in ensuring quality school education. In other words, effective formal education implies effective teacher education.

Teachers can act as trail-blazers in the lives of learners and in the process of education for development. If they acquire the professional competence and attitudes that enable them to effectively perform their multiple tasks in the classroom, in the school and in the community, teachers become the single most important contributing factor in ensuring quality educational provision at primary and secondary levels.

There was a time, in most countries, when teacher education was just a single event that took place at the beginning of a career. But this once-in-a-lifetime model is utterly inadequate for the needs of contemporary societies. In the last decades of the twentieth century, society has witnessed unprecedented technological advancements and economic, political and socio-cultural changes that must be reflected in the school. Indeed, these events have already had a very significant impact on schools around the world. In an effort to better meet the diverse needs of learners and ultimately of society, there have been numerous curricular reforms and the introduction of a range of new approaches and strategies in instruction and evaluation.

A COMPREHENSIVE PARADIGM

Clearly, all of these changes have profound implications for the content and processes of teacher education. If teachers are to be able to provide quality education in the face of these challenges, there is an urgent need for on-going reform of teacher education. Effective teacher education for both elementary and secondary stages of schooling has now to be conceived within a more comprehensive paradigm encompassing a number of interrelated components. These components include the following:

- *Pre-service and initial teacher education:* To be provided in a systematic, professional way to all new teachers entering the teaching profession.
- *Recurrent in-service teacher orientation:* To be offered on a recurrent basis and in an organized manner to practising teachers through orientation seminars, workshops, tele-conferences and other such programmes in response to new professional needs and to ensure continued teacher motivation.

- *Continuing professional self-learning:* According to their individual needs, interests and specific professional responsibilities, teachers should pursue their own self-directed and lifelong learning through books, journals, audio-visual aids and other available information and communication technologies. The establishment of adequately equipped resource centres is essential if teachers are to profit from this type of professional development.

Professional orientation for school principals and department heads: When promoted or newly recruited as principals, supervisors, co-ordinators, etc., they should benefit from:

- Recurrent organized professional development opportunities.
- Continuous self-directed professional learning and enrichment.
- Opportunities for learning about international experiences and contacting colleagues in other countries.

Higher professional education: Planning and provision for increased access to higher education (diploma, bachelor, master and doctoral levels).

- Planning and provision for training of high-level specialists in different fields, such as curriculum development, textbook writing and preparation of other teaching/learning materials, evaluation and monitoring, planning and management, research and statistics, etc.
- Adequate planning and provision to prepare teacher educators and resource persons to provide effective pre-service and in-service teacher education at the elementary and secondary levels.
- Provision of opportunities for international exchange.

Enrichment opportunities for teacher educators.

- Provision of opportunities for periodic enrichment of teacher educators through seminars, workshops, presentations and discussion of papers, etc.
- Facilities and opportunities for research and debate on different aspects of education including teacher education.

- Facilities for publication of different kinds of materials on diverse issues relating to school and teacher education.
- Encouraging international contacts.

A sound programme of teacher education needs to include all of these components in a comprehensive and multi-dimensional manner. The first three components cited above are directly focussed on the education of all teachers. The other three components are essential for improving the quality of teacher education in various ways. An effective pre-service teacher education programme serves as a sound basis for implementing and developing each of these additional components.

THE FIRST THREE COMPONENTS OF THE PARADIGM

Within this comprehensive scenario, the first three components call for further examination in view of the urgent need for revision and renewal of teacher-education curricula. The overall objectives are to ensure that elementary and secondary school teachers are professionally well-equipped so that they may become increasingly autonomous in their professional growth and, ultimately, maintain their motivation and effectiveness in their role and performance in the classroom, school and wider community.

Pre-service teacher education should be the start of a process of transforming lay persons into competent and committed professional educators. The untrained entrants should possess the requisite educational background and stipulated qualifications. The aim of pre-service teacher education is to prepare the uninitiated to become sound professional practitioners. In the light of this definition the new curriculum for teacher education must be competency-based and commitment-oriented.

In-service teacher education refers to a recurrent, organized and needs-based continuing education for practising teachers so that they may update and enrich their professional competence, strengthen their commitment and enhance their role and performance in the classroom, school and wider community. As new developments take place in the curriculum, educational techniques, evaluation procedures, classroom management and other aspects of school education, new needs will constantly arise for in-service training, calling for recurrent provision.

In addition to these organized efforts, all teachers should be encouraged and given opportunities for continuous self-study related to their own professional needs, interests and responsibilities. A successful and dynamic teacher remains a self-motivated and self-directed learner throughout his/her career. It is this self-directed and lifelong learning that supplements and complements the organized sector of teacher education and becomes an important dimension in a comprehensive paradigm of professional education for teachers in contemporary society.

If these three components of teacher education are to be realized, a major review of existing teacher education provision needs to be undertaken. Any such effort must clearly establish and define the basic competencies required by teachers within prevailing social, economic, and cultural conditions, and also foresee the new educational needs and challenges likely to arise within the initial decades of the twenty-first century.

MAJOR DIMENSIONS OF A COMPETENCY-BASED AND COMMITMENT-ORIENTED CURRICULUM

As one of the important dimensions of renewing the curriculum, five performance areas were identified on the basis of job analysis and needs analysis. The purpose was to improve the quality and efficiency of school education.

It may be noted that these performance areas will give rise to a series of practical activities which include the present programme of 'teaching practice', but in a more realistic and effective manner will cover evaluation procedures and classroom management in different situations, such as multi-grade teaching. However, they also include other important educational responsibilities in which the teacher should be given adequate practical training to develop pertinent skills. Thus, in the field of practical training, these performance areas mark a major shift from teaching practice to broader educational practice in which pedagogical skills remain an integral part, but where the development of other significant practical skills are also given their due place.

1. *Performance in the classroom:* includes teaching and learning processes, evaluation techniques and classroom management.

2. *School-level performance:* for greater all-round contribution to school organization and management.
3. Performance in extra-curricula and out-of-school activities
4. *Performance related to parental contact and co-operation:* for improved enrolment, attendance and student achievement.
5. *Performance related to community contact and co-operation:* should focus on improving school-community relationship for mutual development and enrichment.

In order to equip teachers adequately in these performance areas, ten sets of competencies have been identified for curriculum renewal. These are designed to provide adequate theoretical and conceptual understanding, empowering teachers to perform their responsibilities with professional insight and confidence. In essence, these are teacher competencies that should aim at the development of learner competencies and qualities at school. To achieve these multiple goals, teacher competencies include relevant conceptual elements, content elements, contextual aspects, transactional and evaluation aspects, etc. All the ten competency areas thus identified converge on one or more of the performance areas and interrelate theory and practice in a focused manner.

These competencies are first to be developed during pre-service teacher education and further updated and strengthened during recurrent in-service teacher orientation, as well as through continuing and self-directed professional enhancement.

Ten Competency Areas

1. *Contextual competencies.* To provide a wider view of the development of education in society and the teachers' role in it.
2. *Conceptual competencies.* Includes the concepts of education and learning, psychological, sociological and neuro-physiological aspects of education, etc.
3. *Curricular and content competencies.* According to specific levels of education.
4. *Transactional competencies.*
5. *Competencies related to the organization of school level and extra-curricular activities.*

6. *Competencies related to teaching/learning materials.* Includes their selection, use and preparation and should include new information and communication technologies, where applicable, and existing local resources.
7. *Evaluation competencies.*
8. *Management competencies.*
9. *Competencies related to parental contact and co-operation.*
10. *Competencies related to community contact and co-operation.*

While every teacher must master professional competencies, such as those enumerated above, it has been observed that, by themselves, they do not necessarily result in effective performance. The actual performance of trained teachers in the classroom or school is equally dependent—if not more dependent—on their levels of commitment and motivation. The task of fostering professional commitment among teachers must become an integral part of pre-service and in-service teacher education. Towards this end, five commitment areas have been identified.

1. *Commitment to the learner:* Concern for the all-round development of all pupils.
2. *Commitment to the society:* Awareness of, and concern about, the impact of the teaching profession on the development of the community and the nation.
3. *Commitment to the profession:* Development of a professional ethic and sense of vocation.
4. *Commitment to excellence:* In all aspects of a teacher's roles and responsibilities.
5. *Commitment to basic human values:* To become a role model in the classroom and community through genuine and consistent practice of professional values, such as impartiality, objectivity and intellectual honesty.

6

The Potentials and Challenges of Information and Communication Technologies (ICTs) for Education: The Training of Teachers

Jean-Marie Sani

INTRODUCTION

Information and communication technologies (ICTs) have a number of characteristics that permit significant innovation in pedagogical methods at all levels of education. These include the following:

- The digitalized form of information allows storage and dissemination of all types of data on computers (text, sound, still pictures, moving pictures), since all are encoded in the same format—a succession of digits (0,1).
- New possibilities for distance communication, whether simultaneous or deferred, permitting collaborative projects between classes, schools or the development of professional networks among teachers.
- The development of these distance communication techniques has led to a new structuring of information, bringing with it a new way of reading and discovering texts. 'Hypertext' permits the combination of linear reading with rapid connection to related texts, each linked to the other. Multimedia encyclopaedias are a good example of the very useful hyper-textual organization of information.
- Multimedia is becoming the principal way of presenting information, allowing information to be presented in different forms, with the result that it is more diverse and take up less space.

NEW CLASSROOM PRACTICES WITH ICTs

In the classroom, two main tasks may be identified in the use of these technologies:

1. ***Mastering the technology:*** The use of ICTs in education can permit both pupils and teachers to acquire skill in the use of these technologies, particularly in exploiting the computer, as these machines become essential everyday tools in contemporary societies. This objective is in accordance with the aims of vocational education, i.e. to teach skills that will equip learners to enter the job market.
2. ***Tools for learning:*** If used effectively, ICTs can enhance the possibilities for instruction and learning for all pupils at all levels of schooling. They should, however, be seen as supplementing rather than replacing the more traditional but proven educational tools and methods. They may be used in the instruction of all disciplines, as well as for interdisciplinary or cross-curricular projects. ICTs may also enhance learning in the following ways:
   - ***Searching for information:*** The use of CD-ROMs or the Internet requires knowledge of research methods, such as those used in traditional library research, i.e. documentary research. To find the right answers to specific questions, it is necessary to identify keywords, find appropriate sources, and select relevant information from the identified resources. This process requires the development of critical abilities in order to assess the quality and validity of the information retrieved. The teacher's role is essential so as to develop documentary research and critical thinking skills in students. This responsibility cannot simply be delegated to the computer, which remains a tool to be suitably exploited.
   - ***Summarizing the information.*** As with the above, the ability to summarize the information obtained is not a new objective of schooling. Again, it is the technology that allows a new *way* of formulating and presenting the information (as in an HTML file). Developing the ability to summarize and reformulate retrieved information should be one of the major objectives of learning. ICTs are ideal tools for developing this skill.

- *Distance communication.* Several sorts of communication are possible:
  (a) ***Among teachers.*** Using electronic mail, newsgroups or websites, teachers can share their ideas, experience and innovations. This may be a very effective and rewarding means of professional development, as there are already numerous sites, newsgroups and networks devoted to education covering wide geographical areas and subject ranges.
  (b) ***Among classes or pupils.*** This use is less developed because it is more difficult to realize. However, it has tremendous potential for developing the concept and practice of living together, of intercultural communication and international understanding. Classes in diverse geographical locations may collaborate together on a project, with the contribution of not only individual and collective student abilities, but also context-specific knowledge and data from the participating regions/localities. For this type of activity to be carried out successfully, teachers must be properly trained and the technical possibilities of networks and the available software (in particular the transmission speed) must be improved.
  (c) ***Between teachers and groups of learners.*** A new style of communication is being invented between teachers and learners in a distance-learning context. It no longer should be considered sufficient, for instance, to follow a lecture via a video-conferencing system. Nowadays, multimedia materials may be used to enhance instruction. Teaching styles have to evolve to keep pace with these technologies.

However, three principal challenges confront the effective exploitation of these technologies in schools:

- the price (of computers, of software and of communications);
- the capacity of telecommunications—which should exceed 2Mbit/sec;

- the development of software with greater capacities, especially for the sharing of documents.

In conclusion, it should be emphasized that the use of ICTs as a tool, like any other means of instruction, such as books, videos or mere observation:

- may be used for all disciplines: science, technology, languages, history, geography, art, as well as for transdisciplinary projects.
- should supplement—not replace—other tools for teaching and learning. Pupils need to remain in close touch with the reality around them; they need to touch, manipulate, build and experiment with the natural world.

THE CHANGING ROLE OF THE TEACHER

In the traditional teaching-learning relationship, the relation between the teacher and the pupil is frontal—the role of the teacher is to deliver knowledge to the pupil. There is some co-operation among pupils.

Often teachers do not possess adequate knowledge and skills for the effective exploitation of ICTs. In many cases, they are less expert than their pupils. Furthermore, there is likely to be confusion in the teacher's mind about his/her new role in relation to the use of these technologies, i.e. teachers find themselves in a situation where they are no longer the principal

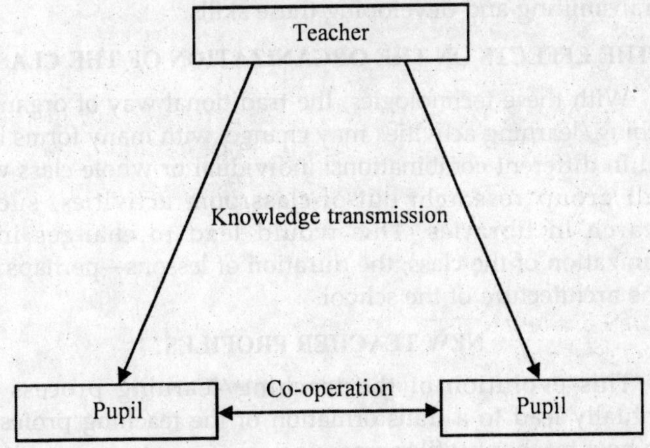

Figure 6.1: The traditional frontal pedagogical relationship

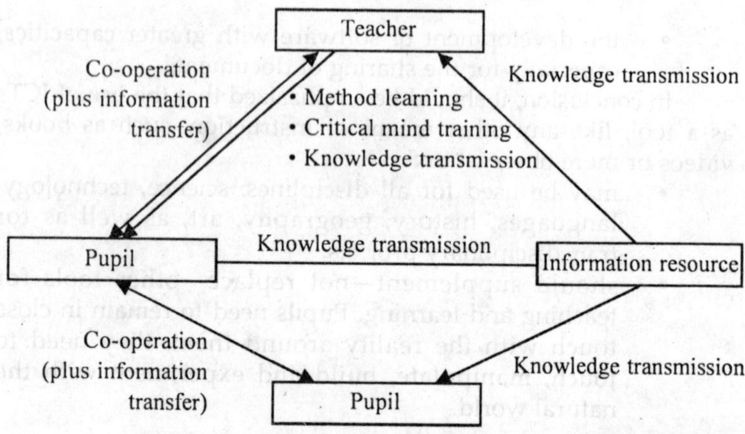

Figure 6.2: The new pedagogical relationship

source and deliverer of information. If teachers are going to find themselves less capable of making effective use of ICTs in the classroom, they need first to be properly trained in pedagogical skills, so as to have adequate knowledge for developing pupils' learning abilities. The skills needed to use computers as tools for learning are similar to those needed to exploit other more traditional tools: how to transform information into knowledge; how to develop a critical mind; how to make links among different sources and types of information; how to find the right answers to the questions rapidly. The teacher's role remains vital in transmitting and developing these skills.

THE EFFECTS ON THE ORGANIZATION OF THE CLASS

With these technologies, the traditional way of organizing teaching/learning activities may change, with many forms being used in different combinations: individual or whole class work; small group research; out-of-classroom activities, such as research in libraries. This would lead to changes in the organization of the class, the duration of lessons—perhaps even in the architecture of the school.

NEW TEACHER PROFILES?

This evolution of the teaching/learning process may eventually lead to a transformation of the teaching profession, with new teacher profiles emerging.

- ***The subject specialist.*** A situation may develop where there are a limited number of specialists in specific subjects in different parts of the world. Schools may enter into distance education relationships with these experts.
- ***The expert in the learning process.*** This second type of teacher will have expertise in the learning process. His/her relationship with the learner will be face-to-face.
- ***The evaluation specialist.*** A third type of teacher profile may evolve: The evaluation specialist. Evaluation is going to become an increasingly complex operation demanding specific skills and the trend may be for pupils to be evaluated by persons other than their teachers.

THE TRAINING OF TEACHERS

Considering these possible important transformations, the adaptation of the teacher-training system becomes a fundamental issue. Two aims of training may be identified:

Preparing teachers for the introduction of ICTs in the classroom:

- It is necessary for teachers to become skilled in operating the new technologies and in exploiting them effectively as educational tools.
- Teachers must master the use of information—skills of research, critical analysis, linking diverse types and sources of information, reformulating retrieved data—if they are to teach their pupils to develop these same skills.
- There needs to be more emphasis placed on training in pedagogy, as opposed to the current trend in many education systems where the major focus is on specialized knowledge in specific curricular subjects. Teachers must be adequately equipped with more didactic competencies so as to assume their new role as experts in the learning process.

  Using ICTs as tools for the training of teachers. As sources of information and expertise, as well as tools for distance communication, ICTs offer many new possibilities for teacher education. Teachers may learn new forms of communication through the regular use of these technologies. Use of new media, new rules of communication—even a new language—have to be learned.

CONCLUSION

A possible economic counterbalance. ICTs are expensive and at present are very unequally distributed around the world, with the developed world having by far the greater share. Nevertheless, it must be recognized that these technologies offer great possibilities for poorer countries, with a reduction in the importance of distance and an increase in the potential for new competitivity. In the coming years, considerable investments in infrastructure, equipment and training are needed if countries are to incorporate the use of ICTs into schools as an integral part of the teaching/learning process.

A real stake in the process of globalization. These technologies offer vast opportunities for the development of contacts and exchanges with the rest of the world. For this to become a possibility for all, however, it will be necessary for prices to fall, a transfer of the necessary technical knowledge, as well as the constant need to keep pace with the evolution of existing technologies. These factors are all realizable and, when they come about, all countries in the worlds may share a major transformation of education—as we know it today.

7

The Potentials and Challenges of Information and Communication Technologies for Education: Globalization of Education, Adaptation of Curricula, New Teaching Materials and the Networking of Schools

M.M. Pant

INTRODUCTION

All new technologies initially provoke reactions ranging from apprehension, to caution, to curiosity, to excitement and expectation. There is considerable scepticism on the part of educators with regard to the potential impact of the new technologies on schooling, since the anticipated impact of earlier communication technologies—such as radio and television—has not been realised.

While communication between learners and mentors is the essence of education, in the traditional learner/teacher paradigm, teachers largely dictate facts (that they themselves may not necessarily comprehend) to pupils who often just reproduce them on paper without interpreting or understanding them. For education to be effective, it is essential that there be more opportunities for student participation in the learning process, more teamwork, more self-study and self-evaluation, as well as more peer-evaluation, and less examination-oriented teaching and learning. It is felt that the new technologies can assist in the promotion of more student-centred and interactive learning.

WHAT ARE ICTs?

ICTs can be briefly summarized as the result of the convergence of technologies—telecommunications and television have merged with informatics. The computer, which can be described as an engine of the mind, has tremendous capacity to store and process data, to produce and disseminate information. With the emergence of diverse multimedia and networking possibilities, computers have increasing potential as tools for innovative teaching and learning.

The new technologies have enormous potential to revolutionize education as we know it, dramatically transforming schools. For the first time in the history of education students are proving more adept at mastering delivery systems than their teachers. It is obvious that the monopolies enjoyed by schools as formal education providers will diminish, with a lot of educational exposure and experience in future being provided outside these institutions. The scope of the new technologies for transforming existing educational provision is enormous and includes the globalization of education, the adaptation of curricula, new teaching materials and the networking of schools.

THE IMPACT OF THE INTERNET ON EDUCATION

The development of the Internet has already had a very significant impact on the structure of higher education provision. Traditional distance education institutions are fast changing to leading-edge 'virtual universities'. Today there are about 800 'cyber colleges' in the United States, with approximately I billion students plugged into 'virtual classrooms' compared to 13 million enrolled in traditional institutions. By the year 2000, this number is expected to triple. It is anticipated that fifty years from now, most of the professions will be taught through the Internet, causing many regional institutions and smaller universities to disappear. Bureaucratic structures are likely to diminish and there will be increasing use of information technologies and resultant networking and partnerships to convey resources for the running of institutions.

Technological developments are leading to accelerating convergence between education and industry and between formal and non-formal education:

- Business organizations are becoming learning organizations;
- Distinctions between formal education and distance education are being reduced resulting in integrated classrooms;

- Lifelong education and training will become increasingly the norm.
- Through technology, education is becoming more closely linked to entertainment.

Some of these changes taking place at the level of higher education are likely to filter down to schools. Already in the United States, 'virtual high schools' are being established. Liberal entry and exit between the school system and the world of work may become increasingly possible.

SYLLABI AND COURSES

It is foreseen that traditional educational institutions, both universities and schools, will have to make important adaptations to the new technologies. Their offerings will become diverse and flexible. Courses of studies in traditional classrooms will no longer meet the public demand for tailored educational services, especially at upper secondary and higher education levels. There is likely to be a move from rigid one-time, full-time, on-campus education to diverse programmes of lifelong learning. The content of courses will need to assume a more global dimension, with cross-cultural variations becoming more evident.

The new educational enterprise will be developed around student outcomes and rigorous evaluation of achievement. The new technologies are also creating a demand for a range of basic competencies, including: improved numeracy skills, communication skills, technical skills and skills in creativity and design, the ability to work with and relate to others, to manage tasks and solve problems, and to foster one's self-development through lifelong learning. In the area of vocational education, there is scope for many new courses including Web design and Web publishing, repair and maintenance of personal computers, office records management and retrieval systems.

These changes are likely to lead to the development of outcomes-based education where each learning outcome will be defined, assessment criteria for each outcome will be clearly stated and guidelines for generating evidence of learning achievements will be laid down.

Teaching methods and strategies, as well as accreditation and certification, will have to change significantly to meet these new demands. The professor/teacher will no longer dominate the teaching/learning process: Students will assume greater autonomy and responsibility for their learning. The teacher will become more of a coach, animator and co-explorer with his/her pupils. The process of learning will be emphasized more than specific detailed subject content, while learning will become more flexible and interactive. In some situations, the technologies make education more affordable.

With the use of a variety of new technologies in instruction, it is imperative to assess the extent to which true learning occurs. Sophisticated assessment methods need to be developed to measure and certify learning competencies.

CHALLENGES TO THE INTRODUCTION OF ICTs IN EDUCATION

Incorporating the technologies successfully into schools requires careful advance planning and preparation. Significant financial and human resources are required, with training an essential component of the process. Redundant and robust systems must be put in place. Innovators have to be prepared to confront bureaucracy and conservative attitudes, including resistance by teachers and other educational staff.

SUMMARY AND CONCLUSIONS

- ICTs will have a deep impact on the way formal education is carried out;
- The rate at which we have to respond to changes affecting education is much faster than ever before;
- It is urgent to realign the curriculum framework to provide outcome-based, flexible learning paths leading to mastery learning;
- Training and orientation of teachers, administrators and students to the new learning technologies is an immediate requirement;
- By sharing courseware and learning resources developed in different parts of the globe, innovative and relevant curricula can be rapidly developed and made available to learners in poor as well as rich countries.

8

Learning to Live Together: The Need for Vocationalizing the Curriculum

Arun K. Mishra

Although the objective of developing employable skills and competencies during upper secondary education is included in Indian national education policy documents and the curriculum framework, in reality skill development courses at this level are virtually non-existent in schools. The concept of 'learning to live together' is likely to be confined to the informal settings of family and local community.

How may a vocational curriculum lead to a better understanding about the culture and ways of life of different countries and the common principles of living in a global village?

The vocational or pre-vocational curriculum lends itself to the possibility of organizing concrete opportunities for students to collaborate for the good of the group, the class in general, the school or the wider community. Many skills may be taught through services that benefit the community as a whole, bringing the school closer to its environs and helping pupils to become aware of their commitments to school and community. Co-operative activities can promote friendship, communal harmony and empathy for others.

All vocational education programmes and activities should stress the concept of sustainable development with a focus on fostering awareness of key environmental concerns and the rights of all to a decent standard of living.

In India, work experience is an integral component of the primary and secondary school curriculum. By the lower secondary stage, skills taught through this programme can be described as pre-vocational. Knowledge and skills taught are expected to be directly relevant to the prevailing developmental and economic activities of the community.

Here are some suggestions for content of pre-vocational curricula or work-experience programmes, particularly within a rural context:

- Management of water for domestic and agricultural purposes;
- Development and use of alternative sources of energy;
- Food production, preservation and storage;
- Machine maintenance and repair;
- Handling and use of fertilizers and other chemicals;
- Information management for rural development.

Part III

Interdisciplinarity, School-Based Management and Non-School Science Education: A Few Topics for Reflection by Curriculum Developers

9

Adaptation of Contents to Address the Principle of Learning to Live Together: The Case for Interdisciplinarity

Ellen-Marie Skaflestad

INTRODUCTION

During the last two or three generations, great changes have taken place in the living conditions of the young. In many countries, both parents spend more time away from the home at their place of work, while their children's links to the world of work and the learning that may be acquired there have waned. The impact of the mass media has grown continuously over the last decade, and schools have become increasingly multicultural.

Social change is no longer an episodic set of events interspersed by periods of stability. Advanced societies are now open to a continuous process of change, difficult to predict scientifically and control socially. These societies are dynamic rather than static, and complex rather than simple entities. Advanced societies are 'risk societies (Elliott, 1998). Social change has ambiguous consequences for the individual. It opens up new possibilities for human fulfilment, but multiplies the risks and hazards that confront the individual in achieving this fulfilment. In these circumstances, responsibility for shaping the conditions of existence in society should be devolved down to the grassroots—to the people themselves. Education can only meet the challenge of social change if it gives all pupils access to appropriate cultural resources in a form that enables them to take responsibility for actively shaping the economic and social conditions of their existence.

The organization of the curriculum in terms of academic subjects for the purpose of systematic instruction is ill-suited to the aim of a general education. More consistent with such an aim is a curriculum that organizes cultural resources in usable forms for the purpose of enabling pupils to deepen and extend their understanding of the problems and dilemmas of everyday life in society, and to make informed and intelligent judgements about how they might be resolved. Such a curriculum will be responsive to the pupils' own thinking and their emerging understandings and insights into human situations.

Educational change implies a focus on both curriculum and pedagogy, and on the development of teachers as experimental innovators. Learning has to be connected with the living experiences of students in a rapidly changing society. Different conceptions of education and of the curriculum presuppose different conceptions of society and the principles governing access to its 'benefits'. Recent trends in curriculum policy-making reflect the dilemmas with which all 'advanced' Western states are wrestling. States feel that they need to steer the curriculum in ways that are consistent with their economic goals, but find that the context of policy implementation is too complex to handle from the centre. The dilemma is illustrated by, on the one hand, attempts at the State level to establish national educational standards through the concept of a core curriculum which will command a public consensus, and on the other, by pressure on schools to respond to the complex changes taking place in their locality.

However, it is clear that the idea of the curriculum having a unifying function in a diverse learning environment, combined with the emergence of the comprehensive common school, has gained strength in many societies. General education can be defined as 'that part of the student's whole education which looks first of all to his life as a responsible human being and citizen' (Harvard Committee, 1945, in Elliott, 1998). The notion of citizenship should be central to defining the character of general education. Without it, general education tends to be defined as a common core of knowledge, skills, and values—a definition that fails to draw attention to basic curricular concerns, such as what knowledge is most worthwhile and what aims and objectives are best suited for schools in democratic societies.

General education is more than a function of the curriculum; it is also an orientation to learning and to curriculum design. Generally speaking, all education justified under general should emphasize a socio-civic content that promotes problem-centred inquiry and group co-operation. General education should also be marked by interdisciplinary subject-matter schemes. The learner in this general education scenario is defined as an autonomously thinking, socially responsible citizen who is able to make decisions. In a broad sense, it is a perspective on learning that emphasizes citizenship priorities. Thus, the consequence of general education is a comprehensive concept of schooling (Hlebowitsh, 1993).

THE CONNECTIVE MODEL OF THE CURRICULUM

Conceptions of knowledge underlie curriculum development and can be defined in either of two ways: 'Curriculum as facts' and 'curriculum as practice' (Young, 1998). 'Curriculum as facts' tends to have a life of its own and obscures the social contexts in which it is embedded. It often results in the curriculum being neither understandable nor changeable. 'Curriculum as practice' does not begin with the structure of knowledge, but with how knowledge is produced by people acting collectively. According to Young, this connective model of curriculum integration does not start with subjects but with the broader notion of curriculum purposes and how subjects can achieve those purposes. It does not start from the requirements of the national curriculum, but with individual schools defining their curriculum purposes and asking how they can be made to correspond to the requirements of the national curriculum.

Schools need to define their purposes in terms of the kind of young person they want to produce, and the kind of adult, worker, citizen and parent roles they wish young people to assume. For schools to move towards a connective model, all staff need to endorse the curriculum criteria and agree to articulate how their subjects or areas of responsibility would be involved—both in supporting shared approaches to teaching and learning and in delivering the agreed outcomes. The model is 'connective' in the sense that subject specialists would be required to connect their subject teaching to: (a) the purposes of the overall school curriculum; and (b) the way other subjects are contributing to the overall school curriculum. According to Young, the role of subjects would need to be made explicit in at least three ways; first, by identifying the specialist skills and knowledge they can offer; second, by showing how any of the specific skills and knowledge of particular subjects can contribute to the broader curriculum goals, such as personal and social education through collaboration with other subject specialists; and third, by identifying the contribution of subject specialists in enabling schools to develop their external links with employers, the community and other education providers.

THE NORWEGIAN NATIONAL CURRICULUM: A CONNECTIVE CURRICULUM MODEL?

Changes in society and the structural changes in education have made it necessary to re-examine the guidelines governing the purpose and content of education. An agreed minimum values framework has been developed in a number of countries, in the search for the common good. Thus, the process of promoting a consensus of values represents a negotiated view of the common good in a given society at a certain period of educational reform. Globalization in education should mean that there is one ultimate goal to be supported by general education in all countries—how to make citizens able to understand both the local and global societies well enough to learn to live together and act as responsible citizens in local and global terms.

When large-scale reforms were being introduced in primary, secondary and higher education in Norway, it seemed natural and fitting to provide a common formulation of the common core of the curriculum, with a view to emphasizing how the stages of education are linked together, not forgetting adult education. Interdisciplinarity has to be considered not only across the curriculum at a certain stage or level of schooling, but in a vertical structure as well. Consequently, curricular reforms are to be seen as a macro-educational planning process. How the reform concept is interpreted is dependent on the social, political, economic, cultural and educational conditions of a country. It is a policy issue to define the context of change, and to view and be willing to develop a holistic reform strategy, where elements of change are linked and provide for interventions enabling curriculum implementation to be realized.

The starting point for the overall work on revised curricula for the different levels of the education system in Norway is to be found in the following Acts of Parliament; the Primary and Lower Secondary Education Act; the Upper Secondary Education Act and the Vocational Training Act; the Adult Education Act and Folk High Schools Act. The main themes in the relevant sections of these acts fell into the following six groups:

- moral outlook;
- creative abilities;
- work;
- general education;
- co-operation;
- natural environment.

The common core of the curriculum expanded on these themes. The introduction to the 'Core Curriculum for Primary, Secondary and Adult Education' in Norway states that the aim of education is to furnish children, young people and adults with the tools that they need to face the tasks of life and surmount its challenges. Education shall provide learners with the capacity to take charge of themselves and their lives, as well as equipping them with the will and determination to stand by others. Education must spur students to diligence and to close collaboration in the pursuit of common goals. It must promote democracy, national identity and international awareness. In short, the aim of education is to expand the individual's capacity to perceive and to participate, to experience, to empathize and to excel.

It was recognized that if education is to achieve these aims, a number of concepts of the human being, illustrating our complex and diverse roles and identities, need to be defined for curriculum development. These include:

- The spiritual human being;
- The creative human being;
- The working human being;
- The liberally-educated human being;
- The social human being;
- The integrated human being.

Education should ultimately form integrated human beings possessing seemingly conflicting capabilities, attitudes, values and skills which permit him/her to lead a full and meaningful life, actively contributing to the common good, yet maintaining his/her own identity and dignity. From this perspective Norwegian education aims to:

- convey the culture's moral values, with its concern for others, while fostering the ability to plot one's own life course;
- provide familiarity with Norway's Christian and humanist heritage, while teaching knowledge of and respect for other religions and faiths;
- teach individuals to overcome self-centredness and belief in the right of the strongest, while inspiring strength to stand alone, to stand up for oneself, dissent and not to acquiesce or submit to the opinions of others unwillingly;
- develop an independent and autonomous personality and, at the same time, to be able to function and work as part of a team.

A number of dual aims are listed in the core curriculum, and the final statement is:

> Education must balance these dual aims. The objective is an all-round development of abilities and distinctive qualities: to conduct oneself according to accepted moral principles; to create and to act, to work with others and in harmony with nature. Education shall contribute to the building of character that gives individuals the strength to take command of their own lives, to assume duties within their society and to take care of the surrounding environment.
>
> When greater knowledge gives greater power, more stress must be placed on the responsibility that accompanies this power. The choices to be made must be based on awareness of consequences and connections, but should also be guided by a scrutiny of values. A distinct precept of education must be to combine greater knowledge, know-how and skills with social awareness, ethical orientation and aesthetic sensibility. The young must be involved in social life, both individually and in a normally coherent way. Education shall promote ethical and critical responsibility in the young for the society and the world they live in.
>
> The ultimate aim of education is to inspire individuals to realize their potential in ways that serve the common good; to nurture humanness in a society in development.

OPERATIONALIZATION OF THE CORE VALUES IN PRIMARY AND LOWER SECONDARY CURRICULUM: THE CASE FOR INTERDISCIPLINARITY

During the 1990s, new curricula have been approved for Norwegian primary and upper secondary education. There are two main differences between the old and new curricula for compulsory education.

The first of these differences is the degree of freedom of choice regarding syllabus content and working methods. The former curriculum allowed local authorities, schools and individual classes a relatively large amount of freedom to decide and allocate the syllabus content at each level. To a great degree, the choices made corresponded to the content of the particular textbooks chosen. Therefore, there were considerable differences in the syllabuses taught from school to school, and district to district. The new national curriculum is based upon a national syllabus. Teaching objectives are set out for each stage (e.g. for age 8 to 10), and instructions about what is going to be taught are formulated.

The other important difference concerns the working methods. The old curriculum recommended pupil-oriented, challenging, working methods. However, more often than not, teaching and learning followed traditional methods. The new curriculum demands that schools, teachers and pupils implement pupil-oriented, challenging, working methods. This is evident in the curricula's general directives, the language used to describe objectives and main features, and the emphasis upon theme-based study and project work.

The compulsory school is a nationwide ten-year school characterized by community and adapted education, which means an all-inclusive school in a co-ordinated school system based on the same curriculum. Adapted education means that it is locally and individually adapted, and takes into account gender equality and linguistic minorities.

The subject curricula are structured according to principles on centrally selected material, local material and adaptation, progression and more subject-specific teaching within a holistic and unified perspective. In the subject curricula, the selection of contents aims at promoting:

- fundamental values, cultural heritage and identity;
- creative abilities and creativity;

- all-round practical skills;
- basic knowledge and broad understanding;
- the ability to co-operate and independence;
- knowledge and awareness of nature, the environment and technology.

It is specified that fields of knowledge across subject boundaries are to be topics related to contemporary society and the individual and society.

Teaching methods must include creative activities and modes of expression, play, practical work, independent work and in-depth study, and project work at all stages.

The structure of the subject syllabi is as follows: (a) introduction; (b) subject-related aims; (c) subject-related objectives for each stage (primary, intermediate and lower secondary). (d) subject-related areas of study for each grade. There are no individual objectives or targets related to the areas of study described. The objectives are an expression of the competence of the pupils to be achieved at each stage of schooling. This provides important possibilities to schools and teachers to teach according to interdisciplinary themes and topics, in particular within the framework of project work. This focus on an interdisciplinary approach is expressed as follows in the curriculum document:

> Although the curriculum is arranged by subjects, it has been designed so those subjects support each other. Aims, objectives and areas of study must be viewed in relation to each other with a view to unity and connections, both within and between subjects. At the primary level, much of the teaching is organized around themes. Structural and thematic connections between subjects and subjects areas should be exploited throughout schooling. This helps pupils to develop overall understandings and to make the most of knowledge and skills across subject boundaries.

Certain fields of knowledge to be taught across subject boundaries have been identified, such as topical issues related to:

- ***Contemporary society***—Nature and environment, international understanding, human rights and peaceful co-existence, technology, information and communication technology, knowledge of the media, working life and vocational guidance, consumer knowledge, road safety.

- ***The individual and society***—The family, sex education, homosexuality, preventive health work, drug abuse, crime prevention, bullying and violence.

The thematic structure of contents shall be based on the experience, interests and understanding of pupils, on connections with the local environment and on topicality. The methods to be used are creative activities and modes of expression in all subjects, practical and independent work, project work and in-depth study. Local work on subject curricula should mainly be decided by individual schools or co-operating schools. It shall establish the foundations for teachers' planning their lessons at all levels. Clear connections between the centrally and locally defined educational content of the teaching must be established, and the aims of the general part of the national curriculum should be borne in mind.

Co-operation at School and with the Wider Community

Co-operation between pupils, teachers and the school administration is essential to the development of the school in terms of creating a learning environment and a place for work. Such co-operation can strengthen the context and possibilities for interdisciplinary perspectives and approaches, helping pupils to see the interconnectedness of learning, and the links between school and the wider community. For the pupils to be included in a social, academic and cultural community, the staff of the school must co-operate, providing a model for the pupils. Co-operation will enable pupils to learn from each other, developing social skills and gaining insight into democratic methods.

It is stressed that the school should develop general social and civic awareness and active involvement in the life of the community. By encouraging pupils to approach new tasks and challenges actively, constructively and deliberately, the school lays the foundations for further learning and helps pupils to master their future work, and their participation in family and social life.

Practical Work and Project Work

Important emphasis is given to practical and project work. Practical work is supposed to form an integral part of all classroom activities and should be designed so as to show pupils the connections between practice and theory and between action and knowledge.

Teachers must provide adequate opportunities for independent work and in-depth study of subjects and subject areas, and in theme and project work. Emphasis should be on teaching research and analytical skills and on developing study methods and working habits which equip pupils for independent work and in-depth study, as well as group work. Teachers are expected to work closely with pupils in realizing projects.

It is stated that project work lends itself to both a single subject and an interdisciplinary approach, and can be linked to specific local themes. Teachers are expected to co-operate with each other to ensure that project work has an interdisciplinary character.

Allocation of Time

School time must be allocated as appropriate to the various activities, bearing in mind the need for unity, continuity and progression in subjects considered separately and together. At lower secondary level, 20 per cent of periods in each year must be devoted to theme and project work.

CONCLUSION

Challenges for curriculum development in our time are to find ways of integrating general education and subject syllabuses—to organize knowledge for teaching and learning into meaningful units. The need for interdisciplinarity is obvious if knowledge is to be made really meaningful in a holistic way for the individual pupil.

The Norwegian model of the curriculum for primary, lower secondary and upper secondary education might be identified as a connective curriculum model. The outcome of the curriculum reform is presently being evaluated. Research-based evaluation will give some feedback on the success of operationalizing the curricula for a better quality of learning.

The former Minister of Education in Norway, Mr. Gudmund Hernes, stated that the main message of the curriculum reform can be summed up like this:

> The most important of all pedagogical tasks is to communicate to children and young people that they are developing continuously in such a way that they can have confidence in their own abilities.

This may be seen as our common curriculum challenge for the next century as well.

REFERENCES AND BIBLIOGRAPHY

Elliott, J. 1998. *The Curriculum Experiment: Meeting the Challenge of Social Change.* Milton Keynes, UK, Open University Press.

Hlebowitsh, P.S. 1993. *Radical Curriculum Theory Reconsidered, a Historical Approach.* New York, Teachers College Press, Columbia University.

Royal Norwegian Ministry of Education, Research and Church Affairs. 1993. *Core Curriculum for Primary, Secondary and Adult Education in Norway.* Oslo.

—. 1997. *Principles and Guidelines for Basic School Education.* Oslo.

—. 1997. *Curriculum for the 10-year Compulsory School in Norway.* Oslo.

Universitetet i Oslo. Institutt for lærerutdanning og skoleutvikling, 1998. *IEA Civic Education Study, Report no 1.* Oslo, university of Oslo.

Universitetet I Oslo. OECD-PISA *Programme for Internaional Student Assessment.* Internet: http://www.ils.uio.no/pisa/

Young, M.F.D. (1998). *Curriculum of the Future: from the 'New Sociology of Education' to a Critical Theory of Learning.* Basingstoke, UK, Falmer Press.

10

Victoria's Schools of the Future

Kenneth Ross

CONTEXT

Victoria is an Australian state with a population of approximately 4.5 million people. In 1996-97, there were 1,700 government schools providing education for 5 to 18-year-olds. The majority of these were primary schools. Another 673 schools were in the non-government sector. Total enrolments were 777,368 students, 517,882 of whom were in government schools. Twenty-five per cent of all Victoria's government schools cater to students from non-English speaking backgrounds, with a small proportion of students from an Aboriginal background.

The formal education system consists of primary schools from preparatory (P) grade to grade six (ages 5 to 12) and secondary schools from grades seven to twelve (ages 12 to 18). Most students (approximately three-quarters) complete thirteen years of schooling and attain the Victorian Certificate of Education (VCE). The total number of government school-teachers in 1996-97 reached 18,159 in primary and 16,902 in secondary schools.

School education is supervised by the 'Office of Schools', which is divided into nine regions for administrative purposes. The Office of Schools is located within the Department of Education[1] and manages the government school system. Several administrative units, including the Office of School Review, in

1 At the inception of the 'Schools of the Future' Programme, the current Department of Education was referred to as the Directorate of School Education (DSE). Both DSE and DOE are used interchangeably in this chapter.

charge of accountability framework, and the Board of Studies, in charge of the curriculum framework, support it. The Office of Schools reports to the Secretary of Education (the public service head) who reports to the Minister of Education. The latter has ultimate responsibility for the school education portfolio.

HISTORICAL DEVELOPMENT

Australia began decentralizing power and authority to schools nearly thirty years ago. In the 1960s, the education system was highly centralized; the states made all key decisions on curriculum, budget and personnel. From the mid-1970s, Victoria became one of the leading states in the move towards a fully decentralized system of education. Between 1980 and 1992, a number of initiatives were taken by successive governments to devolve authority and responsibility progressively to schools. These moves culminated in a policy paper entitled *'Education: Giving Students a Chance'*, published in late 1992 by the Liberal Government. It outlined that quality education can best be achieved by transferring educational decision-making and resource management to the school level. A *'Schools of the Future'* (SOF) Task Force was formed to develop a detailed report outlining the government's objectives and how quality education would be delivered. The task force released the SOF Preliminary Paper in 1993, which revealed that the key to schools' effectiveness would be a 'school charter'. The *'Schools of the Future'* Programme and other legislation has increased the powers and responsibilities of school councils and principals dramatically. Victoria's principals have been placed on limited tenure contracts and local selection of staff was introduced.

The speed with which this reform was institutionalized warrants an explanation. In March 1993, the government asked for applications to place 100 of the state's schools into a pilot programme. The programme aimed at providing 'virtually full authority over the budget and personnel function to the school site'. Within a six-week period, over 700 schools applied, and in July 1993, more than 300 schools entered the first phase of the SOF programme. By early 1994 another 500 schools had entered the programme and an equal number in July 1994. By mid-1995, all Victoria's schools were in the SOF programme. In an attempt to assist schools to understand their self-managing role, the Directorate of School Education published two information kits in 1994. To assist schools to formulate and implement procedures to achieve their respective visions, a Curriculum and Standards, Framework was created. This is a framework within which schools are able to create their own programmes, whilst taking into consideration the identity, aspiration and interests of their teachers. The Board of Studies provided curriculum frameworks for all schools, while the Office of Schools Review developed charter guidelines and an accountability framework.

THE SIGNIFICANCE OF THE VICTORIAN EXPERIENCE

Victoria's model of school-based management (SBM) has two distinguishing dimensions: (a) the involvement of both internal and external constituencies, such as the principal and the representatives of the staff, parents, community and, in the case of secondary schools, students; and (b) the decentralization and devolution to the school level as against the district or the local education authority level, as is the case with other countries, such as in the United States of America or the United Kingdom. Victoria's experience of SBM, through the SOF Programme, is thought to represent one of the most comprehensive strategies at school decentralization attempted anywhere in the world. It is the most sweeping move to decentralization in the history of Australian public education, with nearly 90 per cent of recurrent expenditure distributed to schools within a global school budget. Victoria is the largest system of public education anywhere in the world to have decentralized such a large part of the state budget for school education. The stated objective of the reform was '[to improve] the quality of education for students by moving to schools the responsibility to make decisions, set priorities and control resources'. Accordingly, SOF is intended to 'make more efficient use of resources for the benefit of students, provide a more professional workplace for teachers, and increase the level of community knowledge of, and satisfaction with, schools'. Basically, the reform has four elements, as shown in Figure 10.1.

The curriculum framework: The process of decentralizing the curriculum made standards for student attainment explicit. The framework consists of two elements: the curriculum and

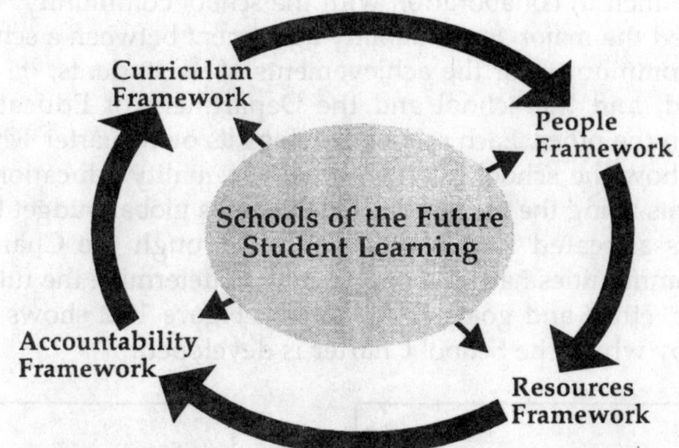

Figure 10.1. The dimensions of schools of the future

standards framework (CSF) for years Prep to 10 (P-10) in eight key learning areas (KLA); and the Victorian Certificate of Education (VCE) for years 11 and 12.

The people framework: The career structures of principals and teachers were addressed in the people framework, consisting of five elements: (1) local selection of staff; (2) full staffing flexibility and workforce planning; (3) performance management for principals and teachers; (4) professional development; and (5) new career structures.

The resources framework: Allocates 90 per cent of the school's recurrent budget directly to the schools; gives schools the flexibility to allocate all resources in accordance with student learning needs; and funds schools in a clear and equitable way through the school global budget.

The accountability framework has three elements: the school charter; the annual report; and the triennial review. 'Quality Assurance in Victoria's Schools' is the main mechanism through which the performance of schools is monitored.

Each of these frameworks has several elements. Whilst most of the elements are in place, some are still to be fully implemented, such as full staffing flexibility in the people framework.

DIMENSIONS OF SOF's REFORMS

The School Charter

As part of the Accountability Framework, the School Charter is the official document produced by the school and the school council in collaboration with the school community. It is considered the major accountability agreement between a school and its community for the achievements of its students, on the one hand, and the school and the Department of Education (DOE), on the other. Each school develops its own charter which outlines how the school intends to deliver quality education to its students using the resources available in its global budget (the resources allocated to it by the DOE). Through the Charter, school communities have the opportunity to determine the future character, ethos and goals of the school. Figure 10.2 shows the process by which the School Charter is developed.

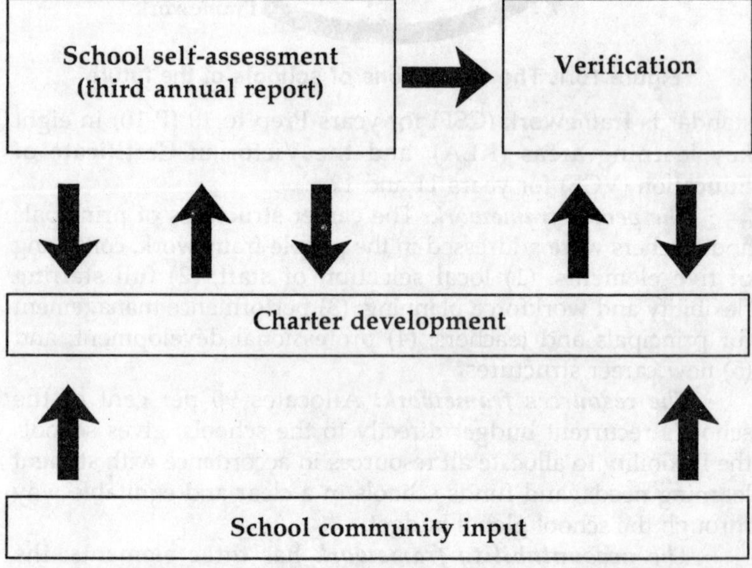

Figure 10.2: The process of charter development

The document of the charter includes: The curriculum profile, codes of practice, students' code of conduct, accountability, budget summary and an agreement to ensure that schools meet their objectives within the limits of available resources. In particular, the document gives:

- A description of the school's philosophy and future directions.
- The school's goals and priorities that are identified as requiring further development.

- How the school intends to deliver the eight mandated curriculum areas and any other special enrichment activities specific to that school.
- Codes of practice for school council members, principals and staff.
- A code of conduct and the discipline approach used for students of the school.
- Details of the processes used for monitoring and reporting on student performance.
- A prediction of student numbers and an indicative budget for the period of the charter.
- A statement that the school agrees to operate within the terms of the charter and to agree to take all reasonable steps to ensure the school meets its goals within the available resources.

The school charter sets the strategic directions for three years. It provides the basis for detailed action plans and allows for the identification of performance measures in meeting the goals and priorities, which relate to curriculum, school environment, management, resource allocation and monitoring performance. Each goal is accompanied by indicators, which enable the achievement of that goal to be measured. The priorities are based on planned and continuous improvement. This places demands on the school to analyse its performance and, using the results of this analysis, to generate priorities for improved student performance. Schools report annually to the DOE and their local community on their performance in achieving their goals and priorities. Every three years a review is conducted at the school, in conjunction with the Office of School Review, to assist with the development of a new charter.

The school charter model adopted in Victoria has a number of features that place it in the category of the world's best practices. First, there is explicit detail concerning the areas identified for improvement and the goals that drive the school; it is not a document that focuses only on improvement, but includes details about the normal operation of the school. Secondly, it is student centred with explicit acknowledgement of the central importance of curriculum and improved student learning. Measurement of both goal and priority outcomes are prominent features. Thirdly, the school charters are firmly located within a broad accountability framework that includes school review and school annual reports. Fourthly, there is detailed specification of the roles of the school community members and a profile of the school. Most importantly, as an accountability instrument, the charter gives parents, via the school council, greater say in the conduct of the school, and increases the requirement to account for the enterprise to the government. The Office of School Review can demand that charters are rewritten, and the objectives not attained in one year are carried over to the next.

CURRICULUM AND STANDARD FRAMEWORK

A second important feature of the SOF programme is the Curriculum and Standard Framework (CSF). This is one of the elements of the curriculum framework noted above. The Board of Studies developed the CSF. There are eight key learning areas in the framework: arts; English; languages other than English; mathematics; sciences; technology, studies of society and the environment; and health and physical education. These guide the development of the curriculum from preparatory year through to year 10. The framework contains two components: (a) the curriculum content in several different levels to be attained over eleven years of study, across the various strands of activity within the key learning areas; and (b) the learning outcomes for students for each of those levels.

The CSF incorporates both content and process standards. Student progress is assessed against the CSF in a programme of state-wide assessment, the Learning Assessment Project (LAP). The LAP assesses students in years 3 and 5, in English and mathematics annually, and in one other key learning area on a five-year cycle. CSF is a teacher assessment of student performance based upon agreed performance levels contained within the CSF documents. In addition, the state has introduced one more test: a state-wide testing of student performance through the Victorian Secondary Assessment Monitor (VSAM) at years 7 and 9.

The introduction of explicit standards by means of the CSF set a yardstick for teachers and the community, and made public what had been the professional concern of individual teachers and staff. The LAP reports to the parents took the locus of information control on student progress out of the teachers' hands. It gave parents 'objective' feedback on their children, gave the school feedback on their performance vis-à-vis other schools, and gave the whole system information on overall attainment. In other words, in government schools, the LAP results became another instrument of accountability when added to school charters.

At the same time, the Victorian Certificate of Education (VCE), which is a two-year (years 11 and 12) school completion certificate, was revised and re-accredited. The VCE provides a wide variety of subjects for students to enable them to undertake studies for either university entrance and/or employment. It contains a series of Common Assessment Tasks (CATs) to be completed by all students undertaking a particular subject to ensure common achievement measures across the system. Some CATs are assessed at the school and others through external examination, but a state-wide moderation system is used to ensure parity for all students' work. All students involved in the VCE are required to sit a General Achievement Test (GAT) to check on the distribution of student grades for school-based CATs within the certificate. Should the school's VCE performance fall within the tolerance range of that school's performance on the GAT, then the results for the VCE assessment will be confirmed. If not, the VCE results will be externally reviewed.

SYSTEMATIC AND LOCAL ACCOUNTABILITY

The third feature of the *'Schools of the Future'* Programme (SOF), which is part of the accountability framework, is the systematic and local accountability processes. These are presented in Figure 10.3. According to this system, the Board of Studies provides curriculum leadership and assistance to schools on a state-wide basis, while the Office of School Review supports the attempts of individual schools to raise the quality of their teaching and learning. The Board is responsible for course development and accreditation, course evaluation and assessment of student performance (including school completion and certification). The Office of School Review is responsible for the co-ordination and management of the accountability processes, particularly as they relate to the development and review of school charters.

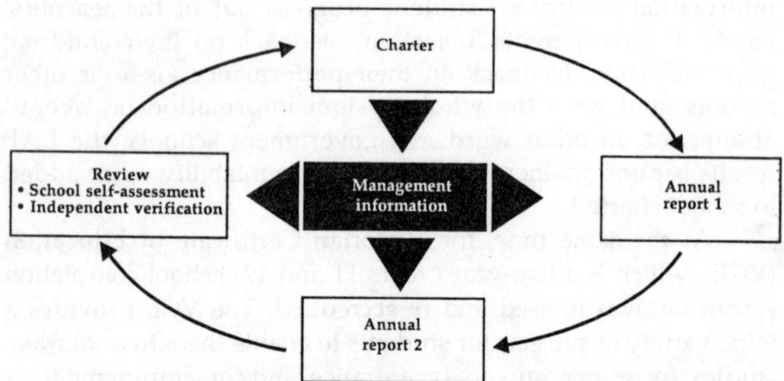

Figure 10.3: Systematic and local accountability framework

As for local or school level accountability, the school councils have the authority to determine the educational policies of the school within the framework of the School Charter. The councils are responsible for maintaining the school premises and grounds, employing non-teaching staff and contracting the services of teachers for particular projects. They are accountable to their local communities, to whom they report through the annual report, and to the Department of Education (DOE), through which independent auditors ensure that the financial dealings of the school conform to the appropriate guidelines.[2]

These elements of the accountability framework serve two main purposes: they satisfy the 'legitimate expectations of government about accountability for the outcomes of schooling', and assist 'schools and teachers to improve standards of student learning'. The framework allows schools to monitor and report on their effectiveness and focus upon improving it. It provides an integrated planning, development and reporting package in which schools develop their own educational plans and priorities within government guidelines (through the school charter), and monitor the progress in meeting these objectives (through annual reports and a school self-assessment). The school's self-assessment is externally monitored through the verification phase of the school review component of the framework. The school charter, annual reports, self-assessment and independent verification are public documents, which are available for

2 Both the LAP and the GAT assessment noted earlier are also used as part of the accountability framework.

community inspection at the school level. The DOE does not allow public access to the accountability documents from schools; this has to be accessed through the schools directly.

SCHOOL REVIEW

This is an element of the accountability framework. School review, as shown in Figure 10.4, is a triennial review based on self-assessment by the school and an independent external verification leading to the development of a new school charter. Community consultation is encouraged at all stages of the process with many schools utilising significant community input in the development of the school self-assessment, including community representation on the verification panel and community involvement in the final development of the new charter. There are three annual reports indicated. Most schools complete two annual reports, with the school self-assessment doubling as both the summary of achievement over three years and the third annual report.

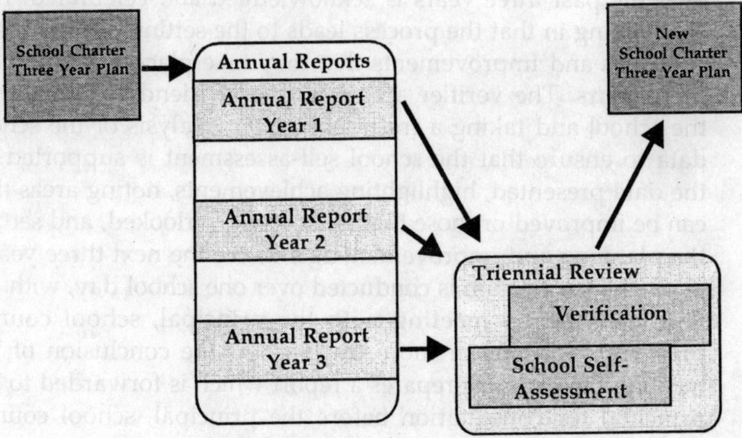

Figure 10.4: School review

SCHOOL SELF-ASSESSMENT

The school self-assessment forms the summary document of the performance of the school over the three-year span of the charter. It is this document that is used in the verification process. It is the school that constructs this document, albeit on the framework provided by the government. It is not until the verification process that external review of this data is conducted.

There are detailed guidelines as to how schools interpret data for self-assessment, as compared to the annual report. Schools are required to present and interpret the data, make judgements and recommendations. The recommendations are focused upon the school's goals and priorities for the next charter. In constructing the self-assessment, schools are encouraged to involve their school community, although the extent of involvement varies. Some schools utilize consultants to help facilitate the process and/or the analysis.

VERIFICATION

An external verifier contracted by the DOE conducts the verification of the school self-assessment. The verification process has been constructed to be both affirming and challenging. It is affirming in that the work of the school and the progress made over the past three years is acknowledged and celebrated. It is challenging in that the process leads to the setting of new goals, priorities and improvements that may take place over the next three years. The verifier acts as a critical friend working with the school and taking a fresh look at the analysis of the school data to ensure that the school self-assessment is supported by the data presented, highlighting achievements, noting areas that can be improved or those that have been overlooked, and setting the planning and improvement agenda for the next three years.

The verification is conducted over one school day, with the verifier typically meeting with the principal, school council president, and one or more teachers. At the conclusion of the meeting the verifier prepares a report which is forwarded to the principal for consultation before the principal, school council president and verifier sign the document. A copy of the document is then sent to the central administration. The end result of the process is that there is a set of firm recommendations on the goals, priorities and improvement focuses to be included in the next charter. The school is in considerable control of the process through its writing of the school self-assessment and has wide representation on the verification panel, not to mention the school principal chairing the verification day.

GLOBAL BUDGETS

The resource framework represents a significant feature of the SOF. Through this framework, the reform has implemented a new basis for funding government schools in Victoria through a well-developed School Global Budget (SGB). The SGB is primarily a formula-based funding model, which consists of a base element for all schools, together with an equity element based on the characteristics of the students enrolled. Hence, individual schools would have the flexibility to allocate all resources in accordance with local needs. Each school receives an SGB, most of which is made on a per student funding basis to reflect the different resource requirements across a range of variables related to learning needs. It provides funding for all school-based costs, including staff salaries, operating expenses and school maintenance.

Schools received support in the introduction of local budgets through increased funding for administrative support and with a software package called 'Computerized Administrative Systems Environment for Schools' (CASES). This assisted the schools to monitor their financial, personnel and administrative functions. Devolving financial management to the local level aimed to empower principals and schools councils to set and allocate resources for local priorities, to separate the purchase of education from its provision, and to decrease the need for a central bureaucracy.

The SGB has its counterpart in other places where there is a high level of school-based budgeting. In every instance, the task of determining the basis for allocation has proved difficult for a range of reasons, including the absence of information about allocations in the past and debate about the relative weightings to be given to the different factors to reflect learning needs.

In 1994 (and again in 1995), an Education Committee was called by the then Minister of Education to advise him on a mechanism by which the DOE could allocate 90 per cent of its state's budget to schools. Most of the committee's recommendations in 1994 and 1995 were implemented. These included a per capita core funding supplemented by need-based allocations for students at educational risks, students with disabilities and impairments, rurality and isolation, students from non-English speaking backgrounds, and priority programmes. The principles underlining these recommendations were detailed as follows:

Pre-eminence of educational considerations. This principle implies that determining what factors ought to be included in the construction of the School Global Budget and what ought to be their relative weighting are pre-eminently educational considerations.

Fairness. This principle implies that schools with the same mix of learning needs should receive the same total of resources in the School Global Budget. In accordance with this principle, SGB should redress the unfair historical allocation of resources, which involved some schools receiving more resources and others receiving fewer resources, when they were otherwise comparable.

Transparency. This principle implies that educational validity and the fairness of the SGB will be apparent only to the extent that the basis for allocations in the SGB is transparent—that is, it is clear and readily understandable by all those concerned. The basis for the allocation of resources to each and every school should be made public.

Subsidiarity. Subsidiarity is the principle that a decision should only be made centrally if it cannot be made locally. It describes the principle of maximizing funds available for school-based decision-making. An implication for the construction of the SGB is that the starting point is to consider included all items of expenditure related to the operation of school. A case must then be made to exclude an item from the SGB. Exclusion of items from the SGB may take place if—and only if: (a) schools do not have control over the expenditure for that item; (b) there is excessive variation in expenditure for the item at the school level from one year to the next; (c) there is unpredictability in expenditure for the item at the school level; (d) expenditure is of a once-off nature; or (e) the item is one for which the school acts simply as a payment conduit.

Accountability. Accountability is a necessary counterpart of the educational focus in the SGB, given that the latter is concerned with matching resources to learning needs. A school which receives resources because it has students with a certain mix of learning needs has the responsibility of providing programmes to meet those needs, and should be accountable for the use of those resources, including outcomes in relation to learning needs.

Strategic implementation. When new funding arrangements are indicated, they should be implemented progressively over several years to eliminate dramatic changes in the funding levels of schools from one year to another.

When implemented in 1995, the SGB consisted of six elements, as follows:

Core funding (based on current staffing and grants formulae with additional funding for administrative support for small schools and early childhood years P-2. This amounted to 80 per cent of the total budget).

1. Additional funding depending on the isolation and rurality (IAR) of schools (depending on the size of the school, and its isolation factor) so as to ensure adequate staffing and a range of curricula in these schools.
2. Additional funding for students from non-English-speaking backgrounds.
3. Additional funding for students with disabilities and impairments (DAI).
4. Additional funding for students at educational risk (SAER).
5. Additional funding for priority programmes such as: physical and sports education; science and technology; instrumental music; professional development; arts in Australia.

USE OF MANAGEMENT INFORMATION SYSTEMS AND TECHNOLOGY

Another important dimension of the SOF is its extensive use of technology and computerized information systems. At the administrative level, and as noted above, CASES was introduced to assist schools to monitor their financial, personnel and administrative functions. This allowed the schools to interface with the central computer system. Schools were issued a standardized computer hardware and software system. CASES stores and processes a range of data including student records (often from teacher input), and financial, physical and human resource data. To enhance the value of information recorded and maintained in CASES, another type of software was designed. This is the CASES Management Information System (CMIS), which is an 'add-on' software package to make better sense of CASES for management purposes. It provides a range of summary reports, often presented graphically, which have been developed in consultation with schools and central personnel. Both CASES and CMIS programmes have been developed in-house by the DOE.

To enhance the system by extending it to include student records, a third software application was introduced. This is a commercial product adapted to Victoria's requirements. The product known as Kidmap provides student assessment and recording, analysis and profiling of student progress/needs, preparation of reports for parents, and access to teaching resources. It allows schools and school systems to access student data, and to analyze and interpret this in a variety of ways.

At the central office and regions, the DOE introduced an Education Management Information System (EMIS), which has some linkages to the CASES/CMIS environment. The basic system in EMIS is the Corporate Information System (CIS) which contains basic school profiles, a diary of events, a phone directory and a range of documents. EMIS also includes a decision support system (DSS) which contains the same databases as CIS, but with additional features. These features provide additional information to allow the construction of an individual school's profile. They also allow the provision of a range of statistical information for downloading to a spreadsheet/word-processing package.

The interface between these information management systems is shown diagrammatically in Figure 10.5. The figure illustrates the main channels of communications and the users of these channels in the SOF management information system. For both the annual report and school review elements, the processes are informed by an extensive array of school and system generated data on student and school achievement. This very much facilitates the collection and analysis of data used in the operation of the accountability framework, and the day-to-day operations of the school.

In terms of curriculum delivery, several curriculum programmes, which make an extensive use of technology, were

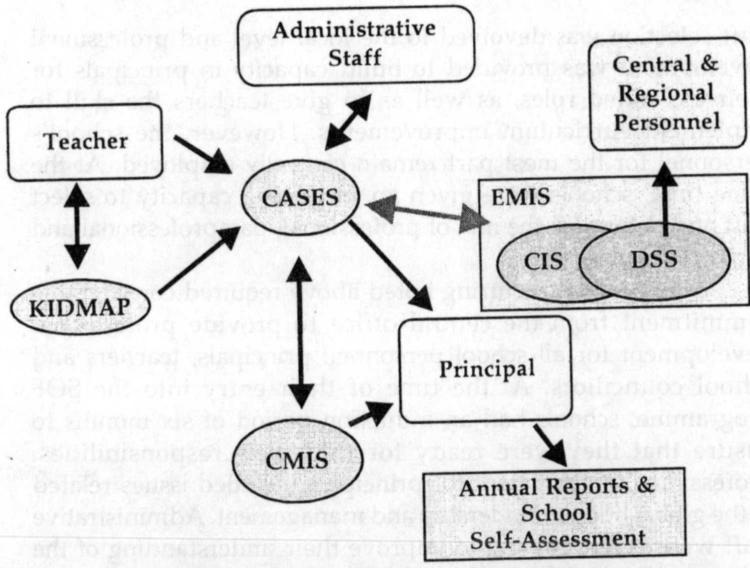

Figure 10.5: Department of Education Management Information System: KIDMAP, CASES, CMIS and EMIS

also introduced. An Interactive Satellite Television (ISTV) programme was established in 1994. Both government and non-government schools installed satellite dishes to receive centrally produced progrmmes. Students could interact directly with the programmes' presenters using either a fax or a telephone. Professional development programmes for teachers and general access for other community groups were also made available through this new technology. Other technological initiatives in the curriculum included programmes which gave all schools access to the Internet and opportunities to develop methods of using the new technologies; and brought people from education and the entertainment arena to work together to develop computer software that both educates and entertains.

PROFESSIONAL DEVELOPMENT AND SUPPORT

As part of the people framework, the career structures of principals and teacher were addressed. This was 'consistent with efforts to restructure the public sector', where 'there has been downsizing of central and regional agencies, with a small but powerful strategic core "steering" the system'. In this framework, staff selection was devolved to the local level and professional development was provided to build capacity in principals for their expanded roles, as well as to give teachers the skill to implement curriculum improvements. However, the school's personnel for the most part remain centrally employed. At the same time, schools were given an increasing capacity to select staff and determine the mix of professional, paraprofessional and support arrangements.

All of the restructuring noted above required considerable commitment from the central office to provide professional development for all school personnel: principals, teachers and school councillors. At the time of their entry into the SOF programme, schools had an induction period of six months to ensure that they were ready for their new responsibilities. Professional development for principals included issues related to the global budget, leadership and management. Administrative staff were given training to improve their understanding of the new computer system and the global budgeting process, including the management of personnel. Teachers were given training in programmes related to curriculum leadership in response to school charters. School councillors were given training to help understand the process of charter development, and the implementation of the SOF programme.

At the central and regional level, special positions were created to deal with local concerns. These positions, known as the District Liaison Principals (DLP) were placed in regions across the state. Two positions were placed in the central office. The role of the DLP was to act as a change agent, providing advice and assistance to principals, assisting with professional development, and ensuring that schools have access to student services and curriculum support staff. In addition, a small number of support staff were located in each region.

Support for principals and teachers is an on-going activity. Areas of support include leadership training, mentoring and coaching, with experienced principals supporting junior ones. These have helped to establish the longer-term future of leadership in schools. The Professional Recognition Programme (PRP) offered teachers the capacity to opt into a system of enhanced pay and career structure, including annual appraisal. The main aims of the programme are:

- To provide a working environment that encourages and rewards skilled and dedicated teachers;
- To encourage the further development of an ethos that values excellence and high standards of achievement; and
- To provide formal feedback on a teacher's performance so that appropriate career development may occur through professional development and other means.

To achieve these goals, the DOE allocated in 1995, the total of A$240 per teacher in each school for professional development. This meant that appraisal could support improvement and provide the basis for promotions based on merit rather than seniority. Local staff selection, appraisal and professional development gave the school greater control over their human resources and greater flexibility in responding to local needs.

AN EVALUATION OF SOF REFORM

The SOF reform has devolved considerable authority and responsibility to the school level. Important features of its success include:

- The framework presents an integrated programme that works at two levels: for school planning and development and for system accountability. It is this dual utility that has been the key to success. Schools value the framework for providing them with a valuable developmental tool. Inspection programmes, such as those used in the United Kingdom, do not offer the same degree of support to schools as that offered by the accountability framework.
- The framework has been supported by the development of a range of performance measures. Some have been developed especially for the framework (staff and parent opinion surveys), whilst others have been developed as part of other elements of the reform (e.g. CSF). Importantly, benchmarks of performance have been created which allow schools to assess their performance against those of both the state average and schools, which have a similar student population.

- The development of the performance measures has been supported by the development of software to facilitate the display and analysis of the data.

This process has provided schools with the tools to monitor performance, a quality assurance framework within which to operate, and a quality control process that meets systemic requirements. In essence, the accountability framework includes the benefits of a supervisor model of school supervision, with an explicit and extensive programme of support for school planning and development.

Appropriate levels of professional development have been used to support the framework implementation. Extensive consultation and trialling occurred in the development of the school charter, the annual report and the triennial review. The Office of School Review also consulted widely with experts throughout the world, and its personnel have been active in gaining experience of best practice. The independence of the Office of School Review from the schools section has enabled it to develop the accountability framework without the constraints that it might have been subject to had it been part of the bureaucratic structure of the Office of Schools.

At the early stages of SOF inception, the DSE published 'The schools of the future information kit' and 'Schools of the future guidelines for developing a schools charter' to assist all schools in their transition to SBM. The documents reinforced the trend for principals to be recognized as true leaders of their schools and to be expected to build and lead their teaching teams by clarifying important responsibilities which are determined at the school level. The unique character of each school and community is reflected in distinctive curricula selected from a broad range of studies established by Victoria's Board of Studies. So as to achieve the overall aim of providing quality education for every student, enabling each to realize his/her full potential, different approaches to learning and teaching are encouraged to optimize the advantages gained from technological progress. The school charter, developed within the guidelines of the DSE/DOE, encapsulates the school's vision and establishes a framework for the allocation of resources. The crucial element for the success of the school would seem to be its ability to respond to the needs of the community and to provide a service which sustains an on-going demand for places within the school, as well as a boost to the employment potential of its graduates.

However, there remain considerable constraints on schools. For example, the leadership role of a principal of a SOF is a demanding one. The tasks carry onerous responsibilities, both 'upwards' to the DOE and the Minister of Education, as well as 'outwards' through the members of the School Council to staff, students, parents and the community.

Moreover, there is evidence of increased teacher workload and time demands, concern over the level of resources, increased reliance on local fund-raising, including the collection of fees, teacher disempowerment and a decrease in school diversity. There is also frustration at the inability of parts of the reform to be fully implemented, especially the promise of school control over staffing and the implementation of the principal performance management plan.

Concern has been expressed that reforms to bring about the decentralization of authority in the current education system are cost-cutting measures, rather than a means of improving school effectiveness. As a matter of fact, this phenomenon of cutting-costs and reducing staff has been a feature of most SBM reforms worldwide.

On the other hand, the new arrangements of decision-making and the increased community involvement give a clear impression that education is a partnership between the staff and the parents. However, much of this seems to have been undermined in favour of more power being given to the principal.

11

Non-School Science Education: A Case Study from France

Jean-Marie Sani

INTRODUCTION

This article attempts to show how a non-school educational institution can be a valuable resource for complementing school-based education, and how it can be a place of innovation and experiment.

A TYPOLOGY OF SCIENCE MUSEUMS

It is possible to place science museums into three categories, which have developed over time:

Collection museums are the most traditional—the museum is based on a collection. A principal disadvantage of such museums is the separation of the visitor from the objects on display. Direct contact and manipulation of real objects have obvious advantages over the simple viewing of artefacts.

Demonstration museums. In the second type of museum, demonstration of facts and mechanisms—'how things work'—is very important. The origins of this type of museum can be found in the eighteenth century, but they are more closely related to universal exhibitions organized in many large cities during the nineteenth century.

Interactive exhibits. The third and most recent category is a type of museum where the emphasis is placed on possibilities for direct interaction between the visitor and the exhibition. Different types of information may be obtained depending on the actions performed on an object by different visitors, who will, in turn, react differently to the results of their actions. Computers may serve as an excellent tool for organizing such interactivity between visitors and exhibitions.

The majority of modern science museums are a combination of these three types. Examples are the Exploratorium in San Francisco, United States of America and the *Cité des Sciences et de l'Industrie*, France.

THE CITÉ DES SCIENCES ET DE L'INDUSTRIE

The main purposes of the *Cité* are:

- to provide visitors with scientific, technological and industrial knowledge and know-how on phenomena and issues affecting their daily lives;
- to provide a forum for debate on current developments in science and technology;
- to operate as a resource centre which relays information from various partner institutions for the benefit of visitors (such as the service provided by the 'Cité des métiers');
- to be a centre for innovation in the fields of communication, education and training in science and technology issues, in partnership with institutions operating in these fields.

Its services consist of:

Exhibitions. These are the most important resources: different exhibitions (permanent or temporary) whose aim is to help visitors understand the effects of the development of science and technology on human life. They combine possibilities for observation, demonstration and hands-on interaction by visitors.

'Explora'. This is the main exhibition: 30,000 square metres on subjects as varied as the environment, communication, health, astronomy, energy, sounds and space.

'Cité des Enfants'. This comprises two permanent exhibitions, for 3 to 5-year-olds, and 5 to 12-year-olds. The subjects include machines, communication, the human body and other aspects of biology. There is also a temporary exhibition on the topic of electricity.

'Techno Cité' is a particular exhibition on technology, intended for teenagers, where the visitors can have real contact with diverse objects, and where teamwork is very important. The exhibition is organized in five sections and each one can be reserved for a class of pupils for ninety-minute sessions.

Its resources are as follows:

'Médiathèque'. This is a multimedia public library with 300,000 books, 3,000 films and a variety of computer software. It offers an excellent complement to the exhibitions.

'Salle science-actualités'. This exhibition is a presentation of current developments in science and technology, assembled by a team of journalists. It is renewed monthly.

'Cité des métiers'. This service space is organized in collaboration with external institutions, where visitors can obtain information on career orientation, vocational education, training, and employment. There is a strong emphasis on the evolution of the job market according to the development of science and technology.

Entertainment theatres. Different places of entertainment complete the complex. These include: (a) the 'Géode': a hemispheric cinema theatre; (b) the 'Planétarium'; and (c) the 'Cinaxe': a dynamic cinema theatre in which the room moves alongside the film.

These places receive from 3.5 to 4 million visitors a year.

THE EDUCATIONAL POLICY OF THE 'CITÉ'

Support to Teachers: A variety of suggestions are provided to teachers relating to project they may wish to carry out with their classes. These are merely suggestions—not directives—and teachers are free to select their own projects. The orientation and documentation services of the *Cité* are available to schools in support of educational projects. The *Cité's* resources facilitate the development of both the content and process of projects:

- *Ccontent:* Work on transversal topics which demonstrate the effects of science and technology on society (daily life, professional life, society's choices, economy);
- *Process:* Access to the diverse media available at the *Cité* in order to learn how to exploit them effectively to answer questions posed by educational staff at the museum.

A Multidisciplinary Approach: All the themes proposed by the *Cité* to classes or teachers are multidisciplinary. This is recognized as a good approach in education, and a necessity for a holistic view of the impact of science and technology on society. To carry out multidisciplinary projects teachers have to:

- use resources which are themselves organized in a multidisciplinary way;
- favour connections between scientific and technological subjects through transversal topics;
- favour treatment of topics using scientific, technological, industrial, social and economic perspectives;
- illustrate the large social debates generated by the evolution of science and technology.

THE ORGANIZATION OF THE EDUCATIONAL PROJECTS

In the museum, school classes can carry out activities that they are unable to do in the school. On the other hand, there are certain important activities that must take place in the school. Thus, the teacher has to use the educational project method, covering a certain time frame, and incorporating activities scheduled before, during and after the visit:

Before the visit. The activities include:

- work on the topic and emergence of a set of questions, originating in the pupils' existing conceptions;
- organization of the project material, budget, transportation and timing. Prior to visiting the museum, pupils know that the visit's aim is to collect information on a particular topic.

During the visit. Activities will include:

- · resolving the original question/problem
- acquisition of methodological tools;
- collection of information in terms of knowledge and know-how;

writing up a brief synthesis;

extending the topic to related subjects that appear relevant during synthesis.

After the visit. In this last phase the following activities are important:

- sharing of acquired information;
- organization and reformulating of the information, for instance in the form of a report, school newspaper or exhibition;
- complementary research, using other resource centres (libraries, museums, factories).

It is very important to involve the pupils in a variety of activities. An example of a possible sequence is as follows:

- determine a research subject;
- identify the pupils' preconceived ideas;
- prepare a set of questions;
- apply a research method, starting from the set of questions;
- evaluate the new knowledge that has been acquired;
- measure the evolution of the pupils' knowledge compared to the initial subject.

This sequence may be adapted to the pupils' level of ability, providing for individualized progression.

THE ROLE OF THE TEACHER

In this type of environment, and in such a didactic relationship, the teacher has a precise role to play. This role is, of course, often evident in the school, but it is probably needed more in a place like the *Cité*. The teacher's role is not to deliver ready-made information to the pupils, but to orient and guide them in the process of structured research, giving them the opportunity to ask their own questions, develop problem-solving techniques, make their own discoveries, and arrive at carefully thought-out conclusions based on scientific evidence. The teacher should help pupils to: formulate questions; organize their research; critically analyze and summarize the results; and formulate new questions based on the outcomes. The teacher must also evaluate the final results.

CONCLUSION

The role of the staff at the *Cité* is to assist teachers in exploiting the museum's resources to the best advantage in the realization of class projects. The most effective approach is a collaborative transdisciplinary project, with several teaches from different disciplines working as a team, and with activities carefully planned before, during and after the visit.

PRESENT POLICY PRIORITIES

Beyond to the *Cité's* general and long-term objectives, specific short or medium-term objectives are formulated (covering a year to several years). These presently include:

- the development of experimental activities;
- the use of information and communication technologies (ICTs) in education;

- work with disadvantaged publics in a strategy of partnership;
- focus on a multidisciplinary and systemic approach;
- provision of information and developing knowledge about training, jobs and careers, and vocational education;
- focus on lifelong learning;
- promotion of international development, especially in the field of science and language projects.

THE ROLE THE CITÉ WOULD LIKE TO PLAY IN THE FORMAL EDUCATIONAL SYSTEM

The *Cité* provides a place where teachers may practise innovation in science and technology education, with the help of tools, training and documentation. It aims to make available to teachers and pupils quality information on the latest developments in science and technology as they relate to the scientific and the industrial communities. The ultimate objective is to foster the introduction of new practices and methods in the whole education system. The museum's close relationship with the formal education system allows it to be a laboratory and place of innovation for new educational practices. A similar role may be played by diverse non-school learning institutions in all countries: museums, libraries, industries, research institutes, art exhibitions. The development of partnerships between the school and non-school learning institutions opens up a wealth of exciting possibilities for innovation and change in the organization and practice of education.

Part IV

Current Trends in the Adaptation of Educational Content in South and South-East Asia

12

Who Should be Doing What in Adapting the Curriculum? The Roles of Various Protagonists with Particular Focus on Policy-Makers, Curriculum Developers and Teachers

Udo Bude

CURRICULUM DEVELOPMENT: THE PROBLEM OF 'SEMANTICS'

The idea of curriculum development has, over recent decades, become firmly established in education systems throughout the world. It took much longer, however, for us to realize that curriculum development is not simply a one-time event during which there was an emphasis on providing training in techniques and skills in order to establish and organize national curricula. For educationalists and practitioners alike, curriculum development is now regarded as an on-going process, with the declared objective being to organize better learning opportunities, and thus focusing on actual interactions at the classroom level.

Continuous efforts to translate educational goals into activities, materials and observable behavioural changes and indicative of the present trend. Learning, teaching and assessment are inextricably linked—it is only in the context of the others that each has a meaning. Without learning, assessment has relatively little value, without assessment, the effectiveness of learning and the accountability of teaching cannot be determined.

Societies advancing into the twenty-first century face the major challenge of developing an open, active, flexible and intercultural curriculum. No longer can a central authority prescribe one curriculum for all primary or secondary schools in a country. What is now required is specific curricula based on a national core curriculum, allowing for local variations and regional diversification according to different ecological and cultural circumstances and learning needs—with the active involvement of all those concerned.

Although the curriculum is generally considered essential and fundamental to the educational process, one quickly discovers that there is no general consensus about the meaning of the term. 'Curriculum' has been defined in different ways by different people in different places. It is meaningless to look for 'right' or 'wrong' definitions. In reality, its meaning seems to be a matter of culture, environment and practicality.

The worldwide application of the term 'curriculum' in the process of organizing education is still a recent phenomenon. In most European languages, one will not find an equivalent to the English term 'curriculum'. The French speak of the *programme scolaire*; the Germans of the *Lehrplan*. However, these terms correspond more to the narrow definition of curriculum indicated in English by the term 'syllabus' (theory of instruction programmes). Analyzing the Latin roots of the term, one finds that it derives from *cúrrere*, meaning 'to run a race course'. This can be translated roughly into educational terms as 'to run a course of subject matters/studies'.

Thus, within the process of globalizing the term 'curriculum', some educators use the term to mean the content or objectives that the schools intend students to learn, while others focus on the means of education—the teaching strategies schools plan to use—that constitute the curriculum. Other educationalists claim that neither the intended learning nor the plans for bringing it about are as important as the actual implementation and learning achievements, which account for the real meaning of the curriculum.

Viewing 'curriculum' in a broader, process-oriented perspective, we may conclude that: *The curriculum is an organized set of intentions, that articulates the relationships among its different elements (objectives, contents, evaluation, etc.), integrating them into a coherent whole. It consists of a continuous chain of activities necessary for translating educational goals into concrete learning opportunities.*

Curriculum development is thus a complicated, dynamic and continuous process. Society is constantly changing. So too do the learners, the teachers and the subjects taught. The curriculum must be responsive to these changing conditions in order to equip the learner to both cope with and contribute to future societal developments. *A permanent search for qualitative improvement, in response to changes in society, is what curriculum development is all about.*

APPROACHES TO CURRICULUM DEVELOPMENT AND THE PARTICIPATION OF STAKEHOLDERS

In most countries, the curriculum for primary or secondary schools is planned at a higher level of authority than that of the classroom. Teachers are regarded as conveyors of the curriculum and are expected to implement the planned curriculum 'to the letter'.

Since it is not possible to teach all knowledge, skills and attitudes, every curriculum consists of selections from the possible range of content that could be taught. Similarly, the teaching methods proposed are also just a sample of the many methods that could possibly be used for teaching that content. The approaches to developing curricula are largely determined by the beliefs and assumptions of those who are actively involved—their philosophy of education or their theoretical perspective. Although some perspectives claim to be 'objective' and 'value-free', in actual fact, no curriculum development can be solely technical.

In a democratic society, the participation of different stakeholders in curriculum development is a fundamental ingredient in the organization of the educational process. However, curriculum development is often not carried out in a systematic way and does not provide regular opportunities for different groups to contribute actively; on the contrary, changes often come about as a reaction to problems or an emergency situation in education or in the larger society.

In such cases, politicians very often react by using the 'panic approach' towards curriculum development, whereby curriculum developers are forced to act as technicians, introducing hastily developed modifications in the curricula. Youth unrest or the boycott of classes by students, for example, are typically counteracted at the political level by demands to change the curriculum and include more teaching on moral values. This neglects the hard facts that young people may be facing a very insecure future on the job market because the curriculum is out of touch with economic reality. In the same way, small groups of 'experts' may impose curricular changes and insist on the introduction of 'fashionable' topics—in many cases, just adding to the burden of an already overloaded curriculum.

A characteristic of the panic approach to curriculum development is the lack of analysis of related situations, with the result that the approach adopted is often piecemeal in nature. As a result, this approach tends to create more problems than it solves.

Curriculum development needs to be carried out in a systematic way, starting with an analysis of prevailing situations (situational analysis). From the very beginning, the main stakeholders have to be involved. This helps to develop realistic procedures, to anticipate problems and to suggest possible solutions. However, only a 'systematic approach' to curriculum development, providing sufficient opportunities for participation by all stakeholders, also enhances the chances of implementing the curriculum changes at the classroom level.

FOCUS ON POLICY-MAKERS, CURRICULUM DEVELOPERS AND TEACHERS

From the Heights of Theory to the Daily Business of Classroom Instruction—National versus School-based Curriculum Development

Notions of curriculum development differ according to educational theories and practices. Some refer to preparing a plan of operation based on an existing syllabus (including the development or selection of textbooks, teacher's guides and other instructional materials). Others regard curriculum development as producing a syllabus and all the accompanying materials needed for teaching/learning in the classroom, and eventually also evaluation instruments for examining the attainment of the programme goals.

Some educationists distinguish between 'local-user development' and 'external development'. Connelly (1991), for example, argues 'that, in theory, a curriculum can be fully developed either by the users or by external developers, although such cases rarely occur in practice' (A. Lewy, *National and school-based curriculum development*, Paris, UNESCO:IIEP, 1991, p. 29).

It seems, however, that in practice only a minority of 'users' will actively take part in the curriculum development process, while most people are happy to take advantage of those curriculum components developed by the 'external developers' (curriculum specialists, publishers, etc.). 'Intelligent teacher-consumers, skilled in the art of selection, will choose the materials they judge suitable, and will alter, augment, process, and transform them to suit their classes' (Lewy, ibid., p. 31).

There is ample evidence for caution about the enthusiasm of teachers for curriculum development. Teachers often tend to be conservative and to resist change. In many instances, they are reluctant to experiment with new ideas concerning the curriculum especially if they do not have a sufficient level of professional training and there is no precedent of their involvement in educational reforms. Nevertheless, teachers will definitely be the most active partners in the curriculum as regards implementation. Even when the scope of their activities may be limited—and this is a quite controversial point of view—their active involvement in curriculum development constitutes a fundamental contribution to the ownership of curricula and the quality of education.

Curriculum development not only takes place at the national level, involving policy-makers, curriculum developers—and some teachers. To a smaller or larger degree, every teacher is also involved in curriculum development at the school level. This fact has been highlighted and elaborated in recent years under the term 'school-based' curriculum development. There are still countries where the nationally prescribed curriculum—especially the schemes of work and textbooks—expects all teachers at the same level to teach the same topics of a subject on the same day of the school year, using the same methods and instructional materials.

The majority of countries, however, tend to provide only a national framework and encourage and foster curriculum development activities at regional, local and school level. This involves different forms of curriculum development initiatives, ranging from selecting and adapting curriculum components according to local circumstances and resources to producing lesson units and instructional materials. Depending on the definition of curriculum development activities, the scope of school-based decisions about the curriculum varies across countries and across schools within a country, covering mostly from 10 to 30 per cent of the total school curricula (Lewy, ibid., p. 36).

The greater the extent of decentralization in curriculum development, the more the participation of schools and teachers in such activities is required. At the same time, such an approach gives practising teachers a chance to contribute as curriculum developers, if they are prepared to assume this professional task.

Who Makes what Choices in Curriculum Development? Clarifying Levels of Decisions and the Actors Involved

The degree and extent to which the different protagonists or stakeholders influence the components of curriculum development vary considerably. Choices are made at different levels of the education system, rarely by the same group of people. The closer curriculum development gets to the school level, however, the more teachers will be involved.

In order to have an idea of the complexity of choice and the reality of power-sharing, we have to clarify who makes what choices and how this influences the instructional practice at school level. The pattern of choice will vary from country to country, but it is easy to demonstrate that curriculum development is not just the business of policy-makers or curriculum specialists. Cooperation and sharing of power are needed among the different groups involved, including the important 'band of middlemen'—teacher trainers, inspectors, local administrators, etc. (See H. Hawes & D. Stephens, *Questions of Quality—Primary Education and Development*, London, Longman, 1990, pp. 66-68).

For the purposes of analysis, the major actors and influences shaping curriculum decisions can be classified as internal and external forces. The internal forces are those legally responsible for curriculum policy and planning, and whose involvement in the curriculum development process is determined through some regular, structured arrangements. On the other hand, the external forces exist outside the government structures and the administrative bureaucracy of education systems. They can influence the curriculum development process through mainly irregular patterns of pressure politics and powers of persuasion. In practice, one will find that these categories of participants and influences in curriculum development overlap significantly.

PROVIDING 'LEARNING OPPORTUNITIES' IN THE CONTEXT OF A MEANINGFUL CURRICULUM

Curriculum Implementation as the Key Element of the Curriculum Process—Getting the Curriculum to the 'Users'

Despite the involvement of the main groups who have a stake in curriculum development, the official (planned, prescribed, intended) curriculum being the result of the joint efforts of the actors involved, differs considerably from the actual organization and type of learning opportunities provided in schools. The prescribed curriculum normally reaches the schools and the teachers in a formal way—as general and pedagogical guidelines, and syllabuses for the different subjects.

However, the taught (implemented, actual, operational, in-use) curriculum is the result of the circumstances, conditions and possibilities affecting each individual school, especially the quality of the school's leadership and its staff. Why is the reality different from the officially planned situation?

In many instances, the official curriculum tends to ignore or neglect the reality of the classroom, particularly in terms of the skills, subject knowledge and motivation of the average teacher. Furthermore, the prevailing conditions in many schools are often not sufficiently considered (overcrowded classes, lack of basic materials and textbooks, etc.).

Successful curriculum implementation lies at the heart of the curriculum development process. It is a continuous process that assists teachers in improving their skills in classroom interaction, providing chances to interpret the official curriculum in the light of local conditions and in a practical way—for example, by developing lesson units within a given framework and trialling them in school.

Curriculum conferences can provide a platform for training workshops in the production of curriculum materials for schools. The main purpose of such conferences is the transformation of nationally or regionally planned curriculum guidelines into practical lesson units that reflect local conditions and concentrate on existing cultural and ecological traits. Educationalists from different levels of primary or secondary education jointly design lesson units for specific subjects for particular classes/standards. If educators learn how to translate curriculum guidelines into structured steps for teaching at the classroom level and develop learning opportunities that create chances for active student involvement, the prescribed core curriculum stands a better chance of actually being implemented.

Only if curriculum conferences are seen as on-going activities within the process of curriculum development can they contribute to empowering teachers, school inspectors, tutors and others involved in education at regional, district or local level. This is a realistic procedure for the curriculum to reach the 'users' and to improve the quality of education provided in schools.

Developing Curricula for and with Teachers to Improve the Quality of Education

Without neglecting the importance of policy-makers and professional curriculum developers in the design and implementation of curricula in primary and secondary education, teachers still play the most crucial role, if we want to see curriculum reforms actually being practised in the schools. One can easily assess the curriculum situation in a school through the following questions:

- How closely does the taught (actual) curriculum follow the official (prescribed) curriculum?
- If there are large deviations, what are the reasons?
- Are there topics or areas of learning that are considered important for the students, but not included in the curriculum?
- Are there parts of the official curriculum not taught in school? What are the reasons for excluding them?
- Does the school offer extra-curricular activities? How and who determines the content? Is there any relationship to the official curriculum?

Although teachers may not have specialist knowledge in many subjects, as 'generalists' they are extremely knowledgeable about the pedagogical situation in schools, for example, about the learning environment, their students and their colleagues. Practising schoolteachers can contribute to curriculum development through their classroom experience, putting ideas and approaches into the true perspective of what works and what does not work. Teachers' roles may vary from describing and explaining the development level and abilities of the students they teach to proposing activities and resources that are feasible and available in their schools and communities. They can also be involved in developing or reviewing draft lesson units and materials based on their experience, and on the matter of trialling or experimenting with newly developed curriculum materials, they are the essential partners in curriculum development.

The ultimate aim of curriculum development should be the improvement of the quality of education through the provision of guidance and assistance to classroom teachers. Effective curriculum materials have to be developed and implemented for and with the teachers. New curricula can only make an impact on the teaching/learning process if they provide guidance and support for using appropriate methods for teaching and assessment. *Teachers are in the ideal position to advise on the appropriateness, relevance and feasibility of curricula. Without consulting, involving and supporting teachers, curriculum reforms stand no chance of succeeding.*

'Schools and classrooms are where the work of policy-makers, planners, curriculum developers, teacher-training institutions and administrators comes together. If the reforms or policies do not make a difference within the classroom, of what value are they?' (A. Hartwell & E. Vargas-Baron, *Learning for all: policy dialogue for achieving educational quality*, Washington, DC, USAID, 1998. Paper presented to the International Working Group on Education, Munich, 23-26 June 1998, p. 22.)

13

An Overview of Country Reports on Curriculum Development in South and South-East Asia

Isabel Byron

As developing nations, the thirteen South and South-East Asian countries which participated in the curriculum development course in New Delhi share many common concerns and problems. However, they also differ considerably in terms of their size and geographical situations, histories and cultures—not to mention levels of development. These individual characteristics will obviously affect the nature and state of development of national education systems and, by extension, the processes of curriculum design, implementation and renewal.

The reform and renewal of school curricula is clearly a priority for all the participating countries. They recognize this process as integral to the improvement of education. Many are presently implementing reforms—Bangladesh, Indonesia, Myanmar, Sri Lanka, Viet Nam—or are preparing for major curriculum change—Thailand, Maldives. Others are monitoring or evaluating the impact of recent reforms: Bhutan, Nepal, India, Philippines. Malaysia refers to its on-going process of curriculum reform and Pakistan to the tradition of four-year curriculum reform cycles.

The focus of reform in a number of countries has been on primary or basic level education (Bangladesh, Indonesia, India, Maldives, Thailand), although most countries are also attempting to improve the quality of secondary education, making it more relevant to the future needs of pupils and reducing an overly academic orientation.

KEY CONCERNS IN EDUCATIONAL CONTENT

Recent curriculum reform in participant countries has often been motivated by factors related to the impact at national or sub-regional levels of changes taking place at international or global levels. Of paramount concern are economic issues. Country reports and profiles express a preoccupation with equipping pupils with the basic knowledge and skills needed to be self-sufficient and productive citizens in an increasingly globalized, rapidly changing, uncertain, competitive and highly technological economic environment. Thus, a principal motivating factor for curriculum reform is the desire to design educational programmes that will more adequately prepare young people for the job market within the existing economic climate, while providing the human resources necessary to ensure sustainable national development.

The concern of keeping pace with rapid global economic change is reflected in the commonly expressed goals of improving the quality and scope of vocational education, strengthening science and technology education, developing competence in information technology (IT) skills by introducing or expanding the use of IT in the classroom, and focusing on the teaching of a wide range of cognitive, social and personality skills so as to develop the capacity for flexibility, problem-solving, creativity, initiative and lifelong learning. An examination of the country reports indicates a variety of efforts being made to incorporate or upgrade these areas of content within the curriculum.

Other key areas of concern in educational content relating to issues of global significance include environmental change and degradation, population control, gender issues, and international understanding and co-operation. Some of these issues are taught as separate subjects, but are often integrated into other existing subjects or treated as cross-curricular areas. The Maldives and Nepal refer to major recent reforms to the social studies curricula so that these issues are properly taken into account.

Importantly, reports indicate a preoccupation with developing curricula fostering respect for, and preservation of, cultural traditions and indigenous values and ways of life, while preparing young people to be part of the modern global society. India, Indonesia, Thailand and Viet Nam all single out the concern for finding a balance between traditional/national and modern/global elements in the curriculum. Nepal refers to the tension existing in the development of the social studies curriculum between these often conflicting values and world views.

Related to the concern to preserve indigenous cultural traditions and lifestyles is the commonly expressed goal of fostering moral values and ethics among pupils through religious education, moral education and other subjects, such as citizenship. Some countries, like India, have advocated the teaching of values through all subjects in the curriculum. Promoting the concept of citizenship is also a key educational goal of many countries, with civics often included as a separate curricular subject. While a priority aim of citizenship is the promotion of national integration and unity, a number of reports additionally cite the objective of promoting democratic values, and respect for, and appreciation of, the cultural diversity which characterizes both their societies and the broader global society.

ORGANIZATIONAL STRUCTURES OF CURRICULUM DEVELOPMENT

The structure of curriculum development in most of the thirteen countries represented is predominantly centralized, with varying degrees of active decision-making at regional, local and school levels. The main pattern is for a central body within, or outside of, the Ministry of Education to be charged with developing curricula based on governmental educational policy, with regional authorities responsible for implementing these policies in schools within their specific province or district.

A number of countries have a section/unit/centre within the Ministry of Education devoted to curriculum issues (Bhutan, Indonesia, Malaysia, Nepal, Pakistan, Philippines, Thailand), while others have a statutory body operating outside of the Ministry, but working closely with it in the development of curriculum (Bangladesh, India, Maldives). More than one national institution or body may share responsibility for developing the curriculum, as in the case of Myanmar where the Department of Educational Planning and Training in the Ministry of Education works with the statutory Basic Education, Curriculum Syllabus and Textbook Committee, or in Sri Lanka, where curriculum objectives and competencies formulated by the

Ministry of Education and the National Education Commission are translated into individual subject curricula and syllabi by the National Institute of Education (NIE).

Wider participation in the curriculum development process takes place in many countries through advisory and subject specialist committees formed within ministries or curriculum development units/centres to assist in curriculum design and implementation. In some cases, technical committees may be formed on an ad hoc basic to support specific reform programmes. Committees are usually composed of a wide representation of educational specialists and professionals from the Ministry of Education, regional education authorities, universities and research institutes, teacher-training institutions and, in some cases, schools.

Indonesia, the Philippines and Thailand report that considerable responsibility is devolved to regional educational authorities, who adapt central curriculum objectives to local contexts where this is felt to be necessary. Indonesia has developed a curriculum network involving all regions of the country in elaborating and implementing the national curriculum according to local realities and needs. Principals and leading teachers from all school levels participate in this network, along with staff from regional educational offices. In India, the country with the most decentralized education system, curricular frameworks formulated at central level with regional input, are translated into curriculum plans by individual states.

While genuine school-based curriculum development does not appear to take place on a significant scale in any of the countries, a number report the participation of teachers in the development of textbooks and other instructional materials. In the Philippines, schools and teachers are encouraged to be innovative in interpreting and adapting the basic curriculum guidelines to suit the local reality and needs. The Maldives express a concern to involve teachers to a greater extent in future reforms in an effort to improve their design and the degree to which they are implemented.

Generally, countries do not refer to existing strategies of broad-based community participation in curriculum development. However, a number (India, Nepal, Philippines, Viet Nam) indicate a growing awareness of the importance of adopting more genuinely participatory approaches to the process.

CHANGES IN TEACHING/LEARNING STRATEGIES AND STUDENT ASSESSMENT

Efforts to develop curricula that will more effectively prepare young people to meet the demands of contemporary and future society are reflected in the growing trend in a number of countries to introduce competency-based programmes at both primary and secondary levels. Under the traditional systems of schooling, large percentages of children fail to acquire even a basic education that would equip them to lead productive and self-sufficient lives once they leave school. This situation becomes all the more critical in countries where vast numbers of children are at best only able to complete the years of primary schooling.

Bangladesh, India, Sri Lanka and the Philippines have all introduced competency-based curricula, while Indonesia is attempting to define 'minimum learning competencies'. A glance at the primary school curricula of Malaysia and Thailand similarly indicates a preoccupation with the teaching of skills, values and attitudes, as opposed to more traditional orientations towards predominantly content-based curricula. Competency-based curricula focus on the attainment by pupils of a stated number of clearly defined skills or competencies at the end of each stage and level of formal education, indicating a concern to change curricula that previously focused on academic achievement and were based primarily on coverage of content. With a competency-based approach, methods of assessment change significantly—continuous and varied evaluation of student progress takes into account many previously ignored areas of competence. Teachers become more accountable for student outcomes, but in principle also have a clearer idea of what pupils are expected to achieve at the end of each stage and level of schooling in each subject area.

Many countries (Bhutan, Indonesia, Maldives, Nepal, Sri Lanka, Thailand, Viet Nam) refer to efforts to introduce more student-centred instructional approaches, in line with world trends. Giving the pupil more autonomy in the learning process is felt to foster greater motivation, creativity and self-sufficiency in the child, developing skills of inquiry, research and problem solving, which present teacher-centred traditions do not encourage.

MAIN PROBLEMS FACING CURRICULUM DEVELOPMENT

A number of problems in the design, implementation and follow-up of curriculum reform were identified, many shared by all countries, although the degree and extent of the problems will vary considerably among the participating nations. Limited financial, human and material resources in all countries constitute a major obstacle to the success of reforms at all levels and stages of the process.

Curriculum Design and Implementation

Countries indicate that ***existing structures and mechanisms for curriculum design and implementation are often inadequate*** in a variety of ways. In addition to the key problem of lack of staff, there are fundamental aspects of decision-making and administrative processes that may create obstacles to successful curriculum development and reform. Top-down patterns of decision-making often do not provide for sufficient situational analysis and research in curriculum design. The failure to involve key stakeholders—and teachers in particular—adequately in decisions about curriculum change is likely to result in resistance and lack of understanding on the part of those who are to be the implementers of the reforms—the teachers.

Inadequate time allocation to the entire curriculum development process was identified as a key problem by Bhutan and is also likely to be an issue in other countries. Once decisions are made, reforms are often hurried through, with too little time devoted to careful implementation and follow-up in relation to vital areas, such as teacher preparation, piloting of reforms, analysis of feedback, subsequent revision and re-testing, regular supervision and evaluation.

Lack of professional expertise is expressly cited by numerous countries as a major obstacle to successful design, implementation and also follow-up of curriculum development (Bangladesh, Bhutan, Maldives, Nepal, Pakistan, Philippines, Viet Nam). Reports refer to the limited number of curriculum specialists available, and, in some cases, to the rapid staff turnover in key areas of curriculum development leading to lack of continuity in implementation and follow-up of change.

A related problem in at least some countries is ***the absence of a strong research base for curriculum change.*** Bangladesh refers to the urgent necessity for a curriculum research wing to be established in order for curriculum development to be able to respond to the nation's real and most pressing educational needs. Aggravating this situation is the limited access to up-to-date information on trends and developments in curriculum reform in other countries, mentioned by Bhutan, Maldives and Viet Nam.

The inadequate preparation of principals and teachers for curriculum change is a challenge for all education systems. Many reports mention the unsatisfactory situation regarding teacher quality and the inadequate provision of suitable pre- and in-service programmes as an integral part of curriculum development and reform. This situation inevitably leads to classroom practices that remain much the same, despite reforms on paper. Countries make reference to the difficulties of implementing major pedagogical innovations in the classroom, such as competency-based curricula and child-centred learning. Resistance to, or inadequate understanding of, these approaches by teachers stems at least partly from their poor preparation for change.

India, Indonesia, Philippines and Viet Nam all cite ***overloaded curricula*** as a major problem. There is clearly a dilemma between the need to update the curriculum, adding important new areas of content, while attempting to reduce overall curriculum load and focus on the teaching of skills and competencies, rather than on mere knowledge. In this regard, countries are faced with the challenge of defining basic subject content and finding effective ways to integrate related subject areas, while identifying and defining minimum learning competencies. However, the competency-based approach is likely to overload the curriculum further, if sufficient attention is not paid to adequately balancing competencies and content. In the Philippines, where the basic education cycle is the shortest among ASEAN countries, and the curriculum is both content/topic based and competency-based, teachers have difficulty covering curriculum content at the different levels, with a 'backlog' of untaught areas building up.

Change is also rendered difficult by ***examination-oriented education systems,*** which pressure teachers to focus on the syllabus rather than implement recommended changes. This problem, cited specifically by Bhutan, was also alluded to by Viet Nam, which referred to the over-dependence on examinations and testing in its education system.

Apart from the limited capacity of many countries in ensuring adequate textbook supply and distribution to large and scattered student populations, there is also the challenge of preparing ***quality teaching materials*** which will effectively serve as tools for changing teaching/learning approaches. Without suitable materials, teachers are all the more likely to continue with traditional methods, particularly where they are overly dependent on textbooks for their teaching.

Follow-up and Evaluation

Several countries (Bangladesh, Bhutan, Indonesia, Maldives, Nepal, Pakistan, Philippines, Thailand) indicate ***the inadequate mechanisms for supervising, monitoring and evaluating change***—a major obstacle to successful reform. This is due to various factors already cited: lack of resources, particularly as regards quality personnel at all levels; inadequate planning, administrative and supervisory structures; as well as large populations, difficult geographical terrain and isolated territories. Bhutan refers to the lack of proper articulation between central administration and provincial/local authorities, while the Philippines describes the failure to staff offices at the divisional level with qualified subject specialists. Failure to carry out systematic evaluation of reforms clearly limits sustained educational development.

Another challenge is that of ***sustaining externally funded reforms,*** a problem specifically referred to by the Philippines and likely to face most developing countries. Once external funding comes to an end, the programme risks being abandoned.

ADVANCES AND SUCCESSES IN CURRICULUM DEVELOPMENT

In spite of the numerous difficulties countries in the region face with regard to curriculum development and renewal, the following reports, indicate a concerted effort by countries to improve their school curricula. There appears to have been considerable innovation in terms of introducing or updating subject matter, as well as in adopting new instructional approaches based on world trends.

Countries describe various major projects—of which curriculum development forms an integral part—undertaken in recent years to improve educational provision. Particular emphasis has been put on providing better quality basic education, with countries such as Indonesia and the Maldives increasing the duration of this cycle in an attempt to improve learning opportunities for all pupils. Reports indicate a growing awareness that the curriculum has to be oriented towards the development of skills which will facilitate and encourage learning throughout life by all pupils, whether or not they have access to secondary or higher education.

Countries such as Bhutan, Maldives and Nepal have gone a long way towards indigenizing curricula inherited from colonial powers or powerful neighbours, recognizing the need to make content relevant to the everyday needs and experiences of pupils. Reforms, such as that of Nepal in social studies, indicate the efforts being made to improve the quality of teaching materials. Countries also testify to a greater awareness of the importance of developing curricula that foster international understanding and the principle of global citizenship.

It is to be hoped that, increasingly in all the countries concerned, the structures for on-going and sustained curriculum development at primary and secondary levels will be systematically strengthened, permitting truly participatory approaches to educational provision and meaningful change in educational practice. In this way, the chances in life for all children and young people will be significantly improved.

Part V

Country Papers on Curriculum Development from Selected States in South and South-East Asia

Bangladesh: Curriculum Planning, Development and Reform for Primary and Secondary Education

Muhammad Abul Hossain and Shawkat Jahan

Estimated population (1995)	118,200,000
Public expenditure on education as percentage of Gross National Product (1995)	2.3
Duration of compulsory education (years)	5
Primary or basic education	
Pupils enrolled (1995)	16,800,000
Teachers (1995)	189,508
Pupil/teacher ratio	71:1
Gross enrolment ratio (1995)	
— Total	63
— Male	72
— Female	53
Net enrolment ratio (1995)	
— Total	84
— Male	89
— Female	78
Estimated percentage of repeaters (1995)	7
Estimated percentage of drop-outs (1995)	55
School-age population out of school (1995)	2,260,000
Secondary education	
Students enrolled (1990)[1]	3,592,995
Gross enrolment ratio (1993)[1]	
— Total	19
— Male	26
— Female	12
Third-level enrolment ratio (1992)[1]	4.0

Estimated adult literacy rate (1995)	
— Total	38
— Male	49
— Female	26

Note:

1. Last year available

Source: *UNESCO Statical yearbook, 1998,* Paris.

INTRODUCTION

The Government of Bangladesh recognizes that education is an important prerequisite for ensuring sustainable development. The country's constitution obligates the State to provide basic education to citizens and eradicate illiteracy within a given time frame. As a signatory to the World Conference on Education For All (Jomtien, Thailand, 1990); World Conference on Children's Rights (New York, 1990); and the EFA Summit Conference of Nine High-Population Countries (New Delhi, 1993), Bangladesh is committed to the eradication of illiteracy by the year 2006.

As a result of both government and private efforts over the last two decades, some important improvements have occurred in the primary education sector. More than 95 per cent of children aged 6 to 10 years are admitted to primary schools and the drop-out rate is now only 38 per cent. The literacy rate for the population over 15 currently stands at 56 per cent, in comparison to 1971 post-liberation figures reflecting a rate of only 22 per cent for the same group. During the past twenty-five years, considerable improvements have also taken place in secondary education.

However, although Bangladesh has experienced *quantitative* education improvement, the *qualitative* aspects of education have become a cause of government concern. Steps have been taken to address educational quality and it is in this larger context that curriculum has come to play a crucial role. In order to place in proper perspective some of the key issues and actions taken to improve educational quality, a brief overview of primary and secondary education structures and of curriculum development is provided below.

PRIMARY EDUCATION IN BANGLADESH

Duration of Primary Education

The current five years for completing primary education is an insufficient amount of time for students to obtain the requisite level of literacy, knowledge, abilities, attitudes and values for solving the problems of everyday life. The New Education Policy, which has yet to be implemented, recommends that primary education be increased from five to eight years in order to enable students to attain minimum ability levels and the capacity for lifelong learning.

Different Systems of Primary Education

A number of different institutions of primary education exist including: kindergarten, general primary, Ebtadayee Madrasah and non-governmental organization schools. The standards and characteristics of these schools vary. Thus, from the very outset, differences in children's abilities, attitudes and values are created. Its has become mandatory to eliminate these existing educational disparities and to introduce a common system of education. Currently, approximately 78,000 primary schools and subsumed under ten or eleven distinct categories. The following tables provide an overview of the current situation.

Table 14.1: Bangladesh primary education institutional enrolment

Year	Total	Boys	Girls
1990	12,051,172	6,662,427	5,388,745
1991	12,635,419	6,910,092	5,725,328
1992	13,017,270	7,048,542	5,968,728
1993	14,067,332	7,525,862	6,541,470
1994	15,180,680	8,084,117	7,132,563
1995	17,284,157	9,094,489	8,189,668
1996	17,580,416	9,219,358	8,361,058
1997	18,031,673	9,364,899	8,666,774
1997	Total number of government primary school students:		11,808,345

Table 14.2: Bangladesh primary education enrolment by gender (%)

Year	Boys	Girls
1990	55.28	44.72
1991	54.69	45.31
1992	54.15	45.85
1993	53.50	46.50
1994	53.02	46.98
1995	52.62	47.38
1996	52.44	47.56
1997	51.94	48.06

Table 14.3: Bangladesh primary education drop-out rate (% of total enrolment)

Year	Rate
1991	59.30
1992	46.60
1993	39.60
1994	38.70
1995	38.00

Beginning in May 1986, the Government decided to merge the primary school classes I and II and, concurrently, to introduce a liberal promotion policy as part of the effort to further reduce drop-out rates. With this liberalized policy, promotion from classes I to II and from classes II to III became automatic. However, assessment of pupils' performance is still supposed to occur throughout the year.

SECONDARY EDUCATION IN BANGLADESH

Current General Secondary Education Scenario

The present secondary education system in Bangladesh encompasses grade levels 6 to 12. Most secondary schools are private. In 1996, only 3.6 per cent of the total number of recognized schools were government schools. Secondary education is divided into three stages: junior secondary (classes VI and VII); secondary (classes IX and X); and higher secondary (classes XI and XII). Higher secondary schools can be intermediate colleges or degree colleges. A significant increase

(16.4%) in the total number of secondary institutions occurred between 1995 and 1996, with secondary and higher secondary schools accounting for most of the increase. Table 14.4 reflects the growth in secondary education from 1991 through 1996.

Table 14.4: Growth of Bangladesh secondary education sector institutions*

Type of institution	1991	1992	1993	1994	1995	1996
Junior secondary	2,000	1,962	1,905	2,136	2,349	3,002
Secondary	8,715	9,038	9,190	9,352	9,363	10,776
(of which Governmental)	(302)	(316)	(317)	(317)	(317)	(317)
Intermediate colleges	366	387	467	603	903	901
(Governmental)	(11)	(11)	(11)	(11)	(9)	(80)
Degree colleges	547	598	603	611	671	786
(of which Governmental)	(204)	(219)	(219)	(219)	(224)	(225)
Total institutions	11,586	11,964	12,085	12,566	13,286	15,465
% increase		3.3	1.0	4.0	5.7	16.4

*Recognized schools only

Noteworthy progress in some areas and stagnation in others characterize secondary education in Bengladesh. Significant progress has been made in improving enrolment, especially among females. Programmes encouraging the enrolment of girls in secondary education have had a profoundly positive effect, with an increase in enrolment from 31.9 per cent in 1991 to 45.6 per cent in 1996. However, in contrast, transition, attendance, completion and pass rates remain low. Table 14.5 reflects the recent enrolment figures for secondary education.

In 1995, 78 per cent of the students completing class V went on to enrol in class VI. The average secondary school attendance rate was about 60 per cent. Completion rates for classes VI to X improved, reaching 56.4 per cent in 1995, in comparison to 48 per cent in 1991. However, these figures belie a highly inefficient system, with drop-out rates remaining high and pass rates in final exams declining. For example, in 1995, the estimated drop-out rates for classes VI to X and classes XI and XII, respectively, were 43.6 per cent (49% for girls) and 38 per cent (36% for girls). In 1997, the pass rate for the class X Secondary School Certificate Examination (SSC) was only 52.14 per cent, compared to 64.95 per cent in 1991. For the class XII Higher Secondary Certification Examination (LISC), the 1997 rate of 52.14 per cent was down

Table 14.5. Bangladesh secondary institutions enrolment, 1991-98

Enrolment	1991	1992	1993	1994	1995	1996
Junior secondary	212,646	284,806	341,975	446,060	494,692	632,211
No. female	75,231	121,174	183,497	227,239	266,811	340,982
Secondary	2,943,473	3,463,236	3,809,515	4,088,742	4,620,769	5,492,114
No. female	994,947	1,478,031	1,680,028,	1,858,222	2,35,973	2,580,578
College	876,756	904,250	936,395	1,127,416	1,267,706	1,246,705
No. female	214,390	248,854	291,566	367,992	422,712	436,624
Total enrolment	4,032,875	4,652,292	5,087,885	5,662,218	6,383,167	7,371,030
Female enrolment	1,284,568	1,848,059	2,155,091	2,453,453	2,825,496	3,358,184
% Female enrolment	31.85	39.72	42.35	43.33	44.26	45.56

from a high of 64.95 per cent in 1991. Furthermore, only 15.8 per cent of the junior secondary school and 36.7 per cent of secondary school teachers have professional qualifications.

With a view to increasing women's participation at secondary level, the Government, the Norwegian Overseas Development Agency (NORAD), the Asian Development Bank (ADB) and the World Bank are supporting stipends for female students. This programme has served to increase the enrolment of women in secondary education institutions.

CURRICULUM DEVELOPMENT IN BANGLADESH

In 1982, the National Curriculum Development Centre merged with the Textbook Board to form the National Curriculum and Textbook Board (NCTB). The NCTB currently serves as the national curriculum agency for the country as a whole and has been entrusted with curriculum and instructional materials development activities from pre-primary to pre-university level. The organizational structure of the NCTB is reflected in Figure 14.1.

The tasks of the National Curriculum and Textbook Board include:

- Completion of curriculum revision for all primary grades;

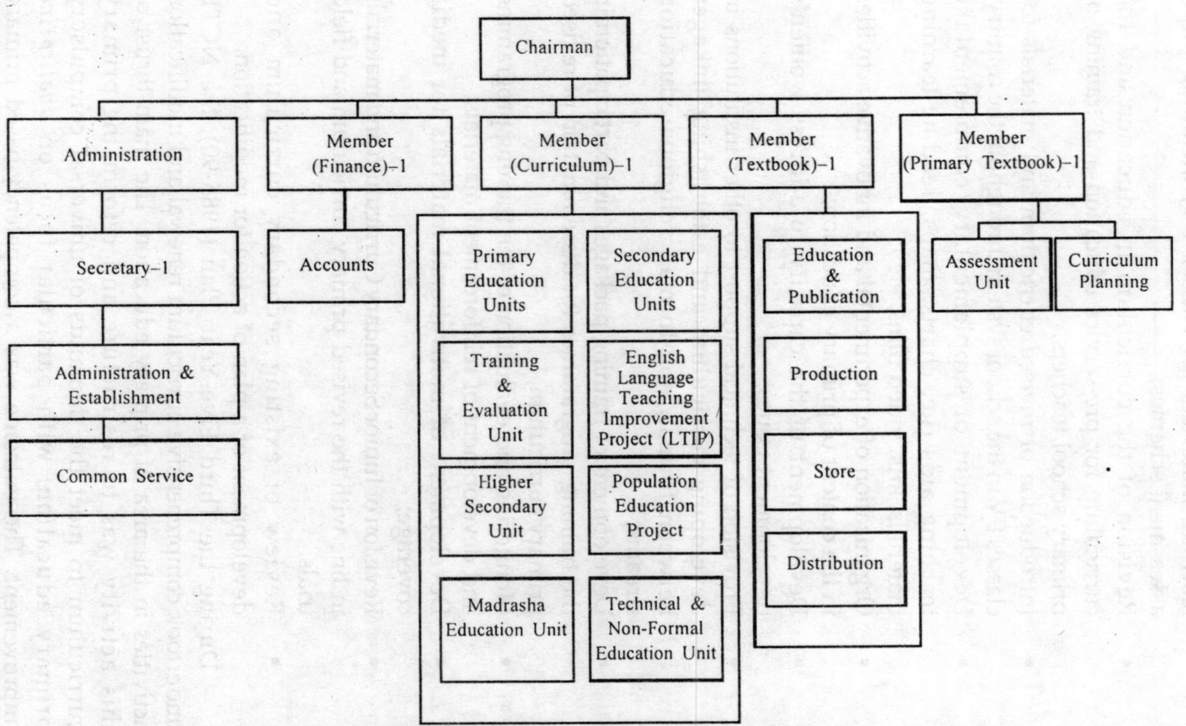

Figure 14.1: The Bangladesh National Curriculum and Textbook Board

- Field trials as well as the production of teaching learning materials for classes I-V, including pupil assessment schemes;
- Revision of the curriculum in education and the curriculum for pre-service education and training of primary school teachers;
- Introduction of revised curriculum and materials for classes I-V in the school system throughout the country;
- Development of supplementary/complementary teaching aids (i.e. charts, maps, reading-learning materials, kits, video films, etc.);
- Organization of experimental and innovative activities in the content of primary education;
- Development of the capacity to address women's development issues;
- Provision of technical support to other institutions in developing curriculum and establishing linkages between formal, non-formal, religious education streams;
- Development of training packages and participation in the training programmes for dissemination of revised primary curriculum;
- Monitoring and evaluation of the training programme and development of reinforcement materials;
- Development of motivational materials for media coverage;
- Revision of Junior Secondary Curriculum and materials in line with the revised primary curriculum and field trials;
- Review of existing secondary curriculum and development of a plan of action for modification.

During the Third Five-Year Plan (1985-90), the NCTB undertook comprehensive curriculum renewal and modification activities in the area of primary education. The main thrust of this activity was to restructure and reform the primary curriculum to meet the demands of universal compulsory primary education, with particular focus on *qualitative* improvement. The outcome was a competency-based primary school curriculum comprised of fifty-three terminal competencies to be achieved by all primary school pupils who successfully complete the five-year primary education cycle.

In this context, the following have been developed: detailed syllabi for eleven subject areas of grades I and II; a trial edition of textbooks; teacher source books; and a continuous pupil assessment scheme, along with a register for keeping assessment records. Improvement of curricula has been undertaken with the help of foreign and domestic experts. Step have been taken to train teachers to be able to use new materials with understanding and skill, with a view to preparing them to introduce the competency-based curriculum. The curriculum emphasizes mastery learning supported by diagnosis of pupils learning difficulties, followed by remedial instruction, appropriate teaching-learning strategies, revised instructional materials, and tools and techniques for continuous assessment of pupil performance.

The new curriculum was implemented during the Fourth Five-Year Plan (1990-95), with new approaches towards the evaluation of pupils' achievements. The primary level curriculum has been formulated on the basis of Bangladesh's socio-economic situation, as well as the children's physical and mental make-up and their cultural awareness. Subjects in primary level include Bangla (the mother tongue), mathematics and environmental studies, social studies, science, arts and crafts, physical education, music. English language and religious studies (with emphasis on the ways of life and moral education) will be introduced in class III. Classes VI to VIII will be gradually incorporated under primary education. At these levels, the subjects to be studied will be Bangla, mathematics, general science, social studies (Islam, Hinduism, Buddhism, Christianity), multi-disciplinary learning (home economics, agriculture, etc.), religion, fine arts, health and physical education. The present curriculum at the primary as well as junior secondary and secondary stages contains elements of population education, including information on the socio-economic impact of the population explosion. The content has been incorporated into the textbooks of classes III to X.

The average amount of weekly teaching hours in each grade at primary level is three hours for Bangla; three hours for mathematics; thirty minutes for music, arts and crafts, and physical education; and two and a half hours for all the other subjects. The medium of instruction is Bengla (mother tongue).

In the present curriculum, summative assessment (in the form of an annual examination in each grade) has been abandoned. Instead, a system of continuous pupil assessment has been introduced. The system requires teachers to assess students regularly for every competency acquired in a particular lesson (through observation, oral/written assessment) and to record achievement on a monthly basis using three scales (namely grades A, B and C). For grade I-II, all the students get promoted to the next classes, and for grades III-V, promotions are given on the basis of students' achievement. No certificate is awarded after completion of the five-years primary schooling, but a scholarship examination is held annually for 20 per cent of the students of grade 5—on the basis of which scholarships are awarded to successful students.

CURRICULUM DEVELOPMENT: AREAS FOR IMPROVEMENT

Some of the major deficiencies in curriculum development in Bangladesh include: (a) lack of professional expertise in the development of modern curriculum, both in the NCTB and nationally; (b) lack of a solid research base providing assessment information about the previous curriculum and the areas needing revision; and (c) insufficient curriculum emphasis on such competencies as understanding, comprehension and application.

CURRICULUM DEVELOPMENT AT THE SECONDARY LEVEL

Under the Secondary Education Development Project of the ADB, the junior secondary school curriculum has been revised and materials prepared on the basis of competencies identified for this stage of education; approximately 150,000 teachers have received short-term training in the new curriculum; ten teacher-training college have been upgraded; and pre-service and in-service training programmes have been conducted. Under the ADB Higher Secondary Education Project, five new higher teacher-training institutes are being established. During 1996-98, the secondary and higher secondary curricula were revised and implemented and textbooks prepared. The monitoring and evaluation systems have been improved and implemented. Table 14.6 provides an overview of the curriculum development and implementation process.

Table 14.6. The curriculum: who makes which choices?

	Central Level	Regional Level	School Level
Aims and Objectives	National Curriculum and Textbook Board prepares curriculum in light of national education philosophy and policy; National Curriculum Co-ordination Committee of Ministry of Education subsequently approves	Local authorities assist with curriculum development by providing suggestions	Teachers prepare lesson plans based on curriculum
Curriculum Plan	Prepares syllabus and distribution of subject marks. Prepares time table.	Monitors the implementation of curriculum	Implements the curriculum at the classroom level
Methods and Approaches to Learning	Field tests and finalizes the intended curriculum	Assists with the finalization of curriculum	Measures competency attainment
Materials	Prepares teachers' guides and learning materials	Distributes teachers's guides and learning materials (local levels: divisional deputy directors, district and Thana education officers)	Utilizes teachers' guides and learning materials in classroom
Evaluation and examiantion	Prepares evaluation and examination plan	Implements secondary level examiantion (various regional boards)	Conducts terminal examinations at primary and secondary levels

Although the three stages of secondary education form part of a concentric curriculum development, during the actual development process there was little co-ordination between these stages. In fact, the greatest part of the work was carried out for the junior secondary stage, and then subsequently reviewed in each of the other stages where attempts were made to address areas of weak emphasis by adding content.

Another problem is the absence of a system of ongoing curriculum review. The syllabus standing committee system within NCTB is not operational. The curriculum section staff should be monitoring the curriculum and textbooks usage and effectiveness in the teaching situation in order to ensure that they are up-to-date and relevant. Unfortunately, at present, this important work is not being done.

The establishment of a curriculum research section is urgently needed for effective curriculum development. The issues affecting the impact of curriculum change in the classroom need to be assessed objectively. Without this research base, curriculum development will continue to be based largely on ad hoc decisions resulting from the views of persons who often lack first-hand knowledge of the overcrowded and poor conditions in most schools. The staff of such a research section will require training and support, as well as increased resources, in order to perform regular classroom visits and assessments.

Lack of expertise is also a fundamental problem. Reviews of the capacity of the NCTB by different institutions in Bangladesh have repeatedly highlighted the lack of trained professional curriculum developers.

CONCLUSION

Although growth in the primary and secondary education sectors in Bangladesh is quite satisfactory, the quality of education is not. However, the country is striving hard to achieve this quality and, in this context, many efforts have been undertaken with the help of domestic and expatriate experts to improve the curriculum.

Bhutan: Curriculum Development for Primary and Secondary Education

Deki C. Gyamtso and Namgyel Dukpa

Estimated population (1995)	1,800,000
Public expenditure on education as percentage of Gross National Product (1995)	—
Duration of compulsory education (years)	—
Primary or basic education	
Pupils enrolled (1994)[1]	60,089
Teachers (1995)	—
Pupil/teacher ratio (1994)[1]	31:1
Gross enrolment ratio (1995)	
— Total	73
— Male	82
— Female	63
Net enrolment ratio (1995)	
— Total	53
— Male	58
— Female	47
Estimated percentage of repeaters (1994)[1]	19
Estimated percentage of drop-outs (1995)	18
School-age population out of school (1995)	100,000
Secondary education	
Students enrolled (1995)	
Gross enrolment ratio (1995)	
— Total	
— Male	
— Female	
Third-level enrolment ratio (1995)	—

Estimated adult literacy rate (1995)	
— Total	42
— Male	56
— Female	28

Notes:

1. Last year available.

- = data not available.

Source: *UNESCO statistical yearbook, 1998,* Paris.

BACKGROUND INFORMATION

The kingdom of Bhutan is a mountainous, Buddhist country with an area of roughly 46,000 square kilometres, located between China in the north and India in the south. It has been an independent nation throughout its history, although it was only in the seventeenth century that Zhabdrung Ngawang Namgyel (1594-1652) unified it into one nation-state. The present King, His Majesty Druk Oyalpo Jigme Singye Wangchuck, is the fourth monarch to rule the country.

Prior to the year 1920, monastic education, which included the study of Buddhist religion, liturgy, philosophy, astrology and the fine arts (painting, sculpture, music and dance), was the only form of formal education available in the country. The first modern school was established in 1920 (Dolkar, 1995). Modernization in Bhutan, however, can be said to have begun only in 1961 with the launching of the Five-Year Development Plan (1961-65). From the outset of the planned development, the number of schools in Bhutan increased to fifty-nine (twenty-nine private and thirty governmental) and a Department of Education was established. With the creation of this institution, the private schools were converted into governmental schools, thereby establishing a centralized and uniform system of education which has largely continue until today, although a few private schools have been opened lately following government encouragement. With the Royal government's heavy investment on education during the Five-Year Development Plans, within a period of three and half decades, Bhutan has been able to create a modern education system from primary to tertiary levels. Enrolment at all levels has grown considerably over the years as a result of the government's commitment to education. By 1998, the total number of students reached 100,355 (Planning Section,1998).

The present pre-tertiary educational structure consists of nine years of basic education, comprising one year pre-primary, six years primary, and two years junior high, followed by four years of secondary education comprising two years of high school and two years of higher secondary level. Access from one level to the other is merit-based and determined by the national and external examinations, as well as by the human resource plans and the space available at the relevant levels of education.

Table 15.1: Summary of education statistics (April 1998)

1. *Number of schools and institutions*

Community schools	115	High schools	18
Primary schools	128	Private schools	7
Junior high schools	44	Institutes	10
		Total	322 and 54 NFE centres

2. *Number of students*

Community school	12,695	Private schools	1,544
Primary schools	41,733	Institutes	2,004
Junior high schools	29,502	Non-formal education	1,842
High schools	11,035		
		Total	100,355

3. *Number of staff*

	Teachers	Others	Total
Community school	305	12	317
Primary schools	1,058	190	1,248
Junior high schools	751	134	885
High schools	386	175	562
Private schools	70	9	79
Institutes	215	161	376
Total	2,785	681	3,466

The National Institute of Education (NIE) offers a three-year Bachelor of Education degree course for grade XII graduates and a one-year teacher training programme—Post-Graduate Certificate in Education (PGCE)—for candidates who already have a Bachelor's Degree either in humanities, commerce or science. The two-year Primary Teacher's Certificate (PTC) course is offered for the grade X graduates at the NIE, as well as the Teacher-Training College (TTC) in Paro. NIE has also started a Distance Education Programme for primary teachers since 1995. An eighteen-month training programme for Dzongkha (national language) teachers is also offered by the Teacher-Training College, Paro.

Education has always been provided free of charge from primary to tertiary level. This includes not only free tuition, but also the provision of textbooks, stationery, meals and boarding facilities where required.

THE CURRICULUM

The school curriculum until the mid-1980s was imported from India and all the teaching materials were those prescribed for Anglo-Indian schools, except for Dzongkha. From the mid-1980s, the then Education Department started *Bhutanizing* the education system so that teaching and learning in schools was in accordance with national needs and aspirations (Second Quarterly Policy Guidelines, 1989, p. 8). The development of a relevant curriculum and curriculum materials for schools throughout the country began following this important policy change.

The first big curricular change came with the introduction of the New Approach to Primary Education (NAPE) project which emphasized activity-based learning, shifting the focus from 'teacher-centredness to child-centredness, as well as from remoteness of content to familiarity of content' (Dolkar, 1995).

The next big change was the introduction of Bhutanese history and geography for grades 6-8 in 1990, and then for grades 9-10 in 1993, whereby students are expected to acquire knowledge, skills, values and attitudes through this change in order to 'develop pride in being Bhutanese;.... a sense of self-discipline and duty;....spiritual, cultural and traditional values and so contribute to national and social cohesion' (National Education policy, 1984). Changes have also been brought about in the other school subjects.

MAIN ORGANIZATIONS INVOLVED IN ADAPTING CURRICULA

The adaptation of curricula takes place at various levels in the country. A brief discussion of the agencies involved is presented here.

Central Level

Since all education institutions are government-financed, educational programmes are largely controlled from the centre. The task of initiating and implementing all educational programmes in the country (in line with national educational policies) is entrusted to the Ministry of Health and Education (MHE) and the Education Division (ED) via its various sections *(see Figure 15.1)*. The Curriculum and Professional Support Section (CAPSS), Educational and Monitoring Support Section (EMSS), Bhutan Board of Examination (BBE), Non-Formal Education Section (NFES) and Youth Guidance and Counselling Section (YGCS) are the key existing professional bodies that are extensively involved in adapting curricula.

Curriculum and Professional Support Section (CAPSS). When the Department of Education (subsequently the Education Division) within the Ministry of Health and Education launched the programme of Bhutanization of the education system in the mid-1980s, a Curriculum and Textbook Development Division (CTDD) was created to develop curriculum and curriculum materials for schools. However, in 1994, the practice of curriculum development had evolved to the point whereby a committee of experienced educators and teachers (formed from various subject committees) took over the implementation. With this development, the CTDD was transformed into the currently existing CAPSS, now the main body involved in curriculum research and development for the formal school system. CAPSS is staffed by curriculum officers, headed by a Director and a Deputy Director, who are responsible for each subject *(see Figure 15.2)*.

The main functions of CAPSS are:

- carrying out curriculum research and development;
- developing and writing syllabi, textbooks and manuals for grades PP-WI in all subjects;
- developing syllabi and textbooks for grades IX and X in consultation with BBE and ISCE Board, Delhi;

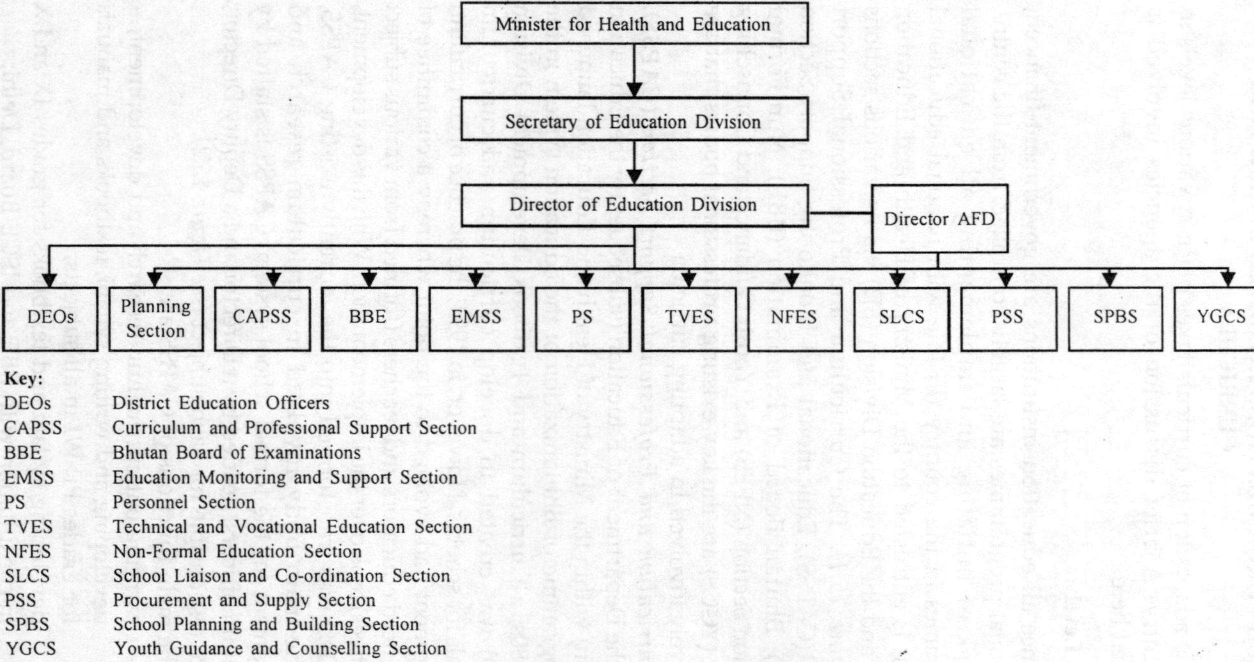

Figure 15.1: Organisation of the Education Division

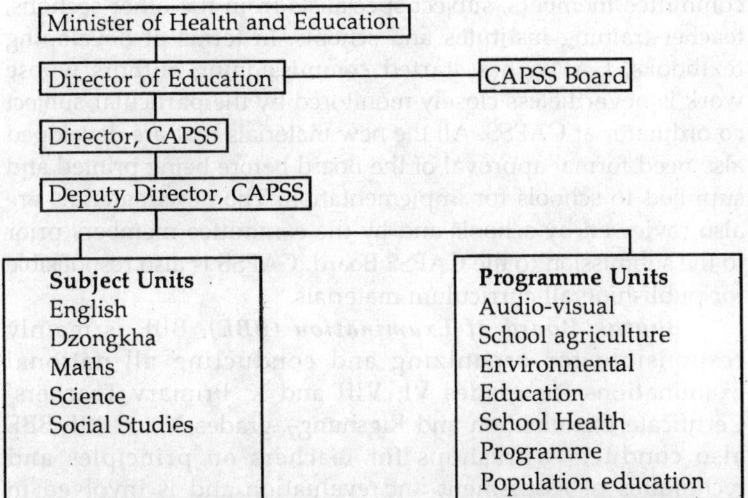

Figure 15.2: The Curriculum and Professional Support Section (CAPSS)

- training of teachers and pilot testling, evaluating, monitoring and revising curriculum materials; and
- conducting in-service workshops for teachers in different subject areas.

For the development of curriculum guidelines and materials at CAPSS, each subject unit has a subject committee (composed of members from various sections, training institutes and schools) with each unit's co-ordinator at CAPSS as the member-secretary. All innovations and changes in the curriculum and curriculum materials are discussed in the subject committee meetings, usually held twice a year. Proposals for initiating the innovations or changes are developed in consultation with the Director and Deputy Director, CAPSS, for further submission to the CAPPS Board, which meets once a year. The CAPSS Board comprises of the Director of Education as the Chairperson, all the section heads of the Education Division, Chairpersons of the various subject committee, the principal of the college, directors of the teacher-training institutes, one principal representing the high schools, and the Director of CAPSS as the member-secretary.

Once the proposal is approved, the curriculum officers of the subject units co-ordinate the work, involving subject committee members, subject specialists from the other sections, teacher-training institutes and schools. In terms of developing textbooks, CAPSS has started commissioning authors whose work is nevertheless closely monitored by the particular subject co-ordinator at CAPSS. All the new materials that are developed also need formal approval of the Board before being printed and supplied to schools for implementation. The new materials are also reviewed by schools and by the committee members prior to the submission to the CAPSS Board. CAPSS is also responsible for publishing all curriculum materials.

Bhutan Board of Examination (BBE). BBE is mainly responsible for organizing and conducting all national examinations for grades VI, VIII and X, Primary Teachers' Certificate Examination and Rigshung—grades X and XII. BBE also conducts workshops for teachers on principles and techniques of assessment and evaluation and is involved in developing general guidelines for assessment with the CAPSS. The subject specialists in BBE are also members of the various subject committees.

Education Monitoring and Support Section (EMSS). EMSS is mainly responsible for reporting on the general quality of education in schools. It monitors compliance with the national education guidelines and reports on the overall school effectiveness in relation to student achievement. This involves reviewing and evaluating school management, maintenance practices, curriculum practices, students' achievement and identification of good teaching practices as well as barriers to learning. EMSS is involved in providing necessary professional support to the District Education Officers and schools on a regular basis through forums such as Dzongkha Based In-service Programmes and School Based In-service Programmes. Some of the officers in EMSS are also members of various subject committees.

Youth Guidance and Counselling Section (YGCS). YGCS is involved in developing and implementing a range of co-curricular and extra-curricular programmes and activities, including Career Education and Guidance Programme, Scout Programme, Adolescent Health Education and Counselling Programme.

Non-Formal Education Section (NFES). NFES is mainly responsible for launching literacy and relevant life-skill programmes for school drop-outs and adults. Its main functions are:

- developing and publishing NFE learning materials;
- training NFE teachers;
- helping districts establish NFE and community learning centres; and
- monitoring NFE programmes and recommending policies.

Technical and Vocational Education Section (TVES). TVES plans, co-ordinates and monitors technical and vocational training programmes. It ensures relevant vocational skills are included in the school learning process and promotes positive attitudes among students regarding technical vocations.

Teacher-training institutes. The two teacher-training institutes, the National Institute of Education and the Teacher-Training College, take an active part in adapting curricula through pre-service teacher training. This is done by: exposing them to various approaches of teaching and learning; assisting them in developing teaching materials from the available resources; equipping them with the necessary skills and teaching techniques; and assessment and evaluation of the teaching/ learning process and the achievement of the students. Both institutes graduate 100 to 150 teachers every year for the primary and secondary schools. The lecturers of the institutes are also involved in the design and implementation of the curriculum as many of them are also members of the various subject committees. They also play an active role in organizing and conducting various in-service curriculum-related workshops for the teachers in the country every year.

District Level

At the district level, the District Education Officers (DEOs) interpret for the benefit of the schools the aims and objectives and the curriculum directives from the Education Division. They provide academic and non-academic supervision, organize teacher and school development activities, monitor development activities and assist with school-based in-service programmes by visiting the schools regularly. They also support schools in obtaining teachers and resources, and developing the school infrastructure.

School Level

At the school level, the head-teachers and teachers interpret the aims and objectives of education through the curriculum. They develop work schemes, lesson plans and teaching materials and practise different methodologies. Teachers also organise various types of co-curricular activities to provide pupils with balanced education. They carry out continuous assessment of the students' performance and set internal tests and examinations. They organize school-based in-service programmes to improve their classroom practices. Many teachers are also involved in developing curriculum materials at the centre and during resource national-based in-service workshops.

PROBLEMS IN ADAPTING CURRICULUM IN BHUTAN

Curriculum specialists in Bhutan face several problems, which are largely due to the short space of time in which the modern system of education has evolved. Discussion will be limited to the main problems faced in the design, implementation and follow-up of reforms.

Curriculum Design

The first main problem is the lack of expertise in curriculum development. There is not much scope for consultation nor the required support and guidance. Curriculum developers are left to cope on their own with their often limited knowledge in the area. Outside experts are often invited to guide and assist, but, in many cases, their expertise and inputs are not relevant to the country's needs and situation. Valid and relevant comments for improvement are rarely received from the field. This largely affects the quality, clarity and appropriateness of the curriculum design.

Lack of resources and access to quality innovations due to financial constraints are other major problems. This limits curriculum developers' access to the wealth of information and ideas on curriculum design that are available in other counties.

Insufficient time for carrying out proper situational analysis and research is another major problem. Final decisions are largely top down (from the administrators who look for immediate action and quick outcomes), leaving staff with no choice but to design the curriculum within the set time frame, which in most cases is inadequate. The problem is further exacerbated by inadequate staffing, since the curriculum developer has other commitments and cannot devote adequate time to curriculum design.

This scenario, however, is rapidly changing with more people becoming specialized in curriculum development and change, thus creating more opportunities for educators to enhance their professional development. With their inputs, education in Bhutan is becoming more relevant and appropriate to the needs of the learners.

Problems in the Implementation of the Curriculum

As with the design process, time and resource constraints (both human and material) are the major problem. Because of the requirement for speedy implementation, inadequate time is allotted for pilot-testing the reforms in terms of their relevance and appropriateness, for carrying out a detailed plan of implementation, and for organizing resources. Furthermore, teachers hardly have the required time for preparation, the shortage of school-teachers being a perennial problem in the system.

More support materials are needed for effective implementation. Teachers do not have much access to reference material, except basic school textbooks. Specialists cannot offer much support to the teachers during the implementation of reforms, since they cannot visit schools regularly due to busy schedules. Thus, in most cases, the initial orientation workshop held in the beginning to familiarize teachers with the reforms and their implementation is the only support and guidance that they receive. Under such circumstances, teachers maintain their old ways of teaching, with which they feel more secure, and new reforms are hardly implemented.

Lack of proper co-ordination among the various sections in the centre, and with the districts and schools, is another problem faced in implementation. Teachers who receive initial training in implementing the reforms are sometimes subsequently given a new assignment or transferred to another school, resulting in the changes not being introduced.

The education system, being examination-oriented also inhibits effective implementation of the curriculum as teaching time is restricted to covering the syllabus in time for the examinations.

The deployment policy of the Education Division can be seen as an additional problem. Very often the teachers who have graduated from NIE and TTC are sent to schools to teach subjects or levels for which they are not trained.

Problems in the Follow-up of the Reforms

The shortage of qualified human resources at all levels has led to a failure in establishing an adequate follow-up mechanism. Because of this, most of the reforms are without proper monitoring and evaluation, leaving the teachers to put the reforms into practice unassisted amidst the numerous difficulties facing them, as already discussed earlier. This has been recognized as a major drawback in the system and measures have been taken to rectify it. An INSET programme has been started to carry out an impact analysis of the 'what after' of the workshops, with a team of educators being given the task of collecting and analysing data from the teachers in the schools. Another positive development is the initiation of regular curriculum reviews of different subjects. These help educators adapt to the changes.

ANALYSIS OF A REFORM IN THE SOCIAL STUDIES CURRICULUM

Reasons for the Reform

The change was brought about in the social studies curriculum from grades 6 to 10, in line with the aims and objectives of the national education policy to make the curriculum relevant to the national aspirations and in keeping with the Bhutanese culture. The old curriculum, being an imported one, was felt to be largely irrelevant to Bhutanese students. The content lacked any input about the history, culture and geography of Bhutan, resulting in the students knowing more about other countries than about their own. The reform thus involved change in content with the inclusion of Bhutanese history and geography and also in the teaching-learning approaches with a focus on activity and inquiry-based learning.

How the Reform was Conducted

The social studies co-ordinator at CAPSS, in consultation with the Director, designed the social studies syllabus with the involvement of subject specialists from the training institutes and school-teachers. Based on the syllabus, textbooks on the history and geography of Bhutan were developed. These went into trial use between 1988 and 1990. A review was then conducted, based on the feedback obtained from teachers. Follow-up workshops were organized for reorienting the teachers to the changes, after which the reform was implemented.

Outcomes and Future Prospects

While the new content was introduced into the curriculum by teachers, few changes were noticed in teaching approaches. The lecture and dictation method was still predominant with hardly any meaningful and experiential learning taking place. This may largely be due to the fact that fundamental change in teaching methods requires a change in belief about practice. To bring this about, a lot of pressure, support and guidance were needed. These elements were largely neglected due to constraints of human resources, resource materials and inadequate communication.

The social studies curriculum envisages a number of additional changes in the future. The most current issue is the integration of social studies up to grade VIII. The students will not be taught history and geography separately, instead they will learn social studies with more focus on development of values and attitudes and social skills. Bhutanese civics will be introduced in 1999 as part of the school syllabus, so as to enable students to develop an awareness of the functioning of the government and their own civic responsibilities. Approaches to the teaching of social studies have gradually improved over the years due to the adoption of the required teaching strategies by newly trained teachers.

REFERENCES

Curriculum and Professional Support Section. 1996. *The Purpose of School Education in Bhutan*. Thimphu, Education Division.

——. 1990. *Social Studies Syllabus*. Thimphu, Education Division.

Dolkar, T, 1995. *Social Studies Curriculum*. Thimphu, CAPSS Education Division.

Education Division. 1989. *Second Quarterly Policy, Guidelines and Instructions*. Thimphu, Education Division.

Education Division. 1993. *Review of Primary Education in Bhutan*. Thimphu, Education Division.

Education Division. 1984. *National Education Policy (Draft)*. Thimphu.

Planning Section. 1998. *Education Statistics*. Education Division, Thimphu.

Youth Guidance and Counselling Section. 1997. *A Profile of Youth Guidance and Counselling Section*. Thimphu, Education Division.

16

India: Education Policies and Curriculum at the Upper Primary and Secondary Education Levels

National Council of Educational Research and Training

Estimated population (1995)	929,000,000
Public expenditure on education as percentage of Gross National Product (1995)	3.4
Duration of compulsory education (years)	8
Primary or basic education	
Pupils enrolled (1995)	109,734,292
Teachers (1995)	1,740,436
Pupil/teacher ratio	63:1
Gross enrolment ratio (1995)	
—Total	100
—Male	110
—Female	90
Net enrolment ratio (1995)	
—Total	87
—Male	98
—Female	76
Estimated percentage of repeaters (1995)	4
Estimated percentage of drop-outs (1995)	38
School-age population out of school (1995)	14,200,000
Secondary education	
Students enrolled (1995)	68,900,000
Gross enrolment ratio (1995)	
—Total	49
—Male	59
—Female	39
Third-level enrolment ratio (1995)	6.5

Estimated adult literacy rate (1995)	
—Total	52
—Male	66
—Female	38

Source: *UNESCO statistical yearbook, 1998*, Paris.

EDUCATION POLICIES AND THE CURRICULUM IN INDIA

Background

India is a union comprised of twenty-five states and seven territories. The Constitution provides directives regarding the development of education throughout the country. The areas in which the respective central and state governments have domain have been identified in the Constitution as the *central* list, *state* list and *concurrent* list. Until the late 1970s, school education had been on the *state* list, which meant that state had the final say in the management of their respective school systems. However, in 1976, education was transferred to the *concurrent* list through a constitutional amendment, the objective being to promote meaningful educational partnerships between the central and state government. Today, the central government establishes broad education policies for school curricula development and management practices. These serve as guidelines for the states.

Education Policies

National policies are evolved through a mechanism of extensive consultations, in which all the states and union territories actively participate. Periodically, the central/state governments appoint commissions and committees to examine various aspects of education. In addition, countrywide debate takes place on various educational issues. The recommendations of various commissions, committees and national seminars, and the consensus that emerges during these national debates, form the basis for India's education policies. During the post-independence period, a major concern of the Government of India and of the states was education as a factor vital to national development. In this context, India's educational reconstruction problems have been periodically reviewed by several commissions and committees. Their deliberations, recommendations and reports have formed the basis for the 1968 National Policy on Education (NPE) and the National Policy on Education Resolution of 1986

The Impact of National Debates on Curriculum

In 1986, extensive deliberations by various national committees on the country's education system and policy culminated with the decision for a national curricular framework containing a *common core* along with *flexible components*. The common core includes the history of India's freedom movement; constitutional obligations and other content essential to nurture national identity. These core elements are intended to cut across subject areas and were designed to promote a number of values (such as India's cultural heritage, egalitarianism, democracy, and secularism, equality of the sexes, protection of the environment, removal of social barriers, observance of the small-family norm and inculcation of the scientific approach). Also, in order to reinforce the view that the whole world is one family, the curriculum would have the objective to promote international co-operation and peaceful co-existence.

With regard to re-orientation of education content and processes, the NPE emphasized the need for bridging the schism between the formal education system and the country's rich and varied cultural traditions. To this end, the preoccupation with modern technologies must not be allowed to sever new generations' ties to India's history and culture. In view of the growing concern over the erosion of essential values and increasing cynicism in society, readjustments in the curriculum are to be carried out so that education becomes a forceful tool for the cultivation of social and moral values. The policy further emphasized the integral role that manual work, sports and physical education should play in the learning process and the need to strengthen science and mathematics education.

THE CURRICULUM DEVELOPMENT PROCESS

The process of curriculum development in India lies between the two extremes of centralization and decentralization. From time to time, the national government formulates the National Policy on Education which includes broad guidelines regarding content and process of education at different stages. These guidelines are further elaborated by the National Council of Educational Research and Training (NCERT).

Using as its foundation the NPEs of 1968 and 1986, two curriculum initiatives have been launched by NCERT: (a) The Curriculum for the Ten-Year School—a framework (1975); and (b) The National Curriculum for Elementary and Secondary Education—a framework (1988). The curriculum framework prepared at the central level provides a broad overview of the school curriculum, including general objectives, subject-wise objectives, suggested scheme of studies, and guidelines for the transaction of the curriculum and the evaluation of pupil outcomes. These detailed curricula, syllabi and instructional materials are developed at the national level. The NCERT has also developed the syllabi and instructional materials used in the school run by organizations.

However, the states consider whether to *adopt* or *adapt* and NCERT syllabi and instructional materials. Thus, the NCERT curriculum framework is always *a suggestion* rather than *prescriptive* and it is not enforceable by law in the states. However, it is readily accepted by the states because of the NCERT's credibility and the participatory development approach it follows. (The NCERT curriculum framework is developed on a consensus basis; all the state and union territories are involved in the curriculum elaboration).

The National Curriculum

The following social, cultural, political, economic and educational parameters have guided the development of the national curriculum framework:

- All citizens of India should have equal access to education. The specific needs of the disadvantaged sections of the society ought to be met through the curriculum;
- Education regarding India's cultural heritage needs to be imparted to students in order to develop national identity and a spirit of togetherness;
- It is essential to impart knowledge of the citizens' duties and rights, and ideals of the Constitution of India to children;

- In view of the erosion of values it is imperative through the curriculum to inculcate moral and social values amongst students;
- Besides national identity and unity, it is also imperative to develop international understanding through the curriculum;
- Protection of the environment and conservation of natural resources should be major objective of school curriculum;
- In view of the increasing population of the country, it is imperative to include suitable content relating to population education in the syllabi of different subjects;
- The curriculum should aim at preparing a child for life, which means that relevant knowledge should be imparted and appropriate skills, competencies and values developed;
- Education plays a significant role in national development by increasing human resources. Therefore, the primary objective of the curriculum ought to be total development of the child's personality;
- All the process of education should be child-centred, with the teacher playing the role of a facilitator during the process of learning;
- The curriculum should aim at developing students' creative potential;
- The curriculum should develop a scientific approach amongst students;
- Work should not be considered as distinct from education. Instead, work should be adopted as a medium for imparting education;
- The process of evaluation should be continuous and comprehensive;
- Media and educational technology ought to be employed to make the transactions of curriculum effective.

An important development since the National Policy on Education was formulated in 1986 has been the acceptance across the country of a common structure of education and the introduction by most states of the 10+2+3 system. There are eight years of elementary education (five years of primary school and three years of upper primary/meddle schooling) and four years of secondary education (two years of general secondary and two years of higher secondary).

The education system seeks to give due recognition and importance to the social organization, traditions, customs and value systems of the various communities, particularly Scheduled Castes and Scheduled Tribes. This is supported among other ways by the development of materials and curricula in their languages.

The main characteristics of the national curriculum, developed in accordance with the above-mentioned principles, are described in the following sections.

General Education

The national curriculum envisages the first ten years of school as the period of general education and that the diversified curriculum should be introduced at the end of general education (i.e. at the beginning of the senior secondary stage.) This plan provides all students with an opportunity to receive instruction in each of the curricular areas considered essential for their overall development.

Undifferentiated Curricula

The national curriculum framework also envisages an undifferentiated curriculum for all children—irrespective of sex and place of residence (i.e. urban or rural).

Minimum Levels of Learning

The 1986 NPE recommended the establishment of Minimum Levels of Learning (MLLs) for the various subject areas at the different school stages. In this context, a Government of India committee (under of Chairmanship of Professor R.H. Dave) elaborated the MLL curriculum concept that designates the *competencies* to be mastered by the primary level pupils in each *subject*, at *specific points in time*. For the first five years of primary schooling, the MLL covers the mother tongue, mathematics, social science and science. The MLL approach implies that the teacher's responsibility is not confined to syllabus coverage. Rather, teachers must be responsible for their pupils mastering designated competencies. This approach has necessitated on-going development of MLL-based textbooks and MLL-based evaluation. It has also introduced a higher concept of *teacher accountability*. Teachers are now held responsible for pupil competency development and not merely for teaching the prescribed syllabus—as was the previous practice.

Common-core Elements

The 1988 National Curriculum Framework (NCF) recommended compulsory core curriculum elements to be taught throughout the country. Most of these core elements are aimed at the development of national identity and a spirit of togetherness leading to national unity. The common core elements recommended in the NCF are: the history of India's freedom struggle; constitutional obligations; content essential for the development of national identity; common cultural heritage of India; democracy, secularism, socialism; gender equality; environmental conservation; removal of social barriers; the small-family norm; and development of a scientific approach. The core elements are not be treated as separate subject areas. Rather, the content is to be interwoven into the different subject areas. Here, it should be noted that, for the first time during India's post-independence period, conscious efforts have been made to place *values* at the centre stage of curriculum.

Continuous and Comprehensive Evaluation

The NCF also considered the limitations of the existing evaluation system, which relies mostly on one-shot, end-of-the-year impact evaluation. This annual examination measures skills attainment and the affective domain is generally ignored (i.e. attitudes development). To remedy this, the NCF recommended that evaluation should be treated as an integral part of the classroom teaching/learning process. Furthermore, evaluation, conducted periodically, should provide the type of feedback on student achievement that enables teachers to improve their methodology, if required.

Interactive Teaching

It is recognized that both the educational curriculum content and process must be re-oriented in order to bring about overall quality improvement. During past few years, successful attempts have been made to re-orient the educational content to current development and demands of both society and the different disciplines. However, this initiative has not been accompanied by a corresponding change in the modes of curriculum transaction, which remains predominantly one of verbal exposition by the teacher. The expository style of teaching, involving mostly one-way communication, puts the learner in the role of a passive recipient—a mere *object* of education. This situation is not conducive to the development of creative, critical and analytical thinking by students. An interactive teaching methodology involving continuing dialogue between the teacher and pupils (discussion, investigation, problem-solving, etc.) could provide an educational environment more conducive to developing certain abstract cognitive skills.

Scheme of Studies

The 1988 National Curriculum Framework has recommended the areas shown in Table 16.1, along with the appropriate time weightage at the upper primary and secondary levels:

Languages. The NCF envisages the study of three languages at the upper primary and secondary stages: first, the mother tongue/regional language; second, Hindi or English (in the case of non-Hindi-speaking states); and third, one of the modern Indian languages (English in Hindi-speaking states; Hindi *or* English in non-Hindi-speaking states).

Mathematics. Functional mathematics are taught at the upper primary stage; arithmetic including commercial mathematics, should be completed, to a very large extent, by the end of the upper primary stage. The secondary stage begins the transition from functional mathematics to the study of mathematics as a discipline.

Science. The upper primary science teaching objectives are to develop an understanding of the nature of scientific knowledge; and certain physical, chemical, biological principles and their relationship to the operation of scientific principles in

Table 16.1: The National Curriculum Framework

Upper primary stage	Time weightage (%)
1. Three languages	32
2. Mathematics	12
3. Social science	12
4. Science	12
5. Health and physical education	10
6. Arts	10
7. Work experience	10
Secondary stage	
1. There languages	30
2. Mathematics	13
3 Social science	13
4. Science	13
5. Work experience	13
6. Health and physical education	9
7. Arts	9

nature, as well as in daily life. The aim of the teaching of science at the secondary level is focused on problem-solving and decision-making through the learning of key concepts, which cut across all the science disciplines.

Social sciences. The study of social sciences at the upper primary stage is comprised of the study of history, geography, civics and contemporary issues and problems. At the secondary stage, it incorporates elements of history, geography, civics and economics to promote an understanding of contemporary India.

Arts. The aim of art education is learner sensitization to the beauty in line, colour, form movement and sound. The upper primary programme incorporates: (i) drawing, painting, printing, collage, clay modelling, puppet construction; (ii) free expression artistic creation; (iii) handling of simple musical instruments; (iv) movement, mime, simple dance forms, community singing; (v) simple concepts of visual and performing arts; (vi) stories of great personalities in the field of arts, and stories connected with other countries. At the secondary stage, it incorporates: (i) study and. exploration of visual and aural resources; (ii) projects leading to creative visual and aural forms; (iii) inter-group, inter-school art activities; (iv) study groups, interaction with community artists; (v) exploration of community/neighbourhood traditional art forms.

Health and physical education. This area focuses on the holistic health of the learner and the community, thereby establishing the important place of mental and emotional, as well as physical health. The first ten years of content focuses on general promotion of healthful living as well as on major health problems of the country. In physical education, sports and games, the emphasis is given to indigenous traditional games. Furthermore, as a system which promotes the integral development of body and mind, yoga receives special attention.

Morals and Values. These areas are treated as an integral curriculum component for which all teachers are responsible.

Work experience. The work experience incorporates purposive, manual work resulting in either goods or services useful to the community. It is an essential component at all stages of education and is to be provided through well-structured, graded programmes. At both the upper primary and secondary stages, work experience emphasizes agricultural and technological processes to facilitate the integration of science, mathematics and technology into community life.

State-level Curriculum Implementation

The available feedback from the states indicates that, for the most part, they have revised their curricula along the lines recommended by the 1986 NPE and 1988 NCF. However, several have made adjustments that respond to specific local needs or socio-political pressure. Following are some highlights that reflect the dynamics of curriculum implementation at the state level.

Languages

All states (except Tamil Nadu) have adopted the three-language formula. Although it was envisaged that only one language (mother tongue or regional language) would be taught at the primary stage, many states have taken the initiative and introduced at second and third language at this level. For example, Punjab state recently decided to introduce English, along with the regional language in class. I. In the state of Sikkim, English is taught as a subject and used as the medium of instruction—beginning right from class I. The policy of using English as the medium of instruction beginning in class I is being implemented in almost all the private, unaided schools throughout the country. (The growth of these private schools has mushroomed during the past few years and this trend is likely to accelerate in the future.)

Another interesting variation encompasses classical language teaching—an area that the NCF did not address. Most of the Hindi-speaking states, and even central school organizations, have made provisions for teaching Sanskrit as a third language. In fact, to accommodate the study of Sanskrit along with other language, some states have even made provisions for the teaching of a fourth language. (In Uttar Pradesh, Sanskrit is compulsory and taught as part of Hindi). It would appear that Sanskrit is in demand because it is associated with ancient Indian culture and is the mother to many modern Indian languages.

Other Scholastic Areas

Several states have modified the NCF science and social sciences recommendations. For examples in place of the integrated science approach, they have opted for the single-subject discipline approach. West Bengal has made provisions for the teaching of history and geography only at the upper primary stage.

Non-scholastic Areas

In India, work experience, arts, and health and physical education are generally catagorized as non-scholastic areas . The curriculum framework has emphasized that these areas are essential for all-around development of the child's personality. However, being non-examination subjects, these areas are not taken seriously by the teachers and students. In some states, arts has not been made a compulsory subject at the secondary stage. In the states and schools where arts is a compulsory or optional area, only the visual arts are taught. Performance arts like music and dance are taught only in a limited number of schools. In the states of Punjab, Haryana and Himachal Pradesh, arts has been clustered with home science and agriculture and the students have been given the option to choose any one of the areas.

Values Education

The NCF also envisaged that values education should permeate all aspects of school life and, therefore, should be integrated into all the curriculum areas. However, states like Haryana, Goa, Himachal Pradesh, Karnataka, Punjab, Sikkim, Tamil Nadu, Andhra Pradesh and Uttar Pradesh have all introduced moral education or moral science as independent subject areas, with distinct time allocations. Apparently, this has occurred in the light of pleas that, given the progressive weakening of the moral fabric of society, the integrated approach does not provide values education with the prominent place it deserves.

Competency-based Textbooks

After the identification of MLL (minimum levels of learning), introductory advocacy programmes were launched to promote the concept as an approach to curriculum development teaching and learning and pupil evaluation. A number of states have since produced primary-state competency-based textbooks in different curricular areas.

CURRICULUM OUTCOMES

The Curriculum Review

The Government of India found it necessary to appoint a National Advisory Committee (NAC) to look into frequent complaints about the *excessive burden* of the curriculum on children. The NAC submitted report findings in 1993. This report, which took note of the widespread perception regarding the heavy load of the school curriculum, also identified the roots of the problem: inability to distinguish between information and knowledge; society's competitive social ethos; the desire to catch up with developed countries; centralized curriculum development processes; non-participation of teachers in the various curriculum development processes; excessive dependence on experts; incomprehensibility of textbook; and absence of an academic ethos in schools. The committee further determined that the academic burden perception is tied to *incomprehension,* a problem which can be addressed (to some extent) by modifying the curriculum development goals, as well as the textbook writing process and by improving the school environment by providing the required infrastructure.

Most of the NAC recommendations were accepted by the Government. All state governments were asked to initiate appropriate follow-up measures to implement the recommendations of the committee, including review of curriculum and textbooks.

Evaluation of Textbooks

The textbooks used in different states and union territories are already subject to periodic evaluation. They are reviewed from the perspective of national integration—the objective being to promote peace and harmony in the country and enable children to learn to live together with people of different religious, linguistic, ethnic and cultural groups. During these reviews any content deemed to have a secular bias or to be destructive to national unity is identified and recommended for removal. Textbooks are also continuously evaluated from the standpoint of gender, scientific temper and relevance, etc.

Improving Quality of Curriculum Transaction

NCERT, the Central Board of Secondary Education (CBSE), the State Councils of Educational Research and Training (SCERTS) and the State Boards of Secondary Education have initiated a number of projects to improve the quality of curriculum transmission in schools. To improve the quality of mathematics education, CBSE has launched the project Operation Mathematics emphasizing the re-orientation of all teachers. The CBSE has also developed source materials in the education areas of: environment, values, consumer and population. Furthermore, a network of training institutions has been established in the country to enhance teacher competencies through in-service education.

Emerging Trends in Curriculum Development

In the light of changing societal needs and aspirations, certain high demand-driven areas are likely to have an impact on the school curriculum. It appears that, among other areas, language education, values education and information technology shall be matters of serious debate in the next cycle of curriculum renewal. Art education is also scheduled for reform in the next cycle of curriculum renewal.

Language education. The introduction of English at the primary stage is one of the demands which school systems are finding difficult to resist. Perhaps this demand is based on the assumption that the study of English can give children an edge in a highly competitive society.

Values education. After the establishment of a National Commission for Human Rights, momentum has been gathering for incorporating this area into the school curriculum. The interim report issued by Justice Verma's Committee (on the fundamental duties of citizens) strongly recommended that fundamental duties should be incorporated at all stages of the school curriculum, as well as into different teacher-education programmes. The incorporation of human rights and citizens' fundamental duties into the curriculum can help pupils *learn to live together,* one of the four pillars of learning identified by the International Commission on Education in the Twenty-first Century.

Computer education is expected to soon occupy a prominent space in India's school curriculum, it being recognized that, in the information age, skills in the use of these technologies are invaluable.

REFERENCES

Arora, G.L. *Child Centred Education for Learning Without Burden.* Gurgaon, Haryana, Krishna Publishing Company, 1995.

Arora, G.L.; Yadav, S.K, *Sefl-learning Materials for Teacher Educators.* Vol. 1. New Delhi, NCERT, 1998.

Arora, G.L.; Ray Shefali. *Towards an Enlightened Human Resource in Education.* New Delhi, State Council of Educational Research and Training, 1993.

Curriculum Board for Secondary Education. *Scheme of Comprehensive School-based Evaluation.* New Delhi, Government of India.

Government of India. *Operationalization of the Suggestions for Teacher Fundamental Duties to the Citizens of the Country. Interim Report.* New Delhi, 1999.

International Commission on Education for the Twenty-first Century. *Learning: the Treasure Within.* Paris, UNESCO, 1996.

Ministry of Human Resource Development, Department of Education. *Minimum Levels of Learning at the Primary Stage.* New Delhi, Government of India, 1991.

Ministry of Human Resource Development. *National Policy on Education*. New Delhi, Government of India, 1986.

National Advisory Committee, Ministry of Human Resource Development. *Learning Without Burden*. New Delhi, Government of India, 1993.

National Council of Educational Research and Training. *National Curriculum for Elementary and Secondary Education. A Framework*. New Delhi, Government of India, 1988.

—. *Some Aspects of the Upper Primary Stage of Education in India: A Status Study* (Draft). New Delhi, Government of India, 1998.

17

Indonesia: Goals and Objectives of Education

Estimated population (1995)	197,500,000
Public expenditure on education as percentage of Gross National Product (1995)	1.4
Duration of compulsory education (years)	6
Primary or basic education	
Pupils enrolled (1995)	29,721,859
Teachers (1995)	131,157
Pupil/teacher ratio	23:1
Gross enrolment ratio (1995)	
—Total	114
—Male	117
—Female	112
Net enrolment ratio (1995)	
—Total	97
—Male	99
—Female	85
Estimated percentage of repeaters (1995)	8
Estimated percentage of drop-outs (1995)	10
School-age population out of school (1995)	800,000
Secondary education	
Students enrolled (1995)	12,200,000
Gross enrolment ratio (1995)	
—Total	48
—Male	52
—Female	44
Third-level enrolment ratio (1995)	11.1

Estimated adult literacy rate (1995)	
—Total	84
—Male	90
—Female	78

Source: *UNESCO Statistical Yearbook; 1998*, Paris

According to Law No. 2/1989, the objectives of the national education system are:

- to develop citizens whose values are based on *Pancasila* (i.e. State ideology, spelled out in the five basic principles of the Republic of Indonesia: belief in one God: just and civilized humanity, including tolerance to all people: the unity of Indonesia; democracy led by wisdom of deliberation among representatives of the people; and social justice for all);
- to support the Indonesian society, people and State. In the broad context of society and national development, the aim of education is, on the one hand, to maintain Indonesia's cultural background and, on the other, to generate the knowledge, skills and scientific progress that will keep the nation abreast of development in the twenty-first century. National education should improve the life of the nation and develop the Indonesian people fully (i.e. intellectually, morally, spiritually, physically and socially).

THE ADMINISTRATIVE STRUCTURE OF CURRICULUM DEVELOPMENT

The Centre for Curriculum and Educational Facilities Development—or Curriculum Development Centre (CDC)—established in 1969, comes under the authority of the Office of Educational and Cultural Research and Development in the Ministry of Education and Culture. It is composed of four divisions each headed by a director: (a) Pre-school, primary and special education; (b) Secondary schools; (c) Higher education; (d) Educational Facilities. The Centre's main functions are: (a) to formulate technical policies on curriculum development and educational facilities; (b) to conduct, co-ordinate and guide the development of curriculum and educational facilities covering institutional objectives, programme structure and basic course outline, teaching learning models and methods; learning materials, etc.; and (c) to formulate suggestions on government policy.

The CDC has established a curriculum network to strengthen professional support for teachers across the country in the area of curriculum development. The network was established in order: (a) to involve different regions in the development of a national curriculum; (b) to improve the level of professionalism in curriculum development at the various levels (national, provincial, district); and (c) to establish a mechanism for curriculum dissemination and development at both national and provincial levels. The network includes professional groups from twenty-seven provinces across the county, each consisting of thirty-five members including leading teachers from primary, junior and senior secondary schools vocational schools, principals, supervisors and staff from the Regional Office of Education and Culture. The network's tasks are: to plan, develop and implement the curriculum according to local conditions and needs; to assist teachers in curriculum development through adjustments, elaboration and analysis based on the students' immediate environment and community needs and resources; and to monitor and evaluate the implementation of both national and local content. The CDC provides advice, assistance and guidance to the network in the elaboration, analysis, monitoring and evaluation of curricula. Some of the accomplishments of the network include: the development and implementation of local content material (course outlines, teaching guides; assessment guides and counselling guides); analysis and modification of the basic course outline; monitoring and evaluation, including development of instruments for the process, and preparation of a report. It is intended to set up networks at the district level, intensify training for network groups at both provincial and district levels, increase production of materials, and improve monitoring and evaluation techniques.

ON-GOING CURRICULUM REFORM

The existing curriculum designed in 1994 was felt to be inadequate in a number of ways: overloaded; too difficult for the pupils to complete; inadequate attention paid to the importance of the natural and social environments; failure to incorporate new areas of content, including education for human rights, moral education, health and nutrition education; and the need to update the content on Indonesian history.

Public hearings were held to find out public opinion concerning curriculum needs. Relevant recent research on the curriculum, especially as related to future needs, was taken into account. An analysis was made of the gaps between the existing curriculum as it was formulated and designed and as it has been implemented. A working team has been set up consisting of staff from universities and teacher training institutions, experienced teachers and relevant experts.

Current Curriculum-related Priorities and Concerns

Major goals include the expansion of compulsory basic education from six to nine years, the improvement of the quality of primary and secondary education, and enhancing learning achievements for all levels. Indonesia's commitment to the further development of education is based on the recognition that development goes hand-in-hand with advancement in science and technology.

The continuous decrease of employment opportunities in the area of agriculture and the increasing demand for knowledge and skills in industry, especially in the high technology and service sectors, as well as the inevitable impact of globalization, have created an urgent need for on-going curriculum reform. In addition to developing student's intellectual capacities, it is recognized that education must foster and promote creativity, the ability to process and utilize information, adaptability and self-training.

Indonesia has recently entered the *Second Twenty Five Year Long-term Development Plan—PJP II* (1994-95 to 2018-19), the emphasis of which is on the development of human resources to sustain the economic evolution of the nation. To respond to the challenges of modernity, the priorities of education for PJP-II include the following:

- The completion of the ***nine-year universal basic education programme*** (which involves adding three years of schooling for those of 12-15 years of age, i.e. at least six years of primary education and three years of lower secondary or equivalent education). The curriculum of the junior secondary school is also to be expanded with skills training, especially for students who are not able to continue to senior secondary education. The implementation of the nine-year basic education programme will cover efforts to develop an improved learning environment at school and in classrooms; efforts to provide and train more teachers; and efforts to provide quality equipment and textbooks.

- ***The relevance of education to development.*** The policy states that education should be related to industry and the business world starting from planning, implementation, assessment, and certification of education and vocational training relevant to economic needs. It requires the expansion and improvement of technical and vocational education for the production of skilled and flexible human resources who master technology. The 'link and match' programme, which involves industry and commerce in vocational education, will continue to be developed and implemented through the dual system. To support the policy, 2,000 commercial and small industrial institutes have been contacted for co-operation and asked to provide training for students. The co-operation includes curriculum development and an examination system that measures the skills and expertise of the participant after completing a certain level.
- ***Improved capacity to master science and technology*** through improved quality of higher education providing training and research, supported by improvements in mathematics and science instruction within the overall education system. Educational programmes, as preparation for employment, are provided through the junior secondary schools with qualified educational content and through vocational secondary education. At pre-tertiary level, science and technology programmes include (i) science and technology for basic education directed towards general basic comprehension and aiming to implant and develop basic learning tools—this covers mastery

in reading, arithmetic, problem-solving, and moral education for the industrial society (discipline, time appreciation, work ethics, self-learning); and (ii) secondary education programmes aiming to master the basics of science and technology.

- the development of a ***monitoring and evaluation system*** of education that is valid, reliable and continuously comprehensive.

PRIMARY AND JUNIOR SECONDARY EDUCATION (BASIC EDUCATION)

Six years of compulsory education for primary school-age children (PS, 7-12 years) were institutionalized in 1984. Currently (as from 1994) the programme has been extended to the 13-15 years age group (i.e. junior secondary school—JSS). The policy has been recognized as nine-year compulsory basic education. The major purpose of the extension is to alleviate the problem of child labour and to keep children in school up to the point where they are able to keep up with the changing demands of society, especially those who cannot afford to pursue a higher level of education.

In addition to primary and junior secondary education, there is also an Islamic primary school administered by the Ministry of Religious Affairs: the Islamic primary school (*Madrasah Ibtidaiyah*), equivalent to primary school, and the Islamic junior secondary school (*Madrasah Tsanawiyah*), equivalent to junior secondary school.

Primary education provides general education. The goal of basic education is to develop the lives of children as individuals and members of society, citizens and members of mankind, as well as to prepare them to pursue their studies in secondary education. The core content of the basic education curriculum consists of: *Pancasila*, religion, civil education, Indonesian language, reading and writing mathematics, introduction to sciences and technology, geography, national and general history, handicraft and arts, sports and health education, drawing, English language, and local content areas. More than one element may be joined in one subject matter: or , *vice versa*, one element may be divided into more than one subject. The 1994 basic education curriculum was implemented in phases until the end of 1996-97 academic year. The average number of weekly periods of teaching by subject are indicated in Table 17.1 below.

Table 17.1: Basic education curriculum (primary and junior secondary school)

Subjects	Number of weekly periods in each grade								
	Primary						Junior secondary		
	1st	2nd	3rd	4th	5th	6th	1st	2nd	3rd
Pancasila education	2	2	2	2	2	2	2	2	2
Religion	2	2	2	2	2	2	2	2	2
Indonesian language	10	10	10	8	8	8	6	6	6
Mathematics	10	10	10	8	8	8	6	6	6
Sciences	-	-	3	6	6	6	6	6	6
Social Sciences	-	-	3	5	5	5	6	6	6
Handicraft and arts	2	2	2	2	2	2	2	2	2
Health and sports	2	2	2	2	2	2	2	2	2
English language	-	-	-	-	-	-	4	4	4
Local content	2	2	4	5	7	7	6	6	6
Total Periods	30	30	38	40	42	42	42	42	42

Source: *Basic Education Curriculum,* MOEC, 1993; (the average length of teaching periods is 30 minutes is grades 1-2 and 40 minutes in grades 3 to 6 of primary education; 45 minutes in junior secondary).

As part of the expansion of educational opportunities at the basic education level and within the initial stage of the nine-year basic education programme, JSS education was developed. However, the number of PS students continuing to JSS has remained low. By intensifying the JSS expansion, it is hoped that with fifteen years, all 13 million of PS graduates will have the opportunity to continue to the JSS. The JSS expansion will be supported by the building of new schools, hiring new teachers, developing more infrastructure and facilities and by the development of the open junior secondary school programme for 13-15-year-old children who are not able to follow the regular JSS.

SECONDARY EDUCATION

Secondary education lasts three years and is available to graduates of basic education. The types of secondary education include: *general secondary education,* which gives priority to expanding knowledge, developing students' skills and preparing them to continue their studies to the higher level of education; *vocational secondary education,* which gives priority to expanding specific occupational skills and emphasizes the preparation of students to enter the world of work and expanding their professional aptitudes; *religious secondary education,* which gives priority to the mastery of religious knowledge; and *service secondary education,* which emphasizes the training of service tasks for civil servants or candidates for civil service.

General secondary education includes general secondary school and Islamic senior secondary school. It is intended to develop the Student's knowledge in accordance with the progress of science, technology and the arts, and enable him/her to continue studies at higher levels of education. It also develops the student's abilities as a member of the community to interact with the social, cultural and natural environment.

General secondary education consists of general and specific teaching programmes. The general education programme is implemented in the first and second grades, while the specific teaching programme starts to be implemented in the third grade. The new curriculum has been implemented in phases, and was extended to all grades in 1996-97. Quality has been improved by introducing a quarter-year academic cycle (instead of the half-year/semester), and a students' streaming division (by discipline) at grade 3 (instead of grade 2). The average number of weekly 45-minute periods of teaching by subject are indicated in Table 17.2 below.

The vocational secondary education programmes are classified into six different groups of vocational fields, namely: Agriculture and Forestry; Technology and Industry; Business and Management; Community Welfare; Tourism, and Arts and Handicraft. Implementation of vocational education is based on the national curriculum. It is adjusted to the local and environmental needs and distinctive features of the vocational education concerned. The curriculum of vocational secondary school consists of general and vocational education programmes.

The quality of vocational education still needs to be improved, its scope expanded and its programmes matched to the employment needs.

Table 17.2: General senior secondary school curriculum

Subjects	Number of weekly periods in each grade				
	General		Specialist		
	Form 1	Form 2	Form 3		
			Language	Science	Social
A. General					
Pancasila education	2	2	2	2	2
Religion	2	2	2	2	2
Indonesian language and literature	5	5	3	3	3
General and national history	2	2	2	2	2
English language	4	4	5	5	5
Sports and health	2	2	(2)	(2)	(2)
Mathematics	6	6	-	-	-
Sciences					
a. Physics	5	5	-	-	-
b. Biology	4	4	-	-	-
c. Chemistry	3	3	-	-	-
Social sciences					
a. Economics	3	3	-	-	-
b. Sociology	-	2	-	-	-
c. Geography	2	2	-	-	-
Arts	2	-	-	-	-
Sub-total	42	42	14 (16)	14 (16)	14 (16)
B. Specialist					
Languages					
Indonesian language and literature	-	-	8	-	-
English language	-	-	6	-	-
Foreign language (s)	-	-	9	-	-
History of culture	-	-	5	-	-
Sciences					
Physics	-	–	-	7	-
Biology	-	-	-	7	-
Chemistry	-	-	-	6	-
Mathematics	-	-	-	8	-

(Contd...)

Table 17.2: Contd...

Subjects	Number of weekly periods in each grade				
	General		Specialist		
	Form 1	Form 2	Form 3		
			Language	Science	Social
Social Sciences					
Economics	-	-	-	-	10
Sociology	-	-	-	-	6
Civics	-	-	-	-	6
Anthropology	-	-	-	-	6
Sub-Total			28	28	28
Total academic hours	42	42	42 (44)	42 (44)	42 (44)

Source: MOEC, 1993; One teaching period lasts 45 minutes.

METHODS AND APPROACHES TO LEARNING

The learning and teaching climate should generate self-confidence, innovative thinking, and should be orientation towards the future. Child-centred, active and co-operative learning is advocated. Teaching is expected to promote higher learning skills, fostering the students' desire and capacity for learning throughout life.

Four types of assessment exist at pre-university level:

1. *Classroom-based continuous assessment* with directions and guidelines provided to teachers on assessment procedures. Assessment may take place after the completion of a small teaching unit at the end of every quarter or semester, or at the end of an academic year.
2. *External assessment* consisting of a school leaving examination at the end of school level.
3. *A survey of student achievement* involving a sample of a student's performance and other relevant variables to be conducted periodically. At present, surveys conducted are not professionally designed and the instruments tend to be of poor quality. A National Assessment Programme is foreseen.
4. *University entrance examination.* These examinations tend to be very difficult due to the interests of top universities to select the best candidates. Many

teachers in senior secondary school focus on preparing students for these exams rather than on the prescribed secondary school curriculum. Students who have a consistently high performance during senior secondary school may be exempted from the examination.

MAIN CHALLENGES FACING CURRICULAR ADAPTATION

Design

- Divergence of opinion with regard to educational philosophy among key stake holders.
- Determining needs for the social, economic, political and cultural environment of the twenty-first century.
- Determining the aims of different levels and types of education.
- Defining minimum basic learning competencies for all level and types of education.

Implementation

- The vast size of Indonesia makes effective country-wide curriculum implementation very difficult.
- The budget for proper piloting of the new curriculum is inadequate.
- The need for comprehensive reform incorporating all aspects of the teaching/learning process: teachers materials and facilities, role of society.

Follow-up

- Socialization of the process in schools.
- Modification following trialling.
- Full-scale implementation following piloting.

18

Malaysia: Curriculum Planning, Development and Reform

Zamrus Bin A. Rahman and Mokelas Bin Ahmad

Estimated population (1995)	20,100,000
Public expenditure on education as percentage of Gross National Product (1995)	5.2
Duration of compulsory education (years)	11
Primary or basic education	
Pupils enrolled (1995)	3,100,000
Teachers (1995)	140,342
Pupils/teacher ratio	19:1
Gross enrolment ratio, (1995)	
—Total	92
—Male	92
—Female	92
Net enrolment ratio (1995)	
—Total	91
—Male	91
—Female	92
Estimated percentage of repeaters (1995)	-
Estimated percentage of drop-outs (1995)	6
School-age population out of school (1995)	280,000
Secondary education	
Students enrolled (1995)	1,640,461
Gross enrolment ratio (1995)	
—Total	62
—Male	58
—Female	66
Third-level enrolment ratio (1995)	10.6

Estimated adult literacy rate (1995)	
—Total	83
—Male	89
—Female	78

Source: *UNESCO Statistical Yearbook, 1998*, Paris.

INTRODUCTION

Legal foundation

The Education Act 1996 (Act 550, Laws of Malaysia) provides the fundamental basis of curriculum policies in Malaysia. It indicates the specific laws and provisions that give direction to curriculum documents. These regulations are mandatory for all schools.

Education Goals and Objectives

The country's educational goals are manifested in the Malaysian National Education Philosophy (NEP) which states that:

> Education in Malaysia is an on-going effort towards further developing the potential of individual in a holistic and integrated manner so as to produce individuals who are intellectually, spiritually, emotionally and physically balanced and harmonious, based on a firm belief in and devotion to God. Such an effort is designed to produce Malaysian citizens who are knowledgeable and competent, who possess high moral standards, and who are responsible and capable of achieving a high level of personal well-being, as well as being able to contribute to the betterment of the family, the society and the nation at large.

The Education System

A uniform system of education in both primary and secondary schools has been established whereby a national curriculum is used in all schools. Common central assessment and examinations at the end of the respective periods of schooling are also being practised. The national language, Malay, is the official language of instruction.

Curriculum Philosophy

The school curriculum is expected to contribute to the *holistic* development of the individual (mental, emotional, physical spiritual) by imparting general knowledge and skills, fostering healthy attitudes and instilling accepted moral values. The aim is to produce Malaysian citizens who are balanced, trained, skilful and cherish and national aspiration for unity.

The general direction for on-going curriculum reform is to improve the quality of education in order to achieve the aims of the National Education Philosophy (NEP). The NEP has been geared towards achieving the nation's vision to prepare children to become knowledgeable, trained and skilled individuals to meet the growing needs of the millennium. It is envisaged that this can be achieved by emphasizing science and technology, use of information technology, and inculcating good moral and work ethics suitable for the Information Age. The school curriculum is designed to achieve the intended learning outcomes for different ability levels.

CURRICULUM DESIGN

The national curriculum promotes unity through the use of a single medium of instruction (the national language) and the provision of the same core subjects all pupils in all schools within the National Education System. However, the cultural diversity of different ethnic groups in Malaysia is preserved through the existence of National Type Schools, which are allowed to use other major ethnic languages as the medium of instruction.

The underlying theoretical principle of national curriculum formulation is that of general education, using an integrated approach in curriculum planning. The curriculum comprises content and skills, with emphasis on the development of basic skills, the acquisition of knowledge and thinking skills. Each subjects must also incorporate the inculcation of moral values and attitudes and the correct use of Malay and other languages, such as English, Chinese and Tamil.

The integrated approach is the main focus in the design of the Integrated Curriculum for Primary School and Integrated Curriculum for Secondary School. The elements of knowledge, skills and values are incorporated so as to bring the integrated development of the intellectual spiritual, emotional and physical aspects of individual.

ORGANIZATIONAL STRUCTURE AND MECHANISMS OF CURRICULUM DEVELOPMENT

Malaysia's system of curriculum development is centralized. The Ministry of Education through its central agency, namely the Curriculum Development Centre (CDC), is responsible for initiating curriculum development. The CDC is responsible for the development of the pre-school, primary school and secondary school curriculum. In the implementation of the curriculum, however, various committees have been set up in the Ministry of Education, State Education Department, Divisions/District Education Offices and schools.

1. ***Educational Planning Committee.*** (Secretariat: Educational Planning and Research Division, Ministry of Education.) Functions:
   - to approve and formulate the major policies in the Ministry of Education;
   - to consider all projects related to policy matters.
2. ***Central Curriculum Committee.*** (Secretariat: Central Curriculum Committee, Ministry of Education.) Functions:
   - to formulate curriculum policies, as well as study their implications;
   - to determine the direction of curriculum development and co-ordinate efforts to achieve this goal;
   - to consider and make recommendations concerning education planning and implementation, as well as to present these findings to the Educational Planning Committee;
   - to study the implications of curriculum programmes and to make decisions accordingly;
   - to determine aspects which require research and study.
3. ***Curriculum Implementation Committee.*** (Secretariat: Central Curriculum Committee, Ministry of Education.) Functions:
   - responsible for co-ordinating all level of preparation in implementing any curriculum programmes approved by the Central Curriculum Committee;

- to ensure co-ordination between the divisions of the Ministry and the State Education Departments in the implementation of the curriculum;
- to gather feedback on curriculum implementation at the State, division/district and school levels;
- to organize forums to discuss successful innovative programmes implemented at the State level.

4. *State Curriculum Committee.* Functions:
   - to monitor, assess and guide principals, headmasters and teachers in the implementation of the curriculum;
   - to gather and analyze information and take the necessary follow-up action;
   - to co-ordinate the implementation of activities between policy makers and implementers at the Ministry, division/district and school levels;
   - to co-ordinate the use of resources for curriculum implementation;
   - to plan, manage and co-ordinate courses organized for teachers by the State Education Department;
   - to plan, manage and implement innovative projects;
   - to improve professionalism among teachers.
5. *Division/District Curriculum Committee.* Functions:
   - to plan activities and curriculum implementation strategies at division/district levels;
   - to arrange and organize courses and seminars to disseminate the latest information to teachers and all personnel concerned;
   - to advise and guide teachers in curriculum implementation;
   - to monitor, supervise and evaluate the implementation of the curriculum;
   - to provide feedback to the higher authorities on the implementation of the curriculum;
   - to plan, manage and implement innovative projects at the division/district levels.
6. *School Curriculum Committee.* Functions:
   - to plan, organize and evaluate teaching/learning activities in schools;

- to work towards increasing the knowledge and competence of teachers and students;
- to study the suitability of the subject content and inform parties concerned;
- to study, evaluate and determine the suitability of textbooks or other materials;
- to assess the scheme of work;
- to assess pupil performance and to identify follow-up action;
- to plan and conduct in-house training programmes for all teachers;
- to disseminate information to all teachers on the latest progress and development in education;
- to produce more education resource materials in schools;
- to identify suitability of educational electronic media programmes in teaching-learning activities;
- to co-operate in the on-going assessment in schools;
- to co-ordinate additional leaning activities.

ASSESSMENT AND EVALUATION

Centralized examinations are held at the end of Year VI, Form III and Form V. The results of these examinations are used to evaluate the effectiveness of the curriculum. Examination results also provide feedback to the planners to improve students' achievement. Based on the analysis of students' scores by items, their achievement in the various skills can be inferred. Should the performance drop below the expected standard, related divisions of the Ministry look into the problem and take prompt action to improve students' future achievement.

The various divisions of the Ministry also carry out related studies to find out the impact of the curriculum on students' achievement. These focus on specific areas of interest, and information is commonly collected using surveys, class observations or test items. Information obtained reflects the effectiveness of the curriculum and the Ministry is required from time to time to take follow-up action, such as reviewing the syllabus, textbooks and other teaching materials, and improving teacher/learning strategies.

CONCLUSION

Careful planning is necessary to ensure the implementation of the curriculum. Thus, the school plays an important role in creating a conducive environment encouraging excellence. In this respect, headmasters and teachers need to understand and internalize the National Education Philosophy, the aims and objectives of the National Education Policy and the integrated approach of the curriculum. Apart from the school and parents, society also plays an important role. The success of the curriculum depends on society's support in assisting the school to develop pupils' personalities and to participate actively in matters relating to education.

19

Maldives: Education Policies, Curriculum Design and Implementation at the Level of Upper Primary and General Secondary Education

Abdul Muhsin Mohamed and Maryam Azra Ahmed

Estimated population (1995)	200,000
Public expenditure on education as percentage of Gross National Product (1995)	6.4
Duration of compulsory education (years)	-
Primary or basic education	
Pupils enrolled (1995)	50,733
Teachers (1995)	-
Pupil/teacher ratio	31:1
Gross enrolment ratio (1995)	
— Total	133
— Male	135
— Female	130
Net enrolment ratio (1995)	
— Total	100
— Male	100
— Female	100
Estimated percentage of repeaters (1992)[1]	18
Estimated percentage of drop-outs (1995)	7
School-age population out of school (1995)	0
Secondary education	
Students enrolled (1995)	40,000
Gross enrolment ratio (1995)	
— Total	49
— Male	49
— Female	49
Third-level enrolment ratio (1995)	—

Estimated adult literacy rate (1995)	
— Total	93
— Male	93
— Female	93

Notes: 1. Last year available.
Source: *UNESCO Statistical Yearbook, 1998,* Paris.

INTRODUCTION

The Republic of Maldives is an archipelago of approximately 1,190 coral islands located south-west of the Indian sub-continent. The islands form twenty-six natural atolls, which are grouped into twenty atolls for administrative purposes. The total population of the Maldives is around 250,000. The population is dispersed among 200 inhabited islands, with 90 per cent of them having a population of less than 1,000. Around 25 per cent of the population is concentrated in the capital, Malé.

HISTORICAL REVIEW OF EDUCATION IN THE MALDIVES

The system of education prevailing in the Maldives today has its roots in a traditional system of schooling that has existed for hundreds of years. These traditional schools, known as *edhuruge, makthab or madhrasa,* are privately owned or operated by the island communities and are usually self-financing. The *edhuruge* is a gathering of children in a private home with the objective of teaching them to read the Quran, to read and write Dhivehi, the mother tongue of Maldivians, and to privide some rudiments of arithmetic. The *edhuruge* is more formal and offers almost the same curriculum, while in the *madhrasa* the curriculum is more far-ranging. These schools have contributed towards achieving many educational objectives, including a high rate of literacy and the preservation of national culture and tradition (Ministry of Education, 1992). However, the present system of education is the result of a merger between the traditional system of schooling and a Western style of schooling introduced since 1960.

The Western style of schooling was introduced in English-medium schools in the capital Malé as part of a conscious effort to prepare individuals for training that they would receive overseas in order to meet the increasing development needs of the country. Thus, the beginning of a public school system was patterned after the British system in terms of organization of curriculum and methods of instruction.

One of the most significant historical developments in education was in 1978 with the decision to unify the national education system. Under this system, schooling in the Maldives was structured on a 5-2-3-2 cycle—five years of primary schooling leading to two years at the middle school level followed the three years of junior secondary school studies and two years of senior secondary school studies. At the end of the 3-year junior secondary cycle and the 2-year junior secondary cycle students sit the London EDEXCEL GCE Ordinary-level and Advanced-level examinations respectively.

THE NATIONAL CURRICULUM

In January 1980, the Ministry of Education (MOE) introduced the first five-year primary curriculum. In 1982, a workshop was held to review this curriculum and to produce another one for the middle-school level. The resulting revisions and re-orientation of the primary curriculum led to the introduction of the first National Curriculum (NC) in 1984.

The NC in Maldives covers the primary and middle school cycles in all subject areas. The secondary curriculum content is designed around the O-level and A-level examinations offered by EDEXCEL. However, in the case of Islamic studies, the Dhivehi language and fisheries science, the curricula are deigned locally—even for the secondary level.

EDUCATIONAL GOALS AND OBJECTIVES FOR MALDIVES

1. The goals[1] of education follow from the national development goals. They are:
   (a) to develop capable individuals with useful occupational skills, knowledge and attitudes for national development with a sense of dignity about labour, and for preserving the nation's environmental resources.

1. Unpublished document of the Ministry of Education, 1997, *Education Sector Master Plan 1997-1998* p. 6.

(b) to promote social justice and equity by ensuring universal primary education and equal educational opportunity for all citizens.

(c) to develop, with an education system based on the principles of Islam, an awareness among all citizens that, as members of the nation, they are also part of the Muslim Ummah.

(d) to promote in individuals a spirit of independence and self-reliance such that they may seek to enhance the quality of life by discovering ways and means of improving their own health, nutrition, and well-being.

(e) to strengthen national consciousness, and to preserve the nation's cultural heritage by promoting desirable cultural values, traditions and the national language.

(f) to provide facilities for lifelong education for all citizens, so that the individual becomes a self-learner and continues to extend his/her intellectual capacity; technical skill and ability to cope with new technologies and discoveries, and develops an appreciation and understanding of changes now occurring in the social and economic life of Maldives.

(g) to develop a sympathetic appreciation of the diversity and interdependence of peoples in the national and international communities.

2. The major objectives[2] of education are as follows:

(a) Expand and strengthen the provision of childhood care and education.

(b) Provide universal basic education (grades 1-7) by the year 2000.

(c) Increase the numbers of trained manpower.

(d) Contribute to the national effort to increase trained manpower.

(e) Improve educational efficiency through quality enhancement.

(f) Improve equity in access and quality of education by gender and location.

(g) Improve curricular relevance to prepare students both for further education and for emerging educational

---

2. Ministry of Education, 1986, *Educational and Human Resource Development Plan 1985-1995*, p. 7.

opportunities and to develop appropriate values and attitudes.

(h) Improve and strengthen the management of the education system.

3. The national philosophy[3] of education includes four main aspects dealing with students' physical and emotional development, cognitive development, social and moral development and skills development. The subjects are developed so as to enhance these various aspects appropriately. Syllabuses are designed by incorporating these elements in an integrated, cohesive and well-defined manner. The teaching material or textbooks are then prepared with the aim of achieving the objectives of the syllabuses.

4. The National Curriculum[4] offers seven subjects namely: mathematics, English, Dhivehi, Islam, environmental studies, practical arts and physical education at the primary level. At middle-school level, environmental studies is replaced by social studies and general science. There are two levels of secondary education comprising a three-year programme of lower secondary education where students are prepared to take the GCE 'O'-Level examinations, and two years of upper secondary, grades 11 and 12, after which students take the GCE 'A'-Level examinations offered by EDEXCEL.

MECHANISMS OF CURRICULUM DEVELOPMENT

The National Curriculum is based on fundamental principles within an Islamic framework. These principles, derived after several stages of consultations, encompass democracy, equity, nationalism, independence, innovation for development and strengthening of the Maldivian society. Based on these fundamental principles, the MOE, in consultation with the National Education Council (NCE), produces national objectives for the education sector (see above). The Educational Development Centre (EDC) is responsible for translating these

3. Educational Development Centre, Ministry of Education, 1984, *Introduction and Guidance for Teachers on the National Curriculum for Primary and Middle Schools*.
4. Ibid.

national objectives into curriculum statements after appropriate consultations. Once the Minister of Education adopts the curriculum statements as policy, EDC draws up the national frameworks for individual subject areas, the syllabi, textbooks, teacher's guides and other relevant resources. Subject panels, consisting of practising teachers and subject specialists from various sectors, including the Department of Public Examinations and the Institute of Teacher Education, help the EDC in the process.

As of 1999, EDC is also responsible for specifying the curriculum materials for the secondary levels. The centre develops teaching materials and resources for Islamic Studies, Dhivehi and Fisheries Science. For the other subjects EDC specifies the materials and resources to be used. Even at the secondary level, subject panels assist EDC.

Table 19.1 shows in detail the interrelationships between the various bodies involved in the adaptation of curricula in the Maldives.

PROBLEMS FACED BY CURRICULUM DEVELOPERS

Curriculum developers at EDC are responsible for the formulation and development of curriculum materials. Each curriculum developers is responsible for a subject area. The curriculum developer has to design the syllabus, decide on the content and prepare textbooks and teacher's guides. In conducting these activities, curriculum developers have to overcome a number of difficulties, such as the following:

At the curriculum design level

- Some curriculum developers need further training, particularly in modern curriculum design techniques.
- There is a particular lack in the region of adequate reference materials, journals and curriculum materials used in other countries.
- There is limited access to the Internet.

At the implementation level

- There is a need for strengthening a process of regular feedback from teachers on implementation.
- There are limited opportunities for curriculum developers to observe classroom teaching.

Table 19.1. The curriculum: who makes which choices?

	Central Level	Regional/Provincial	School Level
	Ministry of Education (MOE) Educational Development Centre (EDC) Department of Public Examinations (DPE) Institute for teacher Education (ITE) Subject Panels (P)	Island Offices (IO) Atoll Education Centres (AEC) Atoll Primary Schools (APS)	Heads Supervisors (SP) Teachers Community Parent/Teacher Associations (PTA)
Aims & Objectives	Sets national aims (EDC MOE) Sets national codes of behaviour (MOE). Trains teachers according to national goals (ITE). Ensures school-based supervision and support (MOE).	Interprets aims to teachers (IO/AEC/APS). Interprets national codes of behaviour for pupils and teachers. Sets local codes of behaviour for Pupils (AEC/APS).	Interprets aims and objectives to Pupils (heads/teachers). Interprets local and national codes of behaviour for teachers and pupils (heads).
Curriculum Plan	Writes national syllabus and allocates appropriate syllabus for secondary level (EDC, P). Decides time allocations (EDC/MOE). Trains teachers for the implementation of the national curriculum (ITE). Ensures achievement of curriculum objectives (MOE)	Teaches according to national syllabus (AEC, APS). Timetable as recommended (AEC, APS) Recommends community participation (IO, AEC, APS)	Makes schemes of work (teachers, SP, heads). Timetable according to recommended time allocations (heads, SP and teachers). Teach according to national syllabus (teachers). Controls co-curricular activities (heads, SP, PTA) Provides assistance to schools (PTA).

Table 19.1. Contd...

	Central Level	Regional/Provincial	School Level
Methods & Approaches to Teaching	Prepare teachers' guides which recommend teaching methodology (EDC, P). Moderate teaching methodology through supervision (MOE) Train teachers in the use of certrain methodology (ITE).	Conduct workshops for teachers on teaching methodology (AEC, APS). Facilitate in conducting field trips and other field work (AO.IO).	Practices recommended methodologies (teachers) Relate methods according to student needs (teachers). Relate teaching to local community (teachers).
Materials	Commissions to write textbooks for the national syllabus (EDC, P) Choose textbooks for secondary schools. Produces or commissions to produce audio-visual materials for the national syllabus (EDC).	Choose educational resources for school use (AEC, APS) Initiate locally relevant resource materials (AEC, APS)	Gives importance to the use of recommended textbooks. (heads, SP, teachers). Procure resource and supplementary materials (heads, PTA) community/parents)
Evaluation and Examination	Set central examinations and expected standards (DPE). Train teachers in assessment and Evaluation (ITE).	Conduct regional workshops for teachers on assessment and evalaution (AEC, APS)	Evaluate and assess all aspects of student achievement (teachers, SP, heads). Sets all internal tests and examinations (teachers, SP, heads). Marks work and keeps records (teachers, SP, heads).

At the evaluation level

- There is a lack of regular interaction with teachers.
- There is a lack of regular feedback from teachers.
- In some instances, teaching is not geared to achieve curriculum objectives.
- In some instances, testing and examiantions are not aligned with curriculum objectives.

A RECENT CURRICULUM REFORM: REVISING THE SOCIAL STUDIES SYLLABI AND TEXTBOOKS

Rationale for Change

As local and global developments influence Maldivian society, there is recognition of the need for the syllabi and teaching materials to reflect these changes. In this respect, many emerging issues need to be incorporated into the social studies syllabus. These include environmental issues (with special focus on the effect of greenhouse gases and the rise in sea-level), overpopulation, the challenges of living in a rapidly developing world, international understanding, tolerance, health and population education, gender prejudice and other social, economic and political issues.

Process of Reform

In 1990, social studies subject-panel meetings were held to discuss the relevance, appropriateness and accuracy of the existing syllabi. Discussions generated the view that most of the topics in social studies needed to be presented in a new perspective since some of the information was outdated. Thus, the textbooks also needed to be changed. Questionnaires were sent out to all schools that taught this subject. These questionnaires were directed at obtaining a wide range of information on the current materials. Feedback from these schools supported the view of the panel members.

Draft syllabi for social studies were prepared under the guidance of subject experts and the subject panel. After discussions with teachers and the panel, the Ministry of Education approved the final syllabi *(see Figure 19.1)* in 1991.

The syllabi were designed to encourage changes in teaching style in order to accomplish the objectives of the National Curriculum. They encourage teachers:

Figure 19.1: Revision of social studies textbooks

One of the reforms recently undertaken in the area of Social Studies can be illustrated the new syllabus designed for use at middle-school level. The new social studies syllabus is designed to try to cope with the rising challenges arising from globalization.

The textbooks based on the previous syllabuses for Grade 6 and 7 social studies were structured as follows:

Grade 6-Social Studies (published in 1986)

Unit 1: The family.
Unit 2: Comparative study
Unit 3: The community.
Unit 4: Internationalism.
Unit 5: Climatic regions.
Unit 6: International links.
Unit 7: Links by commerce to the international world.

Grade 7-Social Studies (published in 1986)

Unit 1: First Maldivians.
Unit 2: Post-conversion.
Unit 3: Independence of Maldives and our neighbours.
Unit 4: The Republic.
Unit 5: Resource development.
Unit 6: Law and order.
Unit 7: International relations.

Component 2: Economic activities and settlement patterns with special reference to Maldives.

Aim of the component: Introducing students to economic geography, human geography and commerce. In introducing these disciplines, students are first required to view the global background and then the Maldivian scene. This component deals with Main Objective No. 5 in the Social Studies syllabus.

Unit 5: Some major economic activities in Maldives
Unit 6: Economic activities in Maldives.
Unit 7: Population, environment and migration in Maldives.
Unit 8: Introduction to commercial activities.

Component 3: Government at home and abroad

Aim of the component:
Understanding modern governments in the global context and viewing the structure and development of government in Maldives.

Unit 9: Modern statecraft.
Unit 10: The structure of Maldivian government.

Component 4: Towards a global outlook

Aim of the component: Understanding the forces that have shaped and are continuing to shape the global outlook of the contemporary world. Global outlook is a major theme for the whole syllabus and in an important feature in the history of humanity.

Unit 11:1. War and Peace in the twentieth century.
Unit 11:2. Interaction and regional organizations.

The textbooks based on the new syllabuses for Grade 6 and 7 social studies are structured as follows:

Grade 6-Social Studies (published in 1993)

Component 1-Earth the living planet.

Unit 1: The universe, the solar system and planet earth.
Unit 2: The atmosphere, lithosphere and hydrosphere.
Unit 3: Weather and climate.
Unit 4: Physical and political geography of the world.

Component 2-Humanity

Unit 5: Astronomy, geography and navigation in the Maldives.

(Contd...)

Unit 6: Origins and development of human cultures. Unit 7: The world civilisations. Unit 8: Western civilization and the industrial revolution. Unit 9: The space age. *Component 3-Human institutions* Unit 10: Society and culture. Unit 11: Languages and scripts Unit 12: The religions of humanity. **Grade 7-Social Studies (published in 1994)** *Component 1: History of Maldives in the perspectives of global development.* *Aim of the component:* Viewing the	history of Maldives against the background of global developments. Teachers should encourage students to always view the history of Maldives against the global background and not as an isolated entity. Unit 1: The South-Asian heritage of early Maldives. Unit 2: Islam and the history of Maldives. Unit 3: Maldives and the advent of colonialism in South Asia. Unit 4: Nationalism in Asia and the Maldives in the Twentieth Century.

- to develop metacognitive skills and understanding;
- to be problem posers and guides rather than problem solvers;
- to present the materials in everyday contexts;
- to encourage wider involvement of community, parents, etc. in the learning process;
- to enhance group work; and
- to encourage process-oriented teaching.

Textbooks and teacher's guides were prepared and pre-tested with the help of practising teachers according to the requirements of the new syllabi. After the trials, the text-books and teacher's guides were published.

The new syllabi and the accompanying textbooks for social studies brought very positive feedback from schools. It was claimed that students' interest in the subject had increased and that the presentation and layout of the new textbooks stimulated interest, generating lively classroom discussions. Teachers also indicated that the curriculum introduced and enhanced research skills in students, fostering their motivation to learn. It also introduced critical thinking skills, making students more active participants in class. However, some schools noted that the level of English used in the texts was higher than in other subjects, which posed some problems for students.

CONCLUSION

The Maldives islands are widely dispersed and, as access to some of them is not very frequent, curriculum developers do not receive regular feedback from schools. At the implementation stage, curriculum developers need contact with teachers to get first-hand information about the syllabuses or teaching material. To gauge the effectiveness of these materials, systematic evaluation is also of utmost importance. Measures need to be taken to ensure that teachers are informed of the evaluation results so that steps can be taken to remedy the situation.

One of the major constraints in curriculum development is the need for further streamlining of the curriculum development process. In the new framework, EDC's role will shift more towards the management of curriculum development, while increasing the involvement of experienced teachers in the process. As teachers are the deliverers of the curriculum at classroom level, effective implementation largely depends on them.

Several reforms have already been introduced into the content and teaching methodologies of the national curriculum that came into effect in 1984. However, this curriculum now needs a major revision to adjust and strengthen it to enable our citizens to face the challenges of the twenty-first century. A major nation-wide curriculum review activity is planned for April 1999.

20

Myanmar: National Aspects of Curriculum Decision-making

U Myint Aye and Daw Tin Kyi

Estimated population (1995)	45,100,000
Public expenditure on education as percentage of Gross National Product (1995)	1.3
Duration of compulsory education (years)	5
Primary or basic education	
Pupils enrolled (1995)	5,530,502
Teachers (1995)	119,942
Pupil/teacher ratio	46:1
Gross enrolment ratio (1995)	
—Total	100
—Male	102
—Female	99
Net enrolment ratio (1995)	
—Total	85
—Male	85
—Female	85
Estimated percentage of repeaters (1995)	-
Estimated percentage of drop-outs (1995)	-
School-age population out of school (1995)	80,000
Secondary education	
Students enrolled(1995)	1,800,000
Gross enrolment ratio (1995)	
—Total	30
—Male	30
—Female	29
Third-level enrolment ratio (1995)	5.2

Estimated adult literacy rate (1995)	
—Total	83
—Male	89
—Female	78

Source: *UNESCO Statistical Yearbook, 1998,* Paris.

The Union of Myanmar is currently in the process of transforming its political, social, economic and administrative systems. In this context, the education system is being transformed in order to meet changing societal needs.

EDUCATIONAL OBJECTIVES

In order to implement an education system that is compatible with the cultural, traditional, and social values of the country, and in keeping with an economic system that will facilitate national development and nation-building, the Government of Myanmar has identified the following main objectives:

- To enable every individual to acquire basic education;
- To base education on raising moral standards;
- To develop knowledge, including the scientific and technical know-how needed for nation-building;
- To produce technicians, skilled workers and proficient intellectuals with practical knowledge, who are loyal to the State and will contribute to nation-building endeavours;
- To train citizens so that they will achieve all-around development;
- To allow those with the requisite intellectual ability, calibre and industriousness to acquire a university education;
- To offer undergraduate and post-graduate courses for those who are working and thereby enable them to study during employment.

THE MYANMAR SCHOOL SYSTEM

The structure of the country's formal school system is 5+4+2, comprising a total of eleven year: five years at the primary level; four at the secondary (middle) level; and two years at the upper secondary (high) level. All schools in Myanmar are State schools and follow the same centrally prescribed curriculum. The admission age is 5 years. Tables 20.1

to 20.3 provide a current statistical overview of the education system.

ORGANIZATIONAL STRUCTURE OF CURRICULUM DEVELOPMENT

Currently, there are nine departments under the Ministry of Education (MOE). In 1998, in order to manage the education system more effectively, the Department of Basic Education (DBE) was expanded into three departments.

Table 20.1: Department of Basic Education, number of schools by type (1998/99)

Department	High	Middle	Primary
DBE (1)	350	853	15,408
DBE (2)	429	1,017	18,207
DBE (3)	157	237	2,214
Total	936	2,107	35,829

Table 20.2: Department of Basic Education, number of teachers by level (1998/99)

Department	High	Middle	Primary
DBE (1)	5,518	31,911	54,013
DBE (2)	6,722	42,915	55,595
DBE (3)	3,368	14,955	11,944
Total	15,608	89,781	121,552

Table 20.3: Department of Basic Education, number of students by level (1998/99)

Department	High	Middle	Primary
DBE (1)	161,375	528,098	1,836,056
DBE (2)	223,542	703,763	2,329,004
DBE (3)	120,348	287,629	521,954
Total	505,265	1,519,490	4,687,014

The Department of Educational Planning and Training (DEPT) has the main responsibility for the administration and management of curriculum and textbooks, teacher education and special projects. The curriculum section under the DEPT is responsible for the organization of curriculum development at the basic education level. A deputy director and assistant director head the curriculum section. The science and the arts curriculum section each have one staff officer and one deputy staff officer. The curriculum section is also responsible for co-operating on and facilitating the task of printing and distribution of textbooks and stationery for students of all levels.

With the enactment of the 1966 Union of Burma Basic Education Law, the Basic Education Council (BEC), chaired by the Minister of Education, came into existence. A basic education curriculum syllabus and textbook committee has been established at the national level under the BEC. This committee is wholly responsible for curriculum development at all levels of basic education. The functions of this national-level curriculum committee include: drawing-up, scrutinizing and revising curricula and syllabi; compiling and writing textbooks; preparing teaching aids; recommending types of performance assessments.

Subject area curriculum committees, headed by appropriate specialist professors, work under the direction of the national curriculum committee. Members of the subject committees include representatives from the Myanmar Education Research Bureau, Institute of Education, teacher education colleges and schools, subject experts and selected teachers. The DEPT deputy staff officers serve as secretaries for these subject curriculum committees which are involved in: Writing and compiling textbooks; preparing prototype teaching aids and materials and teacher's manuals; designing test formats for performance assessment; and revising the curriculum content in conformity with policy changes. Other functions of these committees involve: conducting in-service training for all levels, whenever necessary; responding to inquiries about curriculum and textbook matters; script-writing for educational radio and television.

An overview of curriculum responsibilities—including those at the regional/provisional and school levels—is provided in the chart entitled *The curriculum: who makes which choices* (see Table 20.4).

PRIMARY CURRICULUM REFORM

The main reasons for primary curriculum reform are to develop and strengthen basic language skills and mathematical skills; foster good citizenship; promote social justice in all

Table 20.4: The curriculum: who makes which choices?

Central level	Regional/provisional level	School level
Ministry Basic Education Curriculum Syllabus and Textbook Committee	State/Division Education Officers Inspectors Education colleges	Heads Teachers Communities
Aims and objectives	**Aims and objectives**	**Aims and objectives**
The Myanmar Government has set the following education objectives: (a) to enable every individual to acquire basic education; (b) to base education on the raising of moral standards; (c) to develop knowledge, including the scientific and technical know-how needed for nation-building; (d) to produce technicians, skilled workers and proficient intellectuals with practical knowledge, loyal to the State and who will contribute to nation-building; (e) to train the citizens so that they will achieve all-around development; (f) to permit access to university education to those who possess the requisite intellectual ability, calibre and industriousness; (g) to offer undergraduate and post-graduate courses for those who are working, thereby enabling them to study during employment.	The main objectives of basic education in Myanmar are as follows: – to enable every citizen of the Union of Myanmar to become a well-equipped physical and intellectual worker; with basic education, good health and moral character, – to lay the foundations for vocational education for the benefit of the Union of Myanmar; – to give priority to the teaching of science that is capable of strengthening and developing productive forces; – to give priority to the teaching of arts that is capable of preserving and developing the culture, fine arts and literature of the State; – to lay a firm and sound educational foundation for the further pursuance of university education.	Heads, teachers and pupils will join hand-in-hand to bring about a change of behaviour in schools relating to life skills, morals and civics.

Table 20.5: The curriculum plan for primary and secondary education

Curriculum plan		Curriculum plan	Curriculum plan
Lower Primary-Level Course of Studies		Course contents and syllabi will vary according to the needs of society. Local timetables will be adapted to environmental studies. Teachers, community participants and members of the board of trustees will be involved in the curriculum plan.	The scheme of work will be drawn up by using the bar graph of quarterly course course contents. The syllabus at the school level will be emphasized so that the changes changes in pupil behaviour can be optimized. Aesthetic and physical education will form a regular part of the timetable.
Subjects	**Periods per week**		
Myanmar langage	11		
English	4		
Mathematics	7		
General studies	9		
Aesthetic education	3		
Physical education	4		
School activities	2		
Total	40		

Note: Each period lasts thirty minutes

(Contd…)

Table 5: Contd...

Curriculum plan		Curriculum plan		Curriculum plan	
Upper Primary-Level Course of Studies		**Lower Secondary-Level Course of Studies**		**Upper Secondary-Level Course of Studies**	
Subjects	**Periods per week**	**Subjects**	**Periods per week**	**Subjects**	**Periods per week**
Myanmar language	8	Myanmar langage	5	Myanmar language	5
English	5	English	6	English	5
Mathematics	7	Mathematics	8	Mathematics	5
Basic science	4	General science	4	Co-ordinated science	9
Social studies	8	History	4	Social studies	6
Aesthetic education	3	Moral education	1	Moral education	1
Physical education	4	Physical education	1	Physical education	2
School activities	1	Arts education and co-curricular activities	2	Arts education and co-curricular activities	2
Total	**40**	**Total**	35	**Total**	35
Note: each period lasts thirty-five minutes.		**Note:** each period lasts fourty-five minutes.		**Note:** each period lasts forty-five minutes.	

Table 20.6: Guiding Principles

Methods and Approaches to Learning	Methods and Approaches to Learning	Methods and Approaches To Learning
Learning activities will be focused on concepts and facts rather than on memorization. By using the activity-based learning materials, *learning to learn* will be emphasized in the teaching learning process.	Teachers will be trained to facilitate pupil learning of concepts by using the learning materials. Supervisory Committee will be organized for monitoring the effectiveness of the teaching-learning process.	Different methods—namely role playing, field trips, demonstration and discovery methods—will be used according to the nature of the subjects. The teaching of the course content in environmental or nature study, morals, civics and life skills will be related to local community life.
Materials	**Materials**	**Materials**
The new textbooks(students' texts and teacher's manuals) will be printed continuously.	Students will be motivated to collect learning materials and also to make them. The new subject-relevant materials will be programmed by subject. Reference books will be provided for local schools.	Some pupils buy and use the prescribed textbooks. Teachers *only* use the teachers' manuals. Parents are not urged to buy the textbooks. Learning materials from the environment will be used in the classrooms.

(Contd...)

Table 20.6: Contd...

Evaluation and Examination	Evaluation and Examination	Evaluation and Examination
Regular assessment takes place at Kindergarten and Standard One. From Standard Two to Standard Six, there is an end-of-chapter test. From standards Seven to Nine, for the first semester, there is only an end-of-chapter test; the year-end examination will be held during the second semester. There is also an end-of-chapter test in Standard Ten.	The testing techniques will be distributed in state and division levels. Teacher training will be conducted in testing techniques—especially for end of chapter tests. There are no local examination.	The affective growth of students will be planned so as to determine whether behaviour change have occurred. The end-of-chapter test will be focused on the primary and general secondary level. The year-end examination will be held for Standards Seven, Eight and Nine. The marks obtained by students in academic subjects and in school activities are recorded in a Comprehensive Personal Record.
After finishing the upper secondary-level course, the students need to sit for the matriculation examination. The students who obtain high marks in the matriculation examination can sit the entrance examination to join the professional institutes. In order to enter the professional institutes, marks obtained from the entrance examination, including marks of their school activities, will be considered.		

communities; and develop life skills for healthy-living. To this end, the primary level curriculum has been designed to assure coverage of three key domains: ***cognitive, affective and psychomotor***. Subjects taught at this level include: the Myanmar and English languages, mathematics, basic science, general studies (which includes moral education and civics, life skills and nature study), and social studies (which includes geography, history, moral education, civics and life skills).

To date, new primary-level textbooks and teacher's manuals have been designed, with training workshops for the teachers to be conducted in May 1999. Schools in Yangon and Mandalay have been selected to field test the new teaching/ learning processes developed for general studies, social studies and basic science, using multi-media classroom settings.

Problems Facing Curriculum Reforms

To date, there have been no major problems in the adaptation of curricula (design, follow-up of the reforms) or in developing students' textbooks and teacher's manuals. However, some weaknesses have emerged in the utilization of teaching/ learning materials in the classroom.

21

Nepal: Education Policies, Curriculum Design and Implementation at the General Secondary Level

Estimated population (1995)	21,500,000
Public expenditure on education as a percentage of Gross National Product (1995)	2.9
Duration of compulsory education (years)	5
Primary or basic education	
Pupils enrolled (1995)	2,800,000
Teachers (1994)[1]	81,544
Pupil/teacher ratio	39:1
Gross enrolment ratio (1995)	
—Total	110
—Male	129
—Female	89
Net enrolment ratio (1995)	
—Total	63
—Male	80
—Female	46
Estimated percentage of repeaters (1992)[1]	27
Estimated percentage of drop-outs (1995)	48
School-age population out of school (1995)	1,000,000
Secondary education	
Students enrolled (1995)	91,000
Gross enrolment ratio (1995)	
—Total	37
—Male	49
—Female	25
Third-level enrolment ratio (1995)	5.2

Estimated adult literacy rate (1995)	
—Total	28
—Male	41
—Female	14

Note:
1. Last year available
Source: *UNESCO Statistical Yearbook, 1998*, Paris

A number of high-level government documents guide the development of curricula. They include the Secondary Education Perspective Plan (1996-2011), the Secondary Education Action Plan (1997-2002) and a recent High-Level Education Commission Report (1998).

ORGANISATIONAL STRUCTURE AND MECHANISMS IN CURRICULUM REFORM

The curriculum planned at the central level is modified by all the processes leading to its adoption. In the classroom at the secondary level most institutionalized activity is centrally controlled.

Ministry of Education

The task of initiating educational activities throughout Nepal lies with the Ministry of Education (MOE). The Ministry is responsible for educational planning and management, as well as in improving service delivery systems across the country. MOE is composed of three divisions: Planning; General Administration; and Educational Administration. Educational programmes and services are prepared by: the Curriculum Development Centre; the Secondary Education Development Centre; the Distance Education Centre; the Office of the Controller of Examinations; the National Centre for Educational Development; the regional education directorates and district education offices. From July 1999 a new structure will be implemented with the formation of a Department of Secondary, Education, containing a Primary and Basic Education Division and a Secondary and Higher Secondary Division. This will separate policy-making from Executive actions.

The National Curriculum Council

A high-level National Curriculum Council (NCC), chaired by the Minister of Education, approves all curricula and guides the detailed developmental work of the Curriculum Development Centre (CDC) by setting operational and administrative policy.

Technical Committees

The NCC forms ad-hoc technical committees when additional advice is required. Matters concerning the relevance of curricula drafted by CDC may be referred to such a technical committee if the NCC feels additional advice is necessary.

Curriculum Development Centre

The CDC is responsible for the maintenance, transmission and renewal of the school-level curriculum and is also concerned with pre-primary education. (See below for further discussion of CDC).

The Secondary Leaving Certificate Board

The Secondary Leaving Certificate Board sets policies and makes decisions relating to the School-Leaving Certificate (SLC), which are then implemented by the Office of the Controller of Examinations (OCE). A reformed SLC, with single subject certification is planned for July 2001. The courses leading to this examination are to be implemented from grade 9, beginning in July 1999.

The Secondary Education Development Centre

The Secondary Education Development Centre (SEDEC) is responsible for a range of in-service training activities at the secondary level. SEDEC operates through twenty-five secondary education development units at locations, which allow for national coverage. The training activities also support the work of the Curriculum Development Centre and the OCE.

Janak Educational Materials Centre

The Janak Educational Materials Centre (JEMC), operating as a public limited company, produces and distributes school textbooks throughout Nepal. JEMC's Board of Directors is comprised representatives from concerned ministries and organizations.

THE CURRICULUM DEVELOPMENT CENTRE

Responsibilities and Activities

The CDC (see Figure 21.1) is responsible for the maintenance, transmission and renewal of the school education curriculum. The wide-ranging activities of the centre include developing, revising and disseminating textbooks and teacher's support materials. A programme of seminars and workshops supports these activities. CDC's development and monitoring work is carried out by specialized curriculum subject units, advised by curriculum subject specialist committees. Subject units cover languages, science and maths, social studies, health and physical education. To support CDC's activities, various studies and surveys are conducted on curriculum-related issues and problems. The activities of CDC give rise to a wide range of relationships with other institutions. The most important of these is with the teachers and student in schools, who are the immediate end-users of the centre's products. CDC also incorporates a publishing unit.

The Publishing Unit (PU)

The PU (in collaboration with subject specialist units, their advisory subject specialist committees, subject advisers and consultants) sets textbook specifications. According to prescribed procedures, the PU also selects textbook writers on the basis of: subject knowledge, classroom experience and sample materials reviewed by teachers/subject experts. A PU staff member serves as the managing editor, and is responsible for briefing contracted writers and liaising closely with them to ensure quality and schedule controls, until the camera-ready copy goes to the printers. The PU manager also arranges distribution of the draft materials to be tested in schools and validated by teacher groups. The managing editor, unit specialist, subject committee members, advisers and specialists all visit schools to collect comments which are subsequently relayed to the writers.

Subject Advisory Committees

The role of these committees is to advise on the preparation and revision of the curriculum, and the preparation, revision and evaluation of textbooks and teachers guides. The Curriculum Officers, as members of the committees, also participate in these

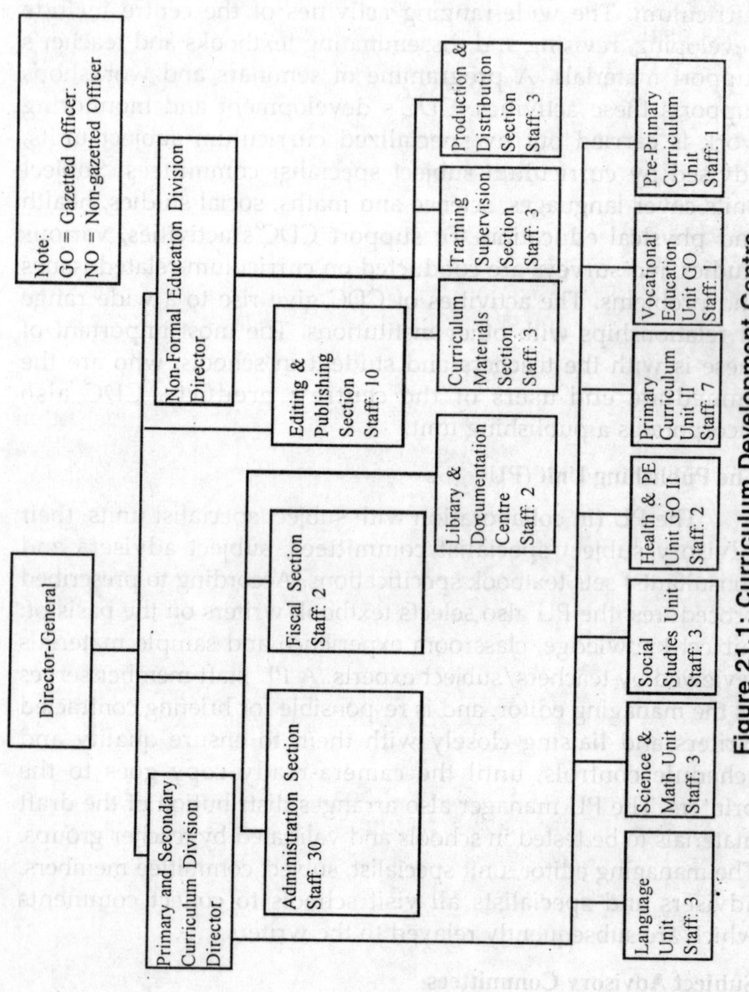

Figure 21.1: Curriculum Development Centre.

activities and provide the secretariat that promotes the committee's work. Sub-committees are established to carry out specialist development inputs. The subject committee cover Nepali, English, science, mathematics, social studies, and health, population and environment (HPE).

Task Committees

Like the publishing unit, the task committees cut across the boundaries of the various specialist units. Examples are the formative assessment committee and the dissemination committee. Task committees are formed to advise the cross-boundary projects managing officer on policy and operational matters.

DEVELOPMENTS IN THE DESIGN, IMPLEMENTATION AND FOLLOW-UP OF CURRICULUM REFORM

Aims of Secondary Education

The general aims of secondary education (framed within the national goals of education, see Figure 21.2) are to: produce healthy citizens who are: familiar with the national tradition, culture, social environment, democratic values; able to use language effectively in daily life; aware of scientific issues;

Figure 21.2: National goals of education.

The National Goals of Education frame all the detailed technical work of curriculum development and should be reflected in all documents and materials developed by the Curriculum Development Centre. The national education goals are to:

- nurture and develop the personalities and inherent talents in each person;
- instill respect for human values and the will to safeguard national and social beliefs so as to help develop a healthy social unity;
- help the individual to socialize, enhancing social unity;
- help the individual keep his or her identity in the national and international context and to help him/her lead a socially harmonious life in the modern world;
- assist the modernization of the country by creating able manpower for its development;
- teach the thoughtful protection and wise use of Nepal's natural resources;
- bring those who are underprivileged into the mainstream of the nation.

creative, co-operative, industrious; able to contribute to economic development.

The Secondary Curriculum

The lower secondary curriculum (grades 6, 7 and 8) and the secondary curriculum (grades 9 and 10) are constructed with *core* subjects and *optional* subjects. The five lower secondary core subjects are Nepali, English, math, science and social studies. The six secondary core subjects are Nepali, English, math, science, social studies, health, population and environment (HPE).

Table 21.1: Secondary school curriculum structure-general secondary school level

Subjects	Classes 9 and 10	
	Weightage	Full Mark
1. Nepali	5	100
2. Mathematics	5	100
3. English	5	100
4. Science	5	100
5. Social studies	5	100
6. Health, Population and Environment Education	4	100
7. Optional Paper I	5	100
8. Optional Paper II	5	100
Total	39	800

A. Option 1 paper subjects (any one)

Languages: Nepali, Arabic, Avadhi, Bhojpuri, Bengali, Chinese, Hindi, Japanese, Maithili, Sanskrit, Tibetan, Persian, Hebrew, Urdit, English, French, German, Greek, Latin, Russian, Spanish. (Other national languages of Nepal will be included in the curriculum, provided that grammar books, teaching materials, etc., are available.)

Humanities, Social Science: history, geography, civics, economics, sociology.

Optional mathematics

B. Option 2 paper subjects (any one)

Interdisciplinary: agricultural education, food science, architectural education, industrial education, office management and accounting, auditing, typing and shorthand, computer science, home science, handicrafts painting, sewing and knitting, bamboo-work, dance, music, Ayurveda naturopathy, health and physical education, Yoga education, photography, journalism, instrumentation.

Secondary Education Development Project

The Secondary Education Development Project (SEDP) began in 1993. SEDP originally aimed at improving and reinforcing three subjects (English, science, mathematics). In 1997, support to Nepali and social studies was added. SEDP also has the goal of reforming the examination system and providing materials and equipment to selected secondary schools. The project has been providing support services through the Secondary Education Development Centre and the twenty-five training centres.

Accomplishments

As of July 1998, CDC had produced ten curriculum booklets outlining the lower secondary curriculum (grades 6–8) and the secondary curriculum (grades 9 and 10) and covering all core subjects therein. The curriculum and textbooks for grades 6 (1996), 7 (1997) and 8 have been printed and distributed (1998). The lower secondary curriculum is being prepared for publication. However, English is the only subject in which teacher's guides have so far been produced.

In July 1998, the National Curriculum Council (NCC) decided to restructure the curriculum of grades 9 and 10, to allow for the inclusion of a new sixth core subject, health, population and environment (HPE). This reduced English, Nepali, mathematics and science from six to five periods each week. The curriculum was adjusted in line with the reduction of periods. The grade 9 production schedule for textbooks and teacher's guides has now been separated from that of grade 10. The grade 9 textbooks for the six core subjects were scheduled for printing by the end of March 1999, while completion of grade 9 teacher's guides was foreseen for the end of April 1999.

CDC has formulated a seven-stage dissemination strategy for the grades 9 and 10 curriculum. The first stage (*approval*) was completed in November 1998. Included in the strategy are:

Figure 21.3: Social studies curriculum

Objectives. There are twenty-eight general objectives listed in the grades 9 and 10 social studies curricula. Highly specific objectives are listed for grade 9 (twenty-seven objectives) and grade 10 (twenty-nine objectives).

Abilities. On achieving the course objectives, students will be able to demonstrate a wide range of abilities. Some examples of the abilities associated with knowledge and understanding and practical abilities are as follows:

Ability 1: Knowledge and understanding.

1. Describing projects in different zones and development regions;
2. Describing the achievements of Nepal in the fields of education, health, transportation, telecommunications, electricity and water supply;
3. Discussing the role of skilled manpower in the development of the infrasctrure of Nepal;
4. Stating the role of international organizations in solving social problems;
5. Explaining different climates and the elements that affect them;
6. Giving a short introduction to the geo-economic activities of continents;
7. Identifying the problems created by population growth and migration and finding out ways to participate in solving them;
8. Describing the physical features of the Earth;
9. Examining the impact of landslides on the physical features of Nepal and participating in the task of avoiding them;
10. Listing the agricultural products grown in the various geographical areas.

Ability 2: Practical abilities.

1. Maintaining national dignity;
2. Drawing a map of Nepal and filling in the main industrial centres and roads;
3. Demonstrating models in class;
4. Analyzing the data presented in charts;
5. Locating historical sites on a map;
6. Investigating the problems of population growth;
7. Appreciating the contributions of our national heroes;
8. Applying the knowledge of social studies for the good of society;
9. Drawing a picture of the Himalayan region;
10. Recording the maximum rainfall of the month;
11. Designing a research proposal to find out about social evils.

Criteria. The new-style social studies textbook must be developed so that students can be thoroughly prepared to take the new SLC examination. The textbook must help them not only to understand the content of the course, but also to develop the necessary skills and ability to think critically about the different topics. Thus, each chapter should have clearly thought-out learning outcomes. These will include not only factual information for knowledge and understanding, but will also include specific skills such as those outlined in the previous section. After studying this lesson the student will be able to:

- interpret a bar chart;
- locate places of historical importance;
- present an argument in favour of a certain course of action;
- collect local data.

The earlier textbooks contained only factual information, followed by recall questions. To prepare for the new SLC examination, the new textbooks must present information in different forms (text, statistics, pictures, photographs, diagrams, graphs and charts, maps, newspaper items, source documents) and provide challenging questions to help students understand, interpret, analyze and evaluate the information.

As 30% of the new SLC marks will be given for project-style work to be carried out during the year, the textbook must include many such activities; starting with simple tasks, and evolving to longer projects. This will mean that most of the material studied has a local component and thus is of relevance to the students. It will also mean that students can undertake genuine research, and gain practical skills in investigating, analyzing and presentating their findings. It will also help them to work together co-operatively in groups. Such tasks are illustrated below.

Examples

- draw up a plan of your VDC, mark the water;
- prepare a poster or a talk to stop young people from experimenting with drugs;
- survey the type and number of animals kept in the local community;
- write and perform a short historical drama;
- keep a record of local weather over a long period and prepare a suitable display, such as a bar chart or graph.

The students should be able to see what skills they are learning in each chapter and gain a sense of achievement and progress towards the SLC. This may be done with a summary of chapters or topics. The exercises given in the textbook will not only test knowledge and understanding, but will require analysis and evaluation of the information and well thought-out answers. They will also need to encourage the practice of practical skills and abilities. This will provide practice for new-type questions on the SLC exam.

detailed planning, package development, training of master facilitators and facilitators, training of head-teachers, cluster-based and school-based workshops. The full programme will be completed by the end of July 1999.

The Publishing Unit's first textbooks are clearly of a better quality than previous CDC/SEDP outputs. The unit has also produced a four-page CDC bulletin of a high standard (February 1998) to publicize its activities, as well as a leaflet in English and Nepali. Textbooks have been monitored for gender, socio-economic and regional equity.

DEVELOPMENT AND CHALLENGES IN THE DESIGN, IMPLEMENTATION AND FOLLOW-UP OF CURRICULUM REFORM

While the capacity of CDC to handle complex tasks is clearly improving, the capacity to plan and to develop human resources has certain constraints due to the rate of staff changeovers. This makes the institutionalization of curriculum development design processes very difficult. These staff changes are at all professional and administrative levels of the organization.

The same changes create problems of continuity and liaison with other organizations connected with the implementation process. These institutions—District Education Offices, Regional Education Offices, OCE and SEDEC—also have high staff instability.

There are very few formalized follow-up activities, except by the SEDP BME Unit. These are mainly concerned with project evaluation.

REFORMING THE CURRICULUM WITH SPECIAL EMPHASIS ON SOCIAL STUDIES

As described earlier all core curricula and many optional subjects have been changed in recent years. In the development of all core curricula there has been tension between traditionalists and modernists, as well as between nationalists and those in favour of globalization or wider horizons—those who speak for 'our values' and those who speak of 'global values'. Balancing these diverse views is a problem for all curriculum developers. No curriculum subject can avoid these controversies. While in science and mathematics the debate is narrower, in English and Nepali there are different dimensions due to the source of each individual language. Both of the two newer subjects—social studies and HPE—have proved to be controversial in different ways (see Figure 21.3).

Translating the goals, general objectives and specific objectives of the social studies curriculum into a useful textbook proved difficult. Although there are many dimensions to the discussion, the fundamental debate concerns whether education is about passing knowledge or learning skills—or what is the balance between the two. Members of the subject committee did not always agree with those who set the framework for developing the new textbook, creating what is hopefully a creative tension.

Those setting the framework for developing the new textbook felt that the following factors are all important in the writing of a new style textbook:

- Each chapter should have clearly thought-out learning outcomes. These will include not only factual information for knowledge and understanding, but specific skills such as interpreting a bar chart, locating places of historical importance, presenting an argument in favour of a certain course of action and collecting local data.
- The earlier textbooks mostly contained only factual information, followed by recall questions. The new textbook must present information in different forms (text, statistics, pictures, photographs, diagrams, graphs and charts, maps, newspaper items, source documents) and provide challenging questions to help students understand, interpret, analyze and evaluate the information.
- The textbook must include many project-related activities, starting the simple tasks, and moving towards longer projects. This will mean that most of the work studied has a local component and thus is of genuine relevance to the students. It will also mean that students can experience research activities, gaining practical investigation and analytical skills, and learning to present their findings. It will also help them to work together co-operatively.
- The students should be able to see what skills they are learning in each chapter and gain a sense of achievement and progress towards the school-leaving certificate (SLC). This may be done with a summary of chapters or topics. The exercises given in the textbook will not only test knowledge and understanding, but will expect analysis and evaluation of the information, and well-thought-out answers. They will also need to encourage the practice of practical skills and abilities. This will provide practice for new SLC-type questions.

Pakistan: Curriculum Design and Development

Estimated population (1995)	136,260,000
Public expenditure on education as percentage of Gross National Product (1995)	2.8
Duration of compulsory education (years)	5
Primary or basic education	
Pupils enrolled (1995)	18,400,000
Teachers (1995)[1]	-
Pupil/teacher ratio	38:1
Gross enrolment ratio (1995)	
—Total	74
—Male	101
—Female	45
Net enrolment ratio (1995)	
—Total	31
—Male	36
—Female	25
Estimated percentage of repeaters (1995)	-
Estimated percentage of drop-outs (1995)	52
School-age population out of school (thousands) 1995	12,700,000
Secondary education	
Students enrolled (1995)	5,300,000
Gross enrolment ratio (1995)	
—Total	21
—Male	28
—Female	13
Third-level enrolment ratio (1991)[1]	3.0

Estimated adult literacy rate (1995)	
—Total	38
—Male	50
—Female	24

Note: 1. Last year available
Source: *UNESCO Statistical Yearbook, 1998*, Paris.

INTRODUCTION

Pakistan—governed under the Islamic, democratic, federal Constitution of 1973—is comprised four autonomous provinces: Punjab, Sindh, North-West Frontier and Balochistan. Education in Pakistan is essentially a provincial affair. However, education is considered to be a vital source of nation-building. Therefore, in order to ensure national cohesion, integration and preservation of the ideological foundation of the State, certain educational functions are the responsibilities of the Federation—via the Federal Ministry of Education. These responsibilities include: curriculum, syllabus, planning, policy and educational standards. The 1976 Act of Parliament authorized the Ministry of Education (MOE) to appoint competent authorities to perform the following curriculum-related functions:

- In connection with the implementation of the education policy of the Federal Government of Pakistan (GOP), prepare or commission; schemes for studies; curricula, textbook manuscripts and strategic schedules for their introduction in various classes of educational institutions;
- Approve manuscripts of textbooks produced by other agencies, before they are prescribed in various classes of an educational institution;
- Direct any person or agency in writing (within a specified period) to delete, amend or withdraw any portion, or the whole, of a curriculum, textbook or reference material prescribed for any class of an educational institution.

Accordingly, a Central/National Bureau of Curriculum and Textbooks (NBCT, commonly known as the *Curriculum Wing*) was appointed to supervise curriculum and textbooks development/approval and to maintain curriculum standards from the primary through to the higher secondary levels. As a logical sequence to this action, four counterpart provincial curriculum centres (one in each province) were established to ensure provincial collaboration and evolve consensus in all activities falling within the purview of the Federation. This initiative was followed by the establishment of four Provincial Textbook Boards (PTTB)—one in each province. Within their respective jurisdictions, these PTTBs are responsible for preparing, publishing, stocking, distributing and marketing school textbooks.

Boards of Intermediate and Secondary Education responsible for conduct of examinations at Secondary (IX-X) and Higher Secondary (XI-XII) levels were also established at each of the divisional headquarters. Subsequently, another institution was established—the Inter-Board Committee of Chairmen (IBCC)—with the following objectives:

- To exchange information among the member boards on all aspects of secondary and higher secondary education;
- To achieve a fair measure of uniformity in academic evaluation standards;
- To promote inter-board curricular and extra-curricular activities;
- To serve as a board chief executives' discussion and consultation forum for all matters relating to secondary and higher secondary education development, and making suitable recommendations to the GOP;
- Advise on and facilitate the exchange of teachers and students;
- Perform such other functions as may be incidental or conducive to the attainment of the above objectives.

THE CURRICULUM DEVELOPMENT PROCESS

Reform

It may be noted that the process of curriculum reform in Pakistan has been introduced as part of the successive series of national education policies (Table 22.1).

Table 22.1: Education policy and curriculum reform cycle

National Education Policy	Curriculum reform cycle
1972	1st Cycle, 1973-76
1979	2nd Cycle, 1982-85
1992	3rd Cycle, 1992-95
1998	4th Cycle, 1998-01

The following six-phase strategy has been adopted for implementing curriculum change:

1. Evolution of curriculum objectives (by level);
2. Development of scheme of studies (by level);
3. Development of syllabus of each subject;
4. Development of textbooks/instructional materials;
5. Review/approval of textual material;
6. Teacher training.

Developing Objectives

Curriculum objectives are basically derived from the recommendations of National Education Policy, national level seminars and other forums (e.g. forums of the Inter-Board Committee of Chairmen (IBCC) and research studies conducted at provincial curriculum centres). The NECT prepares the draft of objectives; which are widely circulated among the provincial institutions responsible for curriculum development, teacher training and examination. Based on their views/comments, these objectives are finalized. They are subsequently translated into the specific teaching objectives for various subjects. Several factors are considered in finalizing curriculum objectives; including the requirements that objectives should; (a) be precise; (b) assist in the selection of teaching strategy.(c) produce (or contribute to) a designated behaviour pattern; (d) enable the teachers to measure or evaluate the quality and effectiveness of learning.

The Studies Scheme

The scheme of studies is based on three key factors: (1) the national education policy; (2) market demand; (3) global issues that relate to new or contemporary education dimensions. Task work in this area is undertaken with active participation of the provincial government, research organizations and experts; as well as feedback from the IBCC.

Development of Syllabi

Based on the objectives and scheme of studies, subject specific syllabi are prepared in consultation with: provincial curriculum centres; subject experts; and psychologists and serving teachers. Collectively, they ensure that the syllabi, in all respects, satisfy the following conditions:

1. They are based on the needs of the learner/child;
2. They take into account the existing knowledge and environmental experience of the learner;
3. The developmental level of the learner is considered in the cognitive, effectiveness and psycho-motor domains;
4. The contents should be focused on attaining the objective(s).

Textbook Development

Provincial Textbook Boards (PTBB) are responsible for development of text-books according to the approved syllabi. Established lists of textbook writers in various subjects are kept. From these lists, invitations are issued to writers to submit draft materials within the prescribed syllabus parameters. Selections are made on the basis of the quality and relevance of materials submitted to local situations. Finally, the selected materials are transformed into textbooks; the final versions of which are sent to the NBCT for approval.

Review and Approval

A National Review Committee, comprising five or six members include: at least one expert from the Syllabus Formulation Committee; two subject experts; two school-teachers (one teaching the relevant material and one from a teacher-training institute). On receipt of textual materials from PTBB, this committee conducts textbook reviews based on the following parameter: (a) the book truly reflects the curriculum; (b) it meets the objectives stated in the curriculum; (c) the book does not contain any material repugnant to Islamic and Pakistani ideology.

In the case of approval, the textbook is sent back for publishing and distribution. In case of objection, the specific complaints are relayed along with revision recommendations.

Teacher Training

Teacher training for curriculum implementation is the exclusive responsibility of the provincial government. However, it is now being stressed that each textbook must have a teacher's guide—also approved by the NBCT. In some cases, assistance in the training of master trainers is provided to provincial governments.

CURRICULUM DEVELOPMENT PITFALLS

Several major obstacles affecting the quality and effectiveness of the curriculum development process in Pakistan are summarized below.

Expertise

Some serving teachers are, of course, involved in curriculum development. But notwithstanding their outstanding subject area expertise, their contribution to the curriculum development is, for all practical purpose, *nominal*. The main reason for this is that they lack the requisite expertise. The existing training programmes provide little exposure in this area, and the teachers' academic qualifications do not necessarily contribute to curricular creativity.

Therefore, at best, the teachers are able to provide opinions about the compatibility between specific concepts or content and the intellectual development level of the children in a specific age group or grade. However, this guidance often reflects, in part, a *particular* situation with which the teachers have been dealing (e.g. children from a rural background) and, in part, their own capability to render a specific concept *comprehensible.*

Textbook Quality

Textbooks often do not reflect the curriculum. Of course, it requires considerable experience and skill to: translate the curriculum in a style that covers the objectives; simultaneously take into consideration the children's language proficiency and background knowledge; and concurrently arrange the content in a logical sequence in a stimulating manner. But the all-important self-assessment questions or activities (especially questions focused on higher order skills) are invariably missing.

Table 22.2: The curriculum: who makes what choices?

	Curriculum Wing Ministry of Education	Regional/Provincial level • Local Authorities • Inspectors • Teacher's Choice	School level • Heads • Teachers • Communities
Aims and objectives	National aims, as reflected in the National Education Policy	Evaluation/study reports of curriculum centres provide change direction	Some aims of effective domains are suggested by teachers
Curriculum plan	National framework: syllabus and weightage	Introduce unique cultural/ regional aspects, including mother tongue	Scheme of work., adjust-ment of time-table, provision for co-curricular activities, exams
Methods and approaches to learning	Teacher-training courses are designed, also in-service teacher training	Teachers' colleges implement training programmes (pre-service)	Practice different methodology
Materials	Provincial textbooks are revie-wed/ approved by Federal Ministry through National Review Committee	Provincial Textbooks Boards Commission writers and select material on merit basis	Representation of teachers. in the National Review Committee.
Evaluation and examination	Inter-Board Committee of Chairmen, co-ordinate activities of the Exams Board	Board of Education holds exams	Trained teaches set the papers and evaluate the script

Implementation and Follow-up

The third problem is that there is lack of follow-up of actual curriculum implementation in classroom practice. The curriculum actually implemented is generally *different* from the official curriculum document. The classroom teacher, who primarily focuses on the textbooks and assessment, does not take into account the educational objectives. No evaluation of the implemented curriculum is carried out; hence no feedback is received to revise the curriculum. In short, each of the steps in the curriculum development process, as outlined above, tends to occur in isolation from the others and there is no visible coherent curriculum development activity.

23

Philippines: Curriculum Development

Bella O. Mariñas Maria Pelagia Ditapat

Estimated population (1995)	67,800,000
Public expenditure on education as percentage of Gross National Product (1995)	2.2
Duration of compulsory education (years)	6
Primary or basic education	
Pupils enrolled (1995)	11,541,570
Teacher (1995)	-
Pupil/teacher ratio	35:1
Gross enrolment ratio (1995)	
—Total	107
—Male	108
—Female	107
Net enrolment ratio (1995)	
—Total	90
—Male	89
—Female	91
Estimated percentage of repeaters (1992)[1]	2
Estimated percentage of drop-outs (1995)	30
School-age population out of school (1995)	1,000,000
Secondary education	
Students enrolled (1995)	4,809,863
Gross enrolment ratio (1995)	
—Total	79
—Male	-
—Female	-
Third-level enrolment ratio (1995)	29.7

Estimated adult literacy rate (1995)	
—Total	95
—Male	95
—Female	94

Note: 1. Last year available.
Source: *UNESCO Statistical Yearbook, 1998,* Paris

INTRODUCTION

Educational Legislation and Policy

The education sector (along with other government agencies) has the task of contributing to the achievement of national development goals espoused in the country's development plan. The general purpose and goals of education in the Philippines have been cited in the national constitution. Section 3(2), Article XIV of the Constitution states that:

> All educational institutions shall inculcate patriotism and nationalism, foster love of humanity, respect for human rights, appreciation of the role of national heroes in the historical development of the country, teach the rights and duties of citizenship, strengthen ethical and spiritual values, develop moral character and personal discipline, encourage critical and creative thinking, broaden scientific and technological knowledge and promote vocational efficiency.

These goals have been translated into educational policies and further elaborated as the basic (elementary and secondary) education framework.

Elementary and Secondary Education

The 1982 Education Act identifies the aims of both elementary and secondary education. For *elementary* education, the aims are: (a) to provide the knowledge and develop the skills, attitudes and values essential to personal development and necessary for living in and contributing to a developing and changing social milieu; (b) to provide learning experiences which increase the child's awareness of the responsiveness to the changes in and just demands of society and to prepare him/her for constructive and effective involvement; (c) to promote and intensify the child's knowledge of, identification with, and love for the nation and the people to which he/she belongs; and (d) to promote work experiences which develop the child's orientation to the world of work and creativity and prepare him/her to engage in honest and gainful work.

The regional level basic education aims and objectives reflect those at the national level, but are modified to suit local conditions and concerns. For *secondary* education the aims are: (a) the provision of general education that was started at the elementary level; and (b) the preparation of students for college and/or the world of work.

Curriculum Policies and Legislation

Curriculum policies are usually set forth by the Department of Education, Culture and Sports through various orders, circulars, memoranda and bulletins. They are aligned with national priorities and contribute to the achievement of development goals. However, several laws passed by the national legislature specifically relate to the school curriculum: Section 3(10), Article XIV of the Constitution mandates the study of the Philippine Constitution; Section 6, Article XIV, designates Filipino as the language of instruction; Section 19(2), Article XIV, states that: 'All educational institutions throughout the country shall undertake regular sports activities in co-operation with athletic clubs and other sectors'. Republic Act No. 4723 mandates music teaching in the schools. The most recent curriculum-specific laws designate: (a) lengthening of the school calendar from 185 to not less than 200 school days per school year; and (b) integration of concepts on human rights, the environment, dangerous drugs and computer education.

The Basic Education System

Basic education in the Philippines is free and compulsory at the elementary level only. The basic education system in the Philippines is composed of six years of elementary and four years of secondary education—a total of ten years. Compared to many countries, this is a relatively short-time period. Filipinos complete their basic education at the age of 16 or 17 years. They then proceed to institutions of higher learning to obtain a post-secondary vocational/technical institution degree or a certificate. Table 23.1 provides a general overview of the country's basic education situation.

Elementary and secondary schools are either government-supported or privately-funded. At the elementary level, the

Table 23.1: The Philippine basic education system

Indicator	Elementary	Secondary	Total
Number of schools	38,631	6,673	45,304
Enrolment	9,354,451	3,940,587	13,295,038
Number of teachers	334,822	146,102	480,924
Number of children per class	34	32	
Enrolment rate	92.70%	62.25%	

Source: Office of Planning Service, 1998.

government schools constitute 92 per cent of the total; at the secondary level, their share is 60 per cent. The school year in the Philippines begins on the first Monday of June and ends on the last Friday of March. The school year for the elementary and secondary levels consist of not less than 40 weeks or 200 days. Class sessions are held Monday to Friday and the school year is divided into four grading periods.

Administrative Structures of Curriculum Development

The education system is decentralized. The central/national office is engaged in policy formulation; while the regional and the division offices are the implementing bodies. Supervision of schools is accomplished at the regional and sub-regional levels.

THE CURRICULUM DEVELOPMENT PROCESS

Administrative Structures of Curriculum Development

Development of the basic education level curriculum is the responsibility of the Central Office Bureau of Elementary and Secondary Education, Curriculum Development Divisions. This bureau defines the learning competencies for the different subject areas; conceptualizes the structure of the curriculum; formulates national curricular policies. These functions are exercised in consultation with other agencies and sectors of society (e.g. industry, socio-civic groups teacher-training institutions, professional organizations, school administrators, parents, students, etc.). The subject offerings, credit points and time allotments for the different subject areas are also determined at the national level. In this sense, a national curriculum exists in the Philippines.

However, while curriculum implementation *guidelines* are issued at the national level, the actual *implementation* is left to school-teachers. They determine the resources to be used;

teaching and assessment strategies and other processes. Furthermore, schools have the option to modify the national curriculum (e.g. content, sequence and teaching strategies) in order to ensure that the curriculum responds to local concerns.

Language of Instruction

A *bilingual policy* is in use whereby both English and Filipino are instructional mediums. At the elementary level, English language, science and health are taught in English; while Filipino, civics and culture, good manners and right conduct (GMRC/character education), home economics, livelihood education, music, are and physical education are taught in Filipino. At the secondary level, English language, science, mathematics, technology and home economics are taught in English; while social studies, values education, physical education, health and music are taught in Filipino.

Curriculum Design

The approach to curriculum design in the country is based on content topic *and* competency. The Department of Education, Culture and Sports (DECS) prescribes competencies for the subject areas in all the grade/year levels. The DECS Bureau of Elementary and Secondary Education develops, publishes and disseminates these learning competencies to the field. Most of the subject/learning areas have a list of learning competencies expected to be mastered by the children at the end of each grade/year level and also at the end of elementary/secondary schooling. Some subject/learning areas have a combination of both (i.e. learning competencies under each content/topic). The curriculum is designed to be interpreted by teachers and implemented with variations. Schools are encouraged to innovate and enrich or adapt, as long as they have met the basic requirements of the curriculum.

In this context, the regional science high schools offer an enriched science and mathematics programme whereby students take additional science and mathematics subjects. In some private schools, English, science and mathematics subjects are taken in lieu of values education; this is because subjects like religion, moral values and ethics already have been incorporated. In addition, students are required to participate in co-curricular activities. These are managed by students with the teacher as facilitator/moderator (see Table 23.2).

Table 23.2: The new elementary school curriculum

Learning Areas	Weekly time allotment (minutes)					
	Class I	Class II	Class III	Class IV	Class V	Class VI
Character-building activities	100-150	100-150	100	100	100	100
Filipino	300	300	300	300	300	300
English	300	300	300	300	300	300
Mathematics	200	200	200	200	200	200
Civics and culture	200	200	200	—	—	—
History/geography/civics	—	—	—	200	200	200
Science and health	—	—	200	200	200	200
Arts and physical education, home economics and livelihood education	—	—	200	200	200	200
Optional	—	—	—	200	300	300
Minutes per week	1,000-1,150	1,000-1,150	1,500	1,700	1,800	1,800
Minutes per day	220-230	220-230	300	340	360	360

Teaching Methods and Learning Activities

The curriculum plan (learning competencies) does not present teaching methods and learning activities that teachers must follow in implementing the curriculum. The guiding philosophy is that the creativity of teachers is stimulated by the option to plan and use the appropriate teaching/learning activities independently. However, teacher's manuals or guides do incorporate higher-level content areas and suggestions for teaching and assessing.

Learning Materials

Until 1987, the government directly managed and supervised the production and distribution of textbooks and manuals through the Instructional Materials Development Council (IMDC). However, this responsibility was transferred to private publishers with the passage of the Book Publishing Industry Development Act (RA 8047). This Act also provided for the adoption of multiple rather than single textbooks. Currently, learning materials and textbooks developed by the private sector are submitted for evaluation to the Instructional

Table 23.3: The new secondary education curriculum

Subjects	1st year		2nd year		3rd year		4th year		Total number of units
	Min.	Unit	Min.	Unit	Min.	Unit	Min.	Unit	
English	200.	1	200	1	200	1	200	1	4
Filipino	200	1	200	1	200	1	200	1	4
Science and technology	400	2	400	2	400	2	400	2	8
Mathematics	200	1	200	1	200	1	200	1	4
Social studies	200	1	200	1	200	1	200	1	4
Physical education, health and music	200	1	200	1	200	1	200	1	4
Values education	200	1	200	1	200	1	200	1	4
Technology and home economics	400	2	400	2	400	2	400	2	8
Minutes weekly	2000	10	2000	10	2000	10	2000	10	40
No. of minutes daily	400		400		400		400		
No. of hours daily	6 hrs. 40'		6 hrs 40'		6 hrs 40'		6 hrs40′		

Materials Council Secretariat (IMCS)—an agency attached to DECS. Approved textbooks are listed in a catalogue from which school-teachers and principals select those that are to be purchased for their respective schools.

Other teaching/learning support materials available in the schools include guides or manuals, teacher support/, workbooks for students, apparatus for science and technology, and home economics, video and cassette tapes, educational computer software, charts, maps and models. All of these must also be submitted for evaluation at the national level before they can be released for purchase for school level use.

Evaluation

At the national level, the National Educational Testing and Research Centre (NETRC) has the task of administering the national achievement tests to students leaving the education sector. For grade VI this means administering the national elementary achievement test and, for year IV, the national secondary assessment test. The tests cover five subject areas and are based on the elementary and secondary level learning competencies. The examinations are administered annually, towards the end of the school year. The results provide the bases for policy formulation and educational reforms. At the regional and division levels, diagnostic and achievement tests are administered to a sample group depending on the availability of funds. No examination is required for admission to public secondary schools.

The purposes of the school-based assessments are: (a) to improve the teaching/learning process; (b) to identify students' strengths and weaknesses; (c) to determine the students' subject area performance and/or achievement level and; (d) to report student progress to parents. Although there are four periods annually at both elementary and secondary levels where students are examined in each subject, formative and summative evaluation are undertaken regularly. Paper and pencil tests are the most common forms of examination in the schools.

ISSUES AND CONCERNS IN CURRICULUM DEVELOPMENT

Issues and concerns abound in almost every aspect of the Philippine curriculum development and implementation process and at every bureaucratic level. Several of these are described below. Table 23.4 provides an overview of the curriculum decision-making process.

Table 23.4: The curriculum who makes which choices?

Aspects of the School Curriculum	Central Level • Department • Bureau (Elementary/Secondary) • NETRC	Regional/Division Level • Supervisors (Region, Division and District)	School Level • Administrators • Teachers • Communities
Aims and Objectives	Formulates and determines educational aims and objectives supportive of national development goals.	Formulates and determines specific vision, mission and objectives of the region/division/district	Formulates the vision, mission and objectives of the school Determines specific cognitive, affective and psychomotor instructional aims and objectives.
Curriculum Plan	Develops national education policies, standards and programmes for curriculum implementation. Formulates learning competencies.	Implements and adapts educational programmes to suit regional/divisional needs and cultures	Implements budget of work based on the Learning Competencies Modifies/adapts the curriculum to learners of different needs and abilities
Methods and Approaches to Learning	Conducts research/studies on innovative approaches and recommends those that are effective.	Conducts teacher training programmes on strategies found to be effective	Uses appropriate methodologies and innovative approaches

(Contd…)

Table 23.4: Contd...

	Recommends strengthening of and continued use of effective methods	Conducts research, field tests and demonstrates new teaching methodologies	Employs activities that enhance lifelong and life-wide competencies.
Materials	Exercises control over evaluation and distribution of text-books and other educational materials	Supervises the selection and distribution of instructional materials to school divisions Ensures availability/adequacy of instructional materials	Supervises the use of instructional materials by learners and teachers Procures materials based on approved list Adopts indigenous learning materials
Evaluation and Examination	Administers national examinations Conducts studies/research on student performance	Conducts supervisory visits Provides technical assistance Administers examinations	Administers formative and summative tests; uses results to improve teaching-learning process Makes report of student performance to parents, school officials.

Design

In addition to the fact that the Philippines has one of the shortest time spans for the completion of basic education, studies point to curriculum *overcrowding*. Everyday, learners must study and do homework in seven of the eight subject areas. When combined with the learning competencies required for each grade/year level, this has proven to be excessive. Reports that science and mathematics content cannot be completed in one school year have confirmed this observation. In this context, a backlog occurs and a carry-over of the previous year's content and competencies to the following school year adversely affects the teaching/learning process. Furthermore, the *scope* and *sequencing* of education (from elementary to secondary level) have also been identified as design defects. Here, content and skills gaps—as well as overlaps and duplications—have emerged. While overlap and duplication further aggravate the curriculum overload, the gaps have helped to produce elementary school graduates who are not entirely ready for secondary school.

Frequently, the inability to limit the number of core or basic subjects has led to curriculum overload. The national examinations are limited to the five subject areas of English, Filipino, science, mathematics and social studies. Very few concepts are included from other subject areas. However, lobby pressure from professional groups to include or increase the time allotments for other subjects has had an impact (i.e. subject area practitioners who demand home economics teaching for both sexes, or an increased time allotment for physical education) with the result that programming problems have occurred.

Implementation

For the nation-wide implementation of the present school curriculum, there has been massive training of school-teachers and orientation of school heads and supervisors. However, the national-level training of trainers' programme was watered down at the regional and division levels and this affected the school implementation.

Another major concern is the availability of instructional materials—most of the time there are none or, if available, they are inadequate. The instructional materials deficit includes not only the students' textbooks and teachers' manuals, but also science and vocational subject facilities, equipment/apparatus and supplementary teaching/learning materials. Other barriers to effective curriculum implementation are large classes, teacher availability (for the specialized secondary subject areas) and quality of instructional supervision

Follow-up

Three main concerns regarding the institutionalization of curriculum reforms are the quality of local leadership, monitoring and evaluation, and sustainability.

Local leadership is critical to a smooth reform implementation. Since the local/field offices are the implementing bodies, institutionalization of the reform is dependent on their priorities and capabilities. Unfortunately, curriculum improvement is often a low priority for local education leaders. The result is curricular reform misimplementation; or misinterpretation of guidelines and procedures.

Monitoring and evaluation of curriculum implementation are also key activities that are not effectively attended to. For example, not all the elementary schools are visited because there are so many of the them. Also, the secondary schools are seldom visited because supervisors are unable to provide technical assistance on specialized subject matter. While supervisors at the regional level are subject specialists, those at the division level are mostly generalists.

Because most reforms are foreign-funded, post-funding sustainability is usually a concern. Sustainability concerns encompass not only the financial aspects—rather more frequently, it is the technical and management aspects that are problematic. In most cases, the success of a reform depends on the quality and feasibility of the proposals/plans for sustainability.

CURRICULAR REFORMS IN THE PHILIPPINES

Reform Rationale

The results of a comprehensive appraisal of the Philippines education system revealed that a great deal was desired as far as the quality of education was concerned. There was a need for students to develop higher critical, logical thinking skills; communication skills, values development and/or general manual skills for higher educational or the world of work.

It was also projected that, due to financial difficulties, students would remain in the government schools and families would begin to move away from the private schools to less expensive public schools. Therefore, the public school sector had to be prepared to accept anyone wishing to complete basic education. The comprehensive appraisal reports became the basic reference documents for improving the quality and efficiency of the education system, and enhancing its utility in terms of access and equity.

Finally, it was recognized that, unless greatly improved, the system's existing capacity would be unable to cope with the educational demands generated by the escalating competitiveness of a growing technological society. Thus, the curricular reforms were also undertaken in order to meet the constant new demands being made on the system.

Reform Implementation

The reforms were implemented after project preparation was undertaken (with the assistance of a foreign-funding organization). Two major initiatives were launched. Both were geared towards *overall quality, access and efficiency* improvements in education sector performance—during and beyond the project cycle. New curricula, with mass training of teachers, were components of the *Program for Decentralized Education* (*PRODED*) and the *Secondary Education Development Programme* (*SEDP*) which focused on the elementary and secondary levels, respectively.

The PRODED was funded with a loan from the International Bank for Reconstruction and Development (IBRD). The project aimed to introduce improvements in policy, management and other sectoral concerns in order to achieve greater efficiency and effectiveness in the operation and administration of the elementary education system. One of the sub-projects was curriculum development.

The SEDP was premised on the fact that the PRODED would bring about higher quality and an increased secondary education student intake. After six years of implementation of the new elementary education curriculum, the 1989 elementary school graduates became the first students for the new secondary education curriculum.

Outcomes

The reforms at the elementary and secondary levels have been implemented over the last fifteen and nine years, respectively. Current indicators are that PRODED and SEDP have indeed succeeded in improving the quality of basic education and in making the sector more effective and efficient in the delivery of basic educational services.

As for outcomes related to the implementation and management of reform, the PRODED and SEDP have meant added responsibilities and accountability for all those involved—from policy-makers to programme implementers and target beneficiaries. Mechanisms and structures needed for the efficient implementation of the reforms have been given priority. Competencies of those involved in curriculum development and implementation are upgraded regularly, so that they may discharge their functions and responsibilities more effectively. Lessons learned from the reform implementation are providing useful baseline information for future reform and development programmes.

The curriculum is continuously undergoing refinement to ensure its relevance to changing needs and demands. The ongoing basic education curriculum review has provided for more indepth indigenization/localization of the curriculum and integration of information technology or multimedia resources in the teaching/learning process. Benchmarking has provided valuable and reliable data about school and student performance. At this point in time, significant improvements in the learners' and schools' performances have been recorded. See Table 23.2 for an overview of curricula.

Future Prospects

In the context of international assessments, the educational performance of the Philippines still needs a lot of improvement. The need for the curriculum to develop students who are *globally competitive* is another factor with which the educational sector will have to contend in the future.

Sri Lanka: Curriculum Design and Implementation for Upper Primary and General Secondary Education

A. Karunasinghe and K.W. Ganasundara

Estimated population (1995)	17,900,000
Public expenditure on education as percentage of Gross National Product (1995)	3.0
Duration of compulsory education (years)	11
Primary or basic education	
Pupils enrolled (1995)	1,962,498
Teachers (1995)	70,537
Pupil/teacher ratio	28:1
Gross enrolment ratio (1995)	
— Total	113
— Male	114
— Female	112
Net enrolment ratio (1995)	
— Total	100
— Male	100
— Female	100
Estimated percentage of repeaters (1995)	2
Estimated percentage of drop-outs (1995)	2
School-age population out of school (1995)	—
Secondary education	
Students enrolled (1995)	2,300,000
Gross enrolment ratio (1995)	
— Total	75
— Male	71
— Female	78
Third-level enrolment ratio (1995)	5.1

Estimated adult literacy rate (1995)	
— Total	90
— Male	93
— Female	87

Source: *UNESCO Statistical Yearbook, 1998*, Paris

ORGANISATIONAL STRUCTURE OF CURRICULUM DEVELOPMENT

In Sri Lanka, the Ministry of Education and the National Education Commission are responsible for all curriculum-related policy-making. However, the National Institute of Education (NIE) has the overall responsibility for curriculum design and development, preparation of syllabi, teacher's guides and textbooks. A council empowered to take all policy and high-level administrative decisions on matters coming under its purview governs the NIE. Its staff consists of curriculum teams of specialist officers and is augmented by teacher educators and teachers seconded for service to the respective teams. Consultation takes place during the curriculum development process and subsequent implementation, with subject specialists and professionals outside of NIE, as well as affected parties, (i.e., teachers, students, parents and prospective employers).

Educational Goals

Sri Lanka's educational goals can be summarized as follows:

- To develop and understand the cultural and religious heritage and the democratic traditions of the country, as well as an appreciation of the contributions made by the different ethnic groups to the national culture;
- To develop a basic understanding of the environment and skills relevant to the needs of life and society;
- To cultivate an appreciation of the arts, literature and science;
- To develop attitudes conducive to harmonious relations among the different ethnic groups;
- To promote moral, spiritual and physical development;
- To inculcate a sense of commitment to national development;

- To develop and promote a system for the acquisition of technical knowledge and vocational skills to meet the manpower needs of the country;
- To promote lifelong education and knowledge renewal through programmes of formal and non-formal education;
- To promote the democratization of education.

PRE-REFORM EDUCATION SYSTEM AND CURRICULUM

Prior to the education reform initiatives in Sri Lanka, the formal pre-tertiary education system consisted of four stages: primary: grades 1-5; junior secondary; grades 6-8; senior secondary: grades 9-11; collegiate: grades 12-13. There was a commonly prescribed curriculum for grades 1 to 11.

The lower primary stage was characterized by a *highly integrated* curriculum (first language, mathematics, environmental studies, creative-aesthetic activities, and religion). In the upper primary stage, the curriculum was *semi-integrated,* with more emphasis on subject competencies. At this level, a beginning science course and English as a second language were introduced.

At the secondary stage, the curriculum consisted of the following subjects, some of which were inter-disciplinary in nature; religion, first language, English, mathematics, integrated science, social studies, and history, aesthetic education and a life-skills/technical subject. At the end of grade 11 (which is also the end of general education) the General Certificate of Education Ordinary Level (GCE 'O'-Level) examination is administered.

EDUCATIONAL REFORM

A New Structure for Primary Education

The Sri Lankan education system is now in the process of being restructured and a number of curriculum reforms are planned or are already underway. A new primary level structure intended to be more suitable to the proposed curriculum reform is being implemented. As a result of this change, grades 1-5 will constitute a *primary stage segment.* Under this framework, the primary cycle is divided into three *key stages* (KS): (KS-1, grades 1 and 2; KS-2, grades 3 and 4; and KS-3; grade-5.

Objectives of the Reform

The new, reformed primary education programme objective listed below have evolved from the largest framework of national education goals, namely:

- development of a child-centred curriculum;
- development of essential as well as desirable competencies during the key stages;
- training of primary grade teachers to implement the revised curriculum;
- provision of adequate facilities and materials to all primary schools;
- establishment of an equal opportunity network throughout the country.

The New Curriculum

The reforms will see the countrywide introduction of a highly integrated curriculum at the lower primary stage, with the subject-area *environment-related activities*, which encompasses several disciplines. Another notable modification related to integration of content is the introduction of activity-based oral English in KS-1. The objective is to create a classroom environment where children and teachers use a mix of mother tongue and conversational English. Appropriate vocabulary will be developed through activity learning and games. Additionally, there is a provision for co-curricular work in all the key stages.

The number of subject areas in the curriculum is now being limited to four: (1) languages; (2) mathematics; (3) religion; and (4) environment-related activities. In the languages component, the formal teaching of English, beginning with KS-2, is being introduced; as well as a second national language (Sinhala/ Tamil)-from KS-3.

In the past, even at primary level, the *content* of subjects received the greatest emphasis. However, nowadays the focus will be placed on *competencies* that children are expected to have acquired at the end of their general education (Table 24.1). This new curriculum establishes a comprehensive set of basic competencies in communications, ethics and religion, environment, learning, enjoyment and leisure. Entry competencies will also be identified, using specially designed assessment instruments, enabling the grade 1 teacher to cater more effectively to children's individual needs. Furthermore, the essential competencies that children are expected to have mastered at the end of each key stage will be identified and this information made available to teachers. Teachers will be encouraged to ensure that at the end of each KS, almost all children in their class have reached the mastery level in the essential competencies, with special emphasis on first language and mathematics. In this content, the practice will be to assign the same class teacher for KS 1 and KS 2.

The new teaching/learning methodology will incorporate an appropriate mix of play, activity and deskwork, the proportion of each component varying gradually with successive grades. The new curriculum also incorporates cross-age play and activity opportunities, where KS-1 children interact with children from grade 6.

There will be continuous classroom-based assessment, with increased emphasis on the use of informal methods. This

Table 24.1: New primary stage competencies

Communications: comprises three subsets: literacy (listening attentively; speaking clearly; reading with understanding; writing accurately and lucidly); numeracy (using numbers: counting, calculating, measuring systematically); graphics (making sense of line, form; expressing and recording details, instructions, ideas, with line, form and colour).

Environment: comprises three parts; social (social awareness and relationships, personal conduct, rights, responsibilities, duties and obligations); **biological** (awareness, sensitivity and skills linked to the living world, man and the ecosystem.); **physical** (awareness, sensitivity and skills relating to space, energy, fuels, matter, materials and their links with human life. Also included are the skills in using tools to shape and form materials for living and learning).

Ethics and religion: comprises *values* and *attitudes* deemed essential for individuals to assimilate, so that they may function in a manner consistent with the ethical, moral and religious modes of conduct.

Play and leisure: related to human emotions that find expression in play, sport and various leisure pursuits essential for mental and physical well-being. They are also connected with such values as: co-operation, team work, healthy life and work competition. Also included are aesthetic and creative activities.

Learning to learn: related to human needs in a rapidly changing, complex, crowded world where learning will require constant review and updating. Includes developing skills of awareness, attentiveness, and perseverance. The information revolution has rendered this competency essential.

represents and attempt to deviate from past assessment techniques (which encourage comparative student achievement) and move towards criterion-referenced assessment techniques. Entry competency tests for grade 1 and terminal competency tests at the end of each key stage will be developed.

The remaining tasks to be carried out include; preparation of a classroom-based evaluation framework; development of training material; and the training of teachers and supervisors. A booklet will be produced on improving testing techniques. Also, a research committee will be set up to carry out action research, surveys and evaluation on special areas, such as group work, multi-grade, multi-level teaching, gifted children and on-going aspects of the curriculum development process, including teacher training.

Resource Materials

A steering committee will decide on the content and layout of textbooks and select the panel of writers who will be registered with the National Institute of Education (NIE). Books will be reviewed by a panel and approved by the steering committee prior to publication, Additional resource materials for primary classes will be prepared at the level of the school, the resource centre and the province with NIE assistance. The Education Publication Department (MOE) and NIE are collectively responsible for printing and distributing teaching and resource materials.

Implementation Schedule

The new primary curriculum was introduced in 1998 as a pilot project in grade 1 of Gampaha district, and introduced to the whole country during 1999. It will be progressively implemented at each grade level, terminating in grade 5 by 2003.

Teacher Education

A profile of the primary school-teacher will be developed based on the vision of the primary school for the twenty-first century. Essential teacher competencies and attitudes will be identified and a new pre-service teacher education curriculum and training materials developed. The staff of colleges of education and other teacher-training institutes will undergo orientation programmes. Each training college will adopt a *problem-solving school* in its catchment area in order to conduct field-based activities. In-service training of primary teacher will be an integral part of the reform. The Master Teacher Programme will be strengthened.

School Managers

As the primary section forms a part of the main school in Sri Lanka, an assistant principal or a senior teacher will be entrusted with responsibility for the primary section of the main school. In schools where there is a large number of pupils, additional adequate supervisory assistance will be provided (i.e. sectional or grade co-ordinators). Anticipated school-level management tasks include: establishing a primary education development committee; strengthening parent/community relations; formalizing collaboration with neighbouring schools through the formation of *school families;* developing and implementing an internal supervisory mechanism for teacher self-evaluation, as well as collegial peer evaluation.

Provincial Administration

The provincial administrative structure will also be strengthened, with a separate primary education division in the Provincial Department (PPEU) being established under a senior officer. The appointment of officers responsible for primary education at the zonal level operating under the PPEU is also foreseen, as well as the creation of primary education development committees in both provinces and zones, comprised principals, teachers, parents and community leaders.

Public Awareness

This will be carried out through a mass media publicity programme involving newspapers, radio and TV and targeting school staff, parents, influential youth groups as well as the general public.

FUTURE PROSPECTS FOR CURRICULAR REFORM

Junior Secondary Stage

In the near future it is projected that grades 6-9 will constitute the junior secondary stage of education, thereby functioning as the upper section of the junior school. Grade 6 will become the *bridging* year between the integrated curriculum

Table 24.2: The curriculum: who makes what choices?

	Central Level • Ministry • National Institute of Education • National Education Commission	Regional/Provincial Level • Provincial Ministry	School Level • Teachers • Communities
National Goals and Basic Competencies	Sets national goals and competencies (Ministry and NEC).		
Aims and Objectives	Sets aims and objectives for subjects (NIE)		
Curriculum Plan	Writes national syllabus and decides time allocations (NIE)		Handles implementation and provides feedback
Methods and Approaches to Learning	Recommends approaches to be used. Trains master teachers in the use of certain methodologies (NIE)	Trains teachers with the help of master teachers. Supervises teachers training	Sets school policy programmes, projects work, activity-room work.
Materials	Writes teacher's guides, textbooks (NIE)		Prepares certain teaching materials.
Evaluation and Examination	Sets central examination grade 5 GCE 'O'-level & 'A'-level	Sets some provincial examinations	Handles on-going assessment. Sets all internal tests and examinations.

Table 24.3: Number of periods for subject/subject areas for grades 6 to 11

Subject	Number of forty-minute periods per week					
	Grade 6	Grade 7	Grade 8	Grade 9	Grade 10	Grade 11
Religion	3	2	2	2	2	2
First language	5	5	5	5	5	5
English	5	5	5	5	5	5
Maths	5	6	6	6	6	6
Environmental studies	9	—	—	—	—	—
Science & technology	—	6	6	6	6	6
Aesthetics	4	3	3	3	3	3
Sinhala/Tamil as a 2nd language	2	2	2	2	—	—
Social studies/history	—	5	5	5	5	5
Life skills	—	3	3	3	—	—
Technical subjects	—	—	—	—	4	4
Groups activities	3	—	—	—	—	—
Additional subjects	—	—	—	—	4*	4*
Total periods per week	40	40	40	40	40	40

of primary school and the subject based curriculum of secondary school. It will serve to inculcate the necessary study skills for secondary education while reinforcing the essential competencies to be acquired during the primary cycle, particularly in language and mathematics. The syllabi in grade 6 will be revised to meet these requirements.

There will be a common curriculum at this stage, comprising 9 subjects: first language, English, mathematics, science and technology, social studies, life skills, religion, aesthetics, health and physical education. (The teaching of a second national language, i.e. Sinhala for Tamil speaking students and Tamil for Sinhala speaking students will also be introduced at this level—when teachers are available.)

At the end of grade 9, a school-based examination, the *Junior School Proficiency Examination,* will test pupil achievement and result in the issue of a certificate.

Teaching methodology will emphasize learning through projects and practical work, as spelt out in the section on practical skills education. Concepts on peace education, conflict resolution, democratic values, human rights and environmental conservation will be integrated into social studies and other relevant subject content.

Implementation Plan

The revision of syllabi and course guides, training of teachers and provision of facilities were scheduled to be for completion at grade 6 level in 1998 and for implementation in January 1999. Reforms will be progressively introduced to grades 7 and 8 with the grade 9 curriculum revised in line with the new GCE "O" level curriculum.

Senior Secondary Stage

The present curriculum at this level is oriented to the GCE (OL or AL) examinations. The programme covers a period of 3 years; namely grades 9, 10, 11 for GCE (OL) and 2 years for GCE (AL) Collegiate level grades 12-13.

At present, the GCE (OL) compulsory curriculum consists of the following subjects (some of which are inter-disciplinary in nature): religion, first language, English, mathematics, integrated science, social studies and history, aesthetic education and life skills/technical subjects. In grade 9, students select 1 vocational type course (out of 53 such courses). In grade 10-11, 1 technical subject may be chosen from several options.

Under the proposed revisions, the time period for senior secondary education will be reduced to 2 years with classes designated grades 10 and 11. The curriculum will be made more flexible with the introduction of a number of core subjects and a number of optional subjects. Core subjects will be: religion, first language, English, mathematics, science and technology, social studies and history, aesthetic studies. Students will be permitted to select up to 3 optional subjects from the following: Sinhala/Tamil as a second language, history, geography, health and physical education, literature (Sinhala/Tamil/English), modern or classical languages, technical subject (from the list of approval technical subjects).

Under the reform, grade 11 remains the end of the general education for all stages. Students sit for the General Certificate of Education Ordinary Level (GCE-OL) examination, which is a centrally planned, national school leaving examination. Only those students who achieve certain prescribed GCE-OL standards are permitted to enter the collegiate stage (about 25-30% of students).

Improving English Language Teaching in Sri Lanka

A special effort is being made to improve English language skills as it is realized that a better knowledge of English will improve employment opportunities and facilitate communication with the outside world. The English language teaching programme will be upgraded to provide opportunities for pupils island-wide to have equal access to English learning for comprehension and communication. Teachers of English will be provided with opportunities and incentives to improve their proficiency in English and skills in teaching English.

25

Thailand: Curriculum Planning, Development and Reform

Kiat Ampra and Chdjane Thaithae

Estimated population (1995)	58,200,000
Public expenditure on education as percentage of Gross National Product (1995)	4.1
Duration of compulsory education (years)	6
Primary or basic education	
Pupils enrolled (1995)	5,961,855
Teachers (1995)	—
Pupil/Teacher ratio	20:1
Gross enrolment ratio (1995)	
— Total	87
— Male	—
— Female	—
Net enrolment ratio (1995)	
— Total	97
— Male	98
— Female	97
Estimated percentage of repeaters (1995)	12
Estimated percentage of drop-outs (1995)	20,000
School-age population out of school (1995)	
Secondary education	
Students enrolled (1995)	3,794,290
Gross enrolment ratio (1995)	
— Total	55
— Male	—
— Female	—
Third-level enrolment ratio (1995)	20.1

Estimated adult literacy rate (1995)	
— Total	94
— Male	96
— Female	92

Source: *UNESCO Statistical Yearbook, 1998*, Paris

INTRODUCTION

The year 2000 will be a year of learning reform in Thailand. In the previous decade, economic growth based on industrial production was rapid, but did not enhance the overall quality of life for the majority of the population. The current economic crisis has heightened the need for educational reform so that citizens are better equipped to cope with present socio-economic demands.

GOALS

Primary education in Thailand aims at developing the *quality of life* of learners so that they can properly serve society, assuming their roles and responsibilities as good citizens under a democratic constitutional monarchy. To achieve this goals, each learner is to be equipped with the basic knowledge and skills necessary for daily living; adjustment to social changes; good physical and mental health; effective work and happy, peaceful living.

THE PRIMARY CURRICULUM

The objectives of the primary curriculum are to provide: (a) basic education for all; (b) experiences useful for daily living; and (c) education for national unity with common purposes. In this context, local authorities are given the opportunity to develop part of the curriculum—rendering it suitable of local conditions and needs.

The curriculum experiences provided for learners comprise five areas:

1. ***Tool subjects:*** Thai language and mathematics;
2. ***Life experiences:*** The process of solving social and daily life problems (with an emphasis on scientific process skills for better living);

3. ***Character development:*** Activities necessary for developing desirable habits, values, attitudes and behaviours leading to an acceptable character.
4. ***Work-oriented experiences:*** General and practical work experiences and basic knowledge for career preparation;
5. ***Special experiences:*** Activities based on learners' interests.

Area 5 is provided for learners in grades five and six only. Experiences provided may include knowledge and skills selected from the other four areas or activities based on learners' interests, i.e. English for everyday life. Schools may select as many activities as desirable. The curriculum aims to develop the following skills, knowledge and attitudes in learners.

1. Basic learning skills, retention of literacy and mathematical skills;
2. Knowledge and understanding about self; the natural environment and social changes;
3. Ability to take care of personal and family health;
4. Ability to identify causes of personal and family problems and to apply scientific reasoning skills to suggest ways and means of solving them;
5. Pride in being Thai, unselfishness, fair-mindedness and the ability to live happily with others;
6. Habits of reading and lifelong learning;
7. Basic knowledge and work skills, good work habits and the ability to work co-operatively with others;
8. Knowledge and understanding of social conditions and changes at home and in the community; ability to carry out the roles as a good family and community member; a sense of responsibility to conserve and develop the environment; to promote religion, arts and culture in the community.

Time Allotments

The total time allotment for the primary curriculum is about six academic years, with not less than forty weeks for each academic year and not less than twenty-five hours or seventy-five periods per week. The time allotted for each period is twenty minutes. Collectively, learning periods cannot be less than 200 days or 1,000 hours. See Table 25.1 for details.

Table 25.1: Approximate time allotments in the primary curriculum

	Approximate time allotments; primary curriculum					
	Grade 1-2		Grade 3		Grades 5-6	
	%	Periods/ year	%	Periods/ year	%	Periods/ year
Tool subjects	50	1,500	35	1,050	25	750
Life experiences	15	450	20	600	25	750
Character development	25	750	25	750	20	600
Education for work	10	300	20	600	30	900
Total	100	3,000	100	3,000	100	3,000
Special experiences	—	—	—	—	—	600

Note: There are three periods of twenty minutes in one hour.

IMPLEMENTATION GUIDELINES

To achieve the educational aims, curriculum orientation guidelines have been established as follows:

1. Organising teaching/learning activities that:
   - Are relevant to conditions and needs of the communities by providing opportunities for local authorities to develop part of the curriculum, as well as instructional aids appropriate to their localities that are learner-centred, making the activities relevant to the learners' needs and living conditions and providing equal opportunities for them to develop according to their respective abilities;
   - Maximize linkages and integraction of subject matter, the learning experiences within each area and between different experience areas;
   - Emphasize learning processes, logical and creative thinking and group processes;
   - Promote learning by doing and emphasize development of concepts in all areas of experience.
2. Organizing research studies, follow-up and continuous remedial teaching;
3. Regularly integrating moral education and desirable values into the teaching/learning process, as well as in extra-curricular activities;

4. Organizing an environment and general climate within the school conducive to learning and to the practical activities of learners.

MEASUREMENT, EVALUATION AND FOLLOW-UP

School administrators and classroom teachers are responsible for measurement, evaluation and follow-up in order to determine mid-year and end-of-year learner promotions. Teachers are expected to carry out formative and summative evaluations periodically based on subject content and experiences, in conformity with the Ministry of Education's prescribed evaluation regulations. However, for the area of *special experiences*, measurement and evaluation are designed to assess learners' involvement in activities and are not used as criteria for class promotion.

SECONDARY EDUCATION

Goals and Aims

Secondary education also aims to improve the learner's quality of life and serves as the basis for further education. It should: (a) help learners discover their own abilities, aptitudes and interests; (b) provide a general education as the basis for securing honest occupations or further education; and (c) respond to the needs of the locations and the nation. Based on these aims, the curriculum is designed to permit learners to develop the following characteristics:

1. Knowledge and skills in general education subjects and the ability to keep up with academic advances;
2. The ability to maintain and enhance personal and community health and hygiene;
3. The ability to analyse community problems and choose suitable alternatives for solving them—taking into account various limitations;
4. Pride in being Thai; ability to live in peace with others and to willingly help others—within the limits of one's capability;
5. Creativity, ability to devise and improve practices which will bring about individual and community progress;
6. Good attitudes towards all kinds of honest occupations; love of work and ability to choose occupations relevant to one's aptitudes and interests;

7. Basic skills for carrying out honest occupations; skills in management and in working co-operatively with others;

Table 25.2: The secondary curriculum

1. Compulsory courses: 57 learning units (credits)	**Credits**
Core compulsories	39 units
Thai	12 units
Sciences	9 units
Mathematics	6 units
Social studies	6 units
Health and physical education	3 units
Art education	3 units
1.2 Elective compulsories: 18 units (credits)	
Social studies	6 units
Health and physical education	6 units
Work education	6 units
2. Free elective courses: To be selected from the following—33 units	
2.1 Languages	
Thai	
Foreign languages	
2.2 Sciences/mathematics	
Sciences	
Mathematics	
2.3 Social studies	
2.4 Personality development	
Health and physical education	
Art education	
2.5 Work and vocation	
Vocational education	
3. Activities in accordance with the Ministry of Education's regulations are to be organized in educational institutions under the responsibility of the Ministry of Education	
3.1 Boy Scouts, Girl Scouts, Red Cross Youth, Girl Guides	
3.2 Extra-curricular activities.	
3.3 Guidance/remedial education or academic development activities	
3.4 Independent activities	

8. Understanding social conditions and changes in one's community; the ability to suggest ways of community development; pride in assuming one's roles, duties as a good community member, knowledge of how to conserve and develop the environment, religious and cultural heritage of the community.

IMPLEMENTATION CRITERIA

This revised edition of the lower secondary school curriculum has the following main features:

Duration

The full course requires approximately three years or six semesters. Each academic year is divided into two semesters, with twenty weeks per semester. A school may offer a summer semester, as deemed appropriate. Each week consists of no less than five learning days, at least seven periods per day (one period is fifty minutes long). At least thirty periods per week are to be allocated to regular teaching/learning in accordance with the curriculum.

Learning units/credits

One unit is given to any subject requiring two learning periods per week per semester. Subjects requiring more or less than two learning periods are assigned appropriate proportional units.

Compulsory courses/free Elective Courses

The learners must take compulsory and free elective courses as specified in the structure of the curriculum. Learners may choose only one foreign language.

Evaluation of Learning

Evaluation of learning and transfer of credits must conform to the Ministry of Education's regulations.

Criteria for Course Completion

The completion requirements are: (a) ninety units of compulsory and free elective courses as specified in the curriculum and satisfactory learning outcomes for all subjects; (b) passing the Thai language and social studies core compulsories; (c) obtaining a minimum of eighty units; (d) participating in the curriculum specified activities with at least 80% attendance and having satisfactorily achieved all the major objectives of the activities.

Organizational Structure

Table 25.3 provides an overview of the key agencies involved in curriculum development and implementation at each level of the system as well as their respective roles and responsibilities.

CURRICULA ADAPTATION: OUTCOMES AND ISSUES

Implementation

The most important problem here is how to change learning and teaching behaviour. According to the curriculum orientation guidelines, teachers should be focusing on: (a) integrating content from daily life, (b) making greater use of activities, rather than textbooks; (c) using different learning materials in a variety of ways: (d) making students the centre of learning activities; and (e) reducing explanation and helping students *construct* knowledge from various sources. But in the real classroom situation, the teacher-centred approach still dominates. Basically, teachers still dictate to students and still place emphasis on textbook content. Few teaching materials are used.

Some of the reasons for this are: (a) teachers are afraid that students cannot obtain the necessary fundamental knowledge through activities; (b) current assessment techniques still emphasize knowledge and understanding; (c) entrance examinations to the secondary level and higher education is still based (mainly) on summative knowledge, as opposed to other abilities; and (d) it takes more time to prepare and teach according to the designated teaching/learning curriculum orientations. It is anticipated that all these problems will be solved in the forthcoming process of reforming curriculum and learning activities.

Follow-up

The follow-up of curriculum implementation is not being adequately undertaken because of an insufficient number of external supervisors. In fact, this task, is poorly implemented,

Table 25.3: The curriculum: who makes which choices?

	Central Level Curriculum Centre	Regional Level Educational Regions	School Level Teachers
Aims and Objectives	Department of Curriculum and Instruction, Ministry of Education, establishes goals and aims.	In each educational region, a local syllabus experience component is designed (for primary level) and work-oriented education (for lower secondary level); the objectives and contents relate to each other, but do not conflict with aims and objectives at the central level.	
Curriculum Plan	Curriculum Development Centre, Department of Curriculum and Instruction Development, Ministry of Education, establishes structure and time allotment		1. All teachers make detailed teaching plans for each semester or each academic year. The contents cover only topics and main points. 2. Teachers make their own lesson plans for each hour to cover: 2.1 Behavioural objectives; 2.2 Detailed contents that relate to objectives and apply the scope of contents to suit local situations (based on the community; environment and dailylife—including local employment).

(Contd...)

Table 25.3: Contd...

	Central Level Curriculum Centre	Regional Level Educational Regions	School Level Teachers
Methods and Approaches to Learning	Curriculum Development Centre, Department of Curriculum and Instruction Development, Ministry of Education, works on guidelines for implementation to achieve the aims.		2.3 Teaching and learning activities are related to objectives and contents; focus on learning by doing as in: experiments, reading, writing, mathematics skills, public speaking and workshops, External trainers are necessary; they can be selected by the community.
Materials	Ministry of Education supports purchase of expensive materials, such as laboratory equipment, reference books, computers, audio-visual and workshop equipment, reading books for students	Some communities support more expensive equipment, such as audio-visual, television and video players	Learning supports: reading books, pictures, word cards, CD-ROMs, cassettes tapes and video are provided by teachers. Some other teaching materials can be produced by the teachers.
Evaluation and Examination	The evaluation regulations are prescribed by the Ministry of Education		School administrators and classroom teachers are responsible for measurement, evaluation and follow-up to improve teaching and learning and to determine mid-year and end-of-year promotion for learners, according to their abilities. They should also carry out periodic formative and summative evaluations of contents and experiences.

because the external supervisors innovations and methods do not relate well to real school situations. There are *some internal* supervision problems as well; the school principals have little understanding of curriculum and teaching methods, and they tend to pay less attention to academic development. Furthermore, most of the teachers have negative attitudes toward internal supervision.

FUTURE PROSPECTS

The Thai primary and lower secondary currricula have now been in use for over twenty years. Since (in principle) curriculum development is related to socio-economic conditions, this means that the curricula in use still relate (for the most part) to the social conditions prevailing in 1978—the date when they were initially adopted. (Not withstanding, there were two subsequent revisions to the lower secondary curriculum in 1990). As mentioned at the outset, it is in this context that Thailand plans to implement major reform of the curriculum and learning activities, beginning in 2000.

The Post-reform Education System

Goals. The reforms will emphasize:

- Providing basic education for all (especially *the equality of being*);
- Providing education for adjusting oneself placidly to the changing society and creating social learning;
- Providing education that embraces *international* norms (i.e. using high technology, respective human rights; being generous to children, women; facing new problems, etc.).

Structure. To achieve the objectives, the structure of the school system will be reorganized, in concert with reforms to the basic education curriculum and learning activities. It is expected that the class levels will be grouped into the following four sections;

- first grade-third grade;
- fourth grade-sixth grade;
- seventh grade-ninth grade;
- tenth grade-twelfth grade.

Curriculum. The basic education curriculum will focus on developing the learners' emotional, physical, social and mental capacities, resulting in the following characteristics for each individual;

- high ethical conduct and values, and the ability to work and live happily in both Thai and global society;
- good health, well-rounded personality and a sense of aesthetics;
- the ability to think, solve problem and adopt a very broad vision;
- knowledge, good skills and capacity for lifelong learning;
- a sense of nationalism and good citizenship (for a system based on a democratic monarchy);
- creativity, ability to participate competently in the global society.

Curriculum orientations. There will be four main orientations: (1) *learning details:* self-development, art education, social studies, Thai language, mathematics, science and technology, work-oriented experiences, foreign languages; (2) *organizing learning details:* focused on basic and selected local needs—also on knowledge, skills ethics and values; (3) *projects*—this is a *key* component consisting of project work for learning and meeting students' interests; (4) *social* activities focused on social development.

Teaching/learning approach. The main thrust will be on effectively using the *child/student-centred* approach. Teachers will design relevant activities by which students can themselves construct and follow-up knowledge. This will include activities designed: (1) to cater to individual students' needs and abilities; (2) to permit students to select options according to their own interests; (3) to organize extra-school and classroom-based teaching and learning activities; (4) to facilitate student-lead learning activities with teachers acting as advisors and facilitators; (5) to evaluate individual student progress (based on authentic assessment, as well as student self-assessments).

26

Viet Nam: Curriculum Planning, Development and Reform

Nguyen Thi Minh Phuong and Cao Thi Thang

Estimated population (1995)	73,800,000
Public expenditure on education as percentage of Gross National Product (1995)	2.7
Duration of compulsory education (years)	5
Primary or basic education	
Pupils enrolled (1995)	10,029,000
Teachers (1995)	288,200
Pupil/teacher ratio	34:1
Gross enrolment ratio (1995)	
— Total	114
— Male	—
— Feamale	—
Net enrolment ratio (1995)	
— Total	—
— Male	—
— Female	—
Estimated percentage of repeaters (1995)	—
Estimated percentage of drop-outs (1995)	—
School-age population out of school (1995)	—
Secondary education	
Students enrolled (1995)	3,794,290
Gross enrolment ratio (1995)	
— Total	47
— Male	—
— Female	—
Third-level enrolment ratio (1995)	4.1

Estimated adult literacy rate (1995)	
— Total	94
— Male	97
— Female	91

Source: *UNESCO statistical yearbook, 1998,* Paris.

BACKGROUND

Rationale for educational reform

In April 1991, the seventh Congress of the Communist Party of Viet Nam put forward a new national programme and strategy for socio-economic stabilization designed to build a prosperous, powerful, just and civilized society for all citizens. This renovation programme is expressed in the shift from a subsidy-based economic mechanism to a market-oriented one, the development of an open, multisector, and socialist-oriented economy, under State management.

Education reform in Viet Nam is intricately linked to this major national initiative whose goals require a supportive, reinforcing education programme. The existing socio-economic, political and cultural climate of the country call for a redesign of educational objectives, contents and methods, in order to meet the human resource needs for the projected industrialization and modernization period. The aim is to complete the basic modernization and industrialization of the country by the year 2020. Viet Nam seeks to join with the international community, while still preserving and developing its national traditions.

Reforms to curricula have come about not only due to recognition by the government of these global pressures, but also due to the demands of teachers, pupils and parents who are aware of the outdated nature of the curriculum and the need for on-going curriculum change. Previously, the curriculum had been designed to endure within a long-term perspective: the primary curriculum for about twenty years; and the secondary for over ten years. This is therefore only the third reform, the first took place in the 1950s, the second, in the 1970-80s.

Educational Aims

In the draft Education Act (submitted to the National Assembly in December 1998) the national educational aims are broadly stated as:

> Forming and fostering the personality, quality and ability of citizens; training working people who: are faithful to the ideal of national independence and building of a fair and civilized society; are moral, dynamic and creative; knowhow to preserve and promote the cultural values of the nation; are receptive and open to all cultures; and have the necessary sense of discipline, organization and industrious behaviour to meet the requirements for building and defending the nation.

Curriculum Development and Reform

Curriculum development in Viet Nam is based on three factors: (1) the vision of the country's leaders concerning the economy and society within the next ten to twenty years; (2) educational achievements and curriculum development experience, based on the country's characteristics; (3) the curriculum development trends and experiences of other countries. Current reforms are based on the following orientations:

- a focus on basic, practical content which can be applied in everyday life;
- an update of content based on scientific, technological and other developments in modern society;
- the renovation of teaching/learning methods in order to help students develop initiative and creativity in learning;
- development of each student's ability, especially the ability and methods for self-learning;
- due consideration for humanistic and international education;
- preservation of the national identity of Viet Nam, while participating in the world community;
- focus on international curriculum goals of *learning to know, learning to do, learning to be, learning to live together.*

THE CURRICULUM ADAPTATION PROCESS

Administrative Structure of Curriculum Reforms

The system of curriculum development in Viet Nam is centrally managed by the Ministry of Education and Training (MOET). However, the Centre for the Development of Curriculum and Methods of Education (CDCME), established in 1961 under National Institute of Educational Sciences (NIES), is the agency

Table 26.1: Who does what in curriculum development?

	Central Level	Regional/Provisional Level	School Level
	• Ministry • Local • Curriculum Centre • Central Examination Board	• Local Authorities • Inspectors • Teacher Colleges	• Heads • Teachers • Communities
Aims and Objectives	Drafts national aims and objectives in detail; Completes aims and objectives after receiving feedback from regional levels.	Makes remarks and comments on pre-defined aims, feedback to the central level; Helps teachers understand the aims.	Makes aims and objectives effective, Interprets aims and objectives to pupils.
Curriculum Plan	Drafts curriculum framework, syllabus and decides timetables, teaching plan; Completes them after receiving feedback from regional level.	Makes remarks and comments upon the drafted curriculum framework, feedback to the central level; Updates, if necessary, the national syllabus; Creates local timetables.	Makes schemes of work; Implements curriculum plan.

(Contd…)

Table 26.1: Contd...

	Central Level	Regional/Provisional Level	School Level
	• Ministry • Local • Curriculum Centre • Central Examination Board	• Local Authorities • Inspectors • Teacher Colleges	• Heads • Teachers • Communities
Methods and Approaches to Learning	Recommends methods and approaches to be used; Training of trainers (key teachers from regional level) in the use of advanced methodologies	Trains teachers in the use of suitable methodologies.	Practises various methodologies. Selects approaches related to local cultural community life.
Materials	Commissions and writes textbooks, teachers' books, exercise books, references; Organizes researching and production teaching.	Gives guidance using materials; Initiates locally relevant materials.	Uses textbooks, teachers' guides; Chooses references, exercise books and recommends parents to buy them; Uses (or ignores) materials from environment
Evaluation and Examination	Produces revision guidance handouts for baccalaureate examinations at secondary school; Makes test questions for baccalaureate examinations at secondary school.	Trains teachers on how to supervise exams and to evaluate; Evaluates school and teachers activities and sets standards; Sets some local examination (for LSS, primary)	Evaluates growth of pupils (knowledge, behaviour, health, etc.); Sets all internal tests and examinations; Marks work and keeps records according to certain principles.

with major responsibility for curriculum research and development (including curricular reform). NIES's extensive curriculum mandate was given when the Minister of Education and Training took the decision to establish a Curriculum Development Board (CDB) and designated NIES as the agency in charge.

- In undertaking the current reforms, the Minister of Education and Training decided to establish the following additional agencies:
- In 1996, a Board of Primary Education Curriculum Development—2000 was set up, consisting of seventy-five members drawn from research agencies or administrations of primary education at central or local level, including universities. During the curriculum reform, this board benefited from a short training course held in Viet Nam by education specialists from the United States of America, Japan and Australia.
- A Council for Curriculum Evaluation consisting of ninety-eight members belonging to twenty-eight education agencies at central and local levels, including universities. In addition, there is a primary education project: 'Evaluation of mathematics and mother tongue teaching at 4th and 5th grades' supported by the World Bank and co-ordinated by the National Institute of Education.
- In 1998, a Board of Junior Secondary Education Curriculum Development, consisting of twenty-five experts drawn from NIES, universities and departments within the MOET. This Board operates within an Asian Development Bank—Viet Namese Government Project, Junior Secondary Education Innovation (1999-2004). During the realization of the project, two groups of curriculum developers were sent to Germany, Thailand and Australia in order to collect documents about curriculum development and to exchange experiences.

The stakeholder method guides curriculum development and implementation in Viet Nam. Various MOET subject experts, university teachers and outstanding general education teachers are selected for CDB membership. During the development process, experts, professors, administrators, teachers and parents are all invited to provide comments on, or evaluate the curriculum. An evaluation council, established by MOET, is comprised representatives drawn from central and local level educational administration and universities.

Process of Curriculum Development

The primary and secondary school objectives are based on the general statement of national aims and the curriculum orientation guidelines. Different subjects are identified and defined for each level by curriculum specialists. Teachers can emphasize or omit parts of subject objectives.

In Viet Nam, the curriculum has traditionally been regarded primarily as a written document which sets up the subject contents, consisting (essentially) of three parts: 1. educational objectives; 2. educational contents for each school

Table 26.2: Primary education teaching periods: number per year

		Grade/Year				
	Subject	1	2	3	4	5
Compulsory	1. Vietnamese language	363	330	297	264	264
	2. Mathematics	132	165	165	165	165
	3. Morals	33	33	33	33	33
	Nature and society	33	33	66		
	-Science	66			66	66
	-History	33			33	33
	-Geography	33			33	33
	5. Technology	33	66	66	66	66
	6. Music	33	33	33	33	33
	7. Art	33	33	33	33	33
	8. Physical education	66	66	66	66	66
	9.Health	33	33	33	33	33
	Total	759	792	825	825	825
Optional	10. Foreign language			66	66	66
	11. Information			66	66	66
	12. Club activities			66	66	66
	Total			Max 132	Max 132	Max 132

year; and 3. curriculum interpretation. However, the view of the curriculum document has changed, resulting in modifications that include: (a) considerations of experiences in foreign countries; (b) the methods, orientation and learning aids; and (c) ways of organizing the assessment and evaluation of student learning outcomes. Also, several new actions are being undertaken: (a) standards for each subject are being designed; (b) some 10-15% of the content now includes a local component (local geography, history, economy, culture); (c) local variations in the national syllabus are being introduced by teachers (i.e. local timetables, local schemes of work, etc.).

Learning Methods and Approaches

Teaching and learning methods are presently being reformed, with the intention of fostering, under the supervision and guidance of the teacher, self-directed discovery learning, based on each student's individual abilities. This is an attempt to transform traditional teacher-centred approaches, in which students play a largely passive role, and which stifle both the puils' and teachers' creativity. The curriculum development centres, educational research centres and teacher-training

Table 26.3: Lower secondary education curriculum: periods per year

	Subject	Grade/Year 6	7	8	9
Compulsory	1. Vietnamese language & literature	132	132	132	132
	2. Mathematics	132	132	132	231
	3. Citizenship education	33	33	33	33
	4. History	99	99	99	99
	5. Natural sciences	99	99	165	165
	6. Technology	66	66	66	66
	7. Music	33	33	33	16.5
	8. Art	33	33	33	16.5
	9. Foreign language	99	99	99	99
	10. Physical education	66	66	66	66
	Total	**792**	**792**	**858**	**858**
Optional			Max. 66	Max. 66	Max. 66

colleges recommend the various learning methods, orientations and approaches to be used. The teacher-training colleges serve as moderators and monitors in the use of certain methodologies (discussion, role play, experimental methods, etc.).

CURRICULUM MATERIALS

The State controls the development and writing of various instructional materials for schools. The government directly supports the compilation and development of textbooks and teachers' guides. Books are sold to teachers and students—they are distributed free of charge only to those students who are particularly disadvantaged. At present, a programme of loaning textbooks to primary students in disadvantaged areas is being developed. Private companies only have the right to develop and print non-compulsory reference books for teachers and students.

- The typical procedure for textbook preparation is as follows:
- The National Institute for Educational Science, the Educational Publishing House and the MOET Councils of Subjects jointly select and introduce a list of authors for approval by the Minister in the MOET;
- The Educational Publishing House (an agency of MOET) organizes the writing of textbooks by giving financial support (creating favourable conditions for consulting professionals and experts);
- The Council for Textbooks evaluates the textbook drafts, then submits them to the Minister for approval. Each level of education usually has only one set of textbooks and there is no option for other textbook use.

Apart from textbooks, teachers' guides, exercise books, reference books for teachers and students, and videocassette tapes are produced for each subject. Other learning aids are reviewed by a committee before being produced and supplied to schools. Learning aids are also made from local materials by teachers and pupils themselves, while others are imported (from China, Germany).

EVALUATION

There is a link between the regular tests carried out in the classroom and the periodic tests set by official regulation. The former occur at the end of each important chapter or textbook subject; the latter at the end of the school term or school year. The purpose of examinations at different levels of education is mainly to consider student promotion to a higher class (which is based on the ability to pass the examination) and to inform the parents about progress. The assessment of students is typically based on percentage calculations, or grading (very good, good, fair, weak).

It is recognized that the current method of evaluating and assessing student learning in Viet Nam should be improved, as the current approach does not take into consideration various categories of student achievement. Due consideration has not yet been given to the inclusion of other assessment information (records, files, practice records in the laboratory, etc.) and to a diagnosis of the individual development of students. This easily results in a biased, one-sided attitude toward teaching/learning and achievement (on the part of both the teachers and the students).

Research is now being conducted on how to evaluate students in more essential ways, that is, by using methods that more accurately reflect the curriculum implementation impact at each level of education.

THE PRINCIPLES OF LEARNING TO LIVE TOGETHER AS CURRICULUM CONTENT

In Viet Nam, there is no specific reform relating only to principles of learning to live together. They may be seen as being part of general reforms. These principles have been translated into objectives for the primary and secondary school based on the national educational goals. These are incorporated into the curriculum through both a cross-curricula approach and extra-curricular activities. The principles are inherent in the content of certain subjects, such as Vietnamese language and literature, civics, history, geography, foreign languages (English, French, Russian). Some aspects of the principles are also included in the sciences (physics, chemistry, biology) and technology.

In both primary and secondary school, the theme of learning to live together is also taught through selected content from several interdisciplinary subjects (i.e. global education, population education, education for environmental protection, technical education, peace education, HIV/AIDS prevention, etc.). Extra-curricular activities can create positive attitudes mutual understanding, responsibility etc. (i.e. class meetings, flag saluting, celebrating public days, cultural galas among schools, song and dance competitions, gymnastics, etc.).

SUCCESSFUL ASPECTS OF REFORM

The following may be cited as advances in the process of curriculum reform: (a) incorporation of new ideas and trends relating to the organization and mechanisms utilized in the curriculum development process; (b) efforts to raise the scientific level and update subject contents; (c) the inclusion of medical and population education, and environmental protection in curriculum contents; and (d) adopting a systematic approach to raising teachers' skills and abilities in subject content and teaching methods.

Progress of Reforms to Date

- New curricula are being developed for primary, junior secondary and streamed secondary education.
- The primary education curriculum is being tested during implementation. Due to financial constraints, only the mathematics, language teaching and ethnic curricula are presently being implemented.
- The curriculum for junior secondary education is being modified on the basis of experts' comments. The Council of Evaluation will be organized in order to exercise the results.
- The curriculum for streamed secondary education is being tested in 180 schools located in all provinces in Viet Nam.

PROBLEMS IN REALIZING REFORMS

- *Resistance to change* shown by administrators, politicians and teachers.
- *Lack of expertise and access to up-to-date information:* Innovation has often been undertaken by staff who have not been specifically trained in the field of curriculum development. Thus, new viewpoints on the curriculum have often been applied ineffectively, due

to a lack of knowledge and skill in using the right approaches. Attempts to define required knowledge and skills within the curriculum present difficulties due to limited expertise in an area where, previously, textbooks were considered to contain all the necessary knowledge for students. There is limited access to up-to-date methodologies on curriculum (definition, function, structure), and on curriculum development processes; for example, in choosing basic contents and life-related contents (social issues) for each subject.

- *Inappropriate content:* The curriculum as it is taught does not meet the stated aims. It is over-loaded, too academic, lacking in practical components and inconsistent with local realities. Therefore, students are not provided with the necessary knowledge and skills they need to enter the world of work.
- *Ineffective teaching methods:* Despite efforts to reform teaching methods, the old methods have remained in use, which prevents changes taking place in student's learning experiences. This is at least partly linked to the fact that teachers are inadequately trained and instructional materials are limited. To date, change in teaching practice occurs mostly in large cities where it is stimulated by high-quality local competitions. There is a need both for more effective organisation of pre and in-service teacher education and for teaching materials that are appropriately designed to meet new teaching/learning approaches.

List of Contributors

Mr Mokelas Bin Ahmad
Assistant Director
Curriculum Development Centre
Ministry of Education
Kuala Lumpur 50604
Malaysia

Ms Maryam Azra Ahmad
Senior Curriculum Developer
Educational Development Centre
Ministry of Education
Male
Maldives

Mr Kiat Amprai
Director
Supervision and Education
Standards Development Office
Office of the National Primary
Education Commission
Ministry of Education
Ratchadanmoen Nok Avenue
Bangkok 10300
Thailand

Dr G.L. Arora
Head, Department of Teacher
Education
National Council of Educational
Research and Training—
NCERT
Sri Aurobindo Marg
New Delhi 110016
India

Mr U Myint Aye
Deputy Staff Oficer (History)
Department of Educational
Planning and Training
Strand Road
Six Storeyed Building
Yangon
Myanmar

Mr G. Balasubramanian
Director (Academics)
CBSE
Shiksha Kendra 2
Community Centre
Preet Vihar
New Delhi 110092
India

Mr Anurag Bhatnagar
Director
Navodaya Vidyalaya Samiti
A-39, Kailash Colony
110048 New Delhi
India

Dr Udo Bude
Deutsche Stiftung für internationale
Entwicklung
Zentralstelle für Erziehung,
Wissenschaft
und Dökumentation
Hans-Bockler-Strasse, 5
53225 Bonn
Germany

Ms Isabel Byron
International Bureau of Education
(UNESCO-IBE)
P.O. Box 199
1211 Geneva 20
Switzerland

Ms Maria Pelagia Cagahastian-Ditapat
Senior Educational Supervisor
Department of Educational and
Sports
DECS
National Capital Region
Misamis St. Bagobantay
Quezon City 1105
Philippines

Ms Thi Thang Cao
Senior Researcher
National Institute for Educational
Science
101, Tran Hung Dao Street
Hanoi
Viet Nam

Dr Bashir Ahmed Dar
Director (Academic) and Head of Curriculum Development and Research Wing
Jammu and Kashmir State Board of School Education Rehari Colony
Jammu Tawi-180005
India

Prof. R.H. Dave
Former Director
UNESCO Institute of Education
Hamburg
Germany

Prof. C.J. Daswani
National Council of Educational Research and Training NCERT
Sri Aurobindo Marg
NEW DELHI-110016
India

Mr Ram Sarobar Dubey
Director
Curriculum Development Centre
Ministry of Education
Secondary Education Curriculum Division
Kesher Mahal, Kantipath
Sanothimi
Nepal

Mr Namgyel Dukpa
Curriculum Officer
C/o Director
Education Division
Ministry of Health and Education
Thimphu
Bhutan

Mr. K.V. Ganasundara
Project Officer
National Institute of Education
High Level Road
Maharagama
Sri Lanka

Mr. S.S. Gauri
Joint Director (Academic)
Navodaya Vidyalaya Samiti
A-39 Kailash Colony
New Delhi-110048
India

Ms Deki C. Gyamtsho
Lecturer
National Institute of Education
C/o Director
Education Division
Ministry of Health and Education
Thimphu
Bhutan

Mr. Jacques Hallak
Assistant Director-General
Director, IBE
P.O. Box 199
1211 Geneva 20
Switzerland

Mr Muhammad Abul Hossain
Project Director
IDA Assisted Primary Education Development Project
Street 59 C
Darus Salam, Mirpur
Dhaka 1216
Bangladesh

Ms Shawkat Jahan
Senior Specialist
National Curriculum and Textbook Board
Ministry of Education
Secretariat, Primary and Mass Education
Dhaka
Bangladesh

Mr. Albert Karunasinghe
Project Officer
National Institute of Education
High Level Road
Maharagama
Sri Lanka

Mr Maqbool Ahmed Khakwani
Deputy Educational Adviser
Curriculum Wing
Ministry of Education
Block 'D' Pak, Secretariat

Islamabad
Pakistan

Prof. B.P. Khandelwal
Director
NIEPA
Shiksha Kendra
2 Community Centre, Preet Vihar
New Delhi-110092
India

Ms Daw Tin Kyi
Deputy Staff Officer (Maths)
Department of Educational
Planning and Training
Strand Road
Six Storeyed Building
Yangon
Myanmar

Mr Faisal Madani
Project Manager of Curriculum
Development
Pusbang Kurrandik Balitbang
Dikbud
JI, Gunung Sahari Raya No. 4
(Eks. Kompleks Siliwangi)
Jakarta Pusat 10002
Indonesia

Ms Bella O. Marinas
Supervising Education Programme
Specialist
Curriculum Development Division
Bureau of Secondary Education,
Culture and Sports
Meralco Avenue
Pasig City 1600
Philippines

Dr Arun Mishra
Director
Central Institute of Vocational
Education-CIVE
Bhopal
India

Dr Abdul Muhsin Mohamed
Director-General
Education Development Centre
Ministry of Education
Male
Maldives

Ms Thi Minh Phuong Nguyem
Vice Director
Curriculum Development Centre
National Institute for Educational
Science
101, Tran Hung Dao Street
Hanoi
Viet Nam

Dr. M.M. Pant
Professor
Indira Gandhi National Open
University
Delhi
India

Ms. Muriel Poisson
UNESCO-IIEP
7-9, rue Eugène-Delacroix
75116 Paris
France

Mr. Zamrus Bin A. Rahman
Assistant Director
Curriculum Development Centre
Ministry of Education
Damansara Town Centre
Kualalumpur-50604
Malaysia

Dr. J.S. Rajput
Chairman
National Council for Teachers
Education
C-2/10, Safdarjung Development
Area
Sri Aurobindo Marg
New Delhi-110016
India

Mr. Kenneth Ross
UNESCO-IIEP
7-9, rue Eugène-Delacroix
75116 Paris
France

Mr. Jean-Marie Sani
Chef du Département Education

Cité des Sciences et de l'Industrie
30, Avenue Corentin Cariou
75930 Paris Cedex 19
France

Dr. Abdul Rahim bin Selamat
Director
Regional Education Centre for Science
and Mathematics—RECSAM
Gelugor, Penang
Malaysia

Prof. A.K. Sharma
Director
National Council of Educational Research and Training—NCERT
Sri Aurobindo Marg
New Delhi-110016
India

Mr. Satya Bahadur Shrestha
Director
Ministry of Education
Curriculum Development Centre
Kesher Mahal, Kantipath
Kathamandu
Nepal

Ms. Ellen-Marie Skaflestad
Norad
P.O. Box 8034
0030 Oslo 1
Norway

Mrs. Chadjane Thaithae
Supervisor
Supervision and Educational Standards Development Office
Office of the National Primary Education Commission
Ratchadammoen Nok Avenue
Bangkok 10300
Thailand

Mr. Erry Utomo
Secretary
Project of Curriculum Development
Pusat Kurikulum
Jin. Gunung Sahari Raya No. 4.
(Eks. Komp, Siliwangi)
Jakarta Pusat 10002
Indonesia